Dr. Grace Wahba
presented by the author
Oskar Ensenwanger.

Developments in Atmospheric Science, 4A

Applied Statistics in Atmospheric Science
Part A, Frequencies and Curve Fitting

Further titles in this series

1. F. VERNIANI (Editor)
Structure and Dynamics of the Upper Atmosphere

2. E.E. GOSSARD and W.H. HOOKE
Waves in the Atmosphere

3. L.P. SMITH
Methods in Agricultural Meteorology

5. G.W. PALTRIDGE and C.M.R. PLATT
Radiative Processes in Meteorology and Climatology

6. P. SCHWERDTFEGER
Physical Principles of Micrometeorological Measurements

Developments in Atmospheric Science, 4A

Applied Statistics in Atmospheric Science

Part A. Frequencies and Curve Fitting

by

O. ESSENWANGER

U.S. Army Missile Command, Redstone Arsenal and University of Alabama, Huntsville, Ala., U.S.A.

ELSEVIER SCIENTIFIC PUBLISHING COMPANY
Amsterdam — Oxford — New York   1976

ELSEVIER SCIENTIFIC PUBLISHING COMPANY
335 Jan van Galenstraat
P.O. Box 211, Amsterdam, The Netherlands

AMERICAN ELSEVIER PUBLISHING COMPANY, INC.
52 Vanderbilt Avenue
New York, New York 10017

With 17 illustrations and 37 tables

ISBN: 0-444-41327-8

Copyright © 1976 by Elsevier Scientific Publishing Company, Amsterdam

All rights reserved. No part of this publication may be reproduced, stored in a retrieval system, or transmitted in any form or by any means, electronic, mechanical, photocopying, recording, or otherwise, without the prior written permission of the publisher,
Elsevier Scientific Publishing Company, Jan van Galenstraat 335, Amsterdam

Printed in The Netherlands

TO MY WIFE KATE
and my daughters
MARIANNE AND ANGELIKA

FOREWORD

It was originally planned to supplement a text on elements of statistical analysis which will be included in the World Survey of Climatology (WSoC), Volume 3 (see Essenwanger, 1974a), by preparation of a manuscript on advanced topics. These two texts would have been published together as a book on statistical analysis for atmospheric science. The enormous material requiring treatment in statistical analysis of atmospheric data and the delay in the publication of Volume 3 (WSoC) made it advisable, however, to separate the elementary and advanced topics. Since it is intended to publish the chapter on statistical analysis from the WSoC as a separate text, the basic knowledge of these elementary topics must be assumed.

This text on advanced topics in statistical analysis builds upon the knowledge which is gained from the elementary text. Atmospheric scientists and statisticians who possess a basic knowledge in statistical analysis should be able to follow, with little difficulty, the course of this book on advanced topics without having the elementary text. To aid their reading, the nomenclature and general descriptions are repeated in a short introduction.

The field of statistical analysis of atmospheric data is very comprehensive and requires treatment of a wide variety of topics. The completion of a manuscript comprising all the subjects of interest would have further delayed the publication. The advanced topics have been split into two parts. While this first part deals with frequency distribution and what can be called "curve fitting procedures", the second part (in preparation) will treat smoothing and filtering, analysis of variance and advanced test methods. A section on special atmospheric topics will round off that part.

It is not the author's intention to add to the number of texts on statistical theory, and the reader may discover that theoretical points will most of the time find short treatise. The major emphasis is placed on problems arising from the application of statistical procedures in practical work. Thus, many examples illustrate difficulties encountered in data analysis rather than the cases of smooth compliance with theory. The applications of several tools to the same set of data should aid in interpretation and evaluation of results from statistical analysis. Statements on cost effectiveness by individual statistical procedures have been added, if appropriate.

It was considered essential to include in this text brief sections on numerical methods for solutions such as for calculation of eigenvalues and eigenvectors. These procedures belong to the topics on empirical polynomial representation and factor analysis. More sophisticated methods may exist than those presented in this text, and the theory had sometimes to be cut short.

Nevertheless, the reader may find it convenient that techniques for calculations of frequency distributions or eigenvalues, etc. are included. This should aid considerably in attaining a quick answer and enabling the fast evaluation of a particular technique for suitability in the analysis of atmospheric data.

Finally the author wishes to express his gratitude to all the persons who have given their support to the establishment of this text. It is not possible to list all of them, but some must be singled out. First of all, I want to thank Prof. Dr. E. Reiter (Colorado State University) for providing the opportunity to teach in a lecture series on statistical analysis at CSU. These lecture notes became the basic material for some of the topics in this text. His consistent encouragement to publish the notes and his keen interest during the writing of this text convinced me that a book on applied statistics in atmospheric science is necessary. My appreciation is extended to Prof. Dr. E. Wahl (University of Wisconsin) for his critical comments on some of the topics. Prof. Dr. H. Flohn (University of Bonn) and Prof. Dr. Landsberg (University of Maryland) deserve my thanks for inviting me to write the chapter on elements of statistical analysis for the WSoC which is expanded with this text.

My thanks go further to my colleagues Dr. J. Stettler, Dr. H. Meyer and Mrs. H. Boyd (Physical Sciences Directorate, Army Missile Command) who took the cumbersome job of editing and reviewing the text. Last but not least Mrs. C. Brooks deserves much credit for her patience in typing the manuscript. I also wish to thank the Army Missile Command for permission to include some of my unpublished work in atmospheric data analysis.

Huntsville, Al., November 1974   O.M. ESSENWANGER

CONTENTS

Foreword. VII
Nomenclature. XII

Chapter 1. INTRODUCTION. 1
    1.1. Expectancy, probability density, cumulative distribution and classes. 1
    1.2. Moments and cumulants. 2
    1.3. Homogeneity and persistence. 5
    1.4. Significance and confidence. 6
        1.4.1. Estimators. 7
        1.4.2. Confidence intervals 9
        1.4.3. Significance. 9
    1.5. The central limit effect. 14
    1.6. Tchebycheff's and Gaussian inequalities. 15

Chapter 2. FREQUENCY DISTRIBUTIONS 17
    2.1. The hypergeometric distribution 17
    2.2. The lognormal distribution 22
        2.2.1. Regular model 22
        2.2.2. Lambert's model. 31
    2.3. The Cauchy distribution. 35
    2.4. The beta or incomplete beta function 41
        2.4.1. Generalized beta distribution 43
        2.4.2. Computations of the beta functions 46
        2.4.3. Comparison with the negative binomial. 48
        2.4.4. A related frequency distribution with beta function 50
        2.4.5. Cosine function with rectangularly distributed phase angle . . . 56
    2.5. Pearson's system of frequencies 57
    2.6. The U-distribution. 63
        2.6.1. The general U-distribution 63
        2.6.2. The symmetric U-distribution 68
        2.6.3. The recomputation of the U-distribution. 69
    2.7. The logistic distribution 76
        2.7.1. Univariate distribution. 76
        2.7.2. The bivariate logistic distribution. 80
    2.8. The bivariate normal distribution 83
        2.8.1. General, multivariate distributions 83
        2.8.2. The bivariate Gaussian distribution 86
        2.8.3. Regression line, maximum frequency, ellipses. 89
        2.8.4. The bivariate circular distribution 91
        2.8.5. Cumulative bivariate Gaussian distribution, integrals 93
        2.8.6. Concluding remarks. 105
    2.9. The exponential distribution 113
    2.10. The logarithmic series distribution 114

- 2.11. The four-parameter Weibull and hyper-gamma distributions. 116
  - 2.11.1. The three-parameter gamma distribution 118
  - 2.11.2. The three-parameter Weibull distribution. 119
- 2.12. Truncated distributions 121
  - 2.12.1. Gaussian distribution. 122
  - 2.12.2. Truncated bivariate and multivariate distributions. 129
  - 2.12.3. Truncated binomial and negative binomial. 133
  - 2.12.4. Truncated Poisson distributions. 136
  - 2.12.5. Truncated gamma distributions 139
- 2.13. Mixed distributions 143
  - 2.13.1. Gaussian univariate mixed distributions. 144
  - 2.13.2. Mixtures of other than Gaussian distributions. 183
- 2.14. Folded (Gaussian) normal distribution. 190

Chapter 3. CURVE FITTING 195

- 3.1. General. 195
  - 3.1.1. Introduction 195
  - 3.1.2. Tchebycheff polynomials. 197
  - 3.1.3. Legendre polynomials 199
  - 3.1.4. Percentage reduction and left variance 214
  - 3.1.5. Miscellaneous polynomial techniques 224
- 3.2. Spectral analysis 226
  - 3.2.1. Power-spectrum of $x \cdot \sin \alpha$. 226
  - 3.2.2. Power-spectrum and periodogram of non-harmonic waves. 234
  - 3.2.3. Estimation of spectra, separation of waves and aliasing. 237
  - 3.2.4. Spectra of meteorological data 241
- 3.3. Bessel functions 244
  - 3.3.1. General. 244
  - 3.3.2. Definition of Bessel functions and recurrence relations. 245
  - 3.3.3. Complex Bessel functions. 246
  - 3.3.4. The zeros of the Bessel functions 247
  - 3.3.5. Fourier-Bessel expansion 248
- 3.4. Empirical orthogonal polynomials and eigenvalues. 252
  - 3.4.1. Empirical orthogonal functions 252
  - 3.4.2. The eigenvalue and eigenvector problem 255
  - 3.4.3. Significance of eigenvectors and eigenvalues. 269
  - 3.4.4. Empirical polynomials and time-series analysis 273
- 3.5. Factor analysis 276
  - 3.5.1. General concepts and problem formulation 276
  - 3.5.2. The statistical (mathematical) model 276
  - 3.5.3. The communality problem. 279
  - 3.5.4. Factor loading or computation of the factors. 281
  - 3.5.5. Summary of factor-analysis procedure 283
  - 3.5.6. Decision on the number of factors 284
  - 3.5.7. Rotation in factor analysis 285
  - 3.5.8. Analysis of covariance 286
  - 3.5.9. Modified correlation input 288
- 3.6. Analysis of time series 289
  - 3.6.1. General representation. 290
  - 3.6.2. The autoregressive model 291
  - 3.6.3. The moving-average model 292
  - 3.6.4. A mixed model. 293

CONTENTS XI

      3.6.5. Moving average and trend 294
      3.6.6. Distribution of residuals in autoregressive–moving-average models . 296
      3.6.7. Spectral relationship 296
      3.6.8. Autocorrelation functions of selected models 298
      3.6.9. Estimation and model identification 302
      3.6.10. Inverse autocorrelation 306
      3.6.11. Godske's model 308
      3.6.12. Time-series and quality control 310
      3.6.13. Multivariate and other atmospheric models 315
  3.7. Transformations 316
      3.7.1. Transformation of special functions 316
      3.7.2. Transformation to Gaussian (normal) distribution 317
            3.7.2.1. Johnson's transformation system 318
            3.7.2.2. Other systems 321
            3.7.2.3. Shenton's system 321
            3.7.2.4. Some original and related transformed distributions 323
            3.7.2.5. Transformation related to square root 323

Chapter 4. CALCULATION OF EIGENVALUES AND EIGENVECTORS 325

  4.1. Matrices and operations 325
  4.2. Types of matrices (Definition) 329
  4.3. Determinants 332
  4.4. Equivalence of matrices 337
  4.5. Adjoints 343
  4.6. The inverse of a matrix 344
  4.7. Similar matrices 348
  4.8. Characteristic equations, eigenvalues and eigenvectors 348
  4.9. Eigenvalues and diagonal matrix 352
  4.10. Largest and smallest eigenvalues by iteration 358
  4.11. Computation of the characteristic polynomial 360
  4.12. The determination of the roots 363
  4.13. Determination of the eigenvectors 371
  4.14. Linear equations 379
  4.15. Conclusions 381

Appendix
  Integral and ordinate of the Gaussian distribution 383
  Table of the Gaussian distribution 384

References 387

Author index 401

Subject index 405

NOMENCLATURE

1. $1/2B \equiv 1/(2B)$, both versions utilized, but always:
   $0.5B \equiv (1/2)B$
   $1/2Bx \equiv 1/(2Bx)$ in contrast to $(1/2B)x$
2. tanh, sinh, cosh are hyperbolic functions.
3. $\tan^{-1}$, $\cos^{-1}$, etc. are $1/\tan$, $1/\cos$, etc.
   arctan or arcsin is spelled out.
4. The square root sign without bar is valid to the end of the line, parenthesis, or equal sign.
   $(1+\sqrt{1}+x) = 1+\sqrt{1+x}$, etc.
   e.g.
   $$\sqrt{\beta_1/\{(c+2)\beta_1 + (c+1)\}} \equiv [\beta_1/\{(c+2)\beta_1 + (c+1)\}]^{1/2}$$
   $$\sqrt{2/\pi} \equiv (2/\pi)^{1/2} \quad \text{but} \quad \sqrt{2}/\pi \equiv (1/\pi)\sqrt{2}$$
5. matrix **M** (capitals)
6. vector *x* (small letters)
7. $\sim$ approximately.

CHAPTER 1

INTRODUCTION

This text requires the knowledge of elementary topics treated by the author in a separate volume (1974a), but the reader who is acquainted with the basic principles in statistical analysis should be able to follow the advanced topics in this book independently of the cited reference. Comprehension of the basic frequencies such as the binomial, Poisson, gamma and Gaussian distribution is assumed. A further prerequisite is familiarity with regression analysis, power spectrum, and some basic test procedures such as the t, F, χ^2 and Kolmogorov-Smirnov tests. These basic topics can be found in the statistical literature (e.g. textbooks quoted in the list of references) although the author (1974a) has illustrated their special application to atmospheric science.

In the Introduction of this first of two books on applied statistics the reader will become familiar with the nomenclature. Chapter 2 treats frequency distributions and Chapter 3 can be called "curve fitting procedures" applied to atmospheric data. A final chapter comprises a brief description of required mathematical techniques. The second volume (Essenwanger, 1974b) will deal with the problems of filtering, analysis of variance, advanced tests and special meteorological topics.

It should be stressed that the author did not intend to prepare a theoretical treatise. Excellent textbooks which achieve this goal are in existence. This text is intended to serve the practitioner and to delineate the practical aspects of statistical analysis with examples from the area of atmospheric science.

1.1 EXPECTANCY, PROBABILITY DENSITY, CUMULATIVE DISTRIBUTION AND CLASSES

It is not the intention of this section to repeat the introduction to elements of statistical analysis (Essenwanger, 1974a). Some of the general definitions and formulae should be repeated, however, to enable the reader to utilize this book as a text of its own.

It was introduced that in general N will be the total number of data points in a sample, the n a partial sample, e.g. an observed frequency in a certain class with determined boundaries. Then:

$$n/N = f \to p = P(A) \text{ for } N \to \infty \qquad [1.1]$$

The p denotes the stabilized relative frequency of an event A. The $P(A)$ and p are expected probabilities for N being very large. We may also call $P(A)$ the population probability. Then:

$$N_e = N \cdot p \qquad [1.2]$$

is the expected number of observations and n_i represents the empirical number of observations for the ith class interval. Relationships of probability will not be repeated here.

$$F(x) = P(A) = P(X \leqslant x) \qquad [1.3]$$

is called the cumulative distribution function (c.d.f.). It represents the probability of an event A with a variate value X being less than or equal to the threshold x.

If we can write:

$$P(X \in A) = \int_A f(x)\,dx \qquad [1.4]$$

then $f(x)$ is called the probability density function (p.d.f.) or, expressed differently:

$$F(a) = P(X \leqslant a) = \int_{-\infty}^{a} f(x)\,dx \qquad [1.5]$$

The class width is defined as:

$$w = x_u - x_1 \qquad [1.6]$$

where x_u and x_1 are the upper and lower boundaries, respectively. The central class value x_c is then:

$$x_1 < x_c < x_u \qquad [1.6a]$$

In many cases:

$$x_c = (x_u - x_1)/2 \qquad [1.6b]$$

but this postulation is not fulfilled if the slopes to both sides of the central class value are not linearly related, e.g. for a logarithmic progression of the variate x. Then x_c must be redefined.

1.2 MOMENTS AND CUMULANTS

Primitive characteristics such as the maximum, minimum, range, or other characteristics based on c.d.f. (e.g. quantiles), or p.d.f. need not be redefined here. Mathematical characteristics can be defined as:

$$E(X^\nu) = \mu_\nu = \int_{-\infty}^{\infty} X^\nu\,dx \qquad [1.7a]$$

which are called the moments of the distribution, with the notation $E(\ )$ standing for the expectancy. The ν denotes the order of the moment. For discrete variables:

MOMENTS AND CUMULANTS

$$E(X^\nu) = \mu_\nu = \Sigma(X^\nu)/N \qquad [1.7b]$$

where it is immaterial whether the left or right expression of [1.7c] below is employed:

$$\Sigma[(X^\nu)/N] \equiv (\Sigma X^\nu)/N \qquad [1.7c]$$

If X can assume large numbers in practical analysis, the left side of the expression may be more accurate due to computer truncation of large numbers for ΣX^ν.

Note. If no boundaries for the summation are specified, it means automatically that the summation should be carried out over all available data.

Equations [1.7] can be modified for inclusion of the p.d.f., namely:

$$E(X^\nu) = \mu_\nu = \int_{-\infty}^{\infty} f(x) x^\nu \, dx \qquad [1.8a]$$

and

$$E(X^\nu) = \mu_\nu = \sum_{1}^{n} f(x) x^\nu \Delta x \qquad [1.8b]$$

Usually the Δx is assumed to be unity, and the n denotes the number of classes. As the reader will notice, the symbol μ stands for the general moments reference zero of the variate.

The central moments are based on the mean value.*

$$E[(X - \bar{X})^\nu] = \nu_\nu = \sum_{1}^{n} f(x_j)(x_j - \bar{x})^\nu = \sum_{1}^{N} (x_i - \bar{x})^\nu/N \qquad [1.9]$$

The following expressions are important:

$$\mu_0 = N = \Sigma X_i^0/N = 1 \qquad [1.10a]$$

$$\mu_1 = \bar{x} = \bar{X} = \Sigma X_i/N = \sum_{1}^{n} f(x_j) x_j \quad \text{(mean)} \qquad [1.10b]$$

$$\mu_2 = \sigma^2 + \bar{x}^2 = \nu_2 + \bar{x}^2 \qquad [1.10c]$$

Furthermore

$$\nu_1 = 0 \qquad [1.11a]$$

$$\nu_2 = \sigma^2 \quad \text{(variance)} \qquad [1.11b]$$

$$\nu_3 = \mu_3 - 3\bar{X}\nu_2 - \bar{X}^3 \qquad [1.11c]$$

$$\nu_3 = \mu_3 - 3\bar{X}\mu_2 + 2\bar{X}^3 \qquad [1.11d]$$

$$\nu_4 = \mu_4 - 4\bar{X}\nu_3 - 6\bar{X}^2\nu_2 - \bar{X}^4 \qquad [1.11e]$$

$$\nu_4 = \mu_4 - 4X\mu_3 + 6\bar{X}^2\mu_2 - 3\bar{X}^4 \qquad [1.11f]$$

---

* The central moments are given for the summation only. The integral form can readily be deduced from a comparison between [1.8a] and [1.8b].

Some characteristics for frequency distributions are derived from the moments. Skewness (γ) and kurtosis (k_u) are defined as:

$$\gamma_1 = \nu_3/\sigma^3 \qquad [1.12a]$$

$$k_u = \gamma_2 = (\nu_4/\sigma^4) - 3 \qquad [1.12b]$$

Karl Pearson (1895) has defined parameters β, namely:

$$\beta_1 = (\gamma_1)^2 = \nu_3^2/(\sigma^2)^3 \qquad [1.13a]$$

$$\beta_2 = (\gamma_2 + 3) = \nu_4/(\sigma^4) \qquad [1.13b]$$

If moments are calculated from grouped data Sheppard's corrections are usually applied:

$$\nu_2' = \nu_2 - w^2/12 \qquad [1.14a]$$

and:

$$\nu_4' = \nu_4 - \nu_2 w^2/2 + (7/240)w^4 \qquad [1.14b]$$

The odd moments do not need a correction for grouping. (Note: 7/240 = 0.02917.).

A mixed moment is defined as:

$$\nu_{xy} = \sum_1^N (X_i - \bar{X})(Y_i - \bar{Y})/N \qquad [1.15a]$$

$$= \sum_1^n f(xy)_j \cdot (x_j - \bar{x})(y_j - \bar{y})/N \qquad [1.15b]$$

The term ν_{xy} or $N\nu_{xy}$ is also called covariance or cross-product depending on the definition.

The integrals of [1.7a] or [1.8a] are very often difficult to solve. When all moments are finite, the integral exists and a "moment generating function $g(t)$" can be defined:

$$g(t) = E(e^{xt}) = \int_{-\infty}^{\infty} e^{xt} f(x)\, dx \qquad [1.16a]$$

By differentiation:

$$\frac{d^\nu g(t)}{dt^\nu} = \int_{-\infty}^{\infty} x^\nu e^{xt} f(x)\, dx \qquad [1.16b]$$

and for $t = 0$:

$$\frac{d^\nu g(0)}{dt^\nu} = E(x^\nu) = \mu_\nu \qquad [1.16c]$$

This procedure indicates that the νth moment follows from the νth derivative of $g(t)$ and substitution of $t = 0$.

Sometimes the "cumulant" generating function $k(t)$ is simpler:

$$k(t) = \ln g(t) = k_1 t + k_2 t^2/2 + \ldots k_\nu t^\nu/\nu! \qquad [1.17]$$

It should be noted that:

$$g(t) \cdot \exp(-\mu_1 t) = g_1(t) = 1 + \nu_2 t^2/2! + \ldots \nu_4 t^\nu/\nu! \qquad [1.18]$$

A useful relationship between moments and cumulants exists as follows:

$$\mu_1 = k_1 \qquad [1.19a]$$
$$\mu_2 = k_2 + k_1^2 \qquad [1.19b]$$
$$\nu_2 = k_2 = \mu_2 - k_1^2 \qquad [1.19c]$$
$$\nu_3 = k_3 = \mu_3 - 3k_2 k_1 - k_1^3 \qquad [1.19d]$$
$$\nu_4 = k_4 + 3\nu_2 k_2 = \mu_4 - 4k_3 k_1 - 6k_2 k_1^2 - k_1^4 \qquad [1.19e]$$
$$\nu_5 = k_5 + 10\nu_3 k_2 \qquad [1.19f]$$

While the non-central moments are a function of the origin, the cumulants are invariants like the central moments.

One final version of the moments may be given. It is sometimes useful to determine the moments from a different reference point than either the mean $\bar{x}$ or zero, let us say for γ. Then:

$$\bar{x} = N^{-1} \Sigma (x - \gamma) + \gamma \qquad [1.20a]$$
$$\sigma^2 = N^{-1} \Sigma (x - \gamma)^2 - (\gamma - \bar{x})^2 \qquad [1.20b]$$
$$\nu_3 = N^{-1} \Sigma (x - \gamma)^3 + 3(\gamma - \bar{x})\sigma^2 + (\gamma - \bar{x})^3 \qquad [1.20c]$$
$$\nu_4 = N^{-1} \Sigma (x - \gamma)^4 + 4(\gamma - \bar{x})\nu_3 - 6(\gamma - \bar{x})^2 \sigma^2 - (\gamma - \bar{x})^4 \qquad [1.20d]$$

For $\gamma = 0$ this set of formulae reduces to the version of [1.11]. These formulae are practical for calculation by hand when e.g. the γ can be substituted as a whole number close to the mean and $(x - \gamma)$ can be expressed in whole numbers. This procedure speeds up calculations of the moments considerably. In electronic computer data handling the γ-version is generally not necessary but may increase accuracy.

1.3 HOMOGENEITY AND PERSISTENCE

These topics have been treated extensively by Essenwanger (1974a). It may be repeated here that the homogeneity of meteorological data sampling is always a major concern in any data collection or analysis work. No formalistic concept exists to guarantee homogeneity. The only assurance is the homogeneity of the physical processes and instrumentation. It is, therefore, always appropriate to investigate the homogeneous background of atmospheric data sampling.

It is self-evident, too, that many meteorological time series display persistence. This is not always a handicap as this property can be very useful, e.g. in the establishment of prediction schemes. The reader is only cautioned to answer the question first whether the outcome of an investigation is influenced by persistence in a manner inconsistent with the applied methods. It may occur that we intend to check for randomness in a statistical parameter, e.g. the amplitude of a periodicity. Then statistical significance of the parameter may be caused by the presence of persistence. This result may not be the actual goal of the employed test.

The problem of homogeneity and persistence is again later debated (Essenwanger, 1974b) in connection with other topics such as the analysis of variance, etc.

1.4 SIGNIFICANCE AND CONFIDENCE

The problem of significance and confidence of the outcome of a statistical data analysis has received extended treatment in statistical tests and in the statistical literature. Since a more comprehensive elaboration on this topic has been prepared by Essenwanger (1974a), only a few additional factors may be called to the attention of the reader. Significance and confidence are tools of sampling evaluation and aid in a conclusion of whether the outcome of a study is a mere product of chance. Two different problems are involved, the evaluation of estimators and the testing of a hypothesis. In the estimation of parameters we speak generally of the confidence which we have that a calculated parameter lies within a specified interval of tolerance. In the task of hypothesis testing we like to know what the probability is that we erroneously reject a true hypothesis (type-I error) or accept a false hypothesis (type-II error). This probability is called the significance level of a test. Individual authors may arbitrarily choose particular levels of significance; the most common thresholds are the 5%, 1%, 2 or 3 σ levels.

In the parameter estimation the problem can be mathematically formulated in the following definition:

$$P[c_l < \theta < c_u] = 1 - \alpha \qquad [1.21]$$

where θ is the estimator, α the significance level (e.g. for 5% the $\alpha = 0.05$), c_l and c_u are the lower and upper bounds, and $c_u - c_l$ the confidence interval.

Some words of caution should be added. Many of the atmospheric time series or data samples do not comprise independent observations. Application of standard test procedures may then lead to significant conclusions caused only by persistence, Markovian or harmonic wave relation. More details follow in subsequent sections.

If the probability of a type-II error is β, then $1 - \beta$ is called the power of the test.

1.4.1 Estimators

It is evident that a small data sample does not necessarily provide the true population parameter, and that random processes may lead to an estimator which is different from the true value of the population. The chances of a large deviation from the true value are higher the smaller the sample and the larger the error variance of the estimated parameter. It is, therefore, important to know the error variance of individual parameters. One and the same parameter, such as the variance, may have a different error variance depending on the method of estimation.

Four properties are desirable for a good estimate. The estimator, sometimes also called a statistic, should be unbiased, efficient, sufficient and consistent. The expression "best" statistics if often applied to unbiased and efficient estimators

Definitions of the four properties can be found in most statistical texts, and have been presented by the author (1974a). No elaborate discussion is intended in this section. Some short remarks may be in order, however.

An unbiased estimator is defined as a statistic whose expectation is equal to the true value of the population parameter. In different words, the estimator approaches the true parameter θ as N goes to infinity. A classical example for an unbiased estimator is the sample mean, while the sample variance is a biased estimator if $\Sigma\,(x-\bar{x})^2/N$ is used. In many cases of biased estimators, correction formulae are available.

The efficiency of an estimator is a continual topic of discussion in the statistical literature. The efficiency can be generally stated in the form:

$$E_f = \mathrm{Var}_{min}/\mathrm{Var}_s \qquad [1.22]$$

where Var_{min} is the minimum variance achievable for a particular parameter θ whose estimator s has the variance Var_s. By definition $\mathrm{Var}_{min} \leq \mathrm{Var}_s$. A most efficient statistic has a ratio of 100%. We may reformulate this statement in mathematical terms:

$$\mathrm{Var}_{min} = \left[N \cdot E\left\{ \left(\frac{\partial \ln f(x,s)}{\partial s} \right)^2 \right\} \right]^{-1} \qquad [1.22a]$$

or

$$\mathrm{Var}_s \geq \left[N \cdot E\left\{ \left(\frac{\partial \ln f(x,s)}{\partial s} \right)^2 \right\} \right]^{-1} \qquad [1.22b]$$

A further rule is that a most efficient statistic, if it exists, can be found by the method of maximum likelihood. Although electronic computers have made it possible to calculate maximum-likelihood estimators in most cases, it is sometimes more economical to obtain an inefficient statistic and then apply some corrections. This procedure works well when the inefficient estimator does not have too low an efficiency. As Kenney and Keeping (1954) have shown:

$$s = s' + \text{Var}_s \left(\frac{\partial L}{\partial \theta}\right)_{s'} \quad [1.22c]$$

where the s' denotes an inefficient statistic such as a moments estimator. The L stands for the likelihood function and:

$$L(\theta, x_1, x_2, \ldots, x_n) = \prod_1^n f(x_i, \theta) \quad [1.23]$$

$$\left(\frac{\partial L}{\partial \theta}\right)_{s'} = \left(\frac{\partial L}{\partial \theta}\right)_s + (s' - s)\left(\frac{\partial^2 L}{\partial \theta^2}\right) + \ldots \quad [1.23a]$$

The economic considerations of inefficient statistics have been treated by McElrath and Bearman (1959) who point out that it is sometimes considerably less expensive to sample for an inefficient statistic. This economic consideration includes electronic data processing. An inefficient statistic means, in other words, that we do not get the maximum information which we could obtain from the data sample by an efficient statistic. A 60% efficiency can thus be interpreted as furnishing the same reliability with 100 observations which we would have obtained from 60 observed data by an efficient statistic. In turn, one can argue that it may sometimes be cheaper to add to a data sample in order to obtain the same information by an inefficient statistic. The gain from an efficient statistic may be disproportionate to the economics involved. Mosteller (1946) has recommended the use of some inefficient statistics such as order statistics, especially the use of the median as a replacement for the mean in a symmetrical distribution.

Various estimation methods have been described by Essenwanger (1974a), and will not be repeated here, since most of them are basic knowledge in statistical analysis which is a prerequisite for this volume.

A statistic is called sufficient when the estimation is independent of the parameter θ. Sufficient statistics do not always exist, but maximum-likelihood estimators are sufficient.

Consistency in an estimator means that the statistic tends towards the true parameter as N increases towards infinity, i.e. a consistent estimator cannot be biased.

In order to distinguish in the text, estimators for moments have been given special symbols such as:

x_m for $\bar{x}$ m_3 for μ_3
s for σ m_4 for μ_4
s^2 for ν_2 M_3 for ν_3
m_2 for μ_2 M_4 for ν_4

SIGNIFICANCE AND CONFIDENCE

1.4.2 Confidence intervals

An evaluation formula for a confidence interval has been given by [1.21], assuming a lower and upper bound. In symmetrical distributions such as the Gaussian distribution it is often only of interest to determine the magnitude of the deviation from the mean. Then we talk about a two-sided test and [1.21] expands to:

$$P[c_l < \theta < c_u] = 1 - 2\alpha \qquad [1.21a]$$

where now the α stands for the fractional exceedance of one boundary.

The problem of the confidence limit would be relatively easy if the parameters of the frequency distribution for θ, especially the mean and variance, were known. In the case where θ follows a Gaussian distribution the boundaries could be written as:

$$c_l = \bar{x} - b \cdot \sigma \qquad [1.21b]$$

$$c_u = \bar{x} + b \cdot \sigma \qquad [1.21c]$$

where b represents the Gaussian threshold for a certain α.

The problem becomes complex because in most cases the x and σ are not known, and must be replaced by estimators which themselves have deviations. A huge literature exists in statistical analysis dealing with this problem. Among others, a fundamental treatment of the problem can be found in an article by Birnbaum and Tingey (1951) for one-sided confidence contours.

The problem of tolerance limits for frequency distributions plays a major role in quality control. The application of the Gaussian distribution in that field has been treated by Jaquez (1962) and Hahn (1970), among others, because the Gaussian law as the fundamental law of the random error has widespread application in quality control. Tolerances and standards in specifications have been recently described again by Kirkpatrick (1970) and Bartlett and Provost (1973).

1.4.3 Significance

As previously mentioned, the level of significance is connected with the chances for accepting a hypothesis when it is wrong or rejecting it when it is true. Significance is, therefore, intimately related with all testing and will be treated in detail in respective sections where applicable. In this text of advanced statistical analysis topics, it is assumed that the reader is familiar with the common test procedures such as the t, F, and χ^2-tests which have been treated by Essenwanger (1974a). Special tests follow (Essenwanger, 1974b). One peculiarity of testing in atmospheric science should be stressed, however, and has been treated by various authors, among others by Nordo (1966).

Many tests of significance are based on the postulation that the observations of the sample are independent or that the errors are not correlated. This assumption is also called "random" sampling. Atmospheric data as a part of geophysical data generally display time and space interrelationships. If the restrictions of the general assumption for most statistical tests were not taken into consideration one would, in principle, too often accept the hypothesis when acceptance would actually be caused by error interrelation or persistence. Some modifications are therefore necessary, largely adjustments in the degree of freedom, in order to compensate for this difference in assumptions.

As will later be pointed out (Section 3.6), atmospheric data samples with observations x can be written in principle as a function of time, space, or both, i.e.:

$$x(t, r) = \sum_{j=0}^{n_g} A_j \phi_j(t, r) + \epsilon(t, r) \qquad [1.24]$$

where t denotes the time and r the space relationship. The A_j signifies coefficients, the $\phi(t, r)$ time and space functions, the $\epsilon(t, r)$ either the residual or an error at time t. For simplification let us restrict the consideration to the most common cause, the time series. Analogous conclusions can be drawn for the space dependency. We modify [1.24] to apply only to time relationship, and write:

$$X(t) = \sum_{j=0}^{n_g} B_j \phi_j(t) + \epsilon_j(t) \qquad [1.24a]$$

It is evident that:

$$\sum_{t=1}^{N} \epsilon_j(t) = 0 \qquad [1.25]$$

where the summation is taken over the total series of the observation time from 1 to N. This assumption would be fulfilled for any random error. It is relatively easy to prove that condition [1.25] also can be achieved for the residuals, e.g. for a least-square solution of the coefficients B_j. We can then state:

$$\sum_{t=1}^{N} \phi_j(t) \epsilon_j(t) = 0 \qquad [1.25a]$$

which transforms into [1.25] if ϕ_0 is considered to be a constant, and the observations are deviations from the mean.

Equation [1.25] implies that the residuals (or errors) are not correlated with $x(t)$. But we cannot automatically conclude that the ϵ is independent of ϕ which is the basis of "random" sampling. In other notation we could write:

$$\text{Var}_\epsilon = (1 - R^2) \text{Var}_x \qquad [1.25b]$$

(see also analysis of time series, Section 3.6, or variance analysis, Essenwanger, 1974b). The R stands for the multiple correlation (see Essenwanger, 1974a), which is zero for random events of a total population. It can immediately be recognized that Var_ϵ is only a fraction of Var_x for $R \neq 0$. It is known that R

SIGNIFICANCE AND CONFIDENCE

can assume high values in small samples by chance. As pointed out, if the goal is an interpolated value for a particular set of observations, this fact may be immaterial. However, if the purpose of an investigation is the confirmation of a theory, the fact of a high correlation R by chance is quite significant. In other words "sporadic relationships" could be misinterpreted as physical laws. To guard against these pitfalls, significance-test procedures have been designed.

If the ϵ_j are not independent or correlated with the ϕ_j a non-zero correlation coefficient R is produced. In most statistical tests, however, this second cause is not included, because it is not a random factor. Consequently, significance of R is not based on true correlation but is caused by correlated errors or non-independent data.

Nordo (1966) has analyzed the influence of a time relationship on the degrees of freedom ν for the residual variance, and derives approximately:

$$\nu = n - \nu_R \qquad [1.26]$$

where n is the number of degrees for a random process, usually the number of classes or observations, and ν_R the contribution by the correlation:

$$\nu_R = n^{-1} \sum_{j=0}^{n_s} \sum_{u,v=1}^{n} r_{u \cdot v} R_{j, u \cdot v} \qquad [1.26a]$$

The symbols r and R stand for correlations. Then:

$$R_{j, u \cdot v} = \frac{\Sigma \phi_j(u) \phi_j(v)}{[\phi_j^2(u) \phi_j^2(v)]^{\frac{1}{2}}} \qquad [1.26b]$$

is the population correlation for the (orthogonal) functions $\phi_j(u)$ and $\phi_j(v)$ (see [1.24a]) which are $(u - v)$ units of time apart. The $r_{u \cdot v}$ is the population correlation coefficient for the residuals (errors) at a lag $u - v$.

For $r_{u \cdot v} = 0$ and $u \neq v$ we find:

$$\nu = n - s - 1 \qquad [1.26c]$$

which was discussed by Essenwanger (1974a), and accounts under s for the reduction in the number of estimators for the parameters of a frequency distribution.

Lorenz (1956) has deduced the following relationships for the variance of an empirical residual e:

$$\text{Var}(e)_R \sim n(n-s-1)^{-1} \cdot S_\epsilon^2 \qquad [1.26d]$$

and:

$$\text{Var}(e)_I \sim (n+s+1)(n-s-1)^{-1} \cdot S_\epsilon^2 \qquad [1.26e]$$

where the subscripts R and I denote the related and independent population variance for the residuals ϵ, and s stands for the number of estimators. The final term:

$$S_\epsilon^2 = n_s \sum_{j=1}^{n_s} \sigma_j^2(\epsilon, t) \qquad [1.26f]$$

represents the expected sample variance. The reduction of the degrees of freedom by s has also been mentioned by Essenwanger (1974a, section 6.7).

Nordo (1966) has reformulated the same variances as:

$$\text{Var}(e)_R = n \cdot (n - \nu_R)^{-1} \cdot S_\epsilon^2 \qquad [1.26g]$$

$$\text{and: } \text{Var}(e)_I = (n + \nu_R)(n - \nu_R)^{-1} \cdot S_\epsilon^2 \qquad [1.26h]$$

For a simple Markov chain (see Essenwanger, 1974b) we obtain:

$$R_{j, u \cdot v} = R_j^{u-v} \qquad [1.27a]$$

$$\text{and: } r_{u \cdot v} = r^{u-v} \qquad [1.27b]$$

According to Nordo, [1.26g] transforms to:

$$\text{Var}(e)_R \sim n(n - n_1 + n_2)^{-1} \cdot S_\epsilon^2 \qquad [1.27c]$$

$$\text{Var}(e)_I \sim (n + n_1 - n_2)(n - n_1 + n_2) \cdot S_\epsilon^2 \qquad [1.27d]$$

$$\text{with: } n_1 = \sum_{j=0}^{n_s} [(1 + R_j r)/(1 - R_j r)] \qquad [1.27e]$$

$$\text{and: } n_2 = 2n^{-1} \sum_{j=0}^{n_s} [R_j r(1 - R_j^n r^n)/(1 - R_j r)^2] \qquad [1.27f]$$

The first term, n_1, becomes 1 for $R_j r$ not close to unity, and the second term, n_2, would go to zero for $n \to \infty$. Thus, the behavior would then resemble the random sampling relationship. This approximation is also valid for $r = 0$. Otherwise, n_1 and n_2 represent the modification of the degrees of freedom.

If a first-order Markov process is not sufficient, ν_R becomes quite lengthy (see Nordo, 1966), but the first-order Markov process is a fairly good approximation for most atmospheric data. It will suffice in most cases, at least to determine the magnitude of a modification by the time relationship. For the higher-order Markov chain the reader is, therefore, referred to the literature such as Nordo (1966).

Assume that the χ^2-coefficient is defined by:

$$\chi^2 = \sum_1^{n_s} \frac{\varphi(R_{j*})}{\varphi(r_x)} (B_j - C_j)^2 \frac{n\sigma_{\phi j}^2}{\text{Var}(e)} \equiv \sum_1^{n_s} \chi_j^2 \qquad [1.28a]$$

Then the degrees of freedom can be expressed as:

$$\nu = \nu_R \cdot \nu_I/(s + 1) \qquad [1.28b]$$

where ν_I is the number of degrees of freedom for independence. The χ^2-modification is then:

$$\chi^2 = \frac{\Sigma [\varphi(R_{j*})\varphi(r_*)](B_j - c_j)^2 \sigma_{\phi j}^2}{\Sigma (B_j - c_j)^2 \sigma_{\phi j}^2} \cdot \chi_I^2 \qquad [1.28c]$$

where the χ_I^2 denotes the independent χ^2-value. The other symbols utilized in [1.28] are as follows:

The C_j denote the coefficients of a representation of the polynomial series:

$$X(t) = \sum_{j=0}^{n_s} C_j \phi_j(t) + e(t) \qquad [1.24b]$$

The definition $\varphi(r_*)$ for the population is:

$$\varphi(r_*) = 1 - R_e^2 \qquad [1.28d]$$

Furthermore:

$$\varphi(R_{j*}) = (1 + a_1^2 + \ldots a_i^2) - 2(a_1 - a_1 a_2 - \ldots - a_{i-1} a_i) R_{j,2}$$
$$- 2(a_2 - a_1 a_3 - \ldots - a_{i-2} a_i) R_{j,2} - \ldots - 2 a_i R_{ji} \qquad [1.28e]$$

where the coefficients come from an autoregressive model (see Section 3.6.2).

Furthermore:

$$e(t+i) = a_1 e(t+i-1) + \ldots a_i e(t) + \delta_{t+i} \qquad [1.28f]$$

where δ_{t+i} is now a random residual error.

The factor $\varphi(R_{j*})/\varphi(r_*)$ can be expressed for a simple Markov chain of data with $R_{j,1} = 0.75$ as:

$$\varphi(R_{j*})/\varphi(r_*) = 1 - 3.43(R_{j,1} - 0.75) \qquad [1.28g]$$

Another definition of the χ^2 is:

$$\chi^2 = [\varphi(r'_*)/\varphi(r)] \cdot \chi_{*I}^2 \qquad [1.29a]$$

and:
$$\nu_* = \nu_I (n - n_R)/(n - s - 1) \qquad [1.29b]$$

with:
$$\varphi(r'_x) \sim \left[\sum_{t=1}^{N} \delta_{t+i} - \sum_{j=0}^{n_s} (B_j - C_j)^2 \varphi(R_{j*}) \sigma_{\phi_j}^2(t) \right] \Big/ \sigma_e^2(t) \qquad [1.29c]$$

Although the calculation of the χ^2 in both equations ([1.28a] and [1.29b]) is tedious, the computation of the degrees of freedom can readily be performed.

The process can be expanded when the function $\phi(t)$ are harmonic waves instead of orthogonal polynomials, namely:

$$\phi_p(t) = \sin(\alpha_p t + \beta_p) \qquad [1.30]$$

Then:
$$R_{p, u \cdot v} = \cos \alpha_p (u - v) \qquad [1.30a]$$

and:
$$\varphi(R_{p*}) = \prod_{m=1}^{i} (1 - 2R_m \cos \alpha_p + R_m^2) \qquad [1.30b]$$

$$\nu \sim \sum_{m=1}^{i} D_m (1 - R_m^2)/(1 - 2R_m \cos \alpha_p \to R_m^2) \qquad [1.30c]$$

The D_m are constants of the lag correlations r_0 through r_{j-1}, namely:

$$r_k = \sum_{m=1}^{j} D_m R_m^k \qquad [1.30d]$$

Then: $\varphi(r_*) = \varphi(R_{p*})\nu$ [1.30e]

More details can be found in the article by Nordo (1966), who has also treated the case of spectral bands. The difference in the $\varphi(R)$ for a band is a limitation of the integration, namely:

$$\varphi(R_{B*}) = \sum_{L}^{U} k_p^2 \varphi(R_{p*}) \qquad [1.31]$$

where L and U are the lower and upper boundaries of the band B. Furthermore:

$$R_{B,t} = \sum_{L}^{U} k_p^2 \cos \alpha_p \cdot t \qquad [1.31a]$$

Finally: $\nu_* = n - n^{-1} \sum_{u,v=1}^{n} r_{u-v}$ [1.31b]

Some of the modifications for atmospheric time series will be treated in further details during the presentation of regular testing procedures (Essenwanger, 1974b).

1.5 THE CENTRAL LIMIT EFFECT

It is stated in most statistical texts that the frequency of the mean values $\bar{x}$ tends towards normality (Gaussian) with increasing N of the data samples from which the mean values have been calculated. This principle, known as the central limit theorem, was considered to be valid irrespective of the distribution form of the original data samples from which the $\bar{x}$ came. Bradley (1973) has recently investigated this question and has concluded that for many populations the approximation towards normality is very slow as N increases. Departure from symmetry is the most damaging effect rather than deviation from the bell shape in the original distribution.

Atmospheric data may, therefore, not always follow well regular statistical principles. This can even go so far that meteorological extreme values follow the Gaussian distribution law and mean values respond very well to curve fitting by extreme value frequencies (see Defrise et al., 1972).

1.6 TCHEBYCHEFF'S AND GAUSSIAN INEQUALITIES

Sometimes we want to know the probability that a random value of a variate x differs from an *expected* value, say α, by as much as K. Tchebycheff (and independently Bienaymé) found that an upper value of this probability exists irrespective of the distribution form of the population provided that the variance σ^2 is finite. The general principle can be expressed as:

$$P[g(x) \geqslant K] \leqslant [E\{g(x)\}]/K \qquad [1.32]$$

This law is known as Tchebycheff's inequality. It can be reformulated for $K = x - \alpha$, and:

$$P(|x - \alpha| \geqslant K) \leqslant \sigma^2/K^2 \qquad [1.32a]$$

or: $P(|x - \alpha| \geqslant h\sigma) \leqslant 1/h^2$

Let us assume $x - \alpha = K = 2\sigma$, or $h = 2$. Then the probability that $|x - \alpha|$ exceeds 2σ is not more than $1/4$; for 3σ it would be $1/9$.

The probability can be given more precisely when the distribution form of $g(x)$ is known. E.g., for $g(x)$ having a Gaussian distribution the maximum probability of a 2σ deviation is 0.0455, for a 3σ deviation only 0.0027. These probabilities are much smaller than the limits from Tchebycheff's law. However, the strong point of the latter is the fact that it is valid for any type of distribution whose form may not be known. Consequently, one should expect that the additional information about the distribution form permits the estimation of the probability more precisely.

Let us assume now that we know the distribution has one single mode x_{mode}. Then we can apply the inequality derived by Gauss:

$$P(|x - x_{\text{mode}}| \geqslant h\tau) \leqslant 4/(9h^2) \qquad [1.33]$$

where: $\tau^2 = \sigma^2 + (x_{\text{mode}} - \bar{x})^2 \qquad [1.33a]$

Again let us assume that $x - x_{\text{mode}} = 2\sigma$. We know that the mode and the mean coincide for the Gaussian distribution, and $\tau^2 = \sigma^2$. Now $|x - x_{\text{mode}}| = 2\sigma \geqslant h\sigma$. Consequently, $h = 2$ and the probability is $4/(9 \cdot 4) = 1/9$; for 3σ the $h = 3$ and $P(...) \leqslant 4/81 \sim 1/20$. Although these probability values are still above the numbers for the Gaussian distribution, they are smaller than from Tchebycheff's law because additional information about the distribution was available.

Various other inequalities for probability estimations are known. The reader is referred to the literature (such as Kenney and Keeping, 1954, Feller, 1966; Kendall and Stuart, 1961, etc.).

CHAPTER 2

FREQUENCY DISTRIBUTIONS

Some basic distributions have been introduced (Essenwanger, 1974a) and will not be repeated here. These basic distributions are the binomial, Poisson, negative binomial, Gaussian, incomplete gamma, extreme value and Weibull distributions. It is assumed that the reader is familiar with these basic models.

In the following sections these basic frequency distributions are supplemented by models which are less frequently found in the literature of statistical analysis of atmospheric data (such as the hypergeometric or the Cauchy distribution) or which are more complicated and difficult (such as the bivariate Gaussian or the beta distribution). Some rare types such as the Pareto distribution (see Johnson and Kotz, 1970a), although gaining in importance in statistical analysis in recent years, have been omitted. The reader may refer to Johnson and Kotz (1969, 1970a, b, 1972a) who have compiled information on most of the frequency models in statistical analysis.

It is considered important to treat mixed and truncated distributions, although mixed models other than the Gaussian type seem to have limited application in atmospheric science. They are theoretically possible, but processes leading to mixed populations such as the mixed binomial or Poisson are very seldom found in practical analysis, even in the general statistical literature. Special references are given on the topics where the author had to be very brief.

Knowledge of the basic test procedure such as standard error, t, F, z, χ^2 and Kolmogorov-Smirnov test (see Essenwanger, 1974a) is also assumed.

2.1 THE HYPERGEOMETRIC DISTRIBUTION

In a dichotomous population and in sampling without replacement the correct distribution is the hypergeometric rather than the binomial distribution. Further, if we select a subgroup for study, especially from a finite population, and the probability p for a certain characteristic is not independent of the history of selection, then the hypergeometric distribution is appropriate. For large finite populations of size N the variation of p would be insignificant and the binomial law can be assumed. Since the hypergeometric distribution may be quite tedious to evaluate one usually finds approximations, one of which is the binomial.

We assume that a finite population of size N (not too large) is given (e.g. different colored balls in an urne). In an event A a subgroup of balls is drawn. We are now interested in a special characteristic x, e.g. the x is counted if we draw a red ball. With every trial the x can either occur or not. If we replace every ball in the sequence of drawings after removal the chances for x remain the same for every trial. If we do not replace every ball the chances of x vary with every trial. We determine now the probability density function p of the x characteristic for the event A. In n trials the number without the property x is $n-x$. The desired probability density function is then:

$$f(x) = \binom{k}{x}\binom{N-k}{n-x} \bigg/ \binom{N}{n} \quad \begin{cases} \text{for } x = 0, 1 \ldots n \\ 0 \text{ elsewhere} \end{cases} \quad [2.1]$$

The $k = Np$, and n is the size of the subgroup or the number of classes. Further:

$0 \leq x \leq n$ for $k \geq n$

$0 \leq x \leq k$ for $k < n$

Thus, sampling from a finite population without replacement leads to the same result as sampling from an infinite population with replacement.
We derive the distribution mean:

$$\bar{x} = kn/N = np \quad [2.2a]$$

This is identical with the mean of the binomial distribution. The variance shows:

$$\sigma^2 = k(N-k)n(N-n)/[N^2(N-1)] = pqn(N-n)/(N-1) \quad [2.2b]$$

against pqn for the binomial. Both are identical if, as a minimum requirement:

$$(N-n)/(N-1) = 1 \quad [2.2c]$$

or: $(1 - n/N)/(1 - 1/N) = 1$

Consequently the binomial distribution is a reasonable approximation for the hypergeometric distribution for large N, say $N > 50$ and $n/N < 0.05$. It can be further seen that for $N > 10$ and $n/N < 0.10$ the left side of [2.2c] is already close to one. In terms of the mean or k we can express this by $\bar{x} < 0.10k$ or in probabilities $p \leq 10\bar{x}/N$. Under these conditions the binomial approximation is good.

The probability density distribution is again more easily computed by its recursion formula:

$$f(x+1) = f(x) \frac{(k-x)(n-x)}{(x+1)(N-k-n+x+1)} \quad [2.3a]$$

The first $f(0)$ of the distribution is:

THE HYPERGEOMETRIC DISTRIBUTION

$$f(0) = \binom{N-k}{n} \Big/ \binom{N}{n} = \frac{(N-k)!}{(N-k-n)!} \frac{(N-n)!}{N!} \qquad [2.3b]$$

for $N \geq (n+k)$

This can be rewritten as:

$$f(0) = \frac{(N-k)(N-k-1)\ldots(N-k-n+1)}{N(N-1)\ldots(N-n+1)} \qquad [2.3c]$$

More details on the binomial approximation and the comparison with the hypergeometric distribution can be found in an article by Brunk et al. (1968).

The higher moments are:

$$v_3 = \frac{n(N-n)(N-2n)\,k(N-k)(N-2k)}{(N-1)(N-2)N^3} \qquad [2.2d]$$

$$v_4 = \frac{N-n}{(N-1)(N-2)(N-3)} \, n\left(\frac{k}{N}\right)\left(1-\frac{k}{N}\right) \Bigg[N(N+1) - 6n(N-n) + 3\left(\frac{k}{N}\right)\left(1-\frac{k}{N}\right)$$

$$\{N^2(n-2) - Nn^2 + 6n(N-n)\} \Bigg] \qquad [2.2e]$$

Equation [2.2d] can also be used to calculate a moments estimator for k. It is better, however, to obtain a maximum-likelihood estimate. As is customary, we need to maximize $f(x)$, i.e. [2.1]. The solution provides the greatest integer not to exceed:

$$\hat{k} = \bar{x}(N+1)/n \qquad [2.4]$$

If $\hat{k}$ is an integer, then $\hat{k} - 1$ is also a maximum-likelihood estimator. The variance of $\hat{k}$ is:

$$\sigma_{\hat{k}}^2 = [\{(N-1)^2(N-n)\}/\{n(N-1)\}]\,(k/N)(1-k/N) \qquad [2.4a]$$

Sometimes the question may arise as to how large the original finite sample N should be, once n and k are known. This question can be answered by [2.2a]. Rewriting the equation renders a maximum-likelihood estimator for N. Again we obtain the greatest integer not to exceed

$$\hat{N} = nk/\bar{x} \qquad [2.2f]$$

and if $\hat{N}$ from [2.2f] is an integer, $\hat{N}$ and $\hat{N}-1$ are both maximum-likelihood estimators. The variance of $\hat{N}$ is:

$$\sigma_{\hat{N}}^2 = [(n+1)(k+1)/(\bar{x}+1)] - 1 \qquad [2.2g]$$

A comparison between binomial, hypergeometric and Poisson distributions is given in Table 2.1.

TABLE 2.1

Comparison between binomial, hypergeometric and Poisson distributions for two selected conditions

	$n = 20, N = 50, k = 2, p = 0.04, q = 0.96$						$n = 20, N = 50, k = 5, p = 0.1, q = 0.9$					
	$f(x)$			$F(x)$			$f(x)$			$F(x)$		
x	binom.	hyperg.	Poisson	binom.	hyperg.	Poisson	binom.	hyperg.	Poisson	binom.	hyperg.	Poisson
0	0.442	0.355	0.449	0.442	0.355	0.449	0.122	0.067	0.135	0.122	0.067	0.135
1	0.368	0.490	0.360	0.810	0.845	0.809	0.270	0.259	0.271	0.392	0.326	0.406
2	0.146	0.155	0.143	0.956	1.0	0.953	0.285	0.364	0.271	0.677	0.690	0.677
3	0.037	0.000	0.038	0.993		0.991	0.190	0.234	0.180	0.867	0.924	0.857
4	0.006		0.008	0.999		0.999	0.090	0.069	0.090	0.957	0.993	0.947
5	0.001		0.001	1.0		1.0	0.032	0.007	0.036	0.989	1.0	0.983
6							0.009		0.012	0.998		0.995
7							0.002		0.004	1.0		0.999
8									0.001			1.0

THE HYPERGEOMETRIC DISTRIBUTION

The c.d.f. can be written as:

$$C_H F(\alpha, \beta, \gamma, 1) = \Sigma f(x) = 1 \qquad [2.1a]$$

with the definition by the infinite series:

$$F(\alpha, \beta, \gamma) = 1 + \frac{\alpha\beta}{\gamma} + \frac{\alpha(\alpha+1)\beta(\beta+1)}{2\gamma(\gamma+1)} + \ldots \text{ (hypergeometric function)}$$

[2.1b]

and: $C_H = [(N-k)!(N-n)!]/[(N-k-n)!N!]$ [2.1c]

2.2 THE LOGNORMAL DISTRIBUTION

It has been observed that some meteorological parameters form an extremely skew probability distribution, and that a transformation into a new variable such as $Y = \ln x$ provides a distribution which may now follow the Gaussian law. A typical example is the visibility whose class coding for recording of synoptic observations follows an approximate logarithmic progression. Guss (1951 and 1955) has pointed out that the frequency distribution of visibility data in a logarithmic scale closely follows a Gaussian law.

2.2.1 Regular model

If we transform the observation X into a new variable by the general form:

$$Y = \frac{y-A}{B} = C \ln \frac{X-D}{E} \quad [2.5]$$

the new variable Y or y may follow the Gaussian law under certain conditions. We speak then of the logarithmic normal distribution, or "lognormal" distribution for short.

In practical applications we can solve the problem by introducing logarithmically progressing class boundaries and renaming these classes by a linear variate (such as Y). The problem is then reduced to a computation of the statistical parameters for Y as outlined in the discussion of the basic frequencies (see Essenwanger, 1974a). Although the frequency distribution, mean, c.p.f. and other information of interest can be obtain by converting the linear variate Y with class boundaries of logarithmic progression in X back to the observation X, this solution may not always satisfy the need. First, a grouping of observations is necessary. After grouping and computation of parameters we may discover that after all we do not have a lognormal distribution. The choice of scale and origin may influence some of the results, and plus-minus sigma boundaries may not always be easy to establish. Thus, knowledge of the statistical representations of parameters in terms of the observed variable X without the detour of transformation and class grouping may be desirable. Transformation first may also create some problems as to the estimation of parameters and similar problems of statistical theory. We should therefore look into the representation of the lognormal distribution in terms of X.

The p.d.f. of a Gaussian distribution for a variate y would be:

$$f(y) = [1/(\sigma_y \sqrt{2\pi})] \exp\left[-\frac{1}{2}\left(\frac{y-\bar{y}}{\sigma_y}\right)^2\right] \quad [2.6]$$

It would therefore be convenient to identify the A in [2.5] with $\bar{y}$ and B

THE LOGNORMAL DISTRIBUTION

with σ_y. We can now rewrite:

$$Y = \frac{y - \bar{y}}{\sigma_y} = C \ln\left(\frac{X - D}{E}\right) \text{ for } X > D \qquad [2.5a]$$

and must estimate the parameters C, D, and E. Before pursuing this goal, let us introduce:

$$C = 1/c \qquad [2.5b]$$

and transform [2.5a] into:

$$\frac{y - \bar{y}}{\sigma_y} = \frac{1}{c} \ln\left(\frac{X - D}{E}\right) \qquad [2.5c]$$

We assume:

$$\bar{y} = (1/N) \sum \ln x \qquad [2.7]$$

and must now also determine $\bar{X}$ and σ_y. The $\bar{y}$ is also called the logarithm of the geometric mean $\bar{x}_g$, and we can express [2.7] by the following:

$$\bar{y} = \ln \bar{x}_g \qquad [2.7a]$$

$$\text{where: } \bar{x}_g = \sqrt[N]{\prod_{i=1}^{N} x_i} \qquad [2.7b]$$

the product of all the x_i for ungrouped data, and:

$$\bar{x}_g = \sqrt[N]{\prod (x_j)^{f_j \cdot N}} \qquad [2.7c]$$

for grouped data into classes with f_j the class frequency (p.d.f.).

We attempt the first parameter estimation of the above constants by moments. The general solution for moments of a logarithmic distribution is:

$$\mu_r = \exp\left[r\bar{y} + (r^2 \sigma_y^2 / 2)\right] \qquad [2.8]$$

Then the mean $\bar{X}$ becomes:

$$\mu_1 = \bar{X} = \exp(\bar{y} + \sigma_y^2 / 2) > 0 \qquad [2.8a]$$

The second moment is:

$$\mu_2 = \exp(2\bar{y} + 2\sigma_y^2) \qquad [2.8b]$$

The moment with reference to the mean is:

$$\nu_2 = \sigma_X^2 = \exp(2\bar{y} + \sigma_y^2)[\exp(\sigma_y^2) - 1] = \bar{X}^2 [\exp(\sigma_y^2) - 1] \qquad [2.8c]$$

We now define: $c_1 = \sigma_X / \bar{X} = [\exp(\sigma_y^2) - 1]^{1/2} \qquad [2.9]$

Hence: $c_1^2 = \exp(\sigma_y^2) - 1 = \sigma_X^2 / \bar{X}^2 \qquad [2.9a]$

From the third moment $\mu_3 = \bar{X}^3 \exp(3\sigma_y^2)$ we derive:

$$\nu_3 = \bar{X}^3 [\exp(3\sigma_y^2) - 3c_1^2 - 1] \qquad [2.8d]$$

This provides the skewness parameter γ_1 as:

$$\gamma_1 = c_1^3 + 3c_1 = c_1(c_1^2 + 3) \qquad [2.8e]$$

If we define $c_2 = c_1^2 + 1 = \exp(\sigma_y^2)$, then c_2 is the only real root of the third-order equation:

$$c_2^3 + 3c_2^2 - 4 - \gamma_1^2 = 0 \qquad [2.8f]$$

The parameters c, D, and E follow (see Yuan, 1933):

$$c^2 = \ln c_2 = \sigma_y^2 \qquad [2.8g]$$

$$E = (1/c_1)(1/c_2)^{1/2} \sigma_X = E' \sigma_X \qquad [2.8h]$$

$$D = \bar{X} - (1/c_1)\sigma_X = \bar{X} - [(c_2+2)/\gamma_1]\sigma_X \qquad [2.8i]$$

The sign of E corresponds with the one for γ_1 in a general logarithmic distribution.

With [2.9] and the definition of $c_2 = \exp(\sigma_y^2)$ we can derive:

$$E = e^{\bar{y}} \qquad [2.8j]$$

and therefore: $\ln E = \bar{y} = (1/N) \Sigma \ln x \qquad [2.7d]$

We restate now [2.5a]:

$$\frac{y - \bar{y}}{\sigma_y} = \frac{1}{\sigma_y} \ln\left(\frac{X - D}{e^{\bar{y}}}\right) \qquad [2.5d]$$

or by reformation:

$$y - \bar{y} = \ln(X - D) - \ln E = \ln(X - D) - \bar{y} \qquad [2.5e]$$

$$y = \ln(X - D) = \ln x \qquad [2.5f]$$

Hence the $x = X - D$. Notice we introduce [2.9] into [2.8i] and obtain the theoretical value $D = 0$, but by estimation $D = \bar{X} - (c_2 + 2)\sigma_X/\gamma_1$. This expression does not render a zero D unless the reference point of X for the lognormal distribution is zero. Otherwise the total frequency curve is shifted, and we need D to fulfill $X > D$ and avoid negative values. The small values of X for $0 \leq X \leq D$ would not exist. If they were real, the $D = 0$ would appear as a boundary condition. (If we have no lognormal distribution, then also $D \neq 0$).

If we substitute c_1 from [2.9] into [2.8e] the skewness of γ_1 leads to:

$$\gamma_1 = \frac{\sigma_X}{\bar{X}} \left(\frac{\sigma_X^2}{\bar{X}^2} + 3\right) \qquad [2.8k]$$

Since $\bar{X}$ and $\sigma_X > 0$, consequently $\gamma_1 > 0$, i.e. $f(x)$ is positively skew, as expected from the distribution.

THE LOGNORMAL DISTRIBUTION

The first three moments give a forced solution. An independent check would be the fourth moment with the γ_2:

$$\gamma_2 = c_3^2[\gamma_1^2 + 2(3c_3^2 + 8)] \qquad [2.8l]$$

where the c_3 has the following meaning:

$$c_3 = [\tfrac{1}{2}\gamma_1 + (\tfrac{1}{4}\gamma_1^2 + 1)^{1/2}]^{1/3} + [\tfrac{1}{2}\gamma_1 - (\tfrac{1}{4}\gamma_1^2 + 1)^{1/2}]^{1/3} \qquad [2.8m]$$

If we compute γ_2 from the observed data and compare it against the one obtained from [2.8l], we can evaluate whether we could be successful in transforming the data to a lognormal law (up to the fourth moment).

It should be pointed out that the lognormal distribution is not determined by its moments. Heyde (1963) has shown that several distributions may exist sharing the same moments. We can conclude, however, that if the observed γ_2 and the one calculated from [2.8l] do not agree by a significant amount, we have no lognormal distribution and the transformation does not lead to a Gaussian curve.

It may be added that the mode should appear at:

$$X_{mode} = e^{\bar{y} - \sigma^2} y + D \qquad [2.10a]$$

The median and geometric mean should coincide with:

$$X_{med} = e^{\bar{y}} + D = \bar{X} \qquad [2.10b]$$

The difference between the mode and median or mean is:

$$X_{mode} - \bar{X} = X_{mode} - X_{med} = e^{\bar{y}}[\exp(-\sigma_y^2) - 1] =$$
$$= -c_1^2/(1 + c_1^2) = -c_1^2/c_2 \qquad [2.10c]$$

A maximum-likelihood solution has been suggested by Cohen (1951). We start with the standardized variable Y. Then:

$$Y = \frac{1}{\hat{c}} \ln\left(\frac{X + \hat{D}}{\hat{E}}\right) \qquad [2.5g]$$

We must estimate the three parameters c, D, and E again, now by maximum-likelihood methods. The solution requires the following steps:

Determine $\hat{D}$ from the equation:

$$[\Sigma\{1/(x_i - \hat{D})\}] \cdot [N \Sigma \ln(x_i - \hat{D}) - N \Sigma \{\ln(x_i - \hat{D})\}^2 + \{\Sigma \ln(x_i - \hat{D})\}^2]$$
$$- N^2 \Sigma \frac{\ln(x_i - \hat{D})}{x_i - \hat{D}} = 0 \qquad [2.11]$$

This equation can be solved by electronic computers and requires elaborate computations. Further:

$$\hat{c}^2 = (1/N) \Sigma [\ln(x_i - \hat{D})]^2 - [(1/N) \Sigma \ln(x_i - \hat{D})]^2 \qquad [2.11a]$$

$$\ln \hat{E} = (1/N) \Sigma \ln (x_i - \hat{D}) \qquad [2.11b]$$

If $\hat{D} = 0$, then $\hat{E}$ is identical with the moments fit [2.7d].

We can simplify the solution of [2.11] by employing the minimum of the observations, x_{min}; and we compute:

$$X_0 = x_{min} + w/2 \qquad [2.12]$$

where w is the class width or distance to the next observation, or the smallest scale interval of reading observations. We further get Y_0, which is the corresponding threshold in a normal distribution by assuming:

$$n_x/N = (1/\sqrt{2\pi}) \int_{-\infty}^{Y_0} \exp(-Y^2/2)\, dY \qquad [2.12a]$$

The n_x is the number of times the x_{min} has been observed. The Y_0 must be obtained from tables of the Gaussian distribution.

With these parameters [2.11] can be reduced to:

$$N \ln (X_0 - \hat{D}) - \Sigma \ln (x_i - \hat{D}) - Y_0 [N \Sigma \{\ln (x_i - \hat{D})\}^2$$
$$- \{\Sigma \ln (x_{min} - \hat{D})\}^2]^{1/2} = 0 \qquad [2.11c]$$

This may be somewhat simpler to solve. Notice that again we establish the reference of X for estimation of D, but $x_{min} > 0$.

The maximum-likelihood estimation is rather tedious without electronic computers.

It is noteworthy, as found by Chow (1954), that the lognormal distribution is related to Gumbel's distribution for $m = 1$ (see Essenwanger, 1974a, section 2.6.3) with the $\gamma_1 = 1.1396$, which provides a $c_1 = 0.364$. Thus the frequency factor k from:

$$x = \bar{x} + \sigma_x k \qquad [2.13]$$

can be computed by approximation as:

$$k = (1/\sigma_z) [\bar{Z} + \ln \{\ln T - \ln (T-1)\}] \qquad [2.13a]$$

The return period T becomes:

$$T = 1.0/P = 1.0/[1.0 - F(z)] \qquad [2.14]$$

The numerical values for $\sigma_z = \pi/\sqrt{6}$ and $\bar{z} \equiv 0.57721$ (as presented by Essenwanger, 1974a, section 2.6.1). This relationship provides the background that:

$$F(z) = \exp(-e^{-z}) \qquad [2.14a]$$

with the transformation $(z - \bar{z})/\sigma_z = (y - \bar{y})/\sigma_y$ can be used as an approximation of the cumulative distribution of the lognormal distribution.

It may be of interest that $y = y_{med}$ is a property of the Gaussian distribution. Pursuing this, we derive the median of X from:

THE LOGNORMAL DISTRIBUTION

$$\bar{x}/x_{med} = \exp(\sigma_y^2/2) \qquad [2.15]$$

$$\text{or: } \ln x_{med} = \ln \bar{x} - \sigma_y^2/2 = \ln \bar{x} - \ln(c_1^2 + 1) \qquad [2.15a]$$

$$x_{med} = \bar{x}/(c_1^2 + 1) \qquad [2.15b]$$

The problem of obtaining "satisfactory" estimates of $e^{\bar{y}}$ and $c_1^2 + 1$ is treated by Laurent (1963), but may be omitted here as these estimators have not found widespread entry into the literature and are tedious to compute.

Severo and Olds (1956) have shown that for testing a hypothesis on the mean of a lognormal distribution (with known variance) the three methods applicable do not yield the same result unless the mean is large compared with the standard deviation. The three test procedures are the normal theory test applied to the $\ln x$, to x and a test based on the Neyman-Pearson (1933) lemma, see also Essenwanger (1974a, section 6.1.1). This means the following: We make a transformation of the variate x to the lognormal distribution. One would assume that now the data have been prepared for testing a hypothesis by a model for which usually a Gaussian distribution is required. The probability level for testing a threshold x by this model (e.g. by [2.13]) would not necessarily provide the true probability level unless the mean is large relative to the standard deviation. This should be expected from the discussion of the various possibilities of estimating the parameters. Consequently some caution should be exercised when the lognormal distribution is employed as a test curve. Further discussions on the estimation of the mean and confidence limits can be found in articles by Thöni (1969) and Koopmans et al, (1964). The lognormal distribution has also found widespread use in hydrology (see Chow, 1964).

Munro and Wixley (1970) have recently developed order statistics for small samples. Broadbent (1956) has described the lognormal approximation to products and quotients. An extensive treatment of the lognormal distribution is also given by Aitchison and Brown (1957).

Example 2.1. Logarithmic normal distribution. Although other applications of the lognormal distribution in meteoreology could be discussed the distribution of the "cold-sums" from 1766–1969 for Berlin, also studied by Teich (1971), has been found suitable to elucidate some important points.

The cold-sum, or sum of cooling degree days, is defined as the accumulation of the temperature difference between zero (centigrade) and the average daily temperature from 1 November through 31 March. Days with a mean daily temperature above zero contribute nothing. The cold-sum is a measure of the winter severity. Both Teich (1971) and Lyness and Badger (1970) have recommended a logarithmic normal distribution for the frequency distribution of the cold-sums. The latter have analysed thresholds from 27° to 47° Fahrenheit at 5° intervals.

Teich has employed a χ^2-test to check the agreement between the analytical and the empirical distribution. Lyness and Badger have plotted the empirical data against a lognormal progression and have obtained a straight line, confirming their assumptions.

TABLE 2.2a

Lognormal distribution of cooling degree days for Berlin, 1766–1969 ($N = 204$)

				Estimation from x	
x_m	226.7	y_m	5.188	$\ln E$	$= 5.23$
s	153.9	σ_y	0.7167	c	$= 0.616$
s^2	23698	σ_y^2	0.5137	$c^2 = \ln c_2$	$= 0.379$
M_3	4238826	M_{3y}	−0.1087		
M_4	$0.22876418 \cdot 10^{10}$	M_{4y}	0.6865		
γ_{1x}	1.16	γ_{1y}	−0.295	$\gamma_{1x} = c_1(c_1^2 + 3)$	$= 2.35$
γ_{2x}	1.07	γ_{2y}	−0.398	$\gamma_{2x} = c_c^2[\gamma_{1x}^2 + 2(3c_3^2 + 8)]$	$= 2.49$

The distribution of cooling degree days for Berlin (1766–1969) comprises 204 observations, whose statistical parameters are given in Table 2.2a and the grouped frequency distribution in Table 2.2b. The first column in Table 2.2a contains the regular moments for the variable x with significant deviations of γ_1 and γ_2 from the ordinary normal distribution, as expected. The average of 226.7 and standard deviation of 153.9 degrees coincide with Teich's findings, as expected. The γ_{1x} is slightly higher than Teich's; this could have been caused by his calculation from grouped data, but the difference is insignificant.

The next column in Table 2.2a lists the transformed data $y = \ln x$, whose distribution is exhibited in Table 2.2b. For this new variable, the γ_{1y} and γ_{2y} are smaller in absolute value but not zero. A simple significance checking procedure is the comparison with the standard error (see Essenwanger, 1974a, section 6.2), $\epsilon(\gamma_1) = \sqrt{6/N} = \sqrt{6/204} = \pm 0.171$. At the 95-% level of significance we would expect $2\,\epsilon(\gamma_1) = \pm 0.343$. The displayed γ_{1y} is therefore within the limitations of a random fluctuation (at the 95-% level of significance). The standard error for γ_2 renders $\epsilon(\gamma_2) = 2\,\epsilon(\gamma_1) = \pm 0.343$. The 95-% level requires a $2\,\epsilon(\gamma_2)$ which is ± 0.686. The given kurtosis from the empirical data stays well within these significance boundaries and we could not conclude from γ_{1y} or γ_{2y} that any significant deviation from the lognormal law would exist with reference to skewness or kurtosis.

One may add that the moments of the lognormal distribution are not unique, i.e. the sequence of moments does not belong only to the lognormal distribution. Other distributions could have the same moments. The standard error test (see Essenwanger, 1974a) is, therefore, a negative rather than an affirmative test. In other words, had we found that γ_1 and γ_2 exceeded the

THE LOGNORMAL DISTRIBUTION

TABLE 2.2b

Frequency distribution and comparison with lognormal

y	n_i	$f(y)$	$F(y)$	Analytical			From x estimation					
				$F_a(y)$	$f_a(y)$	$F(y) - F_a(y)$	F_2	$F_2^y - F_2$	$n_i = n_e$	$n_0 - n_e$	$(n_0 - n_2)^2$	χ_i^2
<3.166	1	0.49%	0.49%	0.25%	0.25%	0.24%	0.04	0.45	0.5			
3.167–3.408	0	0.00	0.49	0.66	0.41	−0.17	0.16	0.33	0.8			
3.409–3.650	3	1.47	1.96	1.58	0.92	0.38	0.52	1.44	1.9	2.1	4.41	0.64
3.651–3.892	5	2.45	4.41	3.44	1.86	0.97	1.46	2.95	3.7			
3.893–4.134	9	4.41	8.82	6.94	3.50	1.88	3.75	5.07	7.1	1.9	3.61	0.51
4.135–4.376	11	5.39	14.21	12.71	5.77	1.50	8.38	5.83	11.8	−0.8	0.64	0.05
4.377–4.618	18	8.82	23.03	21.19	8.78	1.84	16.11	6.92	17.9	0.1	0.01	0.00
4.619–4.860	21	10.30	33.33	32.28	11.09	1.05	27.43	5.90	22.6	1.6	2.56	0.01
4.861–5.102	16	7.84	41.17	45.22	12.94	−4.05	41.68	−0.51	26.4	10.4	108.16	4.10
5.103–5.344	28	13.73	54.90	58.32	13.10	−3.42	57.14	−2.24	26.7	1.3	1.69	0.06
5.345–5.586	26	12.74	67.64	70.54	12.22	−2.90	71.57	−3.93	24.8	1.2	1.44	0.06
5.587–5.828	31	15.20	82.84	81.06	10.52	1.78	83.65	−0.81	21.4	9.6	92.16	4.31
5.829–6.070	10	4.90	87.74	88.88	7.82	−1.14	97.31	−3.57	15.8	5.8	33.64	2.13
6.071–6.312	16	7.84	95.58	94.06	5.18	1.52	95.99	−0.41	10.6	5.4	27.16	2.56
6.313–6.554	7	3.43	99.01	97.13	3.07	1.88	98.38	0.63	6.3			
6.555–6.796	2	0.99	100.00	98.71	1.58	1.29	99.45	0.55	3.2	3.1	9.61	0.80
>6.797	0				1.29				2.6			

$D_a = 1.36/\sqrt{N} \sim 0.97\%$
$D_a = 1.63/\sqrt{N} \sim 1.14\%$
$N = 204$

n_0 = observed
n_e = expected
$\chi_i^2 = (n_0 - n_e)_i^2 / n_{e_i}$

$\chi^2 = 15.13$
$f = 12 - 2 = 10$
$p \sim 87\%$

boundaries of the standard error test, we could deduce that the frequency distribution is not the lognormal law.

The χ^2-test applied by Teich will not be repeated here. The outcome led Teich to the conclusion that no significant deviation from the lognormal frequency can be concluded from the χ^2.

In this next example the Kolmogorov-Smirnov test has been utilized (see Essenwanger, 1974a, section 6.9). Table 2.2b contains the cumulative distribution $F(y)$ and the expected cumulative distribution $F(a)$. Moments estimators for the y have been employed to calculate the $F(a)$. The column $F(y) - F(a)$ delineates the test statistics. We compare these column values against $D_a = 1.36\sqrt{204} \sim 0.97$-% at the 95-% level or $D_a \sim 1.14$-% at the 99-% level of significance. We learn from Table 2.2b that several differences $F(y) - F(a)$ would exceed D_a. Large negative deviations exist around the center of the cumulative distribution and significant positive deviations towards the upper tail. The D_a-values do not even take into account the correction by Lilliefors (1969) (see also Essenwanger, 1974a, section 6.9) which would lead to smaller D_a-values and more significant deviations.

It can be further noticed that recomputation of the analytical lognormal distribution from the estimation in the x-unit (see F_2) would in general make the differences between F_y and F_2, corresponding to $F(y) - Fa(y)$, even larger. The χ^2-test, carried out in the last columns of Table 2.2b would again disclose no significant difference between analytical and empirical frequency.

The estimation for the unconverted x-unit is based on the estimators given in the column of Table 2a, which lead to $y_m = 5.23$ and $s_y = 0.616$.

Why is there a difference in the conclusion about significance between the two tests? It is evident that the Kolmogorov-Smirnov test is more sensitive for smaller N than the χ^2-test. This may explain that significant differences between empirical and analytical frequency are delineated by the Kolmogorov-Smirnov test while results from the χ^2-test indicate that the departures could be non-significant. An example where the test results are reversed is presented in connection with the U-distribution (see section 2.6). This discrepancy points out the necessity for a careful interpretation of test results, and shows that application of the proper test procedure is very complex. In this case efficiency and sufficiency of estimators becomes important and the power of the test should be studied.

Teich (1971) has intuitively felt (although his application of the χ^2-test did not indicate a significant deviation between analytical and empirical frequency) that the data may not be a homogeneous lognormal distribution. He has therefore separated the total frequency distribution into several lognormal components of homogeneous populations. This separation into components will be discussed in Section 2.13.

THE LOGNORMAL DISTRIBUTION

2.2.2 *Lambert's model*

In the preceding section estimators have been calculated for c, D and E for the transformation process $Y = \ln x$. Lambert (1964) has suggested that estimators be calculated for the variate $Y = (y - A)/B$. In his notation it amounts to a three-parameter logarithmic model, which can be obtained by setting $C = 1$ and $E = 1$ in the formula [2.5]. This definition can be interpreted as establishing a standardized variate $X - D = E \cdot x = \sigma \cdot x$, whose variance is unity. The variate Y is then directly proportional to the logarithm of $X - D$. Thus:

$$Y = (y - A)/B = \ln(X - D) \qquad [2.16]$$

or: $X - D = \exp[(y - A)/B] = \exp(Y) \qquad [2.16a]$

The parameters are now A, B and D. In different words we may state that a vector $q(\bar{y}, \sigma_y^2, D)$ should be found by estimation, e.g. $P(\hat{A}, \hat{B}, \hat{D})$ replacing Q.

Lambert (1964) has outlined an iterative procedure for the calculation:

$$P_{i+1} = P_i + \Delta \qquad [2.17a]$$

where P_1 represents an initial set of estimators. According to Aitchison and Brown (1957) this could be an estimation from quantiles. (See primitive characteristics, Section 1.2.)

Another possibility is first the estimation of $\hat{D}$, by quantiles, then the other estimators $\hat{A}_1$ and $\hat{B}_1$ are found from the first two vector elements of M_L, namely $l_1 = 0$ and $l_2 = 0$. A simpler form of this second version is the assumption $D_1 = x_1 - \epsilon$, where ϵ is positive and small, and x_1 the minimum of x.

The vector Δ is defined by:

$$\Delta = M_{IF}^{-1} M_L \qquad [2.17b]$$

with components Δ_A, Δ_B and Δ_D. Further M_{IF} is the information matrix (the negative of the expectations $E(A, B, D)$ of the second derivation of $\ln L$, where L is the likelihood function).

The M_{IF}^{-1} is asymptotically the variance matrix:

$$M_{IF} = \frac{N}{B} \begin{bmatrix} 1 & 0 & \exp(-A + B/2) \\ 0 & 1/(2/B) & -\exp(-A + B/2) \\ \exp(-A + B/2) & -\exp(-A - B/2) & (\beta + 1)\exp[2(B - A)] \end{bmatrix}$$

$$[2.18a]$$

The inverse is:

$$M_{IF}^{-1} = N_B \begin{bmatrix} (\beta + 1)e^B - 2B & -2B & -\exp(A - B/2) \\ -2B & 2B[(B+1)e^B - 1] & 2B\exp(A - B/2) \\ -\exp(A - B/2) & 2B\exp(A - B/2) & \exp(2A - B) \end{bmatrix}$$

[2.18b]

with $N_B = B/[N\{(B+1)e^B - 2B - 1\}]$

Further:

$$M_L = \begin{bmatrix} l_1 \\ l_2 \\ l_3 \end{bmatrix} = \begin{bmatrix} 1/B \; \Sigma[\ln(X_i - D) - A] \\ \dfrac{1}{2B^2} \Sigma[\ln(X_i - D) - A]^2 - N/2B \\ (1 - A/B) \Sigma[1/(X_i - D)] + (1/B) \Sigma[\{\ln(x_i - D)\}/(x_i - D)] \end{bmatrix}$$

[2.18c]

where $l_1 = \partial(\ln L)/\partial A$, $l_2 = \partial(\ln L)/\partial B$, $l_3 = \partial(\ln L)/\partial D$. The n is the number of observations X_i utilized (see below). As Lambert (1964) states, experience supports $P(\hat{A}, \hat{B}, \hat{D}) \to P(A, B, D) \to Q(\bar{y}, \sigma_y^2, D)$. The iteration process [2.17] can be repeated until it can be regarded converged. This is the case when the elements of Δ are below a certain threshold of accuracy (e.g. $< |0.01|$).

It should be noted that the original observations x_i have been replaced by X_i, i.e. quantile estimates.

The maximum-likelihood solution suggested by Cohen (1951) (see [2.11]) requires one equation to be solved by computer iteration, namely [2.11]. The subsequent parameters $\hat{c}$ and $\hat{E}$ are then found by explicit analytical expressions. Lambert's estimation procedure is lengthy, too.

Lambert suggests that for a more rapid calculation modified quantiles be utilized. An x_2 and x_{31} is defined as the lower and upper 3/64-iles, the median $x_{med} = (x_{16} + x_{17})/12$. Then:

$A_1 = \ln[x_{med} - x_2)(x_{31} - x_{med})/(x_2 + x_{31} - 2x_{med})]$ [2.18d]

$B_1 = [0.59662^* \ln\{x_{31} - x_{med})/(x_{med} - x_2)\}]^2$ [2.18e]

$D_1 = (x_2 x_{31} - x_{med}^2)/(x_2 + x_{31} - 2x_{med})$ [2.18f]

Lambert (1970) has later expanded his method to a four-parameter logarithmic model which can be written:

$X = (D + \theta \exp Z)/(1 + \exp Z)$ [2.19a]

or: $Z = \ln[(X - D)/(\theta - X)]$ [2.19b]

---

* $0.59662 = 1/1.6761$

THE LOGNORMAL DISTRIBUTION

Then Z is normally distributed when X is a four-parameter lognormal variate. This type of lognormal variate arises in connection with transformations, which are discussed in detail in Section 3.7. The procedure follows [2.17a] and [2.17b] with the expression $P(\hat{A}, \hat{B}, \hat{D}, \hat{\theta})$ and $\Delta(\Delta_A, \Delta_B, \Delta_D, \Delta_\theta)$.

The (symmetric) information matrix $\mathbf{M}_{IF}$ can be written:

(Symmetric matrix)

$$\mathbf{M}_{IF} = \frac{N}{B} \begin{bmatrix} 1 & \cdot & \cdot & \cdot \\ 0 & 1/2B & \cdot & \cdot \\ \dfrac{1+\exp(-A+B/2)}{\theta - D} & \dfrac{\exp(-A+B/2)}{\theta - D} & M_{33} & \cdot \\ \dfrac{1+\exp(A+B/2)}{\theta - D} & \dfrac{\exp(A+B/2)}{\theta - D} & M_{43} & M_{44} \end{bmatrix} \quad [2.20a]$$

$M_{33} = [1 + 2\exp(-A+B/2) + (1+B)\exp\{2(B-A)\}]/(\theta - D)^2$

$M_{43} = [2 + \exp(-A+B/2) + \exp(A+B/2) - B]/(\theta - D)^2$

$M_{44} = [1 + 2\exp(A+B/2) + (1+B)\exp\{2(B+A)\}]/(\theta - D)^2$

The inverse $\mathbf{M}_{IF}^{-1}$ is asymptotically the variance matrix of the estimators $\hat{A}, \hat{B}, \hat{D}, \hat{\theta}$. This operation is more easily carried out with numerical numbers than with the algebraic expressions.

$$\mathbf{M}_L = \begin{bmatrix} B^{-1} \Sigma[\ln(X_i - D) - \ln(\theta - X_i) - A] \\ -n/2B + (1/2B^2) \Sigma[\ln(X_i - D) - \ln(\theta - X_i) - A]^2 \\ -n/(\theta - D) + \Sigma 1/(X_i - D) + B^{-1} \Sigma[\ln(X_i - D) - \ln(\theta - X_i) - A]/(X_i - D) \\ n/(\theta - D) - \Sigma 1/(\theta - X_i) + B^{-1} \Sigma[\ln(X_i - D) - \ln(\theta - X_i) - A]/(\theta - X_i) \end{bmatrix}$$

[2.20b]

Lambert suggests the following initial values:

$D_1 = X_1 - (X_n - X_1)/N$ [2.20c]

$\theta_1 = X_n + (X_n - X_1)/N$ [2.20d]

$A = n^{-1} \Sigma \ln[(X_i - D_1)/(\theta_1 - X_i)]$ [2.20e]

$B = -A + n^{-1} \Sigma[\ln\{(X_i - D_1)/(\theta_1 - X_i)\}]^2$ [2.20f]

The N is a somewhat arbitrary constant but can be taken as the sample size while n is identified with the number of summations i. From $L_1 = 0$ and $L_2 = 0$ of the vector M_L, it follows that:

$\hat{A} = n^{-1} \Sigma[\ln(X_i - D) - \ln(\theta - X_i)]$

$\hat{B} = n^{-1} \Sigma[\ln(X_i - D) - \ln(\theta - X_i) - A]^2$

The density function can be written:

$$f(x) = [(\theta - D)/\{(x - D)(\theta - x)\sqrt{2\pi B}\}]$$
$$\exp[-(2B)^{-1}\{\ln(x - D) - \ln(\theta - x) - A\}^2] \qquad [2.21]$$

for $D < x < \theta$ and $f(x) = 0$ for $x \leq D$ and $x \geq \theta$.
It should be pointed out that $f(x)$ is bimodal for $B > 2$ and:

$$|A| < B\sqrt{(1 - 2/B)} - 2 \tanh^{-1} \sqrt{(1 - 2/B)} \qquad [2.21a]$$

(The tanh is the hyperbolic tangent.)

Lambert has applied the three- and four-parameter lognormal distribution to precipitation and river floods.

2.3. THE CAUCHY DISTRIBUTION

The Cauchy distribution, although a classical distribution, has not been used very frequently in present-day meteorological statistics. This may be attributed to difficulties in deriving good estimators, and to limited application in meteorology. It will be discussed here for completeness, for possible application to angular variates (especially rectangular distributions), and to familiarize the meteorologist with the existence of the distribution. How to treat angular variates has been discussed by Essenwanger (1974a, section 7.2). Another possible utilization of the Cauchy distribution is the case of bell-shaped curves with heavier tails than the Gaussian distribution.

The Cauchy distribution was derived from the rectangular form, where we have a p.d.f. of:

$$f(\xi) = 1/\pi \qquad [2.22]$$

with the boundaries $-\pi/2 \leqslant \xi \leqslant \pi/2$.
It means that the variable ξ has a constant probability to occur within the given boundaries. The cumulative distribution would then be:

$$F(\xi) = 1/2 + \xi/\pi \qquad [2.23]$$

which is a straight line, the median at $\xi = 0$.
By the transformation:

$$x = a + \alpha \tan \xi \qquad [2.24]$$

we find a distribution for the variable x on a horizontal tangent of a circle where α denotes the radius of the circle. In turn:

$$\xi = \arctan[(x-a)/\alpha] \qquad [2.24a]$$

is the transformed variate for a normalized variable x, with boundaries $-\infty < x < \infty$. This leads to a c.f.d. of:

$$F(x) = (1/2) + (1/\pi)\arctan[(x-a)/\alpha] \qquad [2.25]$$

and a p.d.f. of:

$$f(x) = [1/(\alpha\pi)][1/\{1 + [(x-a)/\alpha]^2\}] \qquad [2.25a]$$

which is a bell-shaped curve and is called the Cauchy distribution. It has heavier tails than the Gaussian distribution. Since α has the function of a standard deviation, it can be found in the literature with the symbol σ. We adopt this change of symbols and [2.25a] can also be written as:

$$f(x) = \sigma \cdot [\pi\{\sigma^2 + (x-a)^2\}] \qquad [2.25b]$$

The theoretical moments of the Cauchy distribution do not exist (they are infinite).

The distribution is symmetrical. In theory median and mean coincide but they may differ for observational data. The parameter a can be determined from [2.24]. With $\xi = 0$ we find:

$$\bar{x} = x_{med} = a \quad [2.24b]$$

because $\xi = 0$ coincides with the median or equivalently the mean. According to various sources the median would be the better choice.

A scheme to derive an "asymptotic relative efficient" estimator has been developed by Bloch (1966). He proposes to use an order statistic of five levels: 13, 40, 50, 60, and 87%. The parameter would then be computed by:

$$a_0 = \sum_1^5 w_i x_i \quad [2.24c]$$

where the weighting functions $w_1 \ldots w_5$ for the five observations $x_1 \ldots x_5$ corresponding to the five levels are -0.052, 0.3485, 0.407, 0.3485, and -0.052. This provides a best linear unbiased estimate.

The other constant α could be estimated from [2.25] with:

$$\pi[F(x) - 0.5] = \arctan[(x-a)/\sigma] \quad [2.25c]$$

$$\text{or: } \sigma = (x-a)/\tan[\pi\{F(x) - 0.5\}] \quad [2.25d]$$

It should be noted that $\sigma = (x - a)$ for $F(x) = 0.75$ or 0.25.

Since [2.25d] can be written for all available points of $F(x)$, we would calculate an average estimator for σ by an order statistic. Let the individual estimator be called s_i, then:

$$s_m = \sum_1^N w_i s_i \text{ or } \sum_1^N s_i \quad [2.25e]$$

The w_i denotes weights, $\Sigma w_i = 1$. They could be all equal (i.e. $w_i = 1/N$). It would be possible to derive a similar scheme as for the location parameter (see [2.24c]).

Instead of the above estimators the maximum-likelihood estimators $\hat{a}$ and $\hat{s}$ could be used. They would be based on the equations:

$$\sum_1^N [(x_i - a)/s]/[1 + (x_i - a)^2/s^2] = 0 \quad [2.26a]$$

$$\text{and: } (2/N) \sum_1^N [1 + (x_i - a)^2/s^2]^{-1} = 1 \quad [2.26b]$$

These equations must be solved by numerical procedures. The usual step is the introduction of the median for a or employing a_0. Then $\hat{s}$ can be computed and checked against [2.26a]. The $\hat{a}$ would be corrected and the process repeated, until a final solution $\hat{a}, \hat{s}$ has been found. Haas et al. (1970) and Barnett (1966) have pointed out that there are several maxima possible in [2.26a] and one would need to find the absolute maximum instead of a

THE CAUCHY DISTRIBUTION

relative maximum. This could be checked by the second derivative (Newton-Raphson method, see Section 4.1.12d). There is no guarantee, according to Haas, that the absolute maximum is the solution closest to the median.

Since the maximum-likelihood estimators have poor efficiency for small samples ($N < 12$), the order statistic may at the present be the better estimate and save elaborate calculation on electronic computers. More theoretical work needs to be done.

Example 2.2. Comparison between Cauchy and Gaussian distribution. This discussion cannot strictly be classified in the group of empirical examples, but it may be very helpful in the computation of the Cauchy distribution and in the study of the difference between the Cauchy and Gaussian distributions.

Two cases were selected for this analysis. Since the distributions with $\sigma \neq 1.0$ would only reflect a difference in the scaling of the x-variate the example was carried out for a value of $\sigma = 1.0$ for the Cauchy distribution. The second parameter $a = 0$ has been arbitrarily selected. As stated previously, it has the function of a location or reference parameter, and the result would only differ in the x-unit by a shift of a.

TABLE 2.3

Cauchy distribution and normal distribution

x	$f(x)$ (in %)			$F(x)$ (in %)		
	Cauchy	Gauss 1	Gauss 2	Cauchy	Gauss 1	Gauss 2
−6	0.86		0.01			
−5	1.22	0.00	0.09	5.3		0.00
−4	1.87	0.01	0.71	6.3	0.00	0.04
−3	3.18	0.44	3.47	7.8	0.13	0.3
−2	6.37	5.40	10.83	14.8	2.3	8.8
−1	15.92	24.20	21.43	25.0	15.9	25.0
0	31.83	39.89	26.91	50.0	50.0	50.0
1	15.92	24.20	21.43	75.0	89.7	75.0
2	6.37	5.40	10.83	85.2	97.7	91.2
3	3.18	0.44	3.47	89.8	99.87	97.9
4	1.87	0.01	0.71	92.2	100.00	99.7
5	1.22	0.00	0.09	93.7		99.96
6	0.86		0.01	94.7		100.00
Σ	90.67	99.99	99.99			

Gauss 1 with $\sigma = 1.0$, Gauss 2 with $\sigma = 1.4827$ to match the quartiles.

The comparison is illustrated in Table 2.3. The column next to the x displays the p.d.f. of the Cauchy distribution with $\sigma_C = 1.0$. This p.d.f. is compared with a p.d.f. of the Gaussian distribution with $\sigma_G = 1.0$. We learn, as expected, that the Cauchy distribution has the longer tail, and within the range of $x = \pm 6$ only about 91% of the total data is contained while virtually all data outside $x = \pm 4$ (corresponding to 4σ) can be neglected for the Gaussian distribution unless the total N is $> 10{,}000$. The comparison of the cumulative distributions follows in the respective columns. Attention is called to the fact that $F(x) \neq \Sigma f(x)$ since $F(x)$ was computed for the given x-value in the first column, while $f(x)$ is valid for a class boundary from $x - 0.5$ to $x + 0.5$. Furthermore, the computation of $F(x)$ by summing $f(x)$ would need smaller class intervals to provide the $F(x)$ accurately enough.

As a last step, the particular Gaussian distribution is calculated whose quartiles coincide. This requires $\sigma_G = 1.4827$ for the Gaussian distribution for $\sigma_C = 1.0$ for the Cauchy distribution. It is obvious that this Gauss 2 has longer tails than the Gauss 1.

This example illustrates further that a comparison between the Cauchy and Gaussian distributions depends on the postulated common requirements. In the first case it was assumed that $\sigma_G = \sigma_C$, in the second case $F_C(x) = F_G(x)$ for the quartiles (i.e. 25, 50 and 75%). Other conditions could have been selected, but the chosen example should suffice to illustrate the property of a longer tail end for the Cauchy distribution than for the Gaussian.

One more remark may be appropriate. Equation [2.25] contains the arctan function. In electronic data processing the question arises whether $(x - a)/\sigma$ may be given in angular units or radians. The reader will notice that the fraction is dimensionless if x, a and σ are given in the same units, such as degrees. The dimensionless fraction represents radians (i.e. arctan $1.0 = \pi/4$, etc.).

Example 2.3. Wind direction distribution, Cauchy and Gaussian representation. One more example with empirical data may reveal properties of the Cauchy distribution. The utilization of the Cauchy distribution for angular variates has been recommended by some authors. The present example is given to illustrate that wind direction data, the major angular data application in climatology, do not necessarily follow the Gaussian or Cauchy distribution. In turn, this example displays the fitting of a Cauchy distribution to empirical data.

The data have been taken from Essenwanger (1974a, table 7.2). The B_0 is the average wind direction in the layer from surface to 5 km. The frequency count in $10°$ class intervals is given in the column n_i in Table 2.4. This frequency count was converted into a cumulative distribution. Technical details on the process of calculating a "true mean" for angular variates and the computation of the cumulative frequency have been outlined in the reference above or by the author's articles in 1961 and 1964. The reader is referred to this literature.

THE CAUCHY DISTRIBUTION

TABLE 2.4
Wind direction model, Cauchy and Gaussian comparison

$f(x)$

Class degrees	n_i	analytical n_i G1	G2	C1	C2	Class degrees	n_i	analytical n_i G1	G2	C1	C2
0—9	11	28.1	25.6	15.0	20.8	180—189	0	12.0	10.3	7.5	8.5
10	21	29.8	34.6	22.4	23.1	190	1	9.2	5.8	5.7	7.2
20	22	30.7	42.7	34.6	24.4	200	1	6.8	3.0	4.5	6.1
30	42	30.5	48.5	50.9	24.1	210	5	4.9	1.4	3.6	5.3
40	63	29.4	50.4	59.6	22.5	220	3	3.4	0.6	2.9	4.6
50	82	27.4	48.1	49.3	20.0	230	6	2.3	0.3	2.5	4.0
60	46	24.8	42.0	33.1	17.2	240	1	1.5	0.1	2.1	3.5
70	13	21.7	33.7	21.5	14.4	250	2	1.0	0.0	1.8	3.1
80	11	18.4	24.7	14.4	12.1	260	1				
90	8	15.1	16.7	10.2	10.1	270	8				
100	8					280	4				
110	4					290	8				
120	0					300	8				
130	2					310	12				
140	0					320	9				
150	2					330	5				
160	2					340	2				
170—179	0					350—359	13				

$F(x)$

Class degrees	%	analytical % G1	G2	C1	C2
0—20	28.4	30.4	23.4	23.9	45.2
20—40	43.4	58.2	44.7	43.8	56.5
40—60	77.2	71.4	67.6	68.9	66.4
60	90.9	82.2	85.3	81.7	73.8
80	95.3	90.0	94.5	87.5	78.9
100	98.1	95.0	98.7	90.6	82.6
120	99.6	97.7	99.8	92.5	88.3
140	99.1	99.1	99.96	93.7	87.3
160	99.5	99.7	100	94.7	88.8
180	99.8	99.89	—	95.3	90.0
200	1.2		—	3.9	10.1
220	3.2	0.1	—	4.4	11.4
240	4.0	0.4	0.0	5.0	13.0
260	6.1	1.0	0.1	5.8	15.1
280	8.9	2.5	0.6	6.9	17.8
300	13.5	5.6	2.8	8.4	21.7
320	16.8	10.9	9.4	10.9	27.1
340—360	21.0	19.1		15.1	34.9

continued on top

G1 = Gaussian, $\phi_m = 28.5$, $\sigma = 55.7$
G2 = Gaussian, $\phi_m = 44.5$, $\sigma = (22.9) \cdot (1.4827) = 33.95$
C1 = Cauchy, $a = 44.5$, $\sigma = 22.9$
C2 = Cauchy, $a = 28.5$, $\sigma = 55.7$
$N = 429$

Four different p.d.f. and c.p.f. have been calculated. The first pair considers a mean of 28.5° and $\sigma = 55.7°$. The result is given in the columns under G1 and C2 for the Gaussian and Cauchy distribution, respectively. The $f(x)$ was converted to a frequency count, while the $F(x)$ is listed in percentages. As expected, the analytical and observed frequencies diverge. The parameters for the Cauchy distribution have been taken from the moments of the Gaussian distribution, an obvious poor choice.

The second pair of frequencies is based on order statistics. A more sophisticated scheme could have been adopted than discussed below. For quick reference, however, only the quartiles were utilized, providing a σ-estimate of 22.9° for the Cauchy distribution (C1). The respective Gaussian distribution (G2) has been calculated with $\sigma = 1.4827$, as previously mentioned. The two resulting p.d.f. and c.p.f. have been placed close together for easier comparison. At first glance one notices a definite improvement in the curve fitting from the previous method of obtaining estimators. This is not only visible for the Cauchy but for the Gaussian distribution, too. The Cauchy distribution (C1) displays better agreement with the observations, although the cumulative distribution indicates that about 8% of the data have been truncated due to a restriction to 360°.

Since the main purpose of this example is the illustration of a difference between moments estimators and order statistics for the Cauchy distribution, and a general demonstration of the computation of the analytical distribution, further discussion on improvements to better fit the empirical data may be omitted. In this particular case the distribution is neither the strictly Gaussian nor the plain Cauchy type.

2.4 THE BETA OR INCOMPLETE BETA FUNCTION

Up to now the beta function or its variation to the incomplete beta function has not played a major role in statistical analysis of meteorological data. The reason can probably be found in the complexity of recomputing the frequency distribution for fractional values of the constants a and b (see [2.27] below). Other estimators than the moment estimators are difficult to compute. This may be another factor why the beta function is little known in meteorology.

The incomplete beta function is very flexible, however, and its form varies from J-shape to bell and U-shape. Thus it may serve for a variety of conditions. The beta function is also related to the cumulative binomial and negative binomial distributions (see Essenwanger, 1974a, section 2.1 and 2.3). In this capacity the beta function has often been applied in statistical analysis problems such as the computation of confidence limits (Essenwanger, 1974a, section 6.1.2).

The beta function is difficult to compute without electronic computers. Pearson (1956) has therefore extensively tabulated the distribution in the form of the incomplete beta function.

The probability density of the beta function is commonly written as:

$$f(x, a, b) = \frac{(a + b + 1)!}{a!b!} x^a (1 - x)^b \begin{cases} \text{for } 0 < x < 1 \\ 0 \text{ elsewhere} \end{cases} \qquad [2.27]$$

The incomplete beta function is sometimes defined by:

$$G(x, a, b) = \int_0^x t^a (1 - t)^b \, dt \qquad [2.28]$$

but more often the ratio $G(x, a, b)/B(a, b)$ can be found, where:

$$B(a, b) = \int_0^1 t^a (1 - t)^b \, dt = (a!b!)/(a + b + 1)! \qquad [2.28a]$$

Pearson (1956) has tabulated this ratio $G(x, a, b)/B(a, b)$. With the above notation we may write:

$$F(x) = [1/B(a, b)] \int_0^x t^a (1 - t)^b \, dt \begin{cases} = 0 \text{ for } x \leq 0 \\ 0 < x < 1.0 \\ = 1.0 \text{ for } x \geq 1.0 \end{cases} \qquad [2.28b]$$

and refer to it in the following abbreviated as the "beta function" or "beta ratio".

The correspondence to the cumulative binomial distribution is:

$$\sum_{x=0}^{x'-1} \binom{n}{x} p^x q^{n-x} = 1 - I(x', n - x' + 1) = 1 - I(a + 1, b + 1) \qquad [2.29a]$$

where $I(a, b)$ is the incomplete beta function as defined below in [2.27a]. The cumulative negative binomial is related to the beta function by:

$$\sum_{x=0}^{x'} \binom{x+k-1}{k-1} p^k q^x = I(a, b) \qquad [2.29b]$$

Pearson (1956) has used the following notation for $p > q$:

$$I(x, p, q) = \left[\int_0^x x^{p-1}(1-x)^{q-1}\,dx\right]\Big/ B(p, q) = F(x) \qquad [2.27a]$$

Consequently the p and q are related to a and b in [2.27] by:

$$p = a + 1 \qquad [2.30a]$$
$$q = b + 1 \qquad [2.30b]$$

It should be further noted that:

$$I(x, p, q) = 1 - I(x', q, p) \qquad [2.31]$$

where: $1 - x' = x$ \qquad [2.31a]

Since: $\Gamma(n) = (n-1)!$ \qquad [2.32]

we can find: $B(a, b) = \dfrac{\Gamma(p)\Gamma(q)}{\Gamma(p+q)}$ \qquad [2.28c]

The moments about zero reference are easily expressed as:

$$\mu_r = \frac{\Gamma(p+r)\,\Gamma(p+q)}{\Gamma(p)\,\Gamma(p+q+r)} \qquad [2.33]$$

If we define: $n^{[r]} = n(n+1)(n+2)\ldots(n+r-1)$ \qquad [2.34]

then: $\mu_r = p^{[r]}/(p+q)^{[r]} = (a+1)^{[r]}/(a+b+2)^{[r]}$ \qquad [2.33a]

Hence: $\mu_1 = \bar{x} = p/(p+q) = (a+1)/(a+b+2)$ \qquad [2.33b]

$\mu_2 = p(p+1)/(p+q)(p+q+1)$ \qquad [2.33c]
$ = (a+1)(a+2)/(a+b+2)(a+b+3)$

$\sigma^2 = v_2 = pq/(p+q)^2(p+q+1)$ \qquad [2.33d]
$ = (a+b+ab+1)/(a+b+2)^2(a+b+3)$

$\mu_3 = p(p+1)(p+2)/(p+q)(p+q+1)(p+q+2)$ \qquad [2.33e]

In the following notation the conversion from p and q to a and b may be omitted:

$v_3 = 2pq(q-p)/(p+q)^3(p+q+1)(p+q+2)$ \qquad [2.33f]

$$\gamma_1 = \frac{2(q-p)}{(p+q+2)}\sqrt{\frac{p+q+1}{pq}} = \sqrt{\beta_1} \qquad [2.33\text{g}]$$

$$\mu_4 = \mu_3(p+3)/(p+q+3) \qquad [2.33\text{h}]$$

$$v_4 = \frac{3pq[2(p+q)^2 + pq(p+q-6)]}{(p+q)^4(p+q+1)(p+q+2)(p+q+3)} \qquad [2.33\text{i}]$$

$$\gamma_2 - 3 = \beta_2 = 3(p+q+1)[2(p+q)^2 + pq(p+q-6)]/[pq(p+q+2)(p+q+3)] \qquad [2.33\text{j}]$$

These moments are based on $0 < x < 1.0$, a normalized scale, or a scaling factor of $R_t = 1.0$. This requirement is generally not fulfilled for regular variates. Then an appropriate scale factor R_t^v must be applied to the moments (see [2.41a] through [2.41c]).

2.4.1 Generalized beta distribution

The generalized beta distribution has the following probability density:

$$f(z) = [1/B(a, b)](z-a_1)^a(a_2-z)^b/(a_2-a_1)^{a+b+1} \qquad [2.27\text{b}]$$

We need four moments to determine the four constants. A general condition for the beta function is:

$$C_B = a + b + 2 = 6(\beta_2 - \beta_1 - 1)/(6 + 3\beta_1 - 2\beta_2) \qquad [2.34]$$

This equation demonstrates that the exponents a and b are functions of the β_1 and β_2 and thus the skewness and kurtosis of the frequency distribution. The parameters p and q can be determined from C_B and β_1 through:

$$p = (C_B/2)[1 + (C_B + 2)\sqrt{\beta_1/\{(C_B+2)^2\beta_1 + 16(C_B+1)\}}] = q' \qquad [2.35\text{a}]$$

$$q = (C_B/2)[1 - (C_B + 2)\sqrt{\beta_1/\{(C_B+2)^2\beta_1 + 16(C_B+1)\}}] = p' \qquad [2.35\text{b}]$$

This set of equations should be used for $p < q$ with $\gamma_1 > 0$. The parameters q' and p' are utilized for $p' > q'$ with $\gamma_1 < 0$.

The constants a_1 and a_2 may be calculated from the mode or the range:

$$z_{\text{mode}} = a_1 + (a_2 - a_1)(p-1)/(p+q+2) \qquad [2.35\text{c}]$$

$$\text{or: } z_{\text{mode}} = a_1 + (a_2 - a_1) \cdot a/(a+b) \qquad [2.35\text{d}]$$

The difference R_z' can be computed from:

$$R_z' = a_2 - a_1 = (\sigma/2)\sqrt{(C_B+2)^2\beta_1 + 16(C_B+1)} \qquad [2.35\text{e}]$$

It can be shown that R_z' is identical with the scale factor R_t by substituting σ, C_B and β_1 which leads to $R_z' = 1.0$ for utilization of [2.33d], [2.34] and [2.33g]. When a_1 is negative, the distribution has its origin at the mode. The R_z' coincides with the range.

The first two moments of the distribution are sufficient for estimation when a_1 and a_2 are known. We have:

$$\bar{x} = a_1 + (a_2 - a_1)p/(p+q) \qquad [2.36a]$$

and: $\sigma^2 = (a_2 - a_1)^2 pq/[(p+q)^2 \cdot (p+q+1)] \qquad [2.36b]$

These formulae are identical with [2.33b] through [2.33d] for $a_1 = 0$ and $a_2 = 1$, for which $f(z)$ of [2.27b] transforms into $f(x, a, b)$ of [2.27]. Thus the mode of the distribution has an x-value of:

$$x_{mode} = (p-1)/(p+q+2) = a/(a+b) \qquad [2.35f]$$

and as such is determined by the exponents a and b. The p can be calculated from

$$p = [(\bar{x} - a_1)/(\sigma^2 R_t)][(\bar{x} - a_1)(a_2 - \bar{x}) - \sigma^2]$$

and with $a_1 = 0$ and $a_2 = 1$ from:

$$p = (x/\sigma^2)[x(1-\bar{x}) - \sigma^2] \qquad [2.36c]$$

The maximum-likelihood estimate is elaborate, but for known boundaries a_1 and a_2 it simplifies to:

$$\psi(\hat{p}) - \psi(\hat{p} + \hat{q}) = (1/N) \Sigma \ln[(z - a_1)/(a_2 - a_1)] \qquad [2.37a]$$

$$\psi(\hat{q}) - \psi(\hat{p} + \hat{q}) = (1/N) \Sigma \ln[(a_2 - z)/(a_2 - a_1)] \qquad [2.37b]$$

where $a_1 < z < a_2$ and ψ denotes the digamma function. It should be noticed that the boundaries are excluded, or else the logarithm becomes infinity. With the notation $a_2 - a_1 = R_z$ and an x-scale with origin at the boundary a_1, the two equations transform to:

$$\psi(\hat{p}) - \psi(\hat{p} + \hat{q}) = (1/N) \Sigma \ln(x/R_z) \qquad [2.37c]$$

and: $\psi(\hat{q}) - \psi(\hat{p} + \hat{q}) = (1/N) \Sigma \ln[(R_z - x)/R_z] \qquad [2.37d]$

or combined into one equation:

$$\psi(\hat{p}) - \psi(\hat{q}) = (1/N)[\Sigma \ln(x/R_z) - \Sigma \ln\{(R_z - x)/R_z\}] \qquad [2.37e]$$

The summation is taken over all x_i (or z_i) from $i = 1, \ldots N$. If the z or x are given in classes, the equation must be modified to have the $f(z)$ or $f(x)$ as the weighting factor such as $\Sigma f(x) \ln(x/R_z)$.

The digamma function, although tabulated by Davis (1963), may be computed simultaneously during the solution by electronic data processing. A proper approximation is:

$$\psi(\xi) \sim \ln(\xi - 0.5) \qquad [2.37f]$$

when $\hat{p}$ and $\hat{q}$ are not too small, i.e. $\xi \geqslant 2.0$. Other approximations can be used such as:

$$\psi(1+s) = -\gamma + \sum_{n=1}^{\infty} s/[n(n+s)] \qquad [2.37g]$$

for $s \neq -1, -2, -3$ etc.
Another approximation is:

$$\psi(1+s) = -\gamma + \sum_{n=1}^{\infty} (-1)^n \zeta_n s^{n-1} \qquad [2.37h]$$

where the ζ is the Riemann zeta function:

$$\zeta(s) = \sum_{n=1}^{\infty} n^{-s} \qquad [2.37i]$$

Formula [2.37h] requires somewhat longer computational time than [2.37g] because of the double approximation process. The γ is Euler's constant, $\gamma = 0.57721$.

The maximum-likelihood equations [2.37c] and [2.37d] become real simple when the approximation [2.37f] is substituted. After some arithmetic we arrive at a first approximation:

$$\hat{p} \sim 0.5[1-B_s]/[1-A_s-B_s] \qquad [2.37j]$$

$$\hat{q} \sim 0.5[1-A_s]/[1-A_s-B_s] \qquad [2.37k]$$

with the abbreviation:

$$A_s = \prod_{i=1}^{N} (x_i/R_z)^{1/N} \qquad [2.37l]$$

$$B_s = \prod_{i=1}^{N} [(R_z - x_i)/R_z]^{1/N} \qquad [2.37m]$$

The exact solution can then be obtained by an iteration process. It should be remembered that [2.37j] and [2.37k] are approximations for $\hat{p}$ and $\hat{q} \geq 2.0$. Gnanadesikan et al. (1967) have given exact numerical solutions in some cases.

For $a_1 = 0$ and $a_2 = 1$ [2.37a] and [2.37b] become very simple, namely:

$$\psi(\hat{p}) - \psi(\hat{p}+\hat{q}) = (1/N) \Sigma \ln x \qquad [2.37n]$$

$$\psi(\hat{p}) - \psi(\hat{p}+\hat{q}) = (1/N) \Sigma \ln(1-x) \qquad [2.37o]$$

If a_1 and a_2 are unknown, a first trial pair of a_1 and a_2 may be employed, e.g. from the moments fit. Then a succession of trials may follow using modifications of a_1 and a_2. The process is time-consuming in most cases, and for a large number of data samples may be expensive, especially when at the same time a variable length of $R_x = a_1 + a_2$ is permitted. However, the process is not more elaborate than a complete ψ-parameter maximum-likelihood solution.

2.4.2 Computations of the beta functions

The approximation of the beta function and computation of the ratio:

$$I(x, p, q) = F(x) = B(x, p, q)/B(p, q) \qquad [2.38]$$

by Pearson is an almost classical approach. His formula is:

$$F(x) = x^p \left[\frac{1}{p} + \frac{1-q}{p+1} x + \frac{(1-q)(2-q)}{2!(p+2)} x^2 \right.$$
$$\left. + \frac{(1-q)(2-q)(3-q)}{3!(p+3)} x^3 + \ldots \right] \qquad [2.38a]$$

for $0 < x \leq 0.5$

and
$$F(x) = F_{0.5} + \frac{1 - r^q}{q \cdot 2^q} + \frac{(1-p)(1-r^{q+1})}{(q+1)2^{(q+1)}} +$$
$$+ \frac{(1-p)(2-p)(1-r^{q+2})}{2!(q+2)2^{(q+2)}} + \ldots \qquad [2.38b]$$

for $0.5 < x < 1.0$

with the abbreviation: $r = 2(1-x)$ $\qquad [2.38c]$

It takes at least three terms, however, and four to five terms are recommended for sufficient accuracy. Double precision in electronic computers will contribute very little to a reduction of the number of terms.

Other approximations of the beta function have been tried, among them approximations by the Gaussian distribution (Peizer and Pratt, 1968):

$$F(x) \sim F(s) \qquad [2.38d]$$

where $F(x)$ is the cumulative beta ratio ([2.28b]) and $F(s)$ the Gaussian cumulative distribution for the parameter s denoting the standard deviation:

$$s = (d_{xq}/|A_q|) \cdot [(2/A_n)(c_q \ln A_1 + c_p \ln A_2)]^{1/2} \qquad [2.38e]$$

The d_{xq} can be replaced by either d_{xq_1} or d_{xq_2}:

$$d_{xq_1} = q - 1/3 - (n + 1/3)(1 - x) \qquad [2.38f]$$

More accurate results are obtained with:

$$d_{xq_2} = d_{xq_1} + 0.2[x/q - (1-x)/p + (x - 0.5)/(p+q)] \qquad [2.38g]$$

The other symbols denote:

$$A_q = q - 0.5 - (p + q - 1)(1 - x) \qquad [2.38h]$$
$$A_n = 1 + 1/[6(p + q - 1)] \qquad [2.38i]$$

THE BETA OR INCOMPLETE BETA FUNCTION

$$c_q = q - 0.5 \qquad [2.38\text{j}]$$

$$c_p = p - 0.5 \qquad [2.38\text{k}]$$

$$A_1 = (q - 0.5)/[(p + q - 1)(1 - x)] \qquad [2.38\text{l}]$$

$$A_2 = (p - 0.5)/[(p + q - 1)x] \qquad [2.38\text{m}]$$

The reader should be cautious as there are several singularities in the computation of s. The approximation cannot be used for the U-distribution.

Gnanadesikan et al. (1967) recommend the calculation of the beta function based on a modified Laplacian expansion developed by Molina (1932). They define:

$$G'(x, p, q) = \int_x^1 t^{p-1}(1-t)^{q-1} \, dt = B(p, q)[1 - I(x, p, q)]$$

$$= B(p, q) \, I(x', p, q) \qquad [2.39]$$

and they approximate:

$$G'(x, p, q) \sim \sum_{i=0}^{6} (A_i/i!)(-\ln x)^{q+i} \gamma(q+i, z) \qquad [2.39\text{a}]$$

The following constants must be utilized:

$$A_0 = 1 \qquad [2.39\text{b}]$$

$$A_1 = A_3 = A_5 = 0 \qquad [2.39\text{c}]$$

$$A_2 = (q-1)/12 \qquad [2.39\text{d}]$$

$$A_4 = (q-1)(5q-7)/240 \qquad [2.39\text{e}]$$

$$A_6 = (q-1)(35q^2 - 112q + 93)/4032 \qquad [2.39\text{f}]$$

$$z = -[p + 0.5(q-1)] \ln x \qquad [2.39\text{g}]$$

$$\gamma(q+i, z) = \int_0^1 s^{q+i-1} e^{-zs} \, ds = e^{-z} \sum_{j=0}^{\infty} \frac{z^j}{q_i(q_i+1)\ldots(q_i+j)} \qquad [2.39\text{h}]$$

where $q_i = q + i$. The formula for $\gamma(q+i, z)$ is due to Wilk et al. (1962). The calculation of the beta function by [2.39h] is not any simpler or less time-consuming than the system employed by Pearson (see Section 2.5).

Gnanadesikan et al. (1967) recommend also a formula to calculate the digamma function:

$$\psi(\alpha) \sim 0.5 \ln[(10 + \alpha_f)(11 + \alpha_f)] + [6(10 + \alpha_f)(11 + \alpha_f)]^{-1} -$$

$$- \left(\frac{1}{\alpha} + \frac{1}{\alpha + 1} + \ldots \frac{1}{\alpha_f + 10} \right) \quad \text{for } 0 < \alpha < 11 \qquad [2.40]$$

and: $\psi(\alpha) \sim 0.5 \ln[\alpha(\alpha-1)] + [6\alpha(\alpha-1)]^{-1}$ for $\alpha \geqslant 11$ [2.40a]

The α_f denotes the fractional part of α if α is not an integer, e.g. $\alpha_f = 0.6$ for $\alpha = 2.6$. Other computations of the beta function for the U-distribution are presented in Section 2.6.3.

2.4.3 Comparison with the negative binomial

The incomplete beta function can be related to the negative binomial distribution. That means the negative binomial distribution is a special case of the beta function for a certain combination of the parameters p and q. The beta function can be related to the negative binomial provided a proper scaling factor is employed.

In order to derive this special combination, we assume:

$$\bar{x}_{NB} = R_t \cdot \bar{x}_B \qquad [2.41a]$$

$$\sigma^2_{NB} = R_t^2 \sigma^2_B \qquad [2.41b]$$

$$v_{3NB} = R_t^3 v_{3B} \qquad [2.41c]$$

The subscripts NB and B refer to the negative binomial and the beta function, respectively, and the R_t is the scaling factor. Comparing the moments of the negative binomial (Essenwanger, 1974a, section 2.3) with the parameters γ and k we derive the following relationships after some lengthy arithmetic:

$$R_t = (1+\gamma)(1+2\gamma) + \bar{x}(3+2\gamma) \qquad [2.42a]$$

$$\text{or: } R_t = v_3/\bar{x} + \bar{x}(3+2\gamma) \qquad [2.42b]$$

$$p = 2\bar{x}(\gamma + \bar{x})/b \qquad [2.42c]$$

$$q = 2(b-\bar{x})(\gamma + x)/b \qquad [2.42d]$$

$$p + q = 2(\gamma + \bar{x}) \qquad [2.42e]$$

The parameter k for the negative binomial is contained in the mean value $\bar{x}$. This simplifies the writing. It can be noticed that the sum of $p + q$ depends on the γ and $\bar{x}$ only (under the assumption that γ and $\bar{x}$ are parameters of the negative binomial). Once the γ and $\bar{x}$ are chosen, the scaling factor R_t must follow from [2.42a]. This can be expressed in terms of p and q as:

$$R_t = [(p+q)(p+q+3) + 2]/[2(1+p)] \qquad [2.42f]$$

If the moments of the beta function are multiplied by this scale factor R_t, the moments of the negative binomial result (see [2.41a] through [2.41c]). The above equations permit conversion of the negative binomial function into the respective beta frequency and the equations below provide for the computation of the parameters of the negative binomial from the beta function. The key is the scale factor R_t:

THE BETA OR INCOMPLETE BETA FUNCTION

$$\gamma = (p+q)/2 - R_t p/(p+q) \qquad [2.42g]$$

and: $\bar{x} = R_t p/(p+q) \qquad [2.42h]$

It should be noticed that the combination of p and q must lead to a value $\gamma > 0$ (see binomial distribution, Essenwanger, 1974a, section 2.1). Not all beta distributions are negative binomials.

A moments fit for the negative binomial via the beta function based on three moments can be obtained as follows (see Essenwanger, 1967):

$$R_t = [M_3(x_m^2 - s^2) - 4x_m s^4]/(M_3 x_m - 2s^4) \qquad [2.43a]$$

$$p = 2(x_m^2 M_3 + x_m^3 s^2 - x_m s^4)/[4x_m s^4 + (s^2 - x_m)M_3] \qquad [2.43b]$$

$$q = s^2 p(p+1)/(x_m^2 - ps^2) \qquad [2.43c]$$

The M_3 can be negative for the beta function, but the R_t, p and q must all be positive. This follows from the condition $x > 0$ and $\gamma > 0$. A negative M_3 is, therefore, not possible for a negative binomial distribution. In this case, the beta function deviates from the negative binomial by its shape. Solving the three equations [2.42a] through [2.42c] for the parameters may sometimes lead to negative estimators of R_t, p or q because of random errors in the observed distribution of the data x. If an approximation of the negative binomial by a beta function is desired, the parameter estimation may be adjusted by setting $M_3 < 2s^4/x_m$.

Details on the moment generating function, applications and tables can be found in Johnson and Kotz (1970b). Two examples of the beta function follow.

Example 2.4. Beta function. Assume that $x = 5.0$ and $\gamma = 4.0$. Then $k = 1.25$ for the negative binomial. This leads to $\sigma^2 = 25$ and $v_3 = 225$. The scaling factor $R_t = 100$. Further $p = 0.9$ and $q = 17.1$, thus $p + q = 18.0$. Assume that we are interested in the median. In Pearson's (1956) tables, $p > q$. Hence we must determine first the $B(p', q')$ for $q' < p'$ by reversing the p and q into p' and q'. Since Pearson's tables progress in units of 0.5, we extract the information for $p' = 17.0$, $q' = 1.0$. This is a sufficiently close approximation. We find $x' = 0.96$ for $F(x') = 0.4996$. Now we must reverse the result ([2.31a]):

$$x = 1 - x' = 0.04$$

This refers to a range from 0 to 1. The actual conditions must be scaled by a scale factor b_x. Notice that the scale factor is not different from R_t except for rounding:

$$b_x = \sigma(p+q)\sqrt{(p+q+1)/pq} \qquad [2.43d]$$

$$b_x = s(18)\sqrt{19/15.4} = 99.7$$

Hence the median would result in:

$$x_{med} = x \cdot b_x = 3.804$$

Example 2.5. Beta function. Let us assume that $p = 5$ and $q = 20$. We should notice that in the negative binomial v_3 is always positive and hence $q > p$ from [2.33f]. We compute $\bar{x} = 11.7$, $\gamma = 0.8$, $k = 14.625$ and $\sigma^2 = 21.06$. The $v_3 = 54.76$ and the $R_t = 58.5$. We extract from the tables for $p' = 20$ and $q' = 5$ the $x' = 0.81$ with $F(x) = 0.5097$. This may be close enough and further interpolation may be unnecessary.

$$x = 1 - 0.81 = 0.19$$

$$b_x = \sqrt{21.06}\,(25)\,\sqrt{26/100} = (114.72)(0.51) = 58.5 = R_t$$

$$x_{med} = (0.19) \cdot 58.5 = 11.11$$

In this case the median is close to the mean value.

2.4.4 A related frequency distribution with beta function

Mielke (1973) has recently introduced a family of frequency distributions for precipitation, which is related with the beta distribution. This family, termed the Kappa distribution by Mielke, has a c.d.f.:

$$F(x) = [\xi^a/(\alpha + \xi^a)]^{1/\alpha} \quad \begin{cases} \text{for } x \geq 0 \\ = 0 \text{ for } x \leq 0 \end{cases} \quad [2.44a]$$

and a p.d.f.:

$$f(x) = c\,\xi^{\theta-1}[\alpha + \xi^a]^{-b} \quad \begin{cases} \text{for } x > 0 \\ = 0 \text{ for } x \leq 0 \end{cases} \quad [2.44b]$$

with: $\xi = x/\beta$ [2.44c]

$a = \alpha\theta$ [2.44d]

$b = 1 + 1/\alpha$ [2.44e]

$c = \alpha\theta/\beta$ [2.44f]

All three parameters α, β, θ must be positive.
Setting $\theta = 1$, a two-parameter model remains. Mielke (1973) has applied this model to precipitation data. Details of this two-parameter model have been described in the precipitation section (Essenwanger, 1974a), and will not be repeated here.

The moments of the three-parameter model can be expressed easily for a reference of zero, namely:

$$\mu_r = \beta^r \alpha^{-w} r B(z_r, w_r) \quad [2.45]$$

THE BETA OR INCOMPLETE BETA FUNCTION

where: $B(z, w) = \Gamma(z)\Gamma(w)/\Gamma(z+w)$ [2.28d]

with: $z_r = (\theta + r)/(\alpha\theta)$ [2.45a]

$w_r = (\alpha\theta - r)/(\alpha\theta)$ [2.45b]

The moments exist only when $w_r > 0$, i.e. $\alpha\theta > r$. It can be readily seen that [2.45] is related to [2.33] and the beta function. While moments estimators for the two-parameter model can readily be obtained via a table of the functions $g(\alpha)$ and $h(\alpha)$ (defined below), the solutions for the three-parameter model are more difficult to derive. The moments (reference zero) are found from:

$\mu_1 = \bar{x} = \beta\alpha^{-w_1}B(z_1, w_1) = \beta h_1(\alpha, \theta)$ [2.45c]

$\mu_2 = \sigma^2 + \bar{x}^2 = \beta^2\alpha^{-w_2}B(z_2, w_2)$ [2.45d]

$\mu_3 = \beta^3\alpha^{-w_3}B(z_3, w_3)$ [2.45e]

The β can be readily eliminated by building the ratio of moments. In fact, $g(\alpha) = \mu_2/\bar{x}^2$ and $h(\alpha) = \beta/\bar{x}$ has been utilized by Mielke (1973) for the two-parameter model. These ratios could be expanded. Two equations with α and θ only can be derived, e.g.:

$\mu_3/\mu_1^3 = \alpha^{3w_1 - w_3}B(z_3, w_3)/B^3(z_1, w_1)$ [2.46]

$= \alpha^2 B(z_3, w_3)/B^3(z_1, w_1) \quad = g_2(\alpha, \theta)$ [2.46a]

and: $\mu_2/\mu_1^2 = \alpha B(z_2, w_2)/B^2(z_1, w_1) = g_1(\alpha, \theta)$ [2.46b]

The separation of α and θ is difficult because the product appears in the beta function. We need a simultaneous solution of the two equations [2.46a] and [2.46b]. Instead of $g_2(\alpha, \theta)$ a more instructive ratio can be substituted. The skewness $\gamma_1 = \nu_3/\sigma^3$ is also independent of β and can be obtained readily from [2.45c] to [2.45e]. The skewness is given in Table 2.5 with θ and α as elements. Simultaneously Table 2.5 illustrates the range of the three-parameter kappa distribution. The second tabulation (Table 2.6) contains $g_1(\alpha, \theta)$, which is an expansion from $g(\alpha)$ given by Essenwanger (1974a).

Inspection of Table 2.5 reveals that the skewness has not been calculated for all possible combinations of α and θ. First we must remember that $\alpha\theta > 3$, otherwise the third moment would not exist. Furthermore, Mielke's (1973) original intent was an application to precipitation data whose skewness is mostly positive. The cut-off at the first value where $\gamma_1 < -0.5$ does not restrict this application. We can further conclude from Table 2.5 that for $\theta \leq 1.0$ the skewness is still positive for $\alpha > 100$. These are very unlikely conditions to occur for precipitation frequencies.

Furthermore, the moments of an empirical distribution cannot be determined with the required precision to determine the α correctly. It can be seen from Table 2.6 that $g_1(\alpha, \theta)$ necessitates already a distinction in the

TABLE 2.5

Skewness γ_1 as function of α and θ

α \ θ	0.25	0.50	0.75	1.0	1.5	2.0	2.5	3.0	4.0	5.0	6.0	7.0	8.0	9.0	10.0	11.0	12.0
0.5																	
1.0																	
1.5																	
2.0					28.9	(30.8)	(32.2)	(33.3)	(34.8)	(35.9)	(36.5)	8.50	4.87	3.61	2.95	2.53	2.24
2.5					4.28	3.53	4.90	2.73	4.28	2.49	1.82	1.46	1.22	1.06	0.94	0.84	0.76
3.0				(26.6)	2.17	1.73	1.89	1.26	1.49	1.00	0.72	0.54	0.41	0.31	0.22	0.16	0.11
3.5				5.74	0.95	1.03	1.01	0.64	0.67	0.30	0.17	0.02	−0.08	−0.16	−0.22	−0.28	−0.32
4.0				3.00	0.49	0.63	0.55	0.27	0.21	−0.03	−0.19	−0.30	−0.40	−0.47	−0.52	−0.58	−0.62
5.0			(25.3)	1.52	0.24	0.37	0.26	0.02	−0.08	−0.29	−0.43	−0.54	−0.62	−0.69	−0.74	etc.	etc.
6.0			3.81	0.98	0.09	0.06	0.06	−0.15	−0.28	−0.47	−0.61	−0.66	etc.	etc.	etc.		
7.0		(24.3)	2.04	0.70	−0.01	−0.11	−0.19	−0.38	−0.43	−0.61	−0.73	−0.73					
8.0		5.46	1.41	0.53	−0.07	−0.23	−0.34	−0.51	−0.63	etc.	etc.	etc.					
9.0		3.03	1.08	0.41	−0.12	−0.30	−0.44	−0.59	etc.								
10.0		2.20	0.89	0.33	−0.16	−0.35	−0.50	−0.65									
12.0		1.77	0.76	0.23	−0.19	−0.39	−0.55	etc.									
14.0	(25.1)	1.34	0.59	0.17	−0.23	−0.44	etc.										
16.0	6.05	1.12	0.50	0.13	−0.25	−0.48											
18.0	3.63	1.00	0.44	0.10	−0.27	−0.50											
20.0	2.82	0.92	0.40	0.08	−0.28	−0.51											
25.0	2.41	0.86	0.37	0.06	−0.30	etc.											
30.0	1.90	0.78	0.33	0.04	−0.31												
35.0	1.75	0.73	0.31	0.03	−0.32												
40.0	1.64	0.71	0.29	0.03	−0.32												
45.0	1.57	0.69	0.28	0.02	−0.33												
50.0	1.53	0.68	0.28	0.02	−0.33												
75.0	1.50	0.67	0.27	0.01	−0.33												
100.0	1.44	0.65	0.26	0.006	−0.34												
500.0	1.41	0.65	0.26	0.003	−0.34												
1000.0	1.39	0.64	0.26	0.000	etc.												
	1.38	0.64	0.26	0.000													

( ) values for $\alpha = 3.1/\theta$

THE BETA OR INCOMPLETE BETA FUNCTION

TABLE 2.6

Function $g_1(\alpha, \theta)$ i.e. [2.46b]

$\alpha \backslash \theta$	0.25	0.50	0.75	1.0	1.5	2.0	2.5	3.0	4.0	5.0	6.0	7.0	8.0	9.0	10.0	11.0	12.0
0.5												1.328	1.223	1.163	1.125	1.099	1.081
1.0											(1.49)	1.073	1.055	1.043	1.034	1.028	1.023
1.5									(1.54)	(1.51)	1.102	1.041	1.031	1.025	1.020	1.017	1.014
2.0					(1.81)	(1.70)	(1.63)	(1.59)	1.273	1.153	1.056	1.030	1.023	1.018	1.015	1.013	1.011
2.5				(2.05)	1.54	1.39	1.39	1.247	1.129	1.081	1.040	1.025	1.019	1.016	1.012	1.011	1.009
3.0			(2.29)	1.82	1.41	1.28	1.24	1.160	1.089	1.058	1.034	1.022	1.017	1.014			
3.5			1.94	1.67	1.34	1.23	1.18	1.124	1.072	1.047	1.030						
4.0		(2.80)	1.77	1.53	1.30	1.20	1.15	1.106	1.063	1.042	1.027						
5.0		2.48	1.68	1.46	1.26	1.18	1.13	1.096	1.057	1.038							
6.0		2.27	1.63	1.43	1.24	1.16	1.12	1.089	1.054	1.036							
7.0		2.15	1.60	1.40	1.225	1.15	1.11	1.081									
8.0		2.07	1.57	1.39	1.217	1.143	1.104	1.077									
9.0		1.98	1.55	1.38	1.211	1.139	1.110										
10.0	(4.36)	1.93	1.53	1.363	1.207	1.136	1.097										
12.0	3.85	1.90	1.52	1.355	1.202	1.134											
14.0	3.52	1.87	1.510	1.350	1.200	1.131											
16.0	3.33	1.86	1.506	1.347	1.197	1.130											
18.0	3.20	1.84	1.500	1.344	1.196	1.128											
20.0	3.04	1.83	1.494	1.340	1.195	1.127											
25.0	2.95	1.820	1.492	1.338	1.193	1.127											
30.0	2.90	1.815	1.490	1.337	1.551												
35.0	2.87	1.811	1.489	1.336													
40.0	2.85	1.809	1.488	1.335													
45.0	2.84			1.335													
50.0	2.804	1.804	1.486	1.334													
75.0	2.793	1.802	1.486	1.334													
100.0																	

( ) values for $\alpha = 3.1/\theta$

TABLE 2.7

Function $h_1(\alpha, \theta)$ i.e. $\bar{x}$ for $\beta = 1$

α \ θ	0.25	0.50	0.75	1.00	1.5	2.0	2.5	3.0	4.0	5.0	6.0	7.0	8.0	9.0	10.0	11.0	12.0
0.5																	
1.0																	
1.5					(1.07)												
2.0					0.97	(1.12)											
2.5					0.91	1.01	(1.16)										
3.0				(0.97)	0.86	0.94	1.07	(1.19)									
3.5				0.90	0.83	0.90	0.98	1.04	(1.22)								
4.0			(0.89)	0.85	0.79	0.87	0.93	0.97	1.11	(1.25)							
5.0			0.79	0.78	0.77	0.85	0.90	0.93	1.01	1.07	(1.26)						
6.0		(0.76)	0.73	0.73	0.75	0.82	0.88	0.91	0.96	1.00	1.05	1.21	1.17	1.14	1.12	1.10	1.09
7.0		0.70	0.68	0.70	0.73	0.80	0.86	0.89	0.94	0.96	0.99	1.03	1.03	1.02	1.02	1.01	1.01
8.0		0.65	0.65	0.68	0.72	0.78	0.84	0.88	0.92	0.94	0.96	0.99	0.99	0.99	0.99	0.99	0.99
9.0		0.61	0.63	0.66	0.72	0.77	0.82	0.86	0.91	0.93	0.95	0.97	0.97	0.97	0.97	0.98	0.98
10.0		0.58	0.61	0.65	0.71	0.764	0.81	0.84	0.90	0.92	0.94	0.95	0.96	0.96	0.96	0.97	0.97
12.0	(0.56)	0.54	0.58	0.63	0.70	0.757	0.80			0.91	0.93	0.94	0.95	0.95			
14.0	0.50	0.51	0.56	0.61	0.69	0.745											
16.0	0.45	0.49	0.55	0.600	0.68	0.737											
18.0	0.42	0.47	0.54	0.593	0.671	0.730											
20.0	0.39	0.46	0.53	0.585	0.666	0.725											
25.0	0.35	0.44	0.51	0.571	0.655	0.721											
30.0	0.33	0.423	0.501	0.562	0.648												
35.0	0.31	0.411	0.492	0.555													
40.0	0.295	0.403	0.486	0.549													
45.0	0.285	0.396	0.481	0.545													
50.0	0.277	0.391	0.477	0.541													
75.0	0.253	0.375	0.463	0.530													
100.0	0.241	0.366	0.456	0.524													

( ) = values for $\alpha = 3.1/\theta$

THE BETA OR INCOMPLETE BETA FUNCTION

third decimal for $\theta = 1$ and $\alpha > 16$.

The maximum-likelihood estimators can also not be obtained readily. This fact can already be concluded from the two-parameter model. We can write:

$$L = \prod_1^n f(x) \qquad [2.47]$$

$$= \left(\frac{\alpha\theta}{\beta}\right)^n \prod_1^n \xi^{\theta-1} \prod_1^n [\alpha + \xi^a]^{-b} \qquad [2.47a]$$

or $L' = \ln L = n \ln(\alpha\theta/\beta) + (\theta - 1) \Sigma \ln \xi - b \Sigma \ln(\alpha + \xi^a)$ [2.47b]

The three equations of the derivatives are consequently:

$$\frac{DL'}{\partial \alpha} = 0 = \frac{n}{\alpha} + \frac{1}{\alpha^2} \Sigma \ln(\alpha + \xi^a) - b \Sigma \frac{1 + \theta \xi^a \ln \xi}{\alpha + \xi^a} \qquad [2.48a]$$

$$\frac{\partial L'}{\partial \beta} = 0 = \frac{-n\theta}{\beta} + \frac{(1+\alpha)}{\beta} \theta \Sigma \frac{\xi^a}{\alpha + \xi^a} \qquad [2.48b]$$

$$\frac{\partial L'}{\partial \theta} = 0 = \frac{n}{\theta} + \Sigma \ln \xi - b \Sigma \frac{\alpha \xi^a \ln \xi}{\alpha + \xi^a} \qquad [2.48c]$$

The solution requires electronic data processing and an iterative procedure similar to the two-parameter model.

Example 2.6. Let us assume that the following moments have been determined from the data: $x_m = 1.08$, $m_2 = 2.30$, $m_3 = 6.53$, $\nu_3 = 1.62$. Then $s^2 = 1.14$ and $s = 1.07$. This leads to $\gamma_1 = 1.33$. We extract pairs of α, θ from Table 2.5 and $g_1(\alpha, \theta)$ from Table 2.6, e.g. the first combination: $\alpha_1 = 12.0$, $\theta_1 = 0.5$, $g_1 = 1.98$, then $\alpha_2 = 7.3$, $\theta_2 = 0.75$, $g_1 = 1.67$, etc. We calculate $m_2/x_m^2 = 2.30/1.17 \sim 1.97$. The first pair of α, θ is closest. Now $\beta = x_m/h_1 = 1.08/0.54 = 2$.

A second set of data may be given by:

$x_m = 2.60$, $m_2 = 8.13$ or $s^2 = 1.34$, $s = 1.16$

Furthermore, $m_3 = 29.16$ or $\nu_3 = 1.0$, then $\gamma_1 = 1.0/1.55 = 0.65$. We select the following pairs:

α	θ	g_1
75.0	0.5	1.80
11.3	0.75	1.56
07.7	1.0	1.42
03.8	1.5	1.30
03.5	2.0	1.20
02.9	2.5	1.15
02.5	3.0	1.12
02.0	4.0	1.09

Now $m_2/x_m^2 = 8.13/6.76 = 1.20$. The solution is therefore $\alpha = 3.5, \theta = 2.0$. The β can be determined by $\beta = x_m/h = 2.60/0.87 = 2.99$ or $\beta \sim 3.0$. In both cases the assumption $\theta = 1$ would not lead to a good fit. Admittedly, the determination of the moments estimators by the set of Tables 2.5 through 2.7 may not be very accurate, and moments estimators for the three-parameter model should be calculated by electronic computers. This rough method may serve, however, for a quick look to judge whether the kappa function may be applicable, and furthermore, whether the two-parameter model could be employed.

2.4.5 Cosine function with rectangularly distributed phase angle

One application of a special type of beta function in physics may be the following situation. We postulate that:

$$y = \cos\theta \qquad [2.49]$$

and: $f(\theta) = a$ (e.g. $a = 1/[2\pi]$)   [2.49a]

The distribution for θ is rectangular.

Since: $F(y) = \int f(\theta)\,d\theta$ [2.49b]

and: $dF(y) = f(y) = a \cdot dy/\sqrt{1-y^2}$ [2.49c]

the distribution leads to a U-shaped frequency, with $y = 0$ for $\cos 90°$ and $y = 1$ or -1 for $\cos 0$ or $\cos 180°$ (i.e. $\cos 0$ or $\cos \pi$), respectively. The U-shaped distribution is a special type of the beta distribution and is treated later in Section 2.6.

2.5 PEARSON'S SYSTEM OF FREQUENCIES

It is rather obvious by now that frequency distributions of observations may assume various types of shapes other than the Gaussian distribution. Although a number of flexible models have been previously introduced, such as the Weibull (Essenwanger, 1974a, section 2.7), or the beta distribution, these types have limitations, too, and a more universal approach would be appropriate. The problem has been recognized and treated by Pearson in 1895. Although in recent times it appears that the Pearson system has been neglected in the statistical literature, it may be worth while for the meteorologist or physicist to become familiar with the variety of descriptive models. It is sometimes difficult to specify in a unique way that only one particular frequency type would apply to a certain set of observational data. (E.g. Essenwanger, 1974a, section 7.5 on precipitation.) The conclusion of the "best fitting model" may be limited to one set of data only. It may, therefore, be beneficial to become acquainted with some universal problems in the selection of suitable models for frequency distributions.

Pearson (1895) developed a system of frequency curves (types) whose main purpose was a mathematical description of frequency distributions regardless of shape. These curves were not originally based on a statistical theoretical background like the preceding frequency distributions but many of the different frequency "types" proved to be identical with some of the commonly known distributions. The Gaussian frequency is a special case of Pearson's systems, the gamma and Weibull distributions belong to type III, and type I of his curves is the generalized beta function. Some theoretical support is also given by Kenney and Keeping (1954). Pearson's system is based on the expression:

$$dy/dx = y(x + a)/(b_0 + b_1 x + b_2 x^2) \qquad [2.50]$$

where y is synonymous with the p.d.f. $f(x)$. The types have been so determined that $dy/dx = 0$ for the mode and large x. The a, b_0, b_1, b_2 are constants related to the moments by:

$$b_0 = \sigma^2(4\beta_2 - 3\beta_1)/[2(5\beta_2 - 6\beta_1 - 9)] \qquad [2.51a]$$

$$b_1 = \sigma\sqrt{\beta_1}(\beta_2 + 3)/[2(5\beta_2 - 6\beta_1 - 9)] = a \qquad [2.51b]$$

$$b_2 = (2\beta_2 - 3\beta_1 - 6)/[2(5\beta_2 - 6\beta_1 - 9)] \qquad [2.51c]$$

The types are based on β_1, β_2, and k, the latter following from the equation:

$$k = \beta_1(\beta_2 + 3)^2/[4(4\beta_2 - 3\beta_1)(2\beta_2 - 3\beta_1 - 6)] \qquad [2.51d]$$

(β_1 and β_2 as defined previously by [1.13a, b]).

Three main types crystallize, depending on the roots of $b_0 + b_1 x + b_2 x^2 = 0$.

Type I b_0 or b_2 negative and the roots real, k negative.
Type IV $b_1^2 < 4b_0 b_2$, roots complex, $0 < k < 1.0$.
Type VI $b_1^2 > 4b_0 b_2$, roots real, $k > 1.0$.

The U-distribution is included in type I and in modified form in type II. We shall analyze this distribution below. More details on Pearson's types can be found in Elderton (1953), Kendall and Stuart (1961) or Craig (1936). Type IV does not correspond to any common statistical distribution. The limit of all Pearson frequency distributions is:

$$\beta_2 - \beta_1 - 1 \geqslant 0 \qquad [2.51e]$$

Outside this limit no Pearson curves exist.
In detail, the k, β_1, β_2 lead to the following Pearson types:

$k = -\infty$ type III (incomplete gamma, Weibull) $2\beta_2 - 3\beta_1 - 6 = 0$

$k < 0$ type I
  special case $5\beta_2 - 6\beta_1 - 9 = 0$ type XII
  special case $g = 0$, $5\beta_2 - 6\beta_1 - 9 < 0$ type VIII
  special case $g = 0$, $5\beta_2 - 6\beta_1 - 9 > 0$;
                          $2\beta_2 - 3\beta_1 - 6 < 0$ type IX

$k = 0$ $\beta_2 = 3$ normal distribution (Gaussian)
          $\beta_2 < 3$ type II
          $\beta_2 > 3$ type VII

$0 < k < 1$ type IV

$k = 1$ type V

$k > 1$ type VI
  special case $g = 0$; $2\beta_2 - 3\beta_1 - 6 > 0$ type XI

$k = \infty$ type III (incomplete gamma, Weibull) $2\beta_2 - 3\beta_1 - 6 = 0$
  special case $\beta_1 = 4, \beta_2 = 9$ type X

$$g = \frac{(4\beta_2 - 3\beta_1)(10\beta_2 - 12\beta_1 - 18)^2 - \beta_1(\beta_2 + 3)^2(8\beta_2 - 9\beta_1 - 12)}{(3\beta_1 - 2\beta_2 + 6)[\beta_1(\beta_2 + 3)^2 + 4(4\beta_2 - 3\beta_1)(3\beta_1 - 2\beta_2 + 6)]}$$

A short summary of the frequency density formulae for the Pearson types follows.

Type I: $f(x) = A_0(1 + x/a_1)^{c_1}(1 - x/a_2)^{c_2}$,   for $a_1 < x < a_2$

More details in Section 2.6.1.

Type II: $f(x) = A_0(1 - x^2/a^2)^c$,   for $-a < x < a$

More details in Section 2.6.2.

PEARSON'S SYSTEM OF FREQUENCIES

Type III: $f(x) = A_0(1 + x/a_1)^B \exp(-c_1 x),$ for $-a_1 < x$

$2\beta_2 = 6 + 3\beta_1$

$B = c_1 a_1 = 4/\beta_1 - 1$

$x_0 (= \text{origin}) = x_{\text{mode}} = 0$

For $\bar{x} = x_0 = 0$ we find:

$f(x) = A_1(1 + x/a_2)^B \exp(-c_1 x),$ for $-a_2 < x$

$a_2 = (B + 1)/c_1$

$A_0 = B^{B+1}/[a_1 \cdot e^B \cdot \Gamma(B + 1)]$

$x_{\text{mode}} = \bar{x} - \nu_3/(2\sigma^2)$

$A_1 = c_1(B + 1)^B/[e^{B+1} \cdot \Gamma(B + 1)]$

(Whenever possible, the gamma or Weibull function should be employed, see Section 2.11 and Essenwanger, 1974a, section 2.5 and 2.7.)

Type IV: $f(x) = A_0(1 + x^2/a_1^2)^{-c_1} \exp(-c_2 \arctan x/a_1),$
   for $-\infty < x < \infty$

With $x/a_1 = \tan \alpha$:

$f(x) = A_0 (\cos \alpha)^{m+2} \exp(-c_2 \alpha)$

$m = 6(\beta_2 - \beta_1 - 1)/(2\beta_2 - 3\beta_1 - 6)$

$c_1 = 1 + m/2$

$c_2 = m(m - 2)\beta_1^{1/2}/[16(m - 1) - \beta_1(m - 2)^2]^{1/2}$
    (opposite sign of ν_3)

$a_1 = (\sigma/4)[16(m - 1) - \beta_1(m - 2)^2]$

The A_0 must be determined by introducing empirical values of $f(x)$.

$x_{\text{mode}} = \bar{x} - \nu_3(m - 2)/[2\sigma^2(m + 2)]$

For $x_0 = \bar{x} = 0$ we can write:

$f(x) = A_0[1 + (x/a_1 - c_2/m)^2]^{-c_1} \cdot \exp[-c_1 \arctan(x/a_1 - c_2/m)]$

The other moments of type IV are:

$\nu_2 = a_1^2(m^2 + c_2^2)/[m^2(m - 1)]$

$\nu_3 = 4a_1^3 c_2(m^2 + c_2^2)/[m^3(m - 1)(m - 2)]$

$\nu_4 = 3a_1^4(m^2 + c_2^2)[(m + 6)(m^2 + c_2^2) - 8m^2]/[m^4(m - 1)(m - 2)(m - 3)]$

Type V: $\quad f(x) = A_0 x^{-B} \exp(-c_1/x) \qquad\qquad$ for $x_0 < x$

$x_0 = \bar{x} - c_1/(B-2)$

$x_{mode} = \bar{x} - 2c_1/[B(B-2)]$

$B = 4 + (8 + 4\sqrt{4+\beta_1})/\beta_1$

$c_1 = (B-2)\sigma\sqrt{B-3}$

$A_0 = c_1^{B-1}/\Gamma(B-1)$

If $x_0 = \bar{x} = 0$, we find:

$f(x) = (B-2)^B/[c_1 e^{B-2} \cdot \Gamma(B-1)]$

Type VI: $\quad f(x) = A_0(x-a_1)^{z_2} x^{-z_1}, \qquad \left(\text{for } \nu_3 \begin{cases} a_1 < x, \\ \text{negative}, -a_1 < x \end{cases}\right)$

We determine the constants in order:

$m = 6(\beta_2 - \beta_1 - 1)/(6 + 3\beta_1 - 2\beta_2)$

$a_1 = (\sigma/2)[\beta_1(m+2)^2 + 16(m+1)]^{1/2}$

$z = (m-2)/2 \pm [m(m+2)/2][\beta_1/\{\beta_1(m+2)^2 + 16(m+1)\}]^{\frac{1}{2}}$

z_1 and z_2 correspond to the positive and negative sign, respectively.

$A_0 = [a_1^{z_1-z_2-1}\Gamma(z_1)]/[\Gamma(z_1-z_2-1)\cdot\Gamma(z_2+1)]$

$x_{mode} = \bar{x} - [\nu_3/(2\sigma^2)][(m+2)/(m-2)]$

For $x_0 = \bar{x} = 0$ we can write:

$f(x) = A_1(1 + x/a_2)^{-z_1}(1 + x/a_3)^{z_2}$

In this case: $(z_1 - 1)/a_2 = (z_2 + 1)/a_3$

$a_2 = a_1(z_1 - 1)/[(z_1 - 1) - (z_2 + 1)]$

$a_3 = a_1(z_2 + 1)/[(z_1 - 1) - (z_2 + 1)]$

$A_1 = \dfrac{(z_2+1)^{z_2}(z_1-z_2-2)^{z_1-z_2}\Gamma(z_1)}{a_1(z_1-1)^{z_1}\Gamma(z_1-z_2-1)\Gamma(z_2+1)}$

A_0 relates to a point before the curve starts.

By setting $c_1 = z_1, c_2 = z_2$ we can transform the moments of type VI to type I (except for a proportionality factor). Consider also that $a_2 \to \infty$ from type I to type VI.

Type VII: $\quad f(x) = A_0(1 + x^2/a_1)^{-c_1}, \qquad\qquad$ for $-\infty < x < \infty$

$x_0 = x_{mode} = 0$

$$c_1 = (5\beta_2 - 9)/[2(\beta_2 - 3)]$$
$$a_1^2 = 2\sigma^2 \beta_2/(\beta_2 - 3)$$
$$A_0 = \Gamma(c_1)/[a_1\sqrt{\pi} \cdot \Gamma(c_1 - 0.5)]$$

This type is a special case of type IV (for $c_2 = 0$) or can evolve from type II.

Type VIII: $f(x) = A_0(1 + x/a_1)^{-c_1}$, \qquad for $-a_1 < x < 0$

For $\bar{x} - x_0 = 0$:

$$f(x) = A_1(1 + x/a_2)^{-c_1} \qquad \text{for } B_1 < x < B_2$$
$$B_1 = -a_1(1 - c_1)/(2 - c_1)$$
$$B_2 = a_1/(c_1 - 2)$$

The constant c_1 lies within the boundaries $0 < c_1 < 1.0$ and must be obtained from the cubic equation:

$$c_1^3(4 - \beta_1) + c_1^2(9\beta - 12) - 24\beta_1 c_1 + 16\beta_1 = 0$$

An approximation can be found from a quadratic equation:

$$c_1 = -2(5\beta_2 - 6\beta_1 - 9)/(3\beta_1 - 2\beta_2 + 6)$$

The two solutions from the cubic and quadratic equations do not completely coincide.

$$a_1 = \pm \sigma(2 - c_1)\sqrt{(3 - c_1)/(1 - c_1)}$$
$\qquad\qquad$ (a_1 has the opposite sign of ν_3)
$$A_0 = (1 - c_1)/a_1$$
$\bar{x} = -a_1/(2 - c_2)$ \quad reference $x_0 = 0$
$\bar{x} = -a_1 + d$ \qquad reference $x_0 = -a_1$
$$d = a_1(1 - c_1)/(2 - c_1)$$

Type IX: $f(x) = A_0(1 + x/a_1)^{c_1}$, \qquad for $-a_1 < x < 0$
$$a_1 = \pm \sigma(c_1 + 2)\sqrt{(c_1 + 3)/(c_1 + 1)}$$
$\qquad\qquad$ (a_1 has the opposite sign of ν_3)

The c_1 must be found from the cubic equation:

$$c_1^3(\beta_1 - 4) + c_1^2(9\beta_1 - 12) + 24c_1\beta_1 + 16\beta_1 = 0$$

$$A_0 = (c_1 + 1)/a_1$$
$\bar{x} = a_1/(c_1 + 2)$ \quad reference $x_0 = 0$
$\bar{x} = -a_1 + d$ \qquad reference $x_0 = -a_1$
$$d = a_1(c_1 + 1)/(c_1 + 2)$$

Type X: $f(x) = \sigma^{-1} e^{-x/\sigma}$, for $0 \leq x \leq \infty$

With $\bar{x} = \sigma$:

$f(\bar{x}) = 1/(e\sigma)$

This type is a special case of the exponential distribution (see Section 2.15). It can also be obtained from type III with $B = 0$ and $\beta_1 = 4$.

Type XI: $f(x) = A_0 x^{-c_1}$, for $d \leq x \leq \infty$

$d = \pm\sigma(c_1 - 2)\sqrt{(c_1 - 3)/(c_1 - 1)}$

$\bar{x} = d(c_1 - 1)/(c_1 - 2)$

Again, the c_1 must be computed from the cubic equation:

$c_1^3(4 - \beta_1) + c_1^2(9\beta_1 - 12) - 24\beta_1 c_1 + 16\beta_1 = 0$

The form for $\bar{x} = 0$ can be written:

$f(x) = A_1(1 + x/a_1)^{-c_1}$

$A_1 = (c_1 - 2)^{c_1}/[d(c_1 - 1)^{c_1 - 1}]$

If we write type XI in the logarithmic form:

$\ln f(x) = C_0 - c_1 \ln x$

the curve can be fitted by the scheme $f(y) = C_0 - c_1 y$.

Type XII: $f(x) = A_0 \left(\dfrac{C_1 + x}{C_2 - x}\right)^z$, for $-C_1 \leq x \leq C_2$

$C_1 = \sigma(3 + \beta_1)^{0.5} + \beta_1^{0.5}$

$C_2 = \sigma(3 + \beta_1)^{0.5} - \beta_1^{0.5}$

$z^2 = \beta_1/(3 + \beta_1)$ (z is the positive root)

$1/A_0 = d\Gamma(c_1 + 1)\Gamma(1 - c_1)$

$\left. \begin{array}{l} c_1^2 = \beta_1/(3 + \beta_1) \\ d = 2\sigma(3 + \beta_1)^{0.5} \end{array} \right\}$ positive roots for negative ν_3

$\bar{x} = 0$

This type is a special transition case of type I for:

$1.5 + 1.125\beta_1 < \beta_2 < 2 + 1.125\beta_1$

The limit of the curve is obtained for $\beta_1 = 0$.

2.6 THE U-DISTRIBUTION

Most of the frequency curves in meteorology display a central hump or have only one boundary (such as wind, precipitation etc.). These curves can easily be represented by some of the previously discussed frequency distributions. Occasionally we find, however, a frequency distribution such as for the cloudiness or sky cover with a two-sided boundary. Since it appears that the sky tends either to be clear or overcast, the terminal conditions occur more often than the transitional classes and the probability density function shows two peaks, one at each boundary. This type of frequency is called a U-shaped distribution, as the histogram looks like the letter U. Sometimes the distribution is referred to as V-shaped (e.g. Chow, 1964). A function will now be introduced whose properties are specially designed to fit the U-distribution. Although the U-distribution has been mentioned in connection with the Pearson system, the U-distribution is a special case of the beta function.

Essenwanger (1974a, section 7.4) has presented the transformation of observed cloudiness data to a Gaussian model. This transformation may not be satisfactory in all analysis cases, and a special treatment of the U-distribution follows now.

2.6.1 The general U-distribution

Pearson's type-I distribution can be written in the following form for the variate z:

$$f(z) = A_0(1 + z/a_1)^{c_1}(1 - z/a_2)^{c_2} \text{ for } a_1 < z < a_2 \qquad [2.52]$$

We introduce: $z = R_z X - a_1$ $\qquad [2.52a]$

and rewrite the probability density with:

$$f(X) = B_0 X^{c_1}(1 - X)^{c_2} \qquad [2.52b]$$

with: $R_z = a_1 + a_2$ $\qquad [2.52c]$

and: $B_0 = A_0 \left(\dfrac{R_z}{a_1}\right)^{c_1} \left(\dfrac{R_z}{a_2}\right)^{c_2}$ $\qquad [2.52d]$

The reader will readily recognize the identity of [2.52b] with [2.28] or [2.28a] except for a proportionality factor. We expand the identity with the beta function by relating c_1 and c_2 to p and q as follows:

$c_1 = p - 1 = a$ $\qquad [2.52e]$

$c_1 = q - 1 = b$ $\qquad [2.52f]$

In the U-distribution c_1 and c_2 must be both negative, and hence p and $q <$ 1.0. We can utilize the moments or maximum-likelihood fit from the beta function and the recomputation of the frequency curve. If we desire to use Pearson's tables of the incomplete beta function, however, we may be disappointed. In the 1956 edition only $I_x(0.5, 0.5)$ is a U-distribution. We could undertake the cumbersome task of calculating I_x-values by the recursion formula:

$$I_x(p, q) = x\,I(p-1, q) + (1-x)\,I(p, q-1) \qquad [2.52g]$$

This elaborate procedure cannot be recommended; neither can the interpolation of Pearson's tables. The following discussion centers on the problems of the U-distribution and some related meteorological questions.

It becomes clear from the relationship of type I with the beta function that maximum-likelihood estimators are difficult to obtain. Only the moments estimators can be calculated with relative ease, but the moments estimators are not efficient. This means we should employ the moments estimators for descriptive purposes only. If the goal is the representation of a population, therefore, other estimators would be necessary.

Another intricate problem is the recomputation of the p.d.f. or c.d.f. with the given estimators. It is quite evident why Brooks and Carruthers (1953) recommend a graphical smoothing of the cumulative distribution and the reading of probabilities from this curve. This solution probably provides one of the best methods to obtain quick results for a complex problem without elaborate work, but it does not resolve the question of representation of a population.

The availability of electronic computers and the accumulation of new sky-cover data by satellites in recent years may encourage a more refined mathematical approach. Thus the problems involved should at least be outlined for those readers unhappy with a graphical solution only.

It should be noticed that [2.52] is written with the mode (antimode!)* as the reference point. We find immediately the relationship:

$$z_{min} = -a_1 \qquad [2.53a]$$

and: $z_{max} = a_2 \qquad [2.53b]$

Hence: $R_z = a_1 + a_2 = z_{max} - z_{min} \qquad [2.52c \text{ expanded}]$

Since further: $k_c = a_1/a_2 = c_1/c_2 \qquad [2.53c]$

we recognize that $k_c = 1$ for symmetric distributions. The distance between z_{min} and the reference point determines the degree of the asymmetry, which is also expressed by the ratio of the exponents. We observe that [2.52] displays a singularity at $z = -a_1$ or a_2, where $f(z)$ becomes ∞. The $F(z)$ is finite, however, as $F(-a_1) = 0$ and $F(a_2) = 1.0$. Recomputation of $F(z)$ and $f(z)$ must therefore give special treatment to the marginal class intervals.

---

* In the U-distribution the lowest frequency density occurs at the reference point and the expression "antimode" is used to indicate this minimum in the frequency density.

THE U-DISTRIBUTION

Details are presented in Section 2.6.3. For $R_z > 8$ a four-point weighted computation of $f(z)$ may provide good approximations in the boundary classes in some cases, but in most cases even this may not suffice.

The following relationship exists between the reference point of the z-scale (reference mode, [2.52]) and the mean of an x-variate:

$$z_0 = a_1 = \bar{x} - 0.5(v_3/\sigma^2)(c+2)/(c-2) \qquad [2.53d]$$

where: $c = c_1 + c_2 + 2 = p + q \qquad [2.53e]$

The $\bar{x}$, σ^2, and v_3 are the first three moments of the observations as usually denoted.

In the beta distribution we have derived C_B in [2.34]. It should be noted that c and C_B are identical.

Hence: $c = 6(\beta_2 - \beta_1 - 1)/(6 + 3\beta_1 - 2\beta_2) \qquad [2.53f]$

Since both c_1 and c_2 must be negative, we conclude that $c < 2.0$ for the U-distribution.

We derive from the similarity with the beta function and the use of [2.53a] and [2.53b] the following expressions, which serve at the same time as estimators of c_1 and c_2 from the moments:

$$c_1 = (c-2)/2 \mp (c/2)(c+2)\sqrt{\beta_1/B_c} \qquad [2.54a]$$

$$c_2 = (c-2)/2 \pm (c/2)(c+2)\sqrt{\beta_1/B_c} \qquad [2.54b]$$

with: $B_c = \beta_1(c+2)^2 + 16(c+1) > 0 \qquad [2.54c]$

The top signs of the equations must be taken for $v_3 > 0$, the lower signs for $v_3 < 0$. Evidently $c = 0$ would characterize a symmetric curve with $c_1 = c_2 = -1$. For reasons discussed below this boundary condition does not exist. We can also eliminate $c = -2$ (see below). This leaves $\beta_1 = 0$ as the condition for a symmetric curve, as expected.

From [2.35e] we deduce the useful equation:

$$R_z = (\sigma/2)\sqrt{B_c} = a_1 + a_2 = z_{\max} - z_{\min} \qquad [2.54d]$$

Since R_z represents the range of the distribution, the B_c must be real and $B_c > 0$ as stated in [2.54c]. Otherwise we would have either a complex root or $R_z = 0$. Introducing the value of B_c we establish the relationship between R_c and the skewness β_1 with:

$$4R_z^2 = \sigma^2[\beta_1(c+2)^2 + 16(c+1)] \qquad [2.54e]$$

We know that $\beta_1 \geq 0$. Consequently a symmetric curve has the range:

$$R_z = 2\sigma\sqrt{c+1} \qquad [2.54f]$$

and $c + 1 > 0$ for any symmetric solution. This eliminates $c = -2$ as a possibility from [2.54a]. Thus the *symmetric* U-distribution must stay within the limits $-1.0 < c < 2.0$.

Whenever $c < -1.0$, the second term in [2.54c] becomes negative. In order to keep the root real, the following conditions must be met:

$$|\beta_1(c+2)^2| > |16(c+1)| \qquad [2.54g]$$

Another limiting condition can be derived by subtracting c_2 from c_1. We obtain:

$$\beta_1 = 4(c+1)(c_1-c_2)^2/[(c+2)^2(c-1+c_1c_2)] \geq 0 \qquad [2.54h]$$

In the symmetric case $c_1 = c_2$, which makes $\beta_1 = 0$, as expected. To keep β_1 positive, we must fulfill:

$$(c+1)/(c-1+c_1c_2) > 0 \qquad [2.54i]$$

This can be rewritten as:

$$(c_1+c_2+3)/[(c_1+1)(c_2+1)] > 0 \qquad [2.54j]$$

As long as c_1 and $c_2 > -1$ but negative, we automatically stay within the limit. When both c_1 or c_2 become smaller than -1.0, the denominator stays positive while the numerator will be negative for $c_1 + c_2 < -3$. Hence only one exponent, either c_1 or c_2 would then be permitted to be smaller than -1.0. This would produce a large β_1 (see [2.54c]) and in practice one may mistake the curve for a J-shaped form. It should be further noticed that β_1 will become infinite for $c = -2$; $c_1 = -1$ or $c_2 = -1$.

The boundary condition for all Pearson's types requires:

$$\beta_2 - \beta_1 - 1 > 0 \qquad [2.54k]$$

This includes the U-distribution. In this sense the β_2 is not independent once the β_1 is chosen and we have actually only three completely independent moments, $\bar{x}$, σ^2 and ν_3.

One may think we need not be concerned about these limiting conditions in practice, as these must be automatically fulfilled whenever the observations display a U-distribution. It should be pointed out, however, that not all distributions in meteorology with U-shaped appearances are exact U-distributions (see later Example 2.7 for more details).

The interrelationship between parameters has also been discussed to foster the understanding that sometimes even maximum-likelihood solutions do not give a good fit to an apparently U-shaped distribution of the observations.

Equation [2.53c] leads to:

$$a_1 = R_z/(1+k_c) \qquad [2.54l]$$

This leaves only the A_0 undetermined [2.52]. Theoretically, the A_0 should be calculated from:

$$A_0 = (1/R_z)(-c_1)^{c_1}(-c_2)^{c_2}(-c_1-c_2)^{-(c_1+c_2)}\Gamma(c)/[\Gamma(c_1+1)\Gamma(c_2+1)] \qquad [2.54m]$$

THE U-DISTRIBUTION

A solution for fractional values of c, c_1 or c_2 may create some problems, although tables of the Γ-value exist (i.e. Elderton, 1953; Davis, 1963). A practical solution to determine A_0 is the following:

$$1/A_0 = \sum_1^n (1 + z_i/a_1)^{c_1} (1 - z_i/a_2)^{c_2} \qquad [2.54n]$$

where the z_i denotes the central class value (see [1.6.b]), and the entire right side provides a summation of the recomputed frequency without the proportionality factor A_0. Since we know $\Sigma f(z) = 1.0$, the proportionality factor can be computed from [2.54n]. This method assumes that $\Sigma f(z)/A_0 = 1/A_0$ can be calculated with sufficient accuracy. Although in empirical distributions the range $R_x = x_{max} - x_{min}$ is easy to get, it may not be the best estimator for R_z. Usually $R_z \sim R_x$ but in practical solutions by [2.54d] the R_z may not even be an integer. This creates another problem: the calculation of the marginal class values. When the observations range from 0 to 10; such as in the cloudiness distribution, the postulation $R_z = 10$ may not even be the best answer for a suitable frequency.

Turning back to the determination of A_0, the frequency may be given for a sample size N instead of the normalized total of $\Sigma f(z) = 1.0$; then $f(z)$ can be multiplied by N, as is shown in Example 2.7.

The maximum-likelihood estimators can be based on [2.37a] through [2.37e] as described for the beta function (Section 2.4). When a_1 and a_2 are not known, the solution requires elaborate computations. One can start with an initial pair of a_1 and a_2 determined from the range R_x and the antimode. It should be pointed out, however, that this original pair governs the ratio k_c and herewith the ratio between p and q or c_1 and c_2, respectively. Thus several trials must be made. This approach is very elaborate and may not lead to a better curve fitting than the moments fit, although it satisfies the requirements from the statistical-theoretical background as representative of a population. It may also be very costly when parameters for numerous data samples must be estimated.

The moments of the U-distribution follow the incomplete beta distribution:

$$\bar{x} = (c_1 + 1)/c \qquad [2.55a]$$

$$\nu_2 = \sigma^2 = (c_1 + 1)(c_2 + 1)[c^2(c + 1)] \qquad [2.55b]$$

$$\nu_3 = 2(c_1 + 1)(c_2 + 1)(c_2 - c_1)[c^3(c + 1)(c + 2)] \qquad [2.55c]$$

$$\nu_4 = \frac{3(c_1 + 1)(c_2 + 1)[2c^2 + (c_1 + 1)(c_2 + 1)(c - 6)]}{c^4(c + 1)(c + 2)(c + 3)} \qquad [2.55d]$$

$$\beta_2 = \frac{3(c + 1)[2c^2 + (c_1 + 1)(c_2 + 1)(c - 6)]}{(c_1 + 1)(c_2 + 1)(c + 2)(c + 3)} \qquad [2.55e]$$

with: $c_1 + c_2 + 2 = c$ [2.55f]

The β_1 is given by [2.54h]. It should be pointed out that the moments as written refer to a scale $R_x = 1.0$, while for any actual length $R_x \neq 1.0$ the four moments must be multiplied by the respective power of R_x, namely R_x^ν, which is similar to computing the scale factor R_t of [2.41a] through [2.41c].

2.6.2 The symmetric U-distribution

The case of $\nu_3 = 0$ with $\beta_1 = 0$ has been introduced previously. This is a symmetrical distribution. Because of its simplification a brief summary will be given here.

From the former discussion we know that for a symmetric U-distribution $k_c = 1$, $c_1 = c_2 = c_s$. Then the symbol c becomes:

$$c = 2c_s + 2 = 2p = 2q \qquad [2.56]$$

We can therefore rewrite the probability distribution in the form:

$$f(z) = A_0(1 - z^2/a^2)^{c_s} \qquad [2.56a]$$

We have further: $R_z = 2a = 2\sigma\sqrt{2c_s + 3}$ [2.56b]

This shows that $c_s > -1.5$ in the symmetric case, and $c > -1.0$. It is easy to verify that:

$$c_s = (5\beta_2 - 9)/(6 - 2\beta_2) \qquad [2.56c]$$

Then: $a^2 = 2\sigma^2\beta_2/(3 - \beta_2)$ [2.56d]

In this form: $A_0 = [1/(a\sqrt{\pi})] \cdot \Gamma(c_s + 1.5)/\Gamma(c_s + 1.0)$ [2.56e]

Equation [2.54n] is also valid in modified form:

$$1/A_0 = \Sigma(1 - z_i^2/a^2)^{c_s} \qquad [2.56f]$$

The moments simplify to:

$$\bar{x} = 0.5 = (c_s + 1)/[2(c_s + 1)] \qquad [2.57a]$$

$$\sigma^2 = 1/[4(2c_s + 3)] \qquad [2.57b]$$

$$\nu_3 = \beta_1 = 0 \qquad [2.57c]$$

$$\nu_4 = 3/[16(2c_s + 3)(2c_s + 5)] \qquad [2.57d]$$

$$\beta_2 = 3(c_s + 3)/(2c_s + 5) \qquad [2.57e]$$

The moments have been written for a range $R_z = 1.0$, and the mode and mean coincide. Hence $a = 0.5$. The standard deviation is normalized, too. Again, the previous remark for $R_z \neq 1.0$ remains valid.

THE U-DISTRIBUTION

2.6.3 *The recomputation of the U-distribution*

The recomputation of an analytical U-distribution from a set of given estimators is not trivial irrespective of whether we deal with moments or maximum-likelihood estimators. In fact, the computation of the entire incomplete beta distribution is a problem. We cannot use the same technique as is customary for most of the other distributions, namely we calculate $f(x)$ by determining the central class value, (in most cases the midpoint), and the resulting frequency is a good approximation for the true class frequency. This technique would work for the classes away from the margins, but the $f(x)$ of the boundary classes cannot be derived in that manner. It should be remembered that $f(-a_1)$ and $f(a_2)$ are infinite, though $F(-a_1) = 0$ and $F(a_2) = 1.0$ are finite.

It is therefore recommended (as in the case of the extreme value distributions) that one obtains $f(x)$ from the difference $F(x_{i+1}) - F(x_i)$, where the x_{i+1} and x_i are the upper- and lower-class boundaries, respectively. Even then, the integral:

$$f(x) = \int_{x_i}^{x_{i+1}} x^{c_1}(1-x)^{c_2} dx = F(x_{i+1}) - F(x_i) \qquad [2.58]$$

is not easy to compute. Several approximations have been suggested in the past (e.g. Aroian, 1941; Hartley and Fitch, 1951; Thomson, 1947; and Wise, 1960; or Johnson and Kotz, 1970b). Some approximations have been discussed in Section 2.4.2.

A classical approach by Pearson (1956) has been previously introduced ([2.38a–c]) which was used by Pearson (1956) to produce the set of tables for the incomplete beta function. Equations [2.38a] through [2.38c] are written for p and q and the conversion from the q, p-notation to the parameters c_1 and c_2 of the U-distribution must be made by [2.52a] and [2.52b], namely $p = c_1 + 1$ and $q = c_2 + 1$.

Another approximation of the U-distribution can be based on trigonometric functions, although this method usually does not give better accuracy. The approximation is much simpler, however. We merely must compute:

$$f(z) = \int_{\theta_i}^{\theta_{i+1}} (\sin \theta)^{2c_1+1} (\cos \theta)^{c_2+1} d\theta = F(\theta_{i+1}) - F(\theta_i) \qquad [2.59]$$

It would be easy to calculate $f(z)$ from:

$$x = (\sin \theta)^2 \qquad [2.59a]$$

and: $dx = 2 \sin \theta \cos \theta \, d\theta \qquad [2.59b]$

with the transformation: $x = (z-a)/R_z \qquad [2.59c]$

relating the observation x to the z-scale of the U-distribution ([2.52]). One may even think of an approximation by:

$$f(z) = 2(\sin \theta)^{2c_1+1}(\cos \theta)^{2c_2+1} d\theta \qquad [2.59d]$$

The $d\theta$ could then be computed by [2.59b]. It is difficult, however, to assess the proper central class value of θ and hence the method by formula [2.59] is more appropriate since it is based on the difference between two values of the cumulative distribution. The θ_i can then be obtained from [2.59a] and $d\theta$ from the difference $\theta_{i+1} - \theta_i$. This may deviate from [2.59b], as we need no assumption for the θ.

Another method suitable for electronic computers is a trivial iteration by [2.52] in small steps omitting the boundaries $-a_1$ and a_2. The steps need to be very small, however, say at least $dx = 0.01$ to 0.001. This is very time-consuming and calculations based on Pearson's formula lead to results much faster. We have the advantage, however, that the central class value x_c can be computed at the same time, namely:

$$x_c = \sum_{i=1}^{n_j} f(x) x_i \qquad [2.58a]$$

This is a by-product of the calculation.

In practical work another problem arises, however. In analytical writing the R_z is an integer, the R_x determined from [2.54d] is not. How should the marginal classes be adjusted? As an example let us examine the cloudiness records. The classes run from zero to ten tenth sky cover. According to [2.54d] the $R_z = z_{max} - z_{min} = 10.0 - 0.0 = 10.0$. The central class values are 0, 1, 2 ..., 10. It is obvious that these class centers under commonly accepted rules (see Section 1.1, [1.6b]) would lead to a class width of 1.0 units. The two border classes, although theoretically running from -0.5 to 0.5 and 9.5 to 10.5, would be limited to 0.5 units, namely from 0.0 to 0.5 and 9.5 to 10.0. This division renders eleven classes, although the boundary classes in practical application are smaller than the other established class intervals. Assume now that R_x has been computed to be 10.24. We could now truncate R_x to an integer of 10 and adapt the class division as given above. We may, however, keep the length of the analytical distribution with $R_z = 10.24$ and compute a distribution, in which we lengthen the boundary class interval by half of the excess from the integer. The boundary class intervals would then comprise 0.62 units and thus be expanded to range from -0.12 to 0.5 and 9.5 to 10.12 for the lower and upper interval, respectively.

This makes the boundary classes different from the intervals of the inside classes, We need not worry about this unequal division as long as we compute the probability density $f(x)$ or $f(z)$ from the cumulative distribution $F(x)$ or $F(z)$, respectively. The adjustment of the boundary classes helps the curve-fitting process by improving the approximation of the empirical distribution

THE U-DISTRIBUTION

by the analytical curve.

If $f(x)$ is however computed from any frequency density formula by approximation assuming a certain central class value, adjustment of the difference in the class width must be taken into consideration. This procedure must be followed independently from the method of estimation.

We had seen that in the case of R_z being an integer the last classes are only half a class-unit. This also would necessitate a correction when $f(x)$ is computed directly. The adjustment is also recommended by Johnson and Kotz (1970b).

In practice it is further advisable to take the U-shape of the distribution into consideration by modifying the central class values of the boundary classes. An empirical formula such as:

$$x_c = x_b \pm (w/2)[1 - (f_l + f_u)^2] \qquad [2.60]$$

could be used. The symbols have the usual meaning, x_c the central class value, x_b stands for either the given class minimum (positive sign) or the maximum (negative sign), w denotes the class width. Attention is called to the fact that f_l and f_u, the frequency of the lower- and upper-boundary class, respectively are fractional units (see the definition of the probability density, Section 1.1). The term in the brackets disappears when all observed values fall into the boundary classes and x_c is then identical with the given value of the observations (i.e. x_b is zero or ten tenths cloud cover).

In the case that 50% of the observations are grouped into the boundary classes, $x_c = x_b \pm 0.375$ for a class width $w = 1.0$. In our previous class interval example, in which we start with $x_b = -0.5$, the $x_c = -0.125$ for the lower-boundary class. We would modify [2.60] to accommodate a smaller interval by replacing $w/2$ by $\delta/2$, where:

$$\delta = R_x - R_z, \text{ i.e.} \qquad [2.60a]$$

the difference between the computed range for the x observations and the integer of the analytical observation z in [2.52].

An adjustment for fewer classes and therefore a different length of the range R_x in [2.60] need not be made. Since f_l and f_u are the frequencies in the classes, fewer classes are likely to increase the class frequency in the boundary class interval and the formula responds to this change.

Example 2.7. Cloudiness distribution. The example illustrates the curve fitting of cloud-cover data for Hampton, Va./Langley AFB, July 1946—1964, between 3 and 5^h LST. The computed parameters for Table 2.8 are as follows: $c_1 = -0.52$, $c_2 = -0.53$, $R_x = 10.76$ with $a_1 = 5.33$, $a_2 = 5.43$. The moments $x_m = 5.43$ (or $x'_m = 5.10$), $\sigma = 3.84$, $\nu_3 = -1.02$, $\beta_1 = 0.0003$, $\nu_4 = 319.9$, $\beta_2 = 1.47$, $x_{mode} = 5.33$. The first column of Table 2.8 is the adjusted scale in x for the length of R_x corresponding to the cloud-cover scale, column 2. The next two columns list the number of observations in the

TABLE 2.8

Cloudiness distribution, Hampton, Va., July 1946—1964, between $3-5^h$, LST

x	Sky cover x'	Observed	$f(x)$ (%)	$f(z)$ (%)	$F(z)^*$ (%)	z_c	$f(x)-f(z)$ (%)	$F(x)^*$ (%)	$F(x)-F(z)$ (%)	Analytical n_z	Maximum likelihood $f(z)$	$F(z)$
0.38	0	359	20.3	18.9	18.9	−5.05	1.4	20.3	1.4	334	18.6	18.6
1.38	1	91	5.1	8.8	27.7	−3.95	−3.7	25.4	−2.3	155	8.8	27.4
2.38	2	151	8.6	6.9	34.6	−2.95	1.7	34.0	−0.6	123	7.0	34.4
3.38	3	140	7.9	6.2	40.8	−1.95	1.7	41.9	1.1	109	6.3	40.7
4.38	4	101	5.7	5.8	46.6	−0.95	−0.1	47.6	1.0	103	5.8	46.5
5.38	5	71	4.0	5.7	52.3	0.05	−1.7	51.6	−0.7	100	5.8	52.3
6.38	6	94	5.3	5.8	58.1	1.05	−0.5	56.9	−1.2	103	5.9	58.2
7.38	7	127	7.2	6.2	64.3	2.05	1.0	64.1	−0.2	110	6.3	64.5
8.38	8	164	9.3	7.0	71.4	3.05	2.3	73.4	2.0	124	7.1	71.6
9.38	9	88	5.0	8.9	80.3	4.05	−3.9	78.4	−1.9	158	9.0	80.6
10.38	10	381	21.6	19.7	100.0	5.15	1.8	100.0	0.0	348	19.4	100.0
	N	1767	100.0									

* The cumulative distribution is summed up to the upper class boundary, i.e., 0.5, 1.5, 2.5, ... etc.

individual classes and the converted $f(x)$. The following columns exhibit the analytical $f(z)$ and $F(z)$ computed from the moments fit (maximum-likelihood fit last two columns). The corresponding class center for the z-scale is given in column 7.

The columns for the deviations of the analytical from the observed probability density and cumulative frequency deserve special attention. If we apply a test for the goodness-of-fit (see Essenwanger, 1974a, section 6), we would discover that the Kolmogorov-Smirnov test does not reveal any significant deviation at the 5-% significance level, but the χ^2-test would. This appears as a contradiction at first glance.

We cannot claim in our case that the χ^2-test would not be applicable for the reason that the moments fit is not very efficient (see Kenney and Keeping, 1954, vol. II). The fit by the maximum-likelihood method displays little difference from the moments method, which indicates a reason other than efficiency of the estimators.

We notice, however, that the large deviations occur at the marginal classes, especially the second and tenth class. This fact has been noted already by Brooks and Carruthers (1953) and seems to be typical of cloud-cover data. The author has observed this in over 100 computed samples. The deficiency in the above-mentioned class intervals may be a misclassification by observers, who tend to place the observations either into the boundary classes or the adjacent ones. Thus the observed cloud frequencies lack observations in these two classes next to the margins, and the distributions do not constitute a rigorous U-distribution in the sense of a statistical (smooth) function. The significance of the deviation in the χ^2-test may thus indicate an observer bias or an inhomogeneous population, the latter being attributed to the multi-layer structure of clouds.

If we establish the empirical cumulative distribution, the deviations from the analytical model become smaller, as positive and negative deviations balance out. The values in the third classes from the boundaries agree. The Kolmogorov-Smirnov test checking the c.d.f. therefore does not reach significance. In a longer period of record (i.e. larger N), the differences would also exceed the significance threshold in the latter test. If we disregard the three classes each from the margins, the χ^2-test would not provide a significant deviation between observed and analytical models. It is quite possible that records of cloud cover by satellite data will eliminate the bias. The only other explanation of the bias under the assumption that sky-cover data should provide an exact U-distribution would be the postulation of a mixture of populations. The period of day, month and climatic conditions in the sample seems homogeneous enough not to create more of a mixture than in other collectives (Essenwanger, 1974a, section 1.5).

Some readers may prefer the graphical smoothing. In the author's opinion, however, it is difficult to pinpoint the starting points of the smooth curve (see Table 2.8) and a graphical method may therefore introduce another bias.

The maximum-likelihood estimate was computed by utilization of [2.37e] and [2.37g and h]. It should be remembered that c_1 and c_2 are negative. The approximation [2.37k and l] does not therefore lead to faster results than the employment of an iterative process.

The solution provided $\hat{c}_1 = -0.51$, $\hat{c}_2 = -0.52$ with $\Psi(\hat{p}) = -2.012$, $\Psi(\hat{q}) = -2.057$ and $\Psi(\hat{p} + \hat{q}) = -0.627$. The right sides of [2.37c and d] are -1.237 and -1.282, respectively. The two sides in [2.37e] (both 0.045) agree therefore up to the third decimal. The comparison between the left and right side in [2.37c] shows a difference of -0.148, namely $-2.012 + 0.627 = -1.385$ against -1.237. Furthermore, [2.37d] provides $-2.057 + 0.627 = -1.430$ against -1.282, i.e. the difference is -0.148. This may be close enough. The maximum-likelihood estimators in this case are not much different from the moments estimators.

Example 2.8. Cloudiness distribution. The moments fit of Table 2.9 (upper part) provides the following solutions (end points -0.10 and 10.10): $c_1 = -0.79$, $c_2 = -0.85$, $R_x = 10.53$ with $a_1 = 5.06$, $a_2 = 5.47$. The moments are $x_m = 6.17$ ($x'_m = 4.91$), $\sigma = 4.40$, $\sigma^2 = 19.36$, $\nu_3 = -30.22$, $\nu_4 = 501.95$, $\beta_1 = 0.126$, $\beta_2 = 1.34$, $x_{mode} = 5.06$. The columns are the same as in Table 2.8, Example 2.7. We see the same features, only this time the Kolmogorov-Smirnov test agrees with the χ^2-test by displaying a significant deviation, too, because the sample is about ten times the previous size.

We obtain the maximum-likelihood estimators by finding a pair $\hat{c}_1$ and $\hat{c}_2$ for which the differences between the left and right side of [2.37c and d] are simultaneously zero. The solution provides $\hat{c}_1 = -0.62$ and $\hat{c}_2 = -0.70$, deviating somewhat from the moments fit. The right sides of these equations are -1.433 and -2.223, respectively; further, $\hat{p} = 0.38$; $\hat{q} = 0.30$; $\hat{p} + \hat{q} = 0.68$. Then $\Psi(p) = -2.714$, $\Psi(q) = -3.502$, $\Psi(p + q) = -1.278$. The differences between the sides in [2.37e. c and d] amount to -0.002; -0.003, and -0.001, respectively. This solution appears to be very accurate. The recomputed $F(z)$ or $f(z)$ of Table 2.9 (lower part) reveals that the curve fitting is not as good as for the moments fit in this case. Of course, the primary goal of a maximum-likelihood solution is not a better curve fitting.

It should be noticed that the ratio $\hat{c}_1/\hat{c}_2 = c_1/c_2$. Hence $\hat{a}_1 = 4.94$ and $\hat{a}_2 = 5.58$. However, in this solution R_x was kept constant, namely 10.53. If R_x is permitted to vary, the costs of the electronic-computer solution skyrocket.

TABLE 2.9

Cloudiness distribution, Hampton, Va., January 1946–1964, all hours

x	Sky cover x'	Observed	$f(x)$ (%)	$f(z)$ (%)	$F(z)$ (%)	z_c	$f(x) - f(z)$ (%)	$F(x)$ (%)	$F(x) - F(z)$ (%)	Analytical n_z
0.26	0	3616	25.6	25.0	25.0	−4.90	0.6	25.6	−0.6	3527
1.26	1	565	4.0	5.3	30.3	−3.80	−1.3	29.6	0.7	750
2.26	2	579	4.1	3.6	33.9	−2.80	0.5	33.7	0.2	509
3.26	3	509	3.6	3.0	36.9	−1.80	0.6	37.3	−0.4	423
4.26	4	381	2.7	2.7	39.6	−0.80	0.0	40.0	0.4	385
5.26	5	297	2.1	2.7	42.3	0.20	−0.6	42.1	0.2	375
6.26	6	381	2.7	2.7	45.0	1.20	0.0	44.8	0.2	395
7.26	7	494	3.5	3.2	48.2	2.20	0.3	48.3	−0.1	445
8.26	8	749	5.3	3.9	52.1	3.20	1.4	53.6	−1.5	553
9.26	9	480	3.4	6.1	58.2	4.20	−2.7	57.0	1.2	854
10.26	10	6074	43.0	41.8	100.0	5.20	1.2	100.0	0.0	5909
N		14125								14125

continued from Table 2.9, Maximum Likelihood Fit.

Sky cover	$f(x)$ (%)	$f(z)$ (%)	$F(x)$ (%)	$F(z)$ (%)	z_c
0	25.6	18.8	25.6	18.8	−4.78
1	4.0	7.5	29.6	26.3	−3.68
2	4.1	5.6	33.7	31.9	−2.68
3	3.6	4.7	37.3	37.8	−1.68
4	2.7	4.5	40.0	41.3	−0.68
5	2.1	4.5	42.1	45.8	0.31
6	2.7	4.7	44.8	50.5	1.31
7	3.5	5.2	48.3	55.7	2.31
8	5.3	6.2	53.6	61.9	3.31
9	3.4	8.9	57.0	70.8	4.31
10	43.0	29.2	100.0	100.0	5.41
N			14125		

2.7 THE LOGISTIC DISTRIBUTION

Among the frequencies which have found very little application in meteorology is the logistic distribution. Feller (1940 and 1966), has pointed out that the "law of logistic growth" has been largely overemphasized in the statistical literature, and most problems could be solved by fitting other distributions. Gumbel (1961) has suggested, however, a bivariate logistic distribution for wind data. Thus, the basic background will be presented here to familiarize the reader with the distribution and leave the application up to his own judgment by presenting an example at the end of the bivariate normal distribution (see Section 2.8).

2.7.1 Univariate distribution

The logistic distribution is best introduced as a cumulative distribution function:

$$F(x) = (1 + e^{-(x-a)/b})^{-1} \quad \begin{cases} 0 \text{ for } x = -\infty \\ 1.0 \text{ for } x - \infty \end{cases} \quad [2.61]$$

with $b > 0$.

Sometimes we find the hyperbolic trigonometric function form of the logistic distribution, namely:

$$F(x) = [1 + \tanh\{(x-a)/(2b)\}]/2 \quad [2.62]$$

with tanh the symbol for a hyperbolic function, $\tanh(u) = (e^u - e^{-u})/(e^u + e^{-u})$. The probability density function can be written as:

$$f(x) = [e^{-(x-a)/b}]/[b \cdot \{1 + e^{-(x-a)/b}\}^2] \quad [2.61a]$$

and by means of the hyperbolic trigonometric function in a much simpler form:

$$f(x) = [\text{sech}^2\{(x-a)/(2b)\}]/(4b) \quad [2.62a]$$

with $\text{sech}(u) = 2/(e^u + e^{-u})$ the symbol for the hyperbolic secant. The distribution is therefore sometimes called the hyperbolic sec-squared distribution.

Although we could generalize [2.61] by writing (see Nelder, 1961):

$$F(x) = A/[1 + e^{(kx-a)/b}]^{-c} \quad [2.61b]$$

in most cases we set A, k and $c = 1.0$, which reduces [2.61b] to the basic form [2.61]. Since the bivariate logistic distribution introduced by Gumbel (1961) makes use of this basic form, it alone will be discussed. Gumbel (1944) derived the one-dimensional logistic distribution from the ratio x_{max}/x_{min}.

THE LOGISTIC DISTRIBUTION

If we set $z = (x - a)/b$, we obtain the standard form:

$$F(z) = (1 + e^{-z})^{-1} \qquad [2.63a]$$

and: $f(z) = e^{-z}/(1 + e^{-z})^2 = [\operatorname{sech}^2(z/2)]/4 \qquad [2.63b]$

Although these are not the only standard forms, both are in widespread use in the statistical literature.

The general moments notation leads to:

$$\nu|z|_r = 2\Gamma(r+1) \cdot \sum_{j=1}^{\infty} (-1)^{j-1} j^{-r} \text{ for } r > 0 \qquad [2.64a]$$

or: $\nu|z|_r = 2\Gamma(r+1) \cdot (1 - 2^{-(r-1)}) \zeta(r) \text{ for } r > 1 \qquad [2.64b]$

where $\zeta(s) = \sum_{n=1}^{\infty} n^{-s}$ is Riemann's zeta function (see also [2.37i]).

For the variate z we derive therefore (with $\bar{z} = 0$):

$$\sigma_z^2 = \pi^2/3 \qquad [2.64c]$$

$$\nu_3 = 0 \qquad [2.64d]$$

$$\nu_4 = 7\pi^4/15 \qquad [2.64e]$$

$$\beta_2 = 21/5 \qquad [2.64f]$$

Should we be interested in the moments of the original observation x, then we notice from the transformation equation that:

$$\bar{x} = a \qquad [2.65a]$$

$$\sigma_x^2 = b^2 \pi^2/3 \qquad [2.65b]$$

and ν_3 and β_2 remain the same. Hence ν_4 can be computed from β_2 with:

$$\nu_4 = 7b^4 \pi^4/15 \qquad [2.65c]$$

Further: $b = \sqrt{3} \sigma_x/\pi \qquad [2.65d]$

The cumulative distribution function can be expressed in terms of $\bar{x}$ and σ^2 as:

$$F(x) = [1 + \exp\{-\pi(x - \bar{x})/(\sigma\sqrt{3})\}]^{-1} \qquad [2.63c]$$

and: $f(x) = \pi[\exp\{-\pi(x - \bar{x})/(\sigma\sqrt{3})\}]/(\sigma\sqrt{3} \cdot [1 + \exp\{-\pi(x - \bar{x})/(\sigma\sqrt{3})\}]^2) \qquad [2.63d]$

At the same time this formula constitutes the moments fit, where $\bar{x} \sim x_m$ and $\sigma^2 \sim s^2$. The maximum-likelihood equations are given by:

$$\sum_{1}^{N} [1 + \exp\{\pi(x_i - \hat{x})/(\hat{s}\sqrt{3})\}]^{-1} = N/2 \qquad [2.66a]$$

$$\sum_{1}^{N} \frac{x_i - \hat{x}}{\hat{s}} \frac{1 - \exp[\pi(x_i - \hat{x})/(\hat{s}\sqrt{3})]}{1 + \exp[\pi(x_i - \hat{x})/(\hat{s}\sqrt{3})]} = \frac{N\sqrt{3}}{\pi} \qquad [2.66b]$$

These equations cannot be solved in explicit form but need iteration methods. As previously mentioned on several occasions, electronic computers aid substantially in finding solutions for maximum-likelihood equations today. For large N the variance of $\hat{x}$ and $\hat{s}$ becomes:

$$N\sigma_{\hat{x}}^2 \sim (9/\pi^2)\sigma^2 \qquad [2.66c]$$

$$N\sigma_{\hat{s}}^2 \sim [9/(3+\pi^2)]\sigma^2 \qquad [2.66d]$$

As initial estimators the moments estimators may be used. Graphical solutions can be found by plotting log $F(x)$ against x. In a recent article Schafer and Sheffield (1973) have stated that the moments estimators are quite efficient. They also have established tables from which confidence intervals for the mean and standard deviation of the logistic distribution can be obtained when both parameters are unknown.

The complexity of the likelihood equations has led to a search for order statistics for estimation; but that problem is not much simpler. When neither $\bar{x}$ nor σ is known, an approximate best linear estimator is given by:

$$x_m^* = [S(k_2) \cdot S_x - S(k_3) \cdot S_y]/\Delta \qquad [2.67a]$$

$$s_m^* = [-S(k_3) \cdot S_x + S(k_1) \cdot S_y]/\Delta \qquad [2.67b]$$

where: $\Delta = S(k_1) \cdot S(k_2) - S^2(k_3)$ \qquad [2;67c]

In order to find the value of the expressions $S(k_i)$ and S_x, S_y we first define:

$$A_i = \xi_i(1 - \xi_i) \ln[\xi_i(1 - \xi_i)] \qquad [2.67d]$$

$$B_i = \eta_i(1 - \eta_i) \ln[\eta_i(1 - \eta_i)] \qquad [2.67e]$$

$$C_i = \xi_i(1 - \xi_i) \ln[\xi_i/(1 - \xi_i)] \qquad [2.67f]$$

$$D_i = \eta_i(1 - \eta_i) \ln[\eta_i/(1 - \eta_i)] \qquad [2.67g]$$

The ξ and η are explained after [2.67 l]. We obtain the $S(k_i)$ by the following computation:

$$S(k_1) = \sum_{i=1}^{k+1} (\xi_i - \eta_i)(1 - \xi_i - \eta_i)^2 \qquad [2.67h]$$

$$S(k_2) = \sum_{i=1}^{k+1} (A_i - B_i)^2/(\xi_i - \eta_i) \qquad [2.67i]$$

$$S(k_3) = \sum_{i=1}^{k+1} (1 - \xi_i - \eta_i)(C_i - D_i) \qquad [2.67j]$$

$$S_x = \sum_{i=1}^{k+1} (1-\xi_i-\eta_i)[\xi_i(1-\xi_i)X_{ni} - \eta_i(1-\eta_i)X_{ni-1}] \quad [2.67\text{k}]$$

$$S_y = \sum_{i=1}^{k+1} [(C_i - D_i)\{\xi_i(1-\xi_i)X_{ni} - \eta_i(1-\eta_i)X_{ni-1}\}]/(\xi_i - \eta_i) \quad [2.67\text{l}]$$

The k denotes the number of selected order statistics $X_{ni} \ldots X_{nk}$ from N independent random variables $x_1 \ldots x_N$ and $\xi_i = n_i/N$, $\eta_i = n_{i-1}/N$. The n_i denotes the order (rank) of the observation, and n_1/N and N_k/N should not be chosen too close to 0 or 1, respectively. The method has been introduced by Gupta and Gnanadesikan (1966) based on general outlines by Ogawa (1951). To sum for all the N requires extensive calculations if N is large, although calculations become relatively easy to carry out by electronic computers. The restriction to k selected thresholds of the $F(x)$ reduces the computer work considerably. The problem is further simplified when $\bar{x}$ or σ is known. More details on order statistics of the logistic distribution can be found in articles by Gupta and Shah (1965) or Gupta et al. (1967), and Chan et al. (1971).

Another order statistic has been given by Kendall and Stuart (1961). It can be applied to the form [2.63b] with $b = 1$, given as:

$$f(z) = e^{-z}/(1 + e^{-z})^2 \quad [2.63\text{e}]$$

The solution can be based on cumulants k_i. For the order statistic x_{n_0}, with $n_0 > (N+1)/2$ and N denoting the number of observations, we can write:

$$k_1 = \sum_{s_0}^{s_1} 1/s \quad [2.67\text{m}]$$

with $s_0 = N - n_0 + 1$; $s_1 = n_0 - 1$;

$$k_2 = \pi^2/3 - \left(\sum_1^{s_1} 1/s^2 + \sum_1^{s_0-1} 1/s^2\right) \quad [2.67\text{n}]$$

$$k_3 = 2\sum_{s_0}^{s_1} 1/s^3 \quad [2.67\text{o}]$$

$$k_4 = 2\pi^4/15 - 6\left(\sum_1^{s_1} 1/s^4 + \sum_1^{s_0-1} 1/s^4\right) \quad [2.67\text{p}]$$

The special form of the logistic distribution ([2.61b]) with $A = 1$, $b = 1$, $c = 1$, $k' = -k$, can be written as:

$$F(x) = [1 + \exp\{-(a + k'x)\}]^{-1} \quad (-\infty \leqslant x \leqslant \infty) \quad [2.67\text{q}]$$

The estimators $\hat{a}$ and $\hat{k}'$ are the roots of:

$$\bar{x} = \frac{1}{k'} + \sum_1^N \frac{x_i \exp(-k'x_i)}{1 + \exp[-(a + k'x_i)]} \frac{2}{Ne^a} \quad [2.67\text{r}]$$

$$e^a = \frac{2}{N} \sum_1^N \frac{\exp(-k'x_i)}{1 + \exp[-(a + k'x_i)]} \qquad [2.67s]$$

2.7.2 The bivariate logistic distribution

Up to this section all distributions have been treated in one dimension only. It is self-evident that many of them can be expanded to more dimensions, but prominence in this field in practical work has presently been given only to the Gaussian normal distribution. The bivariate (Gaussian) normal distribution will be discussed in detail in the next section. The expansion to the bivariate logistic distribution should have been postponed until the bivariate Gaussian distribution had been introduced. Because it is intended to treat the bivariate logistic distribution only briefly, its description has been combined with the one-dimensional logistic curve, and follows now.

We assume that a logistic distribution exists for two variables x_1 and x_2. In accordance with the discussions of the contingency table (Essenwanger, 1974a, section 1.2.1) the margins must have the following probability distributions:

$$F_1(x_1) = F(x_1, \infty) \qquad [2.68a]$$
$$F_2(x_2) = F(\infty, x_2) \qquad [2.68b]$$

where $F_1(x_1)$ and $F_2(x_2)$ both represent logistic distributions. Although the theoretical background for $F_1(x_1)$ and $F_2(x_2)$ requires $F_1(x_1)$ and $F_2(x_2)$ to exist when the other variable is infinity, in practice $F_1(x_1)$ or $F_2(x_2)$ can be found only from the maximum value on. Thus:

$$F_1(x_1) \sim F(x_1, x_{2\max}) \qquad [2.68c]$$
$$F_2(x_2) \sim F(x_{1\max}, x_2) \qquad [2.68d]$$

Since the following discussion deals with two dimensions only, we call the general function $F(x_1, x_2)$ a bivariate function, and since F_1 and F_2 have a logistic form, the $F(x_1, x_2)$ is now called the bivariate logistic function. The probability density is then defined for an areal segment $\partial x \, \partial y$ as:

$$f(x_1, x_2) = \frac{\partial^2 F(x_1, x_2)}{\partial x_1 \partial x_2} \qquad [2.69a]$$

From the fact that $F(x_1, x_2)$ is a cumulative distribution, it is trivial that:

$$F(-\infty, x_2) = F(x_1, -\infty) = F(-\infty, -\infty) = 0 \qquad [2.68e]$$
$$\text{and: } F(+\infty, x_2) = F(x_1, \infty) = F(\infty, \infty) = 1.0 \qquad [2.68f]$$

In practical application the $-\infty$ or $+\infty$ is replaced by the minimum or maximum, respectively.

If the two variables x_1 and x_2 are independent, then:

THE LOGISTIC DISTRIBUTION

$$F(x_1, x_2) = F_1(x_1) \cdot F_2(x_2) \tag{2.69b}$$

This is a necessary condition of independence. If this condition is not met, x_1 and x_2 must be dependent. In this case a coordinate transformation leads to independent variables (see bivariate normal distribution).

A symmetrical system requires the condition:

$$f(x_1, x_2) = f(-x_1, -x_2) \tag{2.69c}$$

and: $f(-x_1, x_2) = f(x_1, -x_2) \tag{2.69d}$

That makes the marginal distributions automatically symmetric about zero, but not all symmetric marginal distributions are symmetric bivariate distributions. An asymmetric case exists when either [2.69c] or [2.69d] is not fulfilled. Furthermore, a bivariate probability function is not uniquely determined by the two marginal distributions unless x_1 and x_2 are independent. We assume $F(x_1)$ and $F(x_2)$ in their basic form ([2.63a]). Then:

$$F(z_1, z_2) = [1 + \exp(-z_1) + \exp(-z_2)]^{-1} \tag{2.70}$$

This function cannot be split into a product like [2.69b]. Hence the z_1 and z_2 are not necessarily independent in this form. Lines of:

$$F(z_1, z_2) = \text{constant} \tag{2.70a}$$

are curves of equal probability in a two-dimensional system.

The density function for the form of [2.69b] is:

$$f(z_1, z_2) = 2F^3(z_1, z_2) \exp(-z_1 - z_2) \tag{2.70b}$$

The marginal probability density functions hence assume the forms:

$$f(z_1) = \exp(-z_1)[1 + \exp(-z_1)]^{-2} \quad (-\infty < z_1 < \infty) \tag{2.70c}$$

and: $f(z_2) = \exp(-z_2)[1 + \exp(-z_2)]^{-2} \quad (-\infty < z_2 < \infty) \tag{2.70d}$

These marginal density functions are symmetrical. It is easy to prove that $f(z_1, z_2)$ in [2.70b] is not symmetrical.

The maximum occurs at $f(0, 0)$ with the density:

$$f(0, 0) = 2/27 \tag{2.71}$$

The curves: $f(z_1, z_2) = c_i \tag{2.71a}$

are lines of equal probability density. Since it may be necessary to draw these lines in a diagram, a brief description to facilitate the task follows.

Along the diagonal $z_2 = -z_1$ we must solve:

$$2[1 + \exp(-z_1) + \exp(-z_2)]^{-3} = c \tag{2.71b}$$

Hence: $\exp(z_1) + \exp(-z_1) = (2/c)^{1/3} - 1 \tag{2.71c}$

Since $[\exp(z_1) + \exp(-z_1)]/2 = \cosh(z_1)$ is the hyperbolic cosine, the solution can be easily found in tables of hyperbolic functions (e.g. Comrie, 1948 and others):

$$2 \cosh(z_1) = (2/c)^{1/3} - 1 \qquad [2.71d]$$

Along the line $z_2 = 2z_1$ we have the same equation [2.71d], which delivers points also on the line $z_2 = z_1/2$.

Along the abscissa $z_1 = 0$ and the line $z_1 = z_2$ we find the solution from:

$$(2/c)\exp(-z_2) = [2 + \exp(-z_2)]^3 \qquad [2.71e]$$

$$\text{and: } (2/c)\exp(-2z_1) = [1 + 2\exp(-z_1)]^3 \qquad [2.71f]$$

These conditions especially furnish enough points to draw the isolines. Should more points be desired, they also can be derived from [2.70b] and the desired relationship between z_1 and z_2.

The correlation coefficient (Essenwanger, 1974a, section 3.2.1) has a value of:

$$r = 0.5 \qquad [2.71g]$$

The correlation ratio (see Essenwanger, 1974a, section 3.4) emerges as:

$$\eta^2 = 3/\pi^2 \qquad [2.71h]$$

Since the parametric form of the logistic distribution is:

$$F(x) = [1 + \exp\{-(x-a)/b\}]^{-1} \qquad [2.71i]$$

or transformed: $F(x) = [1 + \exp\{-(\alpha - \beta x)\}]$ \qquad [2.71j]

with $\alpha = -a/b$ and $\beta = -1/b$, we must relate the basic form with variables z_1 and z_2 to the regular form with variates x_1 and x_2. The transformation would then require:

$$z_1 = (x_1 - a_1)/b_1 \qquad [2.71k]$$

and analogously: $z_2 = (x_2 - a_2)/b_2$ \qquad [2.71l]

Furthermore: $\alpha_i = -a_i/b_i$ \qquad [2.71m]

$\beta_i = -1/b_i \, (i = 1, 2)$ \qquad [2.71n]

The parameters a_i, b_i or α_i, β_i must then be estimated, e.g. from the marginal distribution. The moments fit would provide:

$$a_i = x_{im} \qquad [2.71o]$$

$$b_i^2 = 3s_i^2/\pi^2 \qquad [2.71p]$$

where $s_1^2 \equiv s_x^2$ and $s_2^2 \equiv s_y^2$.

Other estimations have been introduced earlier with the univariate logistic distribution.

An example of the bivariate logistic distribution can be found later in connection with the discussion of the bivariate normal distribution.

2.8 THE BIVARIATE NORMAL DISTRIBUTION

2.8.1 General, multivariate distributions

Multivariate distributions of various types are known in statistical analysis, but have found little entrance into meteorological work. Only the Gaussian distribution has attained importance in application to some vector quantities such as the wind (e.g. Hesselberg and Bjorkdal, 1929; Brooks et al. 1946, 1950; Crutcher, 1957; and Essenwanger, 1959). Since the bivariate (Gaussian) normal distribution belongs to a group with widespread utilization in meteorology, a brief treatment of the background of multivariate distributions will follow. Only basic problems and some examples can be given here, thus the section serves mainly to familiarize the reader with the general problem. Later the bivariate case is discussed. Several sources of reference are available to the reader who is interested in more details (e.g. Kendall and Stuart, 1961; Linder, 1960; Mood, 1950; Hoel, 1966; Hogg and Craig, 1967; Johnson and Kotz, 1972a, and many more).

If we assume any type of a random function of variables $x_1, x_2 \ldots x_n$, say $g(x_1, x_2 \ldots x_n)$ and the probability density $f(x_1, x_2 \ldots x_n)$, then the expected value of g can be expressed by:

$$E(g) = \int_{-\infty}^{\infty} \ldots \int_{-\infty}^{\infty} g(x_1, x_2 \ldots x_n) f(x_1, x_2 \ldots x_n) \, dx_1 \cdot dx_2 \ldots dx_n \quad [2.72]$$

This is the definition of the mean value of g. We can expand the notation and arrive at the moments of g, namely:

$$E(g^\nu) = \int_{-\infty}^{\infty} \ldots \int_{-\infty}^{\infty} g^\nu(x_1, x_2 \ldots x_n) f(x_1, x_2 \ldots x_n) \, dx_1 \cdot dx_2 \ldots dx_n \quad [2.73]$$

This equation is equivalent to [1.8]. When the variables x_i are given in discrete form we replace the integral sign by the summation sign and end up with the summation notation as customary in statistical analysis.

It may be worthwhile to point out a few properties of $E(g)$. It is easy to prove that:

$$E(cg) = cE(g) \quad [2.73a]$$

where c is a constant. If we assume $g_1 \ldots g_i$ to be functions of a set of random variables, it follows from the integral [2.72] that:

$$E(g_1 \ldots g_i) = E(g_1) + \ldots E(g_i) \quad [2.72a]$$

For two functions g_1 and g_2, whose products can be represented by:

$$E(g_1 g_2) = \int_{-\infty}^{\infty} \int_{-\infty}^{\infty} g_1 g_2 \, f_1(g_1) f_2(g_2) \, dg_1 \cdot dg_2 \quad [2.73b]$$

with $f_1(g_1)$ and $f_2(g_2)$ denoting the respective frequency density functions, we can reformulate:

$$E(g_1 g_2) = E(g_1) E(g_2) \qquad [2.73c]$$

It must be stressed that g_1 and g_2 must be independent. Otherwise $E(g_1 g_2)$ cannot be brought into the form [2.73b] and the basic formula [2.73c] is not valid.

The equations above lead to:

$$\mu_x = \sum_{i=1}^{n} \mu_i \qquad [2.74]$$

for: $x = x_1 + x_2 + \ldots x_n$

Further: $E(x - \mu_x)^2 = \sum_{i=1}^{n} \sum_{j=1}^{n} E(x_i - \mu_i)(x_j - \mu_j) \qquad [2.74a]$

For independent variables $x_1, x_2 \ldots x_n$ this expression reduces to:

$$\sigma_x^2 = \sum_{i=1}^{n} \sigma_i^2 \qquad [2.74b]$$

If the variables are not independent, however, the following relationship exists:

$$\sigma_x^2 = \sum_{i=1}^{n} \sigma_i^2 + 2 \sum r_{hk} \sigma_h \sigma_k \qquad [2.74c]$$

where the r_{nk} denotes the correlation coefficient (Essenwanger, 1974a, section 3.1), and the summation must be performed over all permutations up to $n-1$, excluding $h = k$ which comprises the first term. Since the correlation coefficient is zero for independent variables, the second term in [2.74c] disappears and [2.74c] reduces to [2.74b].

Equation [2.74c] can also be interpreted as the general law of error or dispersion for random errors, when x is a compound variate of various factors x_i with random errors for each factor. It should be kept in mind that [2.74a] through [2.74c] are based on the condition of a compound variate x and are not mere additions of samples.

If we are interested in addition of samples, and the mean value of the combined samples is $\bar{X}$, then it would be computed by:

$$\bar{X} = \sum w_j \bar{x}_j \qquad [2.75]$$

and the variance of the combined samples $\sigma_{\bar{x}}^2$ would be:

$$\sigma_{\bar{x}}^2 = \sum w_j \sigma_j^2 + \sum w_j (\bar{x}_j - \bar{X})^2 \qquad [2.75a]$$

with the weighting function: $w_j = n_j/N \qquad [2.75b]$

THE BIVARIATE NORMAL DISTRIBUTION

The n_j denotes the number of observations x_{ij} in the individual samples j and

$$\sum_1^j \sum_1^i n_{ij} = N$$

Formula [2.74a] is different from [2.75a], as expected. Since we have two different types of problems, the reader should not be confused about the two different solutions.

After the general formulation of the multivariate problem for any type of function, let us restrict the following discussions to Gaussian random distributions.

The joint distribution for the multivariate (N-dimensional) Gaussian (normal) distribution can be expressed as follows:

$$f(x_1, x_2 \ldots x_n) = [\exp\{-(2\|C\|)^{-1} \sum_{i,j=1}^{N} C_{ij}(x_i - \mu_i)(x_j - \mu_j)\}]/$$

$$[(2\pi)^{N/2} \sqrt{\|C\|}] \qquad [2.76]$$

The $\|C\|$ symbolizes the determinant of the covariance matrix M_c of the σ_{ij}, and C_{ij} is the cofactor of the matrix M_c, obtained by omitting the ith row and jth column with the proper sign (see Section 4.3). The matrix M_c is defined by:

$$M_c = \begin{bmatrix} \sigma_{11} & \sigma_{12} & \ldots & \sigma_{1n} \\ \sigma_{21} & \sigma_{22} & \ldots & \sigma_{2n} \\ \vdots & & & \\ \sigma_{n1} & \sigma_{n2} & \ldots & \sigma_{nn} \end{bmatrix} \qquad [2.76a]$$

where the σ_{ij} stands for the cross-products when $i \neq j$ and for the variance for $i = j$. The cross-products $x_i x_j$ of the variates x_i for $i = 1 \ldots n_i$ and x_j for $j = 1, \ldots n_j$ are computed by [1.15] and are identical with the mixed moment (see Section 1.2). Since M_c is a symmetric matrix, $n_i = n_j$, with the n_i or n_j denoting the number of variates. In our case the cross-products are:

$$x_i x_j = (1/N) \sum_1^{N_k} (x_{ik} - \bar{x}_i)(x_{jk} - \bar{x}_j) \qquad [2.76b]$$

We can therefore replace σ_{ij} by:

$$\sigma_{ij} = r_{ij} \sigma_{ii} \sigma_{jj} \qquad [2.76c]$$

The (linear) correlation coefficient r_{ij} is covered by the writer (1974a), in section 3.1.

If the variates $x_1, x_2 \ldots x_{ni}$ are independent, the r_{ij} for $i \neq j$ are all zero and the matrix M_c reduces to the diagonal, a mere variance matrix.

Testing of the means of the population in the multivariate case has recently been discussed by Kramer and Jensen (1969). Confidence limits and predictions have been presented in articles by Chew (1966) and by Stein (1962).

The problem of integrating the n-dimensional normal distribution is covered in an article by Ihm (1960). An integration of multivariate normal distributions by the Monte Carlo technique (see Essenwanger, 1974b) has been suggested by Carter (1973). Dutt (1973) has outlined a conversion into a finite sum of multivariate Fourier transforms whose integrals are numerically evaluated by the Gaussian quadrature. The reader may refer to the literature for details.

2.8.2 *The bivariate Gaussian distribution*

Let us assume that the dimension of matrix M_c ([2.76a]) is one. Then [2.76] reduces to the well-known Gaussian distribution formula. The one-dimensional Gaussian distribution is well known and has been treated by the author (1974a, section 2.4) and will not be repeated here.

For two dimensions of M_c we derive the bivariate normal distribution density and can write, after some algebraic operations:

$$f(x_1, x_2) = [\exp\{-z^2/[2(1-r^2)]\}]/[2\pi\sigma_1\sigma_2\sqrt{(1-r^2)}] \qquad [2.77]$$

where:

$$z^2 = (x_1 - \bar{x}_1)^2/\sigma_1^2 - 2r(x_1 - \bar{x}_1)(x_2 - \bar{x}_2)/(\sigma_1\sigma_2) + (x_2 - \bar{x}_2)^2/\sigma_2^2 \qquad [2.77a]$$

The σ_1^2 and σ_2^2 represent the variances of the variates x_1 and x_2, respectively. The margins must have the probability distributions:

$$F_1(x_1) = F(x_1, \infty) \qquad [2.77b]$$

and: $F_2(x_2) = F(\infty, x_2) \qquad [2.77c]$

as introduced by [2.68a and b]. Equations [2.68c] through [2.68f] are also applicable. We can further state:

$$f(x_1, x_2) = \partial^2 F(x_1, x_2)/\partial x \partial y \qquad [2.77d]$$

This is equivalent to [2.69].

In the form of [2.77] the x_1 and x_2 variates are not independent. The interrelationship is expressed by the correlation coefficient r (Essenwanger, 1974a, section 3.1) with $-1 \leq r \leq 1$. The marginal probability density distributions for the above equations are:

$$f(x_1) = [1/(\sigma_1\sqrt{2\pi})] \exp[-(1/2)(x_1 - \bar{x}_1)^2/\sigma_1^2] \qquad [2.77e]$$

and: $f(x_2) = [1/(\sigma_2\sqrt{2\pi})] \exp[-(1/2)(x_2 - \bar{x}_2)^2/\sigma_2^2] \qquad [2.77f]$

Consequently:

$$f(x_1, x_2) = f(x_1) \cdot f(x_2)(1/\sqrt{1-r^2}) \exp\left[-\frac{1}{2}\frac{r}{(1-r^2)}\{rz^2 - 2(1-r^2)(x_1-\bar{x}_1)(x_2-\bar{x}_2)/(\sigma_1\sigma_2)\}\right] \qquad [2.77g]$$

THE BIVARIATE NORMAL DISTRIBUTION

(z^2 see [2.77a].) When x_1 and x_2 are independent, i.e. $r = 0$, then we can write:

$$f(x_1, x_2) = f(x_1) \cdot f(x_2) \tag{2.77h}$$

which is equivalent to [2.69b].

Equation [2.77] reduces then to:

$$f(x_1, x_2) = [\exp(-z_r^2/2)]/(2\pi \sigma_1 \sigma_2) \tag{2.78}$$

with: $z_r^2 = (x_1 - \bar{x}_1)^2/\sigma_1^2 + (x_2 - \bar{x}_2)^2/\sigma_2^2$ [2.78a]

We set: $A_0 = 1/[2\pi \sigma_1 \sigma_2 \sqrt{(1-r^2)}]$ [2.77i]

and can write: $f(x_1, x_2) = A_0 \, e^{-c^2}$ [2.77j]

where c^2 represents lines of equal probability density:

$$c^2 = z_r^2/[2(1-r^2)] \tag{2.77k}$$

These are ellipses in a coordinate system where x_1 and x_2 are the axes. The bivariate Gaussian distribution is therefore also called the bivariate elliptical (Gaussian) distribution.

It is easy to show that the variates with correlation $r \neq 0$ can be transformed into independent variates y_1, y_2 by coordinate transformation.

Let us designate σ_1 and σ_2 for the x_i-system and σ_a, σ_b the respective standard deviations for the y_i-system. Then we have the invariant S_v^2 as:

$$S_v^2 = \sigma_1^2 + \sigma_2^2 = \sigma_a^2 + \sigma_b^2 = k_1^2 + k_2^2 \tag{2.79}$$

Further: $\sigma_a^2 \sigma_b^2 = (1-r^2) \sigma_1^2 \sigma_2^2 = k_1 k_2$ [2.79a]

These are the roots of the matrix equation:

$$\begin{vmatrix} \sigma_1^2 - k & \sigma_1 \sigma_2 r \\ \sigma_1 \sigma_2 r & \sigma_2^2 - k \end{vmatrix} = 0 \tag{2.79b}$$

Hence $k_1 = \sigma_a^2$ and $k_2 = \sigma_b^2$. We can further learn that for $r = \pm 1$ the $\sigma_a^2 \cdot \sigma_b^2 = 0$. This is valid only when at least one of the factors is zero. Consequently the original ellipses of equal probability decompose into straight lines. This conclusion can also be drawn from the correlation coefficient.

We define:

$$y_1 = (x_1 - \bar{x}_1) \cos \varphi + (x_2 - \bar{x}_2) \sin \varphi \tag{2.80a}$$

$$y_2 = (x_2 - \bar{x}_2) \cos \varphi - (x_1 - \bar{x}_1) \sin \varphi \tag{2.80b}$$

Then: $\sigma_a^2 = \dfrac{\sigma_1^2 \cos^2 \varphi - \sigma_2^2 \sin^2 \varphi}{\cos^2 \varphi - \sin^2 \varphi}$ [2.80c]

$$\sigma_b^2 = \frac{\sigma_2^2 \cos^2 \varphi - \sigma_1^2 \sin^2 \varphi}{\cos^2 \varphi - \sin^2 \varphi} \quad [2.80d]$$

$$\tan 2\varphi = 2r\sigma_1\sigma_2/(\sigma_1^2 - \sigma_2^2) \quad [2.80e]$$

or: $\tan \varphi = (\sigma_2^2 - \sigma_1^2 \pm \sqrt{C_x})/(2r\sigma_1\sigma_2)$ [2.80f]

where: $C_x = (2r\sigma_1\sigma_2)^2 + (\sigma_2^2 - \sigma_1^2)^2$ [2.80g]

The φ designates the angle between x_1 and y_1 or x_2 and y_2 (Fig. 2.3, Example 2.9). The new probability density function now reads:

$$f(y_1, y_2) = [1/(2\pi \sigma_a \sigma_b)] \exp(-z_y^2/2) \quad [2.78b]$$

with: $f(y_1, y_2) = f(y_1) \cdot f(y_2)$ [2.78c]

and: $z_y^2 = y_1^2/\sigma_a^2 + y_2^2/\sigma_b^2$ [2.78d]

Notice, by virtue of the transformation, the reference point coincides with the mean, and $\bar{y}_1 = \bar{y}_2 = 0$. The following sign convention should be helpful for [2.80e].

r	denominator	φ	r	denominator	φ
+	$\sigma_1^2 > \sigma_2^2$	0–45°	−	$\sigma_1^2 < \sigma_2^2$	90–135°
+	$\sigma_1^2 < \sigma_2^2$	45–90°	−	$\sigma_1^2 > \sigma_2^2$	135–180°

We could measure the ellipticity by:

$$\epsilon_l = (\sigma_a - \sigma_b)/\sigma_a \quad [2.80h]$$

where the σ_a is the variance of the major axis, $\sigma_a > \sigma_b$. It is difficult, however, to test ϵ_l for significance. Therefore we may check the statistical significance of the ellipticity by comparison with the circular distribution. We need:

$$F_E = \sigma_a^2/\sigma_b^2 \quad [2.80i]$$

or: $F_E' = \sigma_1^2/\sigma_2^2$ [2.80j]

This is Fisher's variance ratio or F-test, discussed in detail by the author (1974a, section 6.5).

Mauchly (1940) has defined another significance test:

$$\epsilon_M = 2\sigma_{x_1}\sigma_{x_2}(1-r^2)^{1/2}/[\sigma_{x_1}^2 + \sigma_{x_2}^2] \quad [2.80k]$$

He has shown that the probability of obtaining a value $\leq \epsilon_M$ is $(\epsilon_M)^{N-2}$ in a sample of N independent observations.

The eccentricity could also be calculated by:

$$\epsilon_x = (\sigma_a^2 - \sigma_b^2)/\sigma_a^2 \quad [2.80l]$$

THE BIVARIATE NORMAL DISTRIBUTION

Examination of the transformation equations [2.80a] through [2.80f] reveals the following with respect to the correlation. If $r = 0$, then $\varphi = 0$ and no transformation is necessary. We could assume $\varphi = 45°$ for $\sigma_1 = \sigma_2$. This special case also requires $r = 0$, otherwise we could not have a bivariate Gaussian distribution. The condition $r = 0$ and $\sigma_1 = \sigma_2$ produces circles instead of ellipses for the lines of equal probability (see Section 2.8.4). In this case no transformation of the axes is necessary, but the center can be shifted to $\bar{x}_1$ and $\bar{x}_2$.

When φ is known the following formula is useful:

$$\sigma_a^2/\sigma_b^2 = (\sigma_1^2/\sigma_2^2 - \tan^2 \varphi)/[1 - (\sigma_1^2/\sigma_2^2) \tan^2 \varphi] \qquad [2.80m]$$

Although the assumption of a bivariate Gaussian normal distribution has been made for the topic presented above, it should be pointed out that not all distributions with Gaussian marginal distributions are Gaussian bivaraite distributions. Examples have been given by Hogg and Craig (1965) and others. The bivariate Gaussian distribution produces marginal Gaussian normal distributions, but this condition alone is not enough. A proof has recently been given by Pierce and Dykstra (1969) that in the multivariate case this condition is not sufficient.

2.8.3 Regression line, maximum frequency, ellipses

Let us assume that we are interested in the conditional probability distribution $f(x_1|x_{2k})$, i.e. x_2 has a constant value, say $c_k = x_2 - \bar{x}_2$. We can rewrite [2.77a] as:

$$z^2 = (x_1 - \bar{x}_1)^2/\sigma_1^2 - 2rc_k(x_1 - \bar{x}_1)/\sigma_1\sigma_2 + c_k^2/\sigma_2^2 \qquad [2.81]$$

If we remember that the conditional probability is:

$$f(x_1|x_{2k}) = f(x_1, x_2)/f(x_2) \qquad [2.81a]$$

we arrive at the following formula after some algebraic manipulations:

$$f(x_1|x_{2k}) = [\exp\{-z_c^2/[2(1-r^2)]\}]/(\sqrt{2\pi}\,\sigma_1\sqrt{1-r^2}) \qquad [2.81b]$$

$$\text{with: } z_c^2 = (x_1 - \bar{x}_1)^2/\sigma_1^2 - 2rc_k(x_1 - \bar{x}_1)/(\sigma_1\sigma_2) + r^2c_k^2/\sigma_2^2 \qquad [2.81c]$$

$$= [(x_1 - \bar{x}_1) - rc_k\sigma_1/\sigma_2]^2/\sigma_1^2 \qquad [2.81d]$$

This means that z_c^2 is a complete square, and the conditional frequency density is a Gaussian distribution, whose mean is at $x_1 + rc_k\sigma_1/\sigma_2$. An analogous equation exists for $f(x_2|x_{1k})$.

The curve of regression is the locus of the mean of a conditional distribution. Hence the curve of regression is:

$$x_1 = \bar{x}_1 + rc_k\sigma_1/\sigma_2 \qquad [2.82a]$$

and similarly: $x_2 = \bar{x}_2 + rc'_k\sigma_2/\sigma_1 \qquad [2.82b]$

with $c'_k = x_1 - \bar{x}_1$. It is easily recognized that for $r = 0$ the regression lines are parallel to the axes through the means $\bar{x}_1$ and $\bar{x}_2$. For $r \neq 0$ the equations [2.82a and b] represent lines with linear progression. The regression lines can also be derived from:

$$\frac{\partial [f(x_1 x_2)]}{\partial x_1} = 0 \text{ and } \frac{\partial [f(x_1, x_2)]}{\partial x_2} = 0 \qquad [2.82c]$$

This is the condition for either a maximum or a minimum. It can be readily shown that in our case it is the maximum. This fact can also be concluded from [2.81b], as the x_1 and x_2 of [2.82a and b] are identical with the means of Gaussian distributions. Mean and mode coincide in the Gaussian distribution, with the mode providing the maximum probability density at the same time.

It may now appear confusing to the reader that on a line, say x_1 or $x_2 = c_1 =$ constant, apparently four points in the line would be designated as modes (see Fig. 2.3, Example 2.9, points A_1 through A_4). A closer perusal of the problem clarifies the discrepancy, however, very quickly.

The maximum (mode) for the line $x_2 = c_1$ occurs at the intersection with the regression line ([2.82a] point A_1 in Fig. 2.3). This can be clearly concluded from the interpretation that the ellipses represent lines of equal probability density, the latter increasing towards the center. The point closest to the center of the line $x_2 = c_1$ is the intersection with the regression line (tangent line $x_2 =$ constant).

The intersection with the major (minor) axis (A_2 and A_3, Fig. 2.3) serves as the mode for a cross-section perpendicular to the major (minor) axis, i.e. in the y-system. This is not the line $x_2 = c_1$. The last point, an intersection with the second regression line ([2.88b] point A_4, Fig. 2.3) represents a maximum for the line $x_1 =$ constant through this point. This is not the line under consideration which is $x_2 = c_1$. Thus the other three intersections with the line $x_2 = c_1$ are maxima, but for other cross-sections.

In essence, all the points of the bivariate Gaussian distribution can be maxima provided we establish an infinite number of coordinate systems. It is obvious that the correlation coefficient is a function of the coordinate system in use. This can be proven simply by the fact that the regression lines and the axes of the ellipse coincide for the y-system, while axes and regression line differ in the x-system. The correlation coefficient for the y-system is zero, and is not zero (Fig. 2.3) for the x-system unless the two coordinate systems coincide.

The angle θ between regression lines (Fig. 2.3) is described by:

$$\tan \theta = (1 - r^2) \sigma_1 \sigma_2 / [r(\sigma_1^2 + \sigma_2^2)] \qquad [2.82d]$$

When $r = 0$, then $\theta = 90°$, which means the regression lines run parallel to the axis (through point x_1, x_2). The lines fall together when $\theta = 0°$ or $180°$. This requires $r = \pm 1$. In this case the configuration is independent of the

THE BIVARIATE NORMAL DISTRIBUTION

coordinate transformation, as φ is expressed by:

$$\tan 2\varphi = 2\sigma_1 \sigma_2 / (\sigma_1^2 - \sigma_2^2) \qquad [2.82e]$$

Furthermore, $r = \pm 1$ is a decayed bivariate distribution consisting of one straight line.

It was previously mentioned that lines of equal probability density are ellipses. The major axis y_1 displays the angle φ with x_1, hence $\varphi = \sphericalangle y_1 x_1$. We know that the major axis of the ellipse has the equation $y_1 = 0$ and the minor axis $y_2 = 0$.

In the x_1, x_2-coordinate system we have:

$$(x_2 - \bar{x}_2)/(x_1 - \bar{x}_1) = \pm \tan \varphi \text{ (major axis)} \qquad [2.83a]$$

$$(x_2 - \bar{x}_2)/(x_1 - \bar{x}_1) = \mp 1/\tan \varphi \text{ (minor axis)} \qquad [2.83b]$$

Introducing the value of $\tan \varphi$ we obtain:

$$x_2 - \bar{x}_2 = [1/(2r\sigma_1 \sigma_2)][\sigma_2^2 - \sigma_1^2 \pm \sqrt{C_x}] \cdot (x_1 - \bar{x}_1) \qquad [2.83c]$$

Analogously the minor axis can be found. (C_x see [2.80g]).
It is obvious that for $r = \pm 1$ the axes are identical with the regression lines. For $r = \pm 1$ we reduce [2.83c] to:

$$x_2 - \bar{x}_2 = [1/(2\sigma_1 \sigma_2)][\sigma_2^2 - \sigma_1^2 \pm (\sigma_1^2 + \sigma_2^2)](x_1 - \bar{x}_1)$$

from which the regression line results. Notice also that $r = \pm 1$ and $\sigma_1 = \sigma_2$ provide $\tan \varphi = 1$ with $\varphi = 45°$. It has been mentioned that $\sigma_a = \sigma_b$ for the Gaussian bivariate distribution necessitates $r = 0$; the condition $r = \pm 1$ produces a degenerated distribution, which is a straight line. In fact, distributions with r either ± 1 represent straight lines, whose angles with the axes depend merely on the ratio σ_1/σ_2 and the sign of r.

A test for checking the statistical significance that the correlation coefficient $r \neq 0$ is given in Essenwanger (1974a, section 6.4.2). Samiuddin (1970) has recently proven that we can test the significance of r against any assumed correlation ρ by the test parameter t, which has a Student distribution with $N - 2$ degrees of freedom:

$$t = [(r - \rho)\sqrt{(N-2)}]/\sqrt{(1 - r^2)(1 - \rho^2)} \qquad [2.83d]$$

For $\rho = 0$ the commonly known test formula is derived. The t-test is discussed in detail by the author (1974a, section 6.4).

The influence of a single truncation upon the correlation coefficient r in the case of a bivariate Gaussian distribution has been treated by Aitkin and Hume (1963).

2.8.4 The bivariate circular distribution

In the previously discussed bivariate distribution the lines of equal probability were ellipses. This can be recognized best from [2.78b], where z_y^2 is synonymous with a certain probability level, determined by [2.78d], and z_y^2 has a particular value. When the y is known, the x-coordinates can be computed by:

$$(x_2 - \bar{x}_2) = y_1 \sin \varphi + y_2 \cos \varphi \qquad [2.80n]$$

$$(x_1 - \bar{x}_1) = y_1 \cos \varphi - y_2 \sin \varphi \qquad [2.80o]$$

Let us assume now $\sigma_1^2 = \sigma_2^2 = \sigma_a^2 = \sigma_b^2 = \sigma^2$. When we postulate the first equalization the others follow from the presented equations. We may drop the index and call the variance σ^2. The ellipses have become circles. We can define a vector standard deviation as previously introduced (Essenwanger, 1974a, section 3.3.4, eq. 3.44a) by:

$$\sigma_v = \sigma_1 = \sigma_2 \qquad [2.84]$$

$$\text{or: } \sigma_1 \cdot \sigma_2 = \sigma_v^2 \qquad [2.84a]$$

Then the bivariate circular distribution has a vector standard deviation of σ_v, where the σ_v is the standard deviation for any arbitrary axis. The above definition is mostly employed in statistical literature.

In meteorology sometimes the vector standard deviation for the bivariate distribution is defined by:

$$\sigma_{v_2} = (\sigma_1^2 + \sigma_2^2)^{1/2} \qquad [2.84b]$$

In this form it has been given by Brooks and Carruthers (1953), Crutcher (1956), Essenwanger (1959) and Court (1957a, 1962). Thus:

$$\sigma_{v_2} = \sigma_v \sqrt{2} \qquad [2.84c]$$

With the definition of σ_{v_2} by [2.84b] the relationship to the standard deviation of the wind speed and the resultant wind vector becomes simple (Essenwanger, 1974a, section 7.2). The reader is cautioned to check which definitions are in use, as the probability levels with reference to σ_v or σ_{v_2} differ and both parameters can be easily mixed up.

For the bivariate circular distribution [2.78] can be transformed to:

$$f(x_1, x_2) = \exp[-z_1^2/(2\sigma_v^2)]/(2\pi \sigma_v^2) \qquad [2.84d]$$

$$\text{or: } f(x_1, x_2) = \exp[-z_1^2/\sigma_{v_2}^2]/(\pi \sigma_{v_2}^2) \qquad [2.84e]$$

$$\text{with: } z_1^2 = (x_1 - \bar{x}_1)^2 + (x_2 - \bar{x}_2)^2 = z_r^2 \sigma_v^2 \qquad [2.84f]$$

If we equate V with z_1 we can write (with $\sigma_{v_2}^2 = \sigma^2$):

$$f(V) = [1/(\pi \sigma^2)] \exp(-V^2/\sigma^2) \qquad [2.85]$$

It can be readily shown that [2.85] can serve as an approximation of the probability density in the non-circular bivariate case, too. We assume the definition [2.84b and c]. Then:

$$2\sigma_v^2 = \sigma_{v_2}^2 = \sigma_1^2 + \sigma_2^2 \qquad [2.86a]$$

We can state $\sigma_1^2 = \sigma_v^2 + \epsilon$ and $\sigma_2^2 = \sigma_v^2 - \epsilon$ with $\epsilon = (\sigma_1^2 - \sigma_2^2)/2$. Thus we find:

$$\sigma_1 \sigma_2 = (\sigma_v^4 - \epsilon^2)^{1/2} \sim \sigma_v^2 \qquad [2.86b]$$

for small enough ϵ.

Now [2.78] can be brought into form [2.84d] by setting:

$$z^2 = (x_1 - \bar{x}_1)^2 \sigma_v^2/\sigma_1^2 + (x_2 - \bar{x}_2)^2 \sigma_v^2/\sigma_2^2 = z_r^2 \sigma_v^2 \qquad [2.86c]$$

The transformation into formulae [2.84e] and [2.85] as the subsequent steps follows analogously. This approximation is valid as long as the postulation $\sigma_v^4 \gg \epsilon^2$ is fulfilled. The definition of the V in the elliptical case is different from the circular one, too, namely:

$$z_1^2 = V^2 = (x_1 - \bar{x}_1)^2 [0.5 + \sigma_2^2/(2\sigma_1^2)] + (x_2 - \bar{x}_2)^2 [0.5 + \sigma_1^2/(2\sigma_2^2)] \qquad [2.86d]$$

In the circular case this "weighting" factor becomes unity ([2.84f]). The standard vector deviation can, therefore, be utilized as a standard length of the dispersion. Since in the non-circular case the lines of equal probability are ellipses, the actual probability depends on the position of the major and minor axes in the coordinate system. The approximation by [2.85] in the elliptic case may be helpful, however, when rough estimates of probability lines and a quick survey of the magnitude of the area of dispersion are required. The result is exact for the bivariate circular distribution.

Since every estimator has a random error, a ratio of the estimators s_1^2/s_2^2 different from 1.0 does not necessarily mean that the population distribution is elliptical. We may test the significance of the difference of two variances by Fisher's F-test (see Essenwanger, 1974a, section 6.5).

2.8.5 Cumulative bivariate Gaussian distribution, integrals

It is relatively easy to obtain $F(x)$ as the integral of $f(x)$ for the one-dimensional Gaussian distribution. Methods have been discussed by Essenwanger (1974a, section 2.4) and a table of $F(x)$ is available in the Appendix (see p. 384).

The task is rendered more difficult for the bivariate distribution. We must consider several types of problems in meteorology. (See Fig. 2.1.)

(1) What is the probability of exceeding a threshold value of x_1 or x_2?

(2) What area do we need to accommodate a certain probability density or what fraction (multiple) of the standard vector deviation would we expect for a certain probability density?

(3) What probability density can we expect for a circle of radius R around the origin?

(4) What is the probability density of a circle segment with the size of the angle α and the circle radius R?

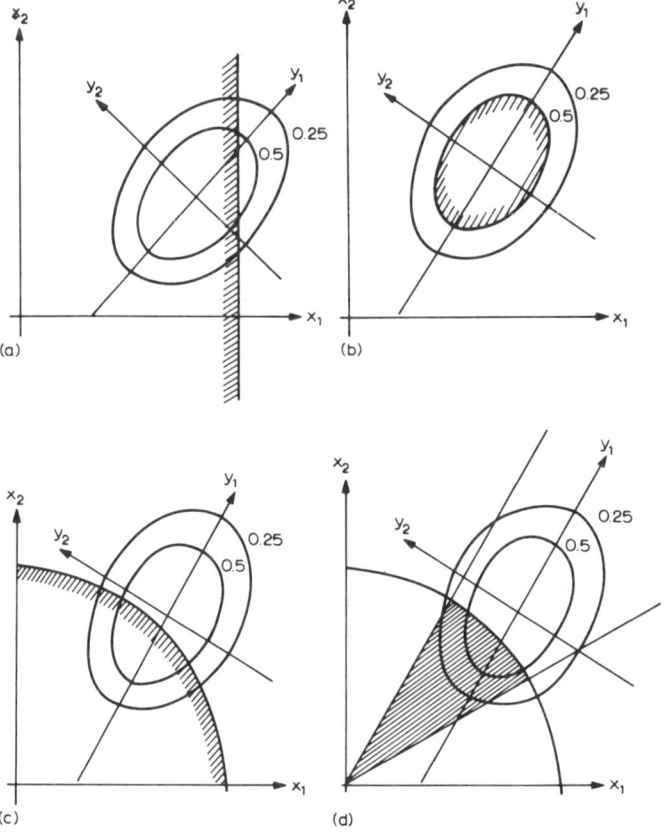

Fig. 2.1. Types of integrations of the bivariate Gaussian distribution in atmospheric science. The probability is needed for:
 (a) exceeding a specified coordinate threshold (e.g. a zonal wind speed of 10 m/sec);
 (b) a particular area (e.g. the area which comprises 50% of the data);
 (c) exceeding a specified circle around the origin (e.g. exceeding a wind speed of 10 m/sec);
 (d) a specified circle segment (e.g. what is the frequency of a NW wind or exceeding a NW wind of 10 m/sec?).

All four problems have one thing in common: They require integration of the bivariate normal distribution. Some of these integrals can be solved directly, but most of them must be approximated. In the following sections some of the solutions are presented although it is impossible in this chapter to discuss all methods.

2.8.5a *The first problem*
It is easily solved by the marginal distribution:

$$F(x_1) = \int_{-\infty}^{x_1} \int_{-\infty}^{\infty} f(x_1, x_2)\, dx_1\, dx_2 \qquad [2.87]$$

THE BIVARIATE NORMAL DISTRIBUTION

where $f(x_1, x_2)$ is defined by [2.77]. Hence the answer is:

$$F(x_1) = \int_{-\infty}^{x_1} f(x_1)\,dx_1 = [1/(\sigma_1\sqrt{2\pi})] \int_{-\infty}^{x_1} \exp[-0.5(x_1 - \bar{x}_1)^2/\sigma_1^2]\,dx_1 \quad [2.87a]$$

This is identical with the solution for a one-dimensional Gaussian distribution, which has been presented by the writer (1974a, section 2.4).

Analogously we find:

$$F(x_2) = \int_{-\infty}^{\infty} \int_{-\infty}^{x_2} f(x_1, x_2)\,dx_1\,dx_2 \quad [2.87b]$$

or:

$$F(x_2) = \int_{-\infty}^{x_2} f(x_2)\,dx_2 = [1/(\sigma_2\sqrt{2\pi})] \int_{-\infty}^{x_2} \exp[-0.5(x_2 - \bar{x}_2)^2/\sigma_2^2]\,dx_2 \quad [2.87c]$$

Very rarely do we need integration over an area bounded between x_{11}, x_{12} in the abscissa and x_{21}, x_{22} in the ordinate. This would be equivalent to integrating over a polygon. This problem is treated by Owen (1956). A simple formula has also been given by Cadwell (1951), based on a comprehensive article by Nicholson (1943), Cadwell first recommends the use of the integral:

$$F(h, k, r) = \frac{1}{2\pi\sqrt{(1-r^2)}} \int_h^{\infty} \int_k^{\infty} \exp[-(x_1^2 - 2rx_1x_2 + x_2^2)/\{2(1-r^2)\}]\,dx_1\,dx_2 \quad [2.87d]$$

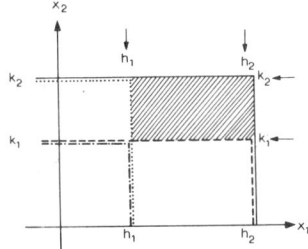

Fig. 2.2. Integration over a bounded rectangular area.
$-\cdot-\cdot-\cdot- = F(h_1, k_1, r)$, $----- = F(h_2, k_1, r)$ $\cdots\cdots\cdots = F(h_1, k_2, r)$,
$\underline{\qquad} = F(h_2, k_2, r)$.

which is tabulated in Pearson's (1931) tables. The probability over a rectangle R_{hk} bound by $x_{11} = h_1, x_{12} = h_2, x_{21} = k_1, x_{22} = k_2$ is then:

$$F(h_1, k_1, r) + F(h_2, k_2, r) - F(h_1, k_2, r) - F(h_2, k_1, r) \quad [2.87e]$$

(see Fig. 2.2). For negative h and k we have:

$$F(h, k, r) = F(-h, -k, r) + F(-h) + F(-k) \qquad [2.87\text{f}]$$

where $F(-h)$ and $F(-k)$ denote the one-dimensional Gaussian integral from 0 to x.

We may notice that:

$$h = (x_{12} - x_{11})/\sigma_{x_1} = (h_2 - h_1)/\sigma_{x_1} \qquad [2.87\text{g}]$$

$$k = (x_{22} - x_{21})/\sigma_{x_2} = (k_2 - k_1)/\sigma_{x_2} \qquad [2.87\text{h}]$$

As proven by Nicholson (1943), a simpler table will suffice with:

$$F(h, q) = \frac{1}{2\pi} \int_0^h \int_0^{qx_1/h} \exp[-0.5(x_1^2 + x_2^2)] \, dx_1 \, dx_2 \qquad [2.87\text{i}]$$

This is the bivariate integral with $r = 0$ over a triangular region with points $(0, 0)$, $(h, 0)$ and (h, q). The definition of q is:

$$q = (k - rh)/\sqrt{1 - r^2} \qquad [2.87\text{j}]$$

Caldwell (1951) makes the following approximation:

$$F(h, q) = \frac{\arctan(q/h)}{2\pi} [1 - \exp\{-hq/2\arctan(q/h)\}] \qquad [2.87\text{k}]$$

For values of $q > h$ we compute $F(q, h)$ instead of $F(h, q)$ and obtain the $F(h, q)$ by the relationship:

$$F(h, q) = F(h) \cdot F(q) - F(q, h) \qquad [2.87\text{l}]$$

Again, the $F(h)$ and $F(q)$ are used as abbreviations for the one-dimensional integral of the Gaussian distribution.

A table for $F(h, k, r)$ has also been published by the National Bureau of Standards (1959).

2.8.5b *The second problem*

This problem arises when we need the area of dispersion from the center point of the bivariate distribution, i.e., we would like to compute the area of the circular probable error where 50% of the frequency is within and 50% outside the area. This problem also has application in turbulence analysis (e.g. Frenkiel, 1951).

For this purpose it is best to transform the bivariate distributions to independent coordinates. Several solutions have been suggested. Hald (1957) solves the problem by an approximation with the χ^2-distribution (Essenwanger, 1974a, section 6.7). The standardized variables can be written:

$$z_y^2 = y_1^2/\sigma_a^2 + y_2^2/\sigma_b^2 = \xi_1^2 + \xi_2^2 = \chi^2 \qquad [2.88]$$

THE BIVARIATE NORMAL DISTRIBUTION

Further: $z_y^2 = z^2/(1-r^2) = \chi^2$ [2.88a]

where z^2 is defined for dependent variables x_1 and x_2 in [2.77a]. (This transformation into the χ^2-distribution is strictly valid for $\sigma_a = \sigma_b = 1$). The $z_y^2 = \chi^2$ is a χ^2-distribution with 2 degrees of freedom (see Hald, 1957). Therefore:

$$F(\chi^2 \leqslant z_y^2) = P(\chi_f^2) = F(y_1, y_2) = \int \int_{z_y^2} f(y_1, y_2) \, dy_1 \, dy_2 \quad [2.88b]$$

In the double integral the z_y^2 symbolizes the area inclosed by the z_y^2. It is self-evident that the conversion to z^2 requires the area to be centered around $\bar{x}_1$ and $\bar{x}_2$.

We compute the probability by the relationship:

$$P(\chi^2) = 1 - \exp(-\chi^2/2) \quad [2.88c]$$

Let us assume $z_y^2 = 1$. Then the coordinate points of the ellipse on the y_1-axes are $y_1 = \pm \sigma_a$, $y_2 = 0$. On the y_2-axes we find the points $y_1 = 0$, $y_2 = \pm \sigma_b$. The probability for any point inside this ellipse can be computed from:

$$\log[1 - P(\chi^2)] = -0.21715 \, \chi^2 \quad [2.88d]$$

After some arithmetic we obtain $P(\chi^2) = 0.3935$. Extracting the P from the χ^2-tables (e.g. Hald 1952) results in the same value.

Equation [2.88d] can be changed to the natural logarithm:

$$\ln[1 - P(\chi^2)] = -0.5\chi^2 \quad [2.88e]$$

Equations [2.88d and e] can also be employed to find a graphical solution. We plot χ_i^2 against $(n - i + 0.5)/n$ on semilogarithmic paper,* where n is the number of observations. We compute the χ_i^2-values from $\chi_1^2 \ldots \chi_n^2$ for the pairs of observations $x_{11}, x_{12}; x_{21}, x_{22} \ldots ; x_{1n}, x_{2n}$ by [2.88] or [2.88a], depending on which form of the observations is used. We rank the χ_i^2 and then the $P(\chi_i^2)$ should correspond to:

$$P(\chi_i^2) = 1 - P(\chi^2) = (n - i + 0.5)/n \quad [2.88f]$$

The points should give a straight line with slope -0.217 and go through the point $(0, 1)$ if a bivariate Gaussian distribution exists. Since, in practice, estimators for $\bar{x}_1, \bar{x}_2, \sigma_a, \sigma_b$ and r are utilized, the points may fluctuate around a straight line.

Assume now that we are interested in finding the line $P(\chi^2) = 0.5$. We determine χ^2 as 1.386 and $z_y^2 = 1.386$. The points on the axes are $\pm 1.18 \sigma_a$ and $\pm 1.18 \sigma_b$ for the y_1- and y_2-axes, respectively. Hence, the circle with

---

* Abscissa in linear scale $x = \chi_i^2$; ordinate in logarithmic scale $y = \ln[(n - i + 0.5)/n]$ or $y = \log(\ldots)$.

radius $1.1774\sigma_1$ or $1.1774\sigma_2$ can be called the 50-% probability circle and the radius is called the circular probable error C_{PE} (see also Essenwanger, 1974a, section 6.2).

In terms of the standard vector deviation $\sigma_{v_2} = \sigma_v \sqrt{2}$ (see [2.84c]), we would obtain the probability 0.5 for $C_{PE} = 1.1774\sigma_v$. Hence, the standard vector deviations as defined by [2.84b] would lead to $C_{PE} = 1.774\sigma_{v_2}/\sqrt{2} = 0.8326\sigma_{v_2}$. A table for the generalized circular error can be found in Weingarten and DiDonato (1961) or Harter (1960). While the theoretical value of the probable error can be readily established from the integration of $F(x_1, x_2)$, the estimation from empirical data generates a different problem. Effects of bias on estimation of the circular probable error C_{PE} have been discussed by Moranda (1960). Kamat (1962) has supplemented these estimates. Morando recommends for $\bar{x}_1 = \bar{x}_2 = 0$ the following estimate:

$$\hat{C}_{PE1} = 1.1774\sqrt{N}\,\frac{\Gamma(N)}{\Gamma\left(\dfrac{2N+1}{2}\right)}\,\sqrt{\frac{1}{2N}\sum_1^N (x_{1i}^2 + x_{2i}^2)} \qquad [2.89]$$

When the mean is greatly different from $\bar{x}_1 = \bar{x}_2 = 0$, he recommends:

$$\hat{C}_{PE2} = 1.1774\sqrt{N}\,\frac{\Gamma(N-1)}{\Gamma\left(\dfrac{2N-1}{2}\right)}\,\sqrt{\frac{1}{2N}\sum_1^N [(x_{1i}-\bar{x}_1)^2 + (x_{2i}-\bar{x}_2)^2]}$$

$$[2.89a]$$

Kamat adds several others, of which:

$$\hat{C}_{EP3} = 1.1774\,\sqrt{\frac{\pi}{2}\left[\frac{1}{2N}\sum_1^N (|x_{1i}| + |x_{2i}|)\right]} \qquad [2.89b]$$

is very simple to compute with variance:

$$\mathrm{Var}(\hat{C}_{EP3}) = (1.1774)^2 (\pi/4 - 1/2)\,\sigma_v^2/N \qquad [2.89c]$$
$$= 0.39564\,\sigma_v^2/N$$

The bar $|\ |$ denotes the absolute value. C_{PE1} and C_{PE2} are not trivial to compute for large N because of the Γ-ratio.

An expansion is:

$$\hat{C}'_{EP3} = C_{EP3}\cdot\sqrt{N/(N-1)} \qquad [2.89d]$$

where $|x_{1i}|$ and $|x_{2i}|$ in C_{EP3} are replaced by $|x_{1i}-\bar{x}_1|$, respectively. The $\mathrm{Var}(\hat{C}_{EP3})$ is not simple, however:

$$\mathrm{Var}(C'_{EP3}) = (1.1774)^2(\pi/4)[N/(N-1)]\,\sigma_v^2\,\mathrm{Var}(m/\sigma) \qquad [2.89e]$$

with: $\mathrm{Var}(m/\sigma) = [(N-1)/N][1/N + (2/\pi)\{\sqrt{(N-2)/N} +$

$$+ (1/N)\arcsin[1/(N-1)] - 1\}] \qquad [2.89f]$$

THE BIVARIATE NORMAL DISTRIBUTION

Weil (1954) has suggested a different solution. He first transforms the x-system with correlation between x_1 and x_2 into the y-system, where the y_1 and y_2 are uncorrelated (see Section 2.8.3). The bivariate distribution is re-written in polar coordinates, $Y_1 = R \cos \theta$; $Y_2 = R \sin \theta$, where the Y_1 and Y_2 run parallel to the y_1 and y_2 with $y_1 = Y_1 - \bar{Y}_1$ and $y_2 = Y_2 - \bar{Y}_2$ (see Fig. 2.3). The end result for the probability $P(R)$ of $R = \sqrt{Y_1^2 + Y_2^2}$ is then:

$$F(R) = P(R)$$

$$= A \cdot R \left[I_0(aR^2) I_0(dR) + 2 \sum_{j=1}^{\infty} I_j(aR^2) I_{2j}(dR) \cos 2j\Psi \right]$$

$$\exp[-R^2(\sigma_a^2 + \sigma_b^2)/4\sigma_a^2 \sigma_b^2] \quad [2.90]$$

The following abbreviations have been used:

$$A = [1/(\sigma_a \sigma_b)] \exp[-(\bar{Y}_1^2 \sigma_b^2 + \bar{Y}_2^2 \sigma_a^2)/(2\sigma_a^2 \sigma_b^2)] \quad [2.90a]$$

$$a = (\sigma_a^2 - \sigma_b^2)/(4\sigma_a^2 \sigma_b^2) \quad [2.90b]$$

$$b = \bar{Y}_1/\sigma_a^2 \quad [2.90c]$$

$$c = \bar{Y}_2/\sigma_b^2 \quad [2.90d]$$

$$d^2 = b^2 + c^2 \quad [2.90e]$$

$$\tan \Psi = \bar{Y}_2 \sigma_a^2 / (\bar{Y}_1 \sigma_b^2) \quad [2.90f]$$

The σ_a^2 and σ_b^2 are variances of the Y_1 and Y_2 variates, respectively, the $\bar{Y}_1$ and $\bar{Y}_2$ denote the mean values of the variates Y_1 and Y_2. The $I_j(x)$ are Bessel functions, such as those tabulated by the British Association for the Advancement of Science (1952).

The solution is relatively simple although lengthy when tables of Bessel functions are available. Computer calculations, especially involving numerous data samples, appear too complex and costly by Weil's system, although it accommodates unequal variances. When routine programs to compute Bessel functions on electronic computers are available, Weil's system would be easy to program, but may still be costly.

The following Table 2.10 is quite useful in determining the radius of a circle for a selected probability (circular distribution) $\sigma_v = \sigma_1 = \sigma_2$; $\sigma_{v_2} = \sigma_v \sqrt{2}$.

2.8.5c *The third problem*

The solution by Weil leads into it. We need only express the variates Y_1 and Y_2 with reference to the origin, i.e., the axes Y_1 and Y_2 parallel to y_1 and y_2, respectively, go through the origin. Then Weil's system gives the answer to the probability density within a circle of radius R around the

TABLE 2.10

Radius of a circle, bivariate distribution

$P(c)$	0.25	0.5	0.75	0.90	0.95	0.99	0.995
$\chi^2 = c_1^2$	0.57536	1.38628	2.77258	4.60516	5.99146	9.21034	10.59663
$c_1 = R/\sigma_v$	0.75852	1.17740	1.66511	2.14596	2.44775	3.03485	3.25525
$c_2 = R/\sigma_{v_2}$	0.53636	0.83255	1.17740	1.51742	1.73082	2.14596	2.30181
c_2^2	0.28768	0.69314	1.38629	2.30258	2.99573	4.60517	5.29832

$R = c_1 \cdot \sigma_v$ or $R = c_2 \cdot \sigma_{v_2}$; $R^2 = c_2^2(\sigma_a^2 + \sigma_b^2)$

$\ln[1 - P(c)] = -\chi^2/2$

$c_1^2 = -2\ln[1 - P(c)]; c_1^2 = 2c_2^2$

THE BIVARIATE NORMAL DISTRIBUTION

origin. Other solutions have been given by Gilliland (1962), Owen (1956), or Ihm (1960) etc., and will be discussed here.

Required is the probability over a circle with radius R around the origin, or in terms of meteorology, the probability of encountering a wind speed of a certain threshold or less. When this probability is known, the chances of exceeding the threshold R can be readily derived. Although a solution can be found by direct representation of the statistical distribution of the wind speed (see Essenwanger, 1968), major scientific studies are based on the vector wind such as Brooks et al. (1946, 1950), (Crutcher 1959), etc. The problem can be restated as such. We like to find:

$$F(R) = P(R) = \int_0^R f(x_1, x_2) \, dx_1 \, dx_2 \qquad [2.91]$$

where $f(x_1, x_2)$ represents the bivariate normal distribution. The equation must be rewritten in polar coordinates, e.g. $y_1 = \sigma_a R \cos \alpha$; $y_2 = \sigma_a R \sin \alpha$ (y-system!). Then:

$$P(R) = \frac{\sigma_a}{2\pi \sigma_b} \int_0^R \int_0^{2\pi} [\exp\{-0.5(\{R \cos \alpha - \bar{y}_1/\sigma_a\}^2$$

$$+ \{(R \sigma_a \sin \alpha - \bar{y}_2)/\sigma_b\}^2)\}] \, d\alpha R \, dR \qquad [2.91a]$$

For $y_1 = y_2 = 0$ and $\sigma_a = \sigma_b = \sigma_v$ (circular!) this equation leads to the known solution:

$$P(R) = 1 - \exp[-R^2/(2\sigma_v^2)] \qquad [2.91b]$$

which is identical with [2.88c].

In polar coordinates [2.91] can also be written in the following form (see also Brooks and Carruthers, 1953; Tucker, 1960, etc.):

$$F(V, \alpha) = P(V, \alpha) = \frac{1}{\pi \sigma^2} \int\int_R \exp(-v_y^2/\sigma^2) \, V \, dV \, d\alpha \qquad [2.91c]$$

The following symbols have been utilized:

$$v_y^2 = y_1^2 + y_2^2 \text{ (independent coordinates)} \qquad [2.91d]$$

$$\sigma^2 = \sigma_1^2 + \sigma_2^2 = \sigma_a^2 + \sigma_b^2 = \sigma_{v_2}^2 \text{ (standard vector deviation)} \qquad [2.84b]$$

Since the definition [2.84b] has been used, the term before the integral is $\pi \sigma^2$ rather than $2\pi \sigma^2$. Notice also that:

$$v_y^2 = V^2 + V_R^2 - 2V_R V \cos \alpha \qquad [2.91e]$$

where: $V_R^2 = \bar{x}_1^2 + \bar{x}_2^2 = \bar{Y}_1^2 + \bar{Y}_2^2 \qquad [2.91f]$

The $\bar{Y}_1$ and $\bar{Y}_2$ denote the coordinates of V_R in the y-system, see text after [2.90f].) The V_R is the resultant wind vector (Essenwanger, 1974a, section

7.2), and α is the angle between the major axes and the individual wind vector whose magnitude is V (see [2.90f]). If we normalize V_R^2 by computing:

$$r_v^2 = V_R^2/\sigma_v^2 = 2V_R^2/\sigma_{v_2}^2 \qquad [2.91g]$$

we can adopt an approximation from the Handbook of Mathematical Functions (Abramowitz and Stegun, 1964, 26, 3.24). We set $R = V/\sigma_v$. For the bivariate circular distribution ($\sigma_v = \sigma_1 = \sigma_2$) we find:

(1) for $R < 1$:

$$P(R^2/2, r_v^2) = [2R^2/(4 + R^2)] \exp[-2r_v^2/(4 + R^2)] \qquad [2.92a]$$

$P(\xi)$ is the probability inside bounded by the circle R.

(2) for $R > 1$:

$P(\xi) = Q(\xi)$ for $\xi < 0$

$P(\xi) = 1 - Q(\xi)$ for $\xi > 0$

$$Q(\xi) = (1/\sqrt{2\pi}) \exp(-\xi^2/2) \qquad [2.92b]$$

$(1/\sqrt{2\pi} = 0.3989)$

with: $\xi = [(R^2/a)^{1/3} - \{1 - 2(1 + b)/(9a)\}]/[2(1 + b)6(9a)]^{1/2}$ [2.92c]

and: $a = 2 + r_v^2$ [2.92d]

$b = r_v^2/a$ [2.92e]

Hence: $R^2 = a[(2/3)\{(1 + b)/a\}^{1/2} \cdot c + 1 - 2(1 + b)/(9a)]^3$ [2.92f]

$c = [-\{\ln P(\xi) + \ln\sqrt{2\pi}\}]^{1/2}$ [2.92g]

$ = [-\{\ln P(\xi) + 0.91844\}]^{1/2}$

(Note, $P(\xi) \geq 0.399$, i.e. $Q \leq 0.601$.)

(3) for $R > 5$ (i.e. R and r_v are both large):
$P(\xi)$ as in [2.92b] with the simplification:

$$\xi = R - \sqrt{r_v^2 - 1} \qquad [2.92h]$$

Therefore: $R = \sqrt{2c} + \sqrt{r_v^2 - 1}$ [2.92i]

Let us assume that $\sigma_v = 1$, then $R = V$ and $r_v = V_R$. We can compare the results from the formulae with tabulations of the circular error probability by setting $r_v = 0$. Two examples are given, $R = 1$ and $R = 1.5$.

Formula [2.92a]:

$P(R^2/2, 0) = [2/(5)] \cdot \exp(-0) = 0.4$

$P(R^2/2, 0) = (4.5)/(6.25) = 0.72$

THE BIVARIATE NORMAL DISTRIBUTION

Formula [2.92b]:

$a = 2, b = 0.$

$\xi = [(1/2)^{1/3} - \{1 - 2(1)/(9 \cdot 2)\}]/[2(1)/(9 \cdot 2)]^{1/2}$

$= [0.794 - (1 - 0.111)]/(1/3) = 3(-0.095) = -0.285$

$Q(\xi) = Q(-0.285)$

Notice, in this case $P(\xi) = (0.3989)(0.9602) = 0.383$.

Second case:

$\xi = [(2.25/2)^{1/3} - (1 - 2/18)]/(2/18)^{1/2} = (1.040 - 0.889)/1/3$

$= 3(0.151) = 0.453; \quad \xi^2/2 = 0.103$

$P(\xi) = 1 - (0.3989)(0.9048) = 1 - 0.318 = 0.682$

From tabulations of the circular distribution we find $P_1 \sim 0.393$ and $P_2 \sim 0.675$. In these cases the approximation may be considered to be sufficient.

It should be noticed that the R cannot be expressed explicitly in terms of P from the formula [2.92a], but the probability $P(R^2/2, r_v^2)$ can be calculated. The boundaries $R = 1$ and $R = 5$ can be computed from both adjacent sides. This permits an evaluation of the approximation errors by comparing both results.

Gilliland (1962) has employed a different approach. His computations are based on [2.91a], with the first approximation like [2.91a and b]. We are interested in the case where $\bar{y}_1 \neq \bar{y}_2 \neq 0$ and $\sigma_a^2 \neq \sigma_b^2$.

Thus, after some arithmetic, Gilliland (1962) obtains:

$$P(R \leq z) = [A/(2\pi a)] \sum_{j=0}^{\infty} B_j P_j(\beta z^2) \qquad [2.93a]$$

The following symbols have been used:

$$P_j(\beta z^2) = 1 - \exp(-\beta z^2) \sum_{k=0}^{j} (\beta z^2)^k /k! \qquad [2.93b]$$

This function is tabulated in a report by the General Electric Company Defense Systems Department (1962), and is a modified Poisson distribution:

$z = V/\sigma_{y_1}$ [2.93c]

$A = \exp[-(b^2 + c^2)/2]$ [2.93d]

$a = \sigma_{y_2}/\sigma_{y_1}$ [2.93e]

$\alpha = (1 - a^2)/(4a^2)$ [2.93f]

$\beta = (1 + a^2)/(4a^2)$ [2.93g]

$$b = \bar{y}_1/\sigma_{y_1} \qquad [2.93\text{h}]$$

$$c = \bar{y}_2/\sigma_{y_2} \qquad [2.93\text{i}]$$

$$d = c/a \qquad [2.93\text{j}]$$

$$B_j = [j!/(2\beta^{j+1})] \sum_{i=0}^{j} D_{j,i} \qquad [2.93\text{k}]$$

$$D_{j,i} = (\alpha^i/i!) \sum_{k=0}^{j-1} [b^{2k} d^{2(j-i-k)}/(2k)!\, 2(j-i-k)]\, G[i, 2k, 2(j-i-k)] \qquad [2.93\text{l}]$$

$$G(m, n, p) = \int_0^{2\pi} \cos^m 2\theta \cos^n \theta \sin^p \theta\, d\theta \qquad [2.93\text{m}]$$

This integral is zero if either n or p is odd. The given solution may lead to elaborate computations for the individual data sample, but in special cases can be simplified. Gilliland has given some of the simplified examples. He has also performed an error analysis, where the absolute error of R after the first $N + 1$ terms is:

$$\epsilon_R \leqslant \frac{Az^2 \exp(-\beta z^2)(n+3)\, C^{n+1}}{2a(n+3-\beta z^2)(n+2-C)(n+1)!} \qquad [2.93\text{n}]$$

with: $0 < C/(n+2) < 1$ [2.93o]

Owen (1956) has considered probabilities over rectangles for the correlated bivariate normal distribution. Although important in problems of the bivariate distribution, the solution is rarely needed in meteorology. The requirement is not among our four questions. For general interest, the following may be stated. Owen's fundamental formula provides the probability:

$$P(h, k, r) = \tfrac{1}{2} G(h) + \tfrac{1}{2} G(k) - T(h, a_h) - T(k, a_k) - c \qquad [2.94]$$

where $c = 0$ for $hk \geqslant 0$ with $h + k \geqslant 0$, otherwise $c = 0.5$. The a_h and a_k denote:

$$a_h = [k/(h\sqrt{1-r^2})] - r/\sqrt{1-r^2} \qquad [2.94\text{a}]$$

$$\text{and: } a_k = [h/(k\sqrt{1-r^2})] - r/\sqrt{1-r^2} \qquad [2.94\text{b}]$$

with r standing for the correlation between x_1 and x_2. $G(h)$ is the univariate normal distribution with zero mean, and $T(h, a_h)$ has been tabulated by Owen (1956).

Another set of tables for integrals of the bivariate normal distribution over offset circles has been established by Rosenthal and Rodden (1961).

Asymptotic properties of some estimators of quantiles of circular error have been discussed by Blischke (1966). In order to keep this section in bounds the reader is referred to the quoted literature.

2.8.5d *The fourth problem*

This problem arises in the construction of a wind rose from the given vector wind data. Again we must integrate [2.91a], this time between boundaries V_1 and V_2 for the first integration and between α_1 and α_2 for the second integration. Brooks et al, (1946) have employed [2.91c] and start with:

$$F(V, \alpha) = P(V, \alpha) = (\pi\sigma)^{-1} \int_{\alpha_1}^{\alpha_2} \int_{V_1}^{V_2} \exp(-v_y^2/\sigma^2)\, V\, dV\, d\alpha \qquad [2.95]$$

This integral also has no analytical solution. Brooks et al, (1946) have attempted to solve the problem via the aid of the persistence of the wind vector. The reader is referred to the original literature by Brooks et al. (1946) and a modification by Knighting (1954).

2.8.6 *Concluding remarks*

Several other topics could be included in this section on the multivariate or bivariate distribution. The reader is referred to the respective literature dealing with the problems.

The problem of estimation for the bivariate lognormal distribution has been treated by Mostafa and Mahmoud (1964).

A thorough analysis of the confidence limit for the mean of a multivariate normal distribution has been made by Stein (1962), while the central tolerance region is covered by John (1968). Probability contents of regions under spherical normal distributions have been described by Ruben (1960). Ihm (1960) has analyzed the numerical integration of n-dimensional normal distributions.

Chew (1966) has published formulae for confidence, prediction and tolerance regions for the multivariate normal distribution. A test for multivariate normality has been introduced by Wagle (1968). Crutcher and Moses (1962) have expanded the bivariate concept to three dimensions, which is very seldom used in meteorology and may be omitted here.

These are only a few examples of the various problems of recent research in the area of multivariate analysis.

Example 2.9. Bivariate elliptical distribution (Gaussian). This first example treats the bivariate distribution. It illustrates the various parameters and the regression lines (Fig. 2.3). The Thule (Greenland) January, 25 km altitude wind data from the period 1956—1963 serve as an appropriate data sample. It was selected for its display of an elliptical shape of the isoline of equal frequency density, its large angle between the major and x_1-axis (abscissa),

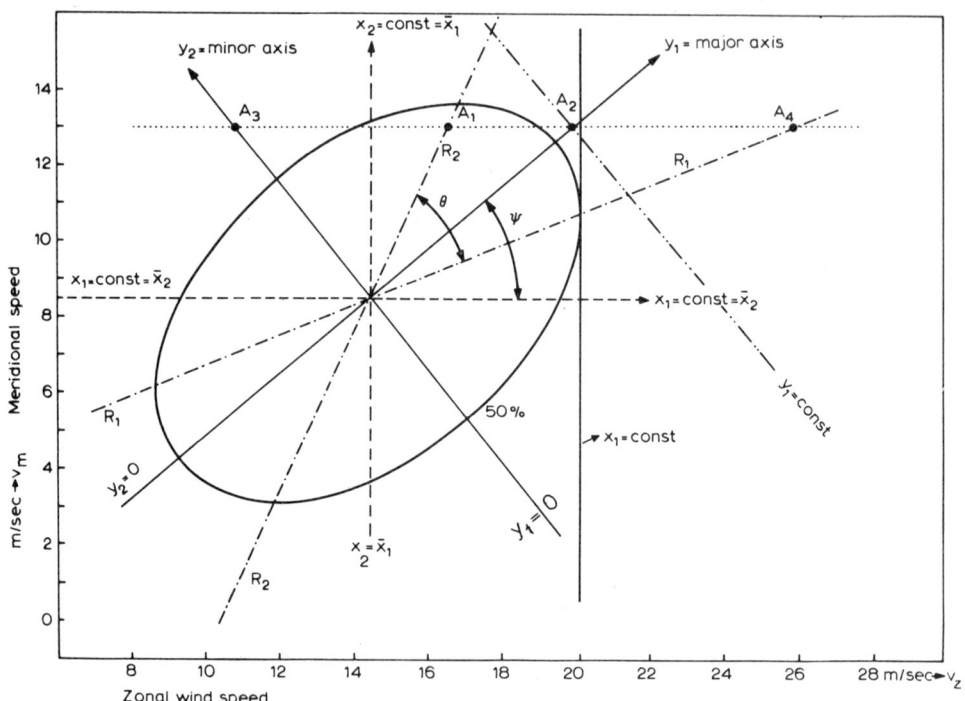

Fig. 2.3. Bivariate distribution for the wind velocity at Thule (Greenland) in January (1956—1963) at 25 km altitudes. Illustration of parameters and regression lines (Example 2.9). $R_1: x_2 = 2.7 + 0.39 x_1$; $R_2: x_1 = 10.5 + 0.47 x_2$.

and its relatively large displacement of the center of the bivariate distribution from the origins of the coordinate system x_1 and x_2. The x_1-component is identical with the zonal wind component, with west being positive; the x_2 coincides with the meridional wind component, south being positive.

The following statistics have been computed from the empirical data: $\bar{x}_1 = 14.5$ m/sec, $\bar{x}_2 = 8.4$ m/sec, $s_{x_1}^2 = 23.22$, $s_{x_2}^2 = 19.72$. It is evident that the two variances would not lead directly to a conclusion of a bivariate elliptical distribution, because both variances are too close to distinctly prove non-circularity. The calculation of the (linear) correlation coefficient between zonal and meridional wind components provided $r = 0.428$. The variances of the major and minor axes are the roots of the matrix [2.79].

$$\begin{vmatrix} 23.22 - k & 9.16 \\ 9.16 & 19.72 - k \end{vmatrix} = 0$$

THE BIVARIATE NORMAL DISTRIBUTION

which result in $\sigma_a^2 = 30.80$ and $\sigma_b^2 = 12.14$. Thus $S_{v_2}^2 = 42.94$ which equals $\sigma_{x_1}^2 + \sigma_{x_2}^2$ and $\sigma_a^2 + \sigma_b^2$. Now the non-circular shape of the lines of equal frequency density can be readily recognized by the ratio $\sigma_a^2/\sigma_b^2 \sim 2.5$.

The formula:

$$\tan 2\varphi = 2 \cdot (0.428) \cdot (4.82)(4.44)/(23.22 - 19.72)$$

provides $\varphi = 39.6°$. Consequently the transformed frequency density can be written:

$$f(y_1, y_2) = [1/\{2\pi(5.55)(3.48)\}] \exp(-Z_y^2/2)$$

and: $Z_y^2 = y_1^2/30.80 + y_2^2/12.14$

$\sigma_a^2 \sigma_b^2 = (1-r^2)\sigma_1^2 \sigma_2^2 = 374.0$

The ellipticity measures: $\epsilon_l = 0.372$, $F_E = 2.537$, $\epsilon_x = 0.778$

The regression lines follow from:

$$x_1 = 14.5 + 0.428(x_2 - 8.4) \cdot (4.82/4.44) = 14.5 + 0.468(x_2 - 8.4) =$$
$$= 10.5 + 0.47 x_2$$

and:

$$x_2 = 8.4 + 0.428(x_1 - 14.5) \cdot (4.44/4.82) = 8.4 + 0.392(x_1 - 14.5) =$$
$$= 2.7 + 0.39 x_1$$

The angle θ between regression lines can be calculated from:

$$\tan \theta = (0.817)(4.82)(4.44)/[(0.428)(23.22 + 19.72)] = 17.47/18.38$$
$$= 0.950$$
$$\theta = 43\tfrac{1}{2}°.$$

The transformed variate equations in a coordinate system y_1, y_2 with major and minor axis render:

$$y_1 = (x_1 - 14.5) \cdot 0.771 + (x_2 - 8.4) \cdot 0.637$$
$$y_2 = (x_2 - 8.4) \cdot 0.771 + (x_1 - 14.5) \cdot 0.637$$

The major and minor axes have the following forms:

Major axes: $(x_2 - 8.4) = \pm(x_1 - 14.5) \tan \varphi = (x_1 - 14.5) \cdot 0.827$

or: $x_2 = 0.827 x_1 - 3.6$.

Minor axes: $(x_2 - 8.4) \tan \varphi = -(x_1 - 14.5)$

$$0.827 x_2 = -x_1 + 14.5 + 6.9$$

or: $x_2 = -1.209 x_1 + 25.9$

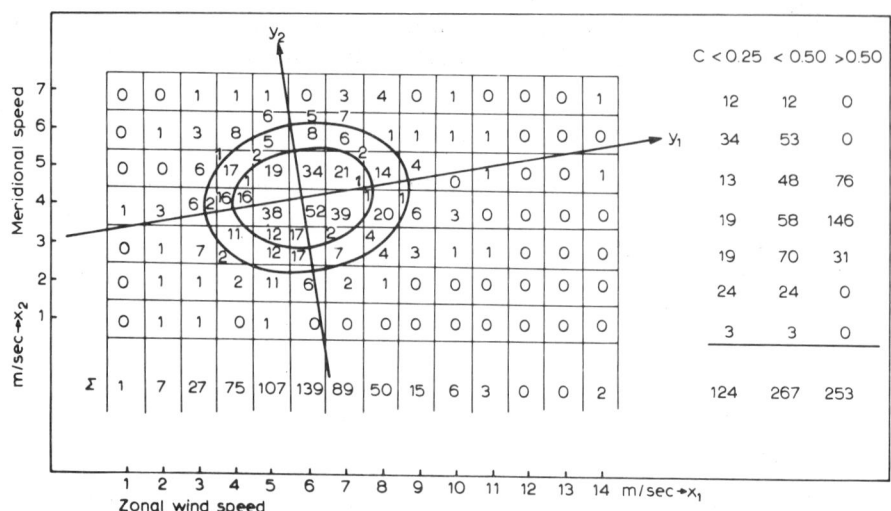

Fig. 2.4. Cumulative Gaussian bivariate distribution. Comparison between analytical and empirical wind data for Montgomery (Alabama) in July (1956–1964).

Example 2.10. Cumulative bivariate distribution. An example has been selected for the comparison between the cumulative analytical and the empirical frequency distribution. The wind observations at 20 km altitude for Montgomery (Alabama) in July (1956–1964) serve this purpose. This sample was selected because the variance ratio $s^2(x_1)/s^2(x_2)$ with $N = 521$ already indicates that the distribution is non-circular and the illustration was intended to be given for an elliptical shape of the density lines. With a transformation angle of $\varphi = 8.8°$ the variances of the major and minor axes ($s_a^2 = 2.91$ and $s_b^2 = 1.20$) do not change much from $s^2(x_1)$ and $s^2(x_2)$.

Lines of equal density $f(x_1, x_2) = A \exp[-z^2/\{2(1-r^2)\}] = A \cdot C$ were calculated for $C = 0.10, 0.25, 0.50, 0.75$ and 0.90, but only the 0.25- and 0.50-ellipse isolines of equal frequency are displayed in Fig. 2.4. It may be pointed out that the ellipse of $C = 0.5$ will correspond to the 50% cumulative probability since $\int_{-\infty}^{\infty} F(x_1, x_2)$ equals $2\pi \sigma_{x1} \sigma_{x2} (1-r^2)^{1/2} = 1/A$. Hence C corresponds to the fraction of the areal integration (see [2.88]). The mean values at the margins of the contingency table indicate a slight increase towards the positive x_1- and x_2-coordinate which should be expected from the correlation $r = 0.137$ between the zonal (x_1) and meridional (x_2) wind components.

In Fig. 2.4 the contingency of the wind frequency in 1 m/sec class units is placed at the respective center of the square, and the isolines $C = 0.5$ and 0.25 have been utilized to analyze the agreement between analytical and empirical cumulative frequencies. First, the data falling outside the ellipse

THE BIVARIATE NORMAL DISTRIBUTION 109

$C = 0.5$ were counted. Since class units in the bivariate elliptical model and the contingency fields do not readily coincide, it is self-evident that the particular square fields which were divided by the $C = 0.5$ ellipse line had to be prorated. The split within one square field has been carried out as displayed in Fig. 2.4, e.g. the 21 observations in the field $x_1 = 5$, $x_2 = 5$ are divided so that 19 observations fall inside the ellipse, and 2 lie outside these numbers. The respective column at the right-side margin provides the total numbers count within a line. On this basis 267 data points have been determined to fall outside the boundaries of the $C = 0.5$ ellipse. This number compares favorably with an expected number of 260. The count could have been determined by a computerized check, comparing the exact data in 0.1 m/sec units and separating data falling within the ellipse and outside. By coincidence this count resulted in 259. The example and graphical solution was chosen for illustration and a check by the reader.

The comparison between the analytical and empirical count was expanded to the $C = 0.25$ ellipse. The counts are assumed along the line and listed in the respective columns at the right margin. Again, the agreement of 124 empirical data with an expected value of 130 must be rated as very favorable. The computerized check in 0.1 m/sec units resulted in 125 cases, which is even closer to the expected value. The difference must be attributed to random fluctuations and imperfections of the prorating method.

One would conclude from the illustrated example that the bivariate (elliptical) distribution may serve as a good approximation of the empirical data. This conclusion has been confirmed by other authors (such as Brooks and Carruthers, 1953; Crutcher, 1956, etc). Nevertheless, some readers may object to generalizing the result. Admittedly, there may be distributions where the bivariate model would not provide such excellent agreements.

Example 2.11. Logistics and Gaussian (elliptical) bivariate distribution. A comparison between the bivariate elliptical (Gaussian) distribution and the bivariate logistic distribution could have been made on a theoretical basis. Instead, an actual example was chosen, namely the Albrook (Canal Zone) wind data at 20 km altitude in January (1956—1964). The bivariate parameter estimators were as follows: $\bar{x}_1 = 10.5$ and $\bar{x}_2 = 7.3$ m/sec, $\sigma^2(x_1) = 18.41$ and $\sigma^2(x_2) = 3.04$, $\varphi = -0.34°$ and $\sigma_a^2 = 18.41$ and $\sigma_b^2 = 3.04$. Thus the major and minor axes for the bivariate elliptical distribution run approximately parallel to the zonal and meridional wind coordinates, since the transformation angle φ is negligible for all practical purposes (see Fig. 2.5). The variance ratio of 6:1 signifies a definite non-circular case. Inspection of the displayed contingency distribution (in class units of 1 m/sec square fields) reveals that two modes exist, one with center $x_1 = 7$ and $x_2 = 7$ m/sec. The second peak appears around $x_1 = 14$ and $x_2 = 7$ m/sec. This bimodality causes the approximation of the data by the bivariate Gaussian distribution to appear elliptical. It cannot be decided from the data on hand whether the displayed bimodality is a significant trait of tropical upper air wind data or is in this case due to the short period of record, i.e. random sampling. Since this question was of secondary interest in this example, the topic was not further pursued. Lines for $C = 0.1$ and 0.5 are contained in Fig. 2.5 for the bivariate Gaussian distribution. The meaning of the lines was explained in Example 2.10.

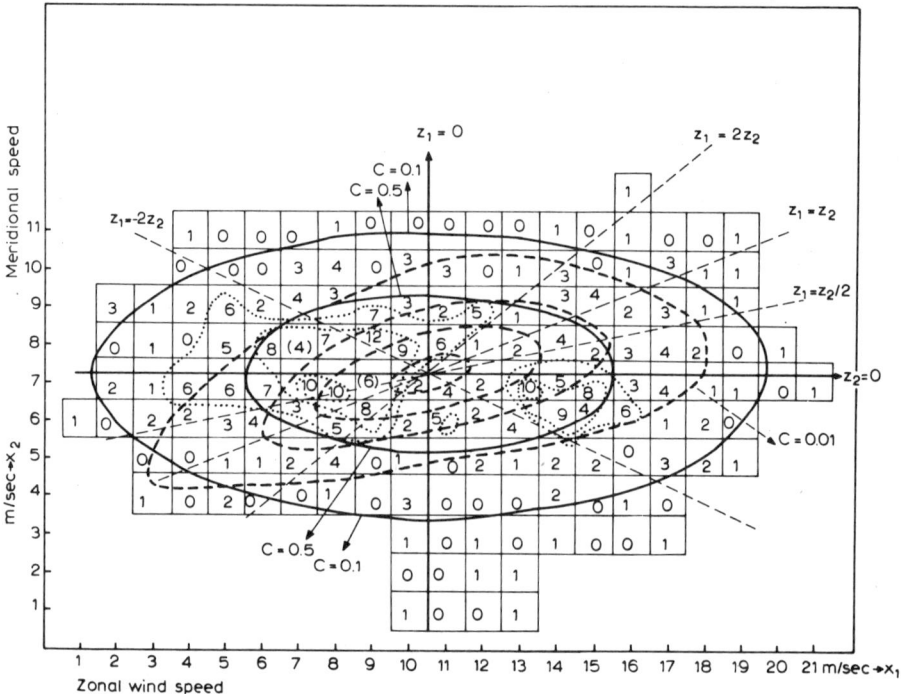

Fig. 2.5. Comparison between bivariate Gaussian distribution and bivariate logistic distribution for wind velocity data at Albrook (Canal Zone) in January (1956—1964) at 20 km altitude. ——— = bivariate Gaussian, — — — = bivariate logistic, · · · · · = empirical.

The lines for the logistic bivariate distribution were drawn based on [2.71]. Solutions for the intersections with several first-order equations such as $z_1 = z_2$, $z_1 = \pm z_2$, etc., are indicated in Fig. 2.5 by the drawn straight lines. The differences between the Gaussian and the logistic lines are self-explanatory. Some analysis as to whether the logistics curve fitting is better for meteorological wind data may be postponed until Example 2.12.

THE BIVARIATE NORMAL DISTRIBUTION

Example 2.12. Elliptical Gaussian and logistic bivariate fitting of data. Although the data from the previous example could have been selected for the purpose of illustrating curve fitting of empirical data by the two models, the wind data for Montgomery (Alabama), 20 km in July (1956—1964) were found more appropriate to exemplify the differences in curve fitting because the data sample was larger and the variance smaller than in the sample for Albrook. The empirical isolines of equal frequency density can thus be drawn with better accuracy.

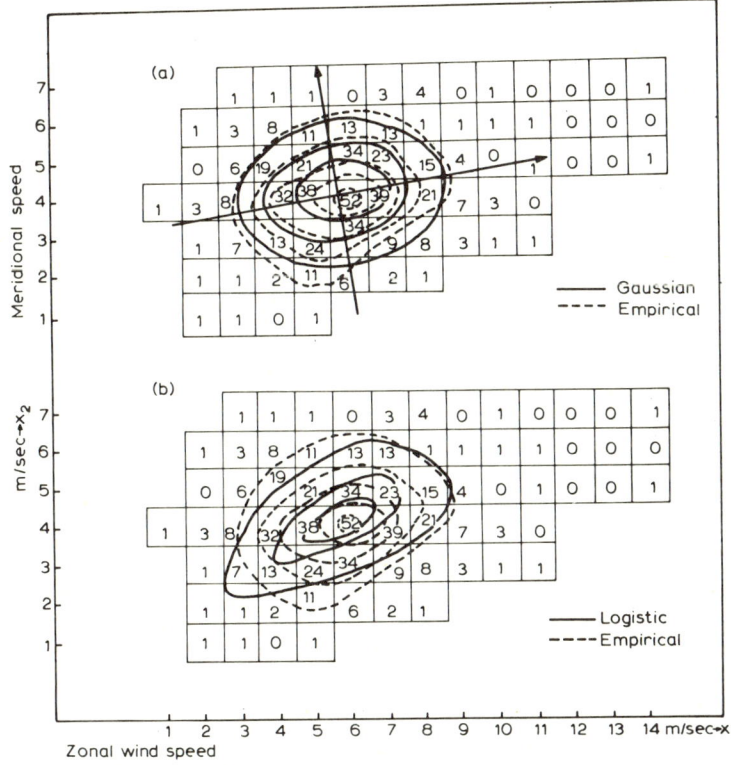

Fig. 2.6. Bivariate elliptical Gaussian (a) and Logistic (b) distribution for wind data at Montgomery (Alabama) in July (1956—1964) at 20 km altitude.

Figure 2.6a illustrates the probability lines for the bivariate elliptical Gaussian distribution. These lines appear to parallel the empirical data lines fairly well. It should be emphasized that again the data sample is not large enough to pin the discrepancies to a deficiency of the bivariate model. Brooks and Carruthers (1953) and the author (1959) have shown other examples of empirical isolines of equal density, which are neither rigid circles or ellipses nor logistic curves.

The logistic model fitting is exhibited in Fig. 2.6b. In conclusion, at least this data sample does not support a general utilization of the logistic model. Figure 2.6 would not indicate that the logistic model would be superior to the Gaussian model. More study would be necessary to render final judgment. The reader will find, however, the information for consideration of both the Gaussian and the logistic models in the two sections presented. As previously mentioned, however, more information is available for the Gaussian model on areal integration, curve fitting, and tables than for the logistic model.

2.9 THE EXPONENTIAL DISTRIBUTION

Some distributions display an exponential decrease of the frequency density with an increase of the variable. We may therefore write:

$$f(x) = a \exp[-(x-b)/c] \quad (x \geqslant b, a > 0) \quad [2.96]$$

This is called the exponential distribution. Other types of distributions whose frequency density function contains an exponential term are sometimes classified in the literature as exponential types. The exponential distribution as defined here is strictly a frequency whose density follows an exponential decline as given by [2.96].

We find the moments for the modified form, in which we replace:

$$a = 1/\sigma = 1/c \quad [2.96a]$$

therefore: $f(x) = \sigma^{-1} \exp[-(x-b)/\sigma]$ [2.96b]

The moments become now:

$$\bar{x} = b + \sigma \quad [2.97a]$$

$$\mu_2 = \sigma^2 \quad [2.97b]$$

$$\mu_3 = 2\sigma^3 \quad [2.97c]$$

$$\mu_4 = 9\sigma^4 \quad [2.97d]$$

$$\beta_1 = 4 \quad [2.97e]$$

$$\beta_2 = 9 \quad [2.97f]$$

The moments estimators are:

$$c^2 = \sigma^2 = s^2 \quad [2.98a]$$

$$b = x_m - s \quad [2.98b]$$

$$a = 1/s \quad [2.98c]$$

The median is $b + \sigma \ln 2$.
The maximum-likelihood estimators can be found from:

$$\hat{s} = \bar{x} - \hat{b} = N^{-1} \Sigma(x_i - \hat{b}) \quad [2.99]$$

Since $x_i \geqslant b$ for any x_i, $\hat{b}$ is the minimum of the x_i-values. In empirical applications the minimum observed value is not always known or at least not precisely. It is worthwhile to consider that the variance of $\hat{b}$ is the error of $\hat{b}$, namely:

$$\epsilon_{\hat{b}}^2 = \sigma^2/N^2 \quad [2.99a]$$

$$E(\hat{b}) = b + \sigma/N \quad [2.99b]$$

then: $\hat{\sigma} = \sigma(1 - N^{-1})$ [2.99c]

with variance: $\epsilon_{\hat{\sigma}}^2 = \sigma^2[N^{-1} + N^{-2} - 2N^{-3}]$ [2.99d]

Let us assume, $b = 0$, then:

$$f(x) = \frac{1}{c} e^{-x/c} = \alpha e^{-\alpha x}$$ [2.100]

It follows from the moments fit [2.98a—c] that:

$x_m = s$ [2.101a]

and: $\alpha = 1/s = 1/x_m$ [2.101b]

or: $x_m = 1/\alpha$ [2.101c]

Further details for the exponential distribution such as estimation by quantiles, unbiased maximum-likelihood estimators etc. may be referred to the respective literature (e.g. Johnson and Kotz, 1970a, etc.). The role of the exponential distribution in life testing has been analyzed by Epstein (1958), whose article applies to the form of [2.100].

2.10 THE LOGARITHMIC SERIES DISTRIBUTION

The logarithmic series distribution, or short logarithmic distribution, has been introduced briefly in Essenwanger (1974a) during the discussion of precipitation data. Bowman and Shenton (1970) have found its usefulness for the duration of storms, but application to ecological problems (Williams, 1944) and in biology (e.g. Fisher et al., 1943) may indicate other suitability in atmospheric science.

The density function can be written:

$f(x) = \alpha \theta^x / x$ [2.102]

with $0 < \theta < 1$ and $x = 1, 2 \ldots n$.

The α is a function of θ and as such not a new parameter, namely:

$1/\alpha = -\ln(1 - \theta) = \ln[1/(1 - \theta)]$ [2.102a]

The distribution has the following moments:

$\bar{x} = \alpha \theta / (1 - \theta)$ [2.103a]

$\mu_2 = \bar{x}/(1 - \theta)$ [2.103b]

$\sigma^2 = \alpha\theta(1 - \alpha\theta)/(1 - \theta)^2 = \bar{x}(1 - \alpha\theta)/(1 - \theta)$ [2.103c]

$\mu_3 = \mu_2(1 + \theta)/(1 - \theta)$ [2.103d]

$\mu_4 = \mu_3[(1 + \theta) + 2\theta/(1 + \theta)]/(1 - \theta)$ [2.103e]

THE LOGARITHMIC SERIES DISTRIBUTION

The third- and fourth-order moments have been expressed with reference to zero since the central moments have no simple form.

The maximum-likelihood estimator for θ can be obtained from:

$$\hat{\theta}/[(1-\hat{\theta})\ln(1/(1-\hat{\theta}))] = \bar{x} \quad [2.104]$$

which is the same as [2.103a] with α being substituted. As Patil (1961) has stated, an asymptotic variance of $\hat{\theta}$ is:

$$\sigma_{\hat{\theta}}^2 = \theta^2/(N \cdot \sigma^2) \quad [2.104a]$$

Patil has prepared a table to solve [2.104]. The entry in the table is the mean. Otherwise an iterative method is needed. Patil has also studied the efficiency of other estimators. He has deduced the following three solutions:

$$\hat{\theta}_1 = 1 - f_1/\bar{x} \quad [2.104b]$$

$$\hat{\theta}_2 = 1 - \bar{x}/\mu_2 \quad [2.104c]$$

$$\hat{\theta}_3 = \sum_{x=2}^{\infty} f(x, \theta) \cdot x/(x-1) \quad [2.104d]$$

The f_1 represents the first-class frequency (n_1/N), and the $f(x, \theta)$ the frequency density. Patil found, however, that $\hat{\theta}_2$ and $\hat{\theta}_3$ have a poor efficiency, while $\hat{\theta}_1$ is reasonably efficient.

2.11 THE FOUR-PARAMETER WEIBULL AND HYPER-GAMMA DISTRIBUTIONS

The three-parameter model of the Weibull distribution was introduced by Essenwanger (1974a, section 2.7). A generalized four-parameter model has been given by Harter (1967), which is not only related to the previously discussed Weibull model but is also associated with the incomplete gamma function. We can write:

$$f(x) = \frac{\delta}{\beta^{\alpha+1}\Gamma(\alpha+1)}(x-\gamma)^{\delta(\alpha+1)-1}\exp\left[\frac{-(x-\gamma)^\delta}{\beta}\right] \begin{cases} x \geq \gamma \\ = 0 \text{ elsewhere} \end{cases}$$ [2.105]

For $\delta = 1$ we find:

$$f(x) = [1/\{\beta^{\alpha+1}\Gamma(\alpha+1)\}](x-\gamma)^\alpha \exp[-(x-\gamma)/\beta]$$ [2.105a]

which is the Pearson type III or incomplete gamma distribution. For $\gamma = 0$ we obtain the two-parameter gamma distribution (Essenwanger, 1974a, section 2.5).

Although the four-parameter model is very versatile and flexible in curve fitting, the estimation of the parameters is difficult. Let us first substitute a location reference γ:

$$X = x - \gamma$$ [2.106]

which simplifies the writing, although it assumes the knowledge of γ. We may further replace:

$$\delta(\alpha+1) = c$$ [2.106a]

$$1/\beta = b$$ [2.106b]

$$\delta = a$$ [2.106c]

and: $\alpha + 1 = c/a$ [2.106d]

Then: $$f(x) = \frac{ab^{c/a}}{\Gamma(c/a)} X^{c-1} \exp(-bX^a)$$ [2.105b]

which is called the hypergamma distribution (Suzuki, 1964).

The moments (with reference zero) can be expressed by:

$$\mu_k = [\Gamma\{(k+c)/a\}/\Gamma(c/a)] \cdot b^{k/a}$$ [2.107]

In specific terms we find:

$$\mu_1 = \bar{X} = [\Gamma\{(1+c)/a\}/\Gamma(c/a)] \cdot b^{1/a}$$ [2.107a]

$$\mu_2 = \sigma^2 + \bar{X}^2 = [\Gamma\{(2+c)/a\}/\Gamma(c/a)] \cdot b^{2/a}$$ [2.107b]

$$\mu_3 = [\Gamma\{(3+c)/a\}/\Gamma(c/a)] \cdot b^{3/a}$$ [2.107c]

$$\mu_4 = [\Gamma\{(4+c)/a\}/\Gamma(c/a)] \cdot b^{4/a} \qquad [2.107d]$$

$$\sigma^2 = b^{2/a} \cdot [\Gamma(c/a)\Gamma\{(2+c)/a\} - \Gamma^2\{(1+c)/a\}]/\Gamma^2(c/a) \qquad [2.107e]$$

No trivial and explicit solution for all parameters exists. Even solutions by electronic computers are not trivial. Suzuki (1964) has prepared a graphical solution from:

$$\bar{X}^2/\mu_2 = \Gamma^2[(1+c)/a]/[\Gamma(c/a) \cdot \Gamma\{(2+c)/a\}] \qquad [2.108a]$$

$$\bar{X}^3/\mu_3 = \Gamma^3[(1+c)/a]/[\Gamma^2(c/a)\Gamma\{(3+c)/a\}] \qquad [2.108b]$$

$$ab^2 = \Gamma[(1+c)/a]/\Gamma(c/a) \qquad [2.108c]$$

The parameter b has been eliminated in the first two equations. Subsequently, b is calculated from [2.108c].

If the γ is unknown, the $\bar{x}$ only can be determined from the observations. The equations by Suzuki would not suffice. Cohen (1969) suggests a solution via computation of Pearson's parameters β_1 and β_2, namely $(\nu_3/\sigma^3)^2$ and ν_4/σ^4. These measures are independent of the location parameter γ. Unfortunately the expressions for ν_3 and ν_4 contain three and four $\Gamma(\ldots)$ terms, respectively, because the total lengthy formula $\nu_3(\mu_3, \sigma, \bar{X})$ cannot be readily condensed. It may be easier to postulate:

$$\gamma = x_{\min} \qquad [2.109]$$

and employ Suzuki's equations, with $\bar{X} = \bar{x} - x_{\min}$ and substitution into [2.108a and b] of μ_2' and μ_3' (reference $x_{\min}$). These μ_2'- and μ_3'-terms can be calculated (see [1.20b, c]). It will lead to estimators for a, b and c. Equation [2.107d] may serve as a check and an iterative correction towards the real γ-estimator.

The maximum-likelihood estimators for $\gamma = 0$ can be obtained from:

$$a \ln x_g - \Psi(c/a) + \ln b = 0 \qquad [2.110a]$$

$$ab \sum x_i^a - Nc = 0 \qquad [2.110b]$$

$$ab \sum x_i^a \ln x_i - N - Nc \ln x_g = 0 \qquad [2.110c]$$

and: $\ln x_g = (\sum \ln x_i)/N \qquad [2.110d]$

It is readily seen that x_g is the geometric mean. The $\Psi(c/a)$ denotes the di-gamma function (e.g. see Abramowitz and Stegun, 1964 or [2.37f etc.]).

Sometimes initial guesses of $\hat{a}_0$, $\hat{b}_0$ and $\hat{c}_0$ may lead more quickly to a solution. Suzuki has suggested an iterative procedure by the Newton-Raphson method with the ith approximation $\hat{a}_i = \hat{a}_{i-1} + h_i$, $\hat{b}_i = \hat{b}_{i-1} + k_i$ and $\hat{c}_i = \hat{c}_{i-1} + l_i$. The corrections h_i, k_i and l_i can be calculated from the matrix equation below:

$$\begin{bmatrix} h_i \\ k_i \\ l_i \end{bmatrix} = \begin{bmatrix} \ln x_g + (c/a^2)\Psi'(c/a) & b^{-1} & -(1/a)\Psi'(c/a) \\ b \Sigma x_i^a + ab \Sigma x_i^a \ln x_i & a \Sigma x_i^a & -N \\ b \Sigma x_i^a \ln x_i + ab \Sigma x_i^a (\ln x_i)^2 & a \Sigma x_i^a \ln x_i & -N \ln x_g \end{bmatrix}_{i-1}$$

$$\times \begin{bmatrix} -a \ln x_g + \Psi(c/a) - \ln b \\ Nc - a \Sigma x_i^a \\ N + Nc \ln x_g - ab \Sigma x_i^a \ln x_i \end{bmatrix}_i \quad [2.111]$$

The suffixes $i-1$ to the matrix symbolize that the parameters without index should be replaced by the index, e.g. $a = a_{i-1}$, $b = b_{i-1}$ and $c = c_{i-1}$ for all elements of the matrix. Furthermore, $\Psi'(c/a)$ is the trigamma function. (See Abramowitz and Stegun, 1964). In electronic data processing the computations by the matrix equation may sometimes be faster than by programming of the original [2.110]. For $\gamma \neq 0$ one additional equation must be added to the set [2.110a–d], namely:

$$(1-c) \Sigma (1/x_i) + ab \Sigma x_i^{a-1} = 0 \qquad [2.110e]$$

Further, the x_i must be replaced by $x_i - \gamma$ in all four equations. Since the task of finding sumultaneous solutions for the estimators of $\hat{a}, \hat{b}, \hat{c}, \hat{\gamma}$ is very tedious (only the b can be eliminated by substituting [2.110b] into the remaining set) the assumption of a $\hat{\gamma}$, and the utilization of Suzuki's iterative procedure ([2.111]) with check by [2.110e] may lead to a faster solution in some cases.

Harter and Moore (1965) have outlined an iterative technique for the determination of maximum-likelihood estimators based on the first m order statistics of a sample of size $N(m \leq N)$ for complete and censored samples for the three-parameter gamma and Weibull distributions. This method was later expanded by Harter (1967) to the four-parameter model. Another recent article by Hassanein (1971) considers estimation of location and scale parameters for three-parameter Weibull distribution by the use of quantiles when the shape parameter is known.

The c.d.f. for the four-parameter model is the integral:

$$F(x) = 1/[\Gamma(\alpha+1)] \int_0^z z^\alpha e^{-z} dz \qquad [2.105c]$$

with: $z = (x-\gamma)^\delta / \beta$ $\qquad [2.105d]$

2.11.1 The three-parameter gamma distribution

Setting $\delta = 1$ in [2.105] leads to the three-parameter gamma distribution ([2.105a]). The moments can be found by substitution of $a = 1$ into [2.107a–e], and the subsequently presented methods for estimators can be modified accordingly.

The maximum-likelihood equations for the three-parameter gamma model can be derived from the set of [2.110a, b, e] by setting $a = 1$. Then:

$$\ln x_g = \Psi(c) + \ln b = 0 \qquad [2.111a]$$

$$b \Sigma x_i - Nc = 0 \qquad [2.111b]$$

$$(1 - c) \Sigma (1/x_i) + bN = 0 \qquad [2.111c]$$

Greenwood and Durand (1960) have established tables for the solution of this three-parameter gamma model.

2.11.2 The three-parameter Weibull distribution

Some details of this distribution were presented by Essenwanger (1974a). Since this distribution model is very adaptable to various shapes of frequency distributions the basic equations may be repeated here. The four-parameter model transforms into the three-parameter form by setting $\alpha = 0$. Then:

$$f(x) = \frac{\delta}{\beta}(x - \gamma)^{\delta - 1} \exp\left[-\frac{(x - \gamma)^\delta}{\beta}\right] \quad \text{for } x \geqslant \gamma \qquad [2.112a]$$

We substitute $\beta = \theta^\delta$ and we derive the form:

$$f(x) = \frac{\delta}{\theta}\left(\frac{x - \gamma}{\theta}\right)^{\delta - 1} \exp\left[-\left(\frac{x - \gamma}{\theta}\right)^\delta\right] \qquad [2.112b]$$

which has been covered by Essenwanger (1974a):

$$F(x) = 1 - \exp\left[-\left(\frac{x - \gamma}{\theta}\right)^\delta\right] \quad \begin{cases} x \geqslant \gamma \\ 0 \text{ elsewhere} \end{cases} \qquad [2.112c]$$

Moments estimators can be determined with the aid of Table 2.11. We calculate first the skewness parameter $\gamma_1 = M_3/s_3$ (see [1.12a]). From Table 2.11 we read off δ. Subsequently, we find:

$$\theta^2 = s^2/D \qquad [2.112d]$$

where the denominator D is given in Table 2.11. Finally:

$$\gamma = x_m - \theta \cdot A \qquad [2.112e]$$

where x_m is the mean and A is given in Table 2.11. Linear interpolation of Table 2.11 is accurate enough for $\delta > 0.5$. A graphical solution will be presented by Essenwanger (1974b).

The maximum-likelihood equations are difficult to solve since γ cannot be readily eliminated. The likelihood equations are listed below. A similar procedure as suggested for the three-parameter gamma model would be developed. Maximum-likelihood estimators require the solution of the system:

$$-\frac{N}{\theta^\delta} + \frac{1}{\delta^2} \sum_1^N (X_i - \gamma)^\delta = 0 \qquad [2.112\text{f}]$$

$$\frac{N}{\delta} - \frac{1}{\theta^\delta} \sum_1^N (X_i - \gamma)^\delta \ln(X_i - \gamma) + \sum_1^N \ln(X_i - \gamma) = 0 \qquad [2.112\text{g}]$$

$$\sum_1^N \left(\frac{1-\delta}{X_i - \gamma}\right) + \frac{\delta}{\theta^\delta} \sum_1^N (X_i - \gamma)^{\delta-1} = 0 \qquad [2.112\text{h}]$$

More details can be found in Essenwanger (1974a). A special three-parameter Weibull model was also developed by Peto and Lee (1973) for cancer-research experiments.

TABLE 2.11

Calculation of moments estimators for the three-parameter Weibull distribution

γ_1	δ	D	A
69900	0.1	$0.2433 \cdot 10^{19}$	$0.3629 \cdot 10^7$
60.09	0.25	$0.3974 \cdot 10^5$	24.0
6.62	0.50	20.0	2.0
3.12	0.75	2.595	1.191
2.0	1.0	1.0	1.0
1.430	1.25	0.5621	0.9314
1.072	1.5	0.3757	0.9027
0.8207	1.75	0.2759	0.8906
0.6311	2.0	0.2146	0.8862
0.3586	2.5	0.1441	0.8873
0.1681	3.0	0.1053	0.8930
0.02511	3.5	0.08107	0.8997
0.00056	3.6	0.07730	0.9011
−0.08723	4.0	0.06466	0.9064
−0.1783	4.5	0.05294	0.9126
−0.2540	5.0	0.04423	0.9182
−0.3181	5.5	0.03756	0.9232
−0.3732	6.0	0.03232	0.9277
−0.4212	6.5	0.02812	0.9318
−0.4634	7.0	0.02470	0.9354
−0.5008	7.5	0.02188	0.9387
−0.5342	8.0	0.01952	0.9417

2.12 TRUNCATED DISTRIBUTIONS

In some observational programs instrumental limitations may lead to a systematic cut-off at some boundary. Beyond that terminus all observations are missing. A typical case involves the observations of upper-air wind data. Only in recent times have improvements in the measuring techniques decreased the number of gaps in the data caused by strong winds in the lower troposphere. Since radar cannot pick up the signal at the low angles which are caused when the balloon is rapidly displaced due to strong winds in the troposphere, the frequency distribution discloses a truncation of extreme observations at higher altitudes. (Other data gaps which may introduce a bias in the frequency distributions at higher altitudes, although of importance, are not of primary interest in this section.)

Another well-known example is the "motor boating" effect in earlier times, in which the actual observed relative humidity could not be determined except when it was above a certain threshold. In other data processing tasks truncation may even be desirable in order to eliminate effects of outliers (wild observation points). (See quality control, Essenwanger, 1974a, section 6.10.) Other examples can be found in life testing of materials when the observation time is not enough to provide detailed information for long time intervals. This problem may arise in atmospheric observations, when time sequences are studied and the material is divided into time intervals such as months, etc. Events of long durations may not be known in detail.

Truncated distribution methodology is also needed in atmospheric physics and meteorology for separation of population mixtures (see Section 2.13).

It exceeds the scope of this section to describe all presently existing methods of truncations for various types of distributions. The Gaussian distribution will be more extensively treated here, and references or shorter descriptions will be given for other types of distribution forms.

In statistical terminology the expression "truncated distribution" is generally employed when the number of observations beyond a certain threshold is not known, while the sample is called "censored" when the number of observations is known, but the only information about the variate is the fact that it lies above or falls below a certain threshold. In the subsequent sections no direct distinction between truncation and censoring is made except that the methods of approach differ. We may distinguish three different types:

(1) The truncation point is known, but the number of observations above or below must be determined (truncated sample), type A.

(2) The number of observations and the truncation point is available (censored sample), type B.

(3) The number of observations is known, the terminus is variable, type C.

The truncation is illustrated in Fig. 2.7 for the Gaussian distribution. Sneyers (1962) has recently discussed the use of the truncated normal distribution in climatology.

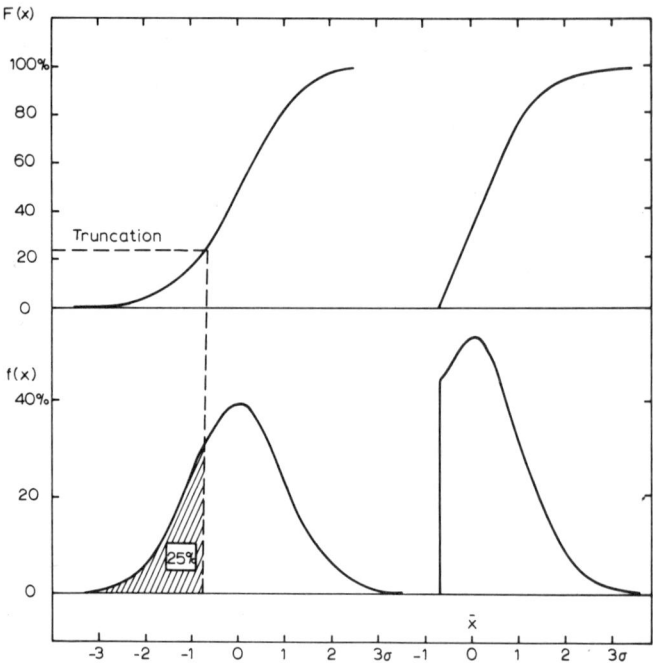

Fig. 2.7. Truncated Gaussian distribution.

2.12.1 *Gaussian distribution*

2.12.1a One-sided truncation. We assume a Gaussian distribution for the variate x with parameters $\bar{x}$ and σ^2, and a truncation at point x_0. Then we introduce a new variable for the truncated distribution:

$$X = x - x_0 \qquad [2.113]$$

Observations are missing above (or below) x_0. Let us call the portion $\emptyset(x_0)$ of missing data the degree of truncation (e.g. $\emptyset(x_0) = 10\%$). The cummulative distribution for X with truncation at the lower end can then be expressed as:

$$F(X) = [F(x) - \phi(x_0)]/[1 - \phi(x_0)] \qquad [2.113a]$$

with $F(x) \geqslant F(x_0)$. Assume, $F(x) = 0.55$, then $F(X) = 0.45/0.90 = 0.50$. For the truncation at the upper end:

$$F(X) = F(x)/[1 - \phi(x_0)] \qquad [2.113b]$$

with $F(x) \leqslant F(x_0)$. In case of $F(x) = 0.55$, this would amount to $F(X) = 0.55/0.90 = 0.61$. The frequency density is then:

$$f(X) = (1/\sigma) f(x)/[1 - \phi(x_0)] \qquad [2.113c]$$

TRUNCATED DISTRIBUTIONS

In order to determine a truncation of less than 10—20%, the number of observations must be large. Otherwise the deviation of fractiles will fall within the error limits (see Essenwanger, 1974a, section 6.2 on standard error).

We can transform $F(X)$ into $F(x)$ if $\emptyset(x_0)$ is known. From [2.113a] we obtain:

$$F(x) = F(X)[(1 - \emptyset(x_0))] + \emptyset(x_0) \qquad [2.113d]$$

for the lower truncation and:

$$F(x) = F(X)[(1 - \emptyset(x_0))] \qquad [2.113e]$$

for the upper truncation.

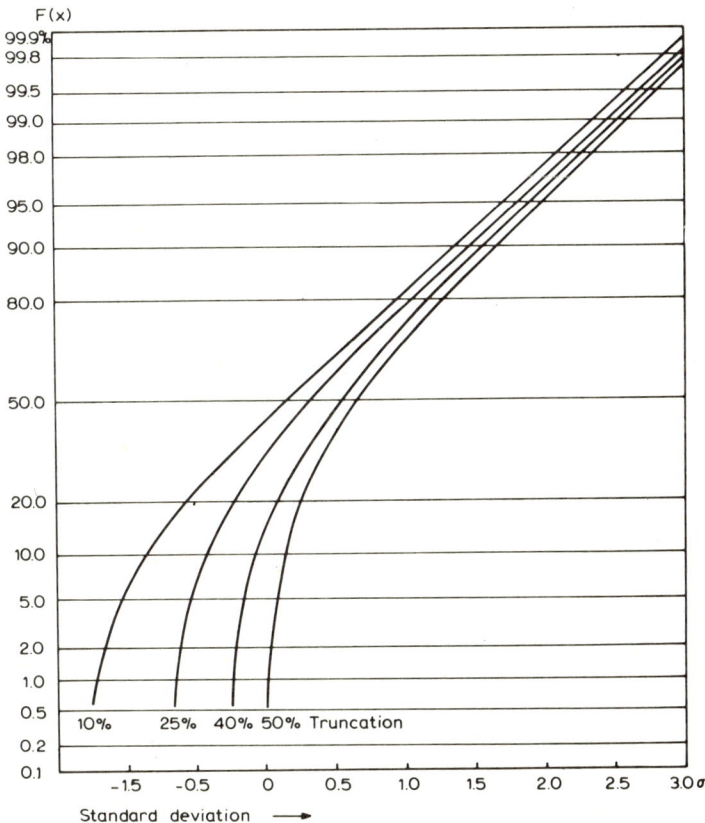

Fig. 2.8. Fractile diagram for various degrees of truncation of the Gaussian distribution.

If we construct a fractile diagram (Fig. 2.8), we may be able to guess the degree of truncation from the plotting of observations on probability paper, although some caution should be exercised, as not all curvatures may indicate truncation.

The following estimation method for $\bar{x}$ and σ from the truncated distribution X is due to Fisher (1931) for a known truncation point, but undetermined number of observations (type A).

First we take the point of truncation as the origin. The new variable corresponds to X, i.e. for the temperature T with truncation at T_0, $X = T - T_0$. We then calculate:

$$Y = N \Sigma X_i^2 / [2(\Sigma X_i)^2] \qquad [2.114]$$

Hald (1952) has established tables to estimate:

$$z = f(y) \sim \bar{X}/\sigma \qquad [2.114a]$$

Furthermore we calculate σ. Again, from Hald (table IX) we obtain $g(z)$. This provides:

$$s = [(1/N) \Sigma X_i] \cdot g(z) \approx \sigma \qquad [2.114b]$$

Knowing s, we readily find:

$$\bar{X} = -s \cdot z \sim -z \cdot \sigma \qquad [2.114c]$$

When the class unit $\neq 1.0$, then the classwidth w enters, namely:

$$s = [(1/N) \Sigma X_i] \cdot w \cdot g(z) \approx \sigma \qquad [2.114d]$$

The corrected mean value for the temperature would then finally be:

$$\bar{T} = T_0 + \bar{X} \qquad [2.114e]$$

2.12.1b One-sided truncation for known $\emptyset(x_0)$, censored sample, type B. In the set of equations [2.114] no knowledge of the degree of truncations was assumed. If this degree is known, we can call the total number of observations N and N_t the number of truncated observations, $N_t = \emptyset(x_0)N$.

We denote $n = N - N_t$, then:

$$y = n \sum_1^n X_i^2 / [2(\sum_1^n X_i)^2] \qquad [2.115]$$

The estimate of y is then selected from table X of Hald's tables, with $h = \emptyset(x_0)$, hence $z = f(y, h)$. A similar modification is made in [2.114b or d] by replacing N by n and taking $g(h, z)$. All other parts of the procedure remain the same.

2.12.1c One-sided truncation with unknown x_0, type C. The solution for estimators in this case is very elaborate, as the point of truncation enters as the third unknown parameter. The point of truncation may sometimes be estimated by graphical methods and then the previous methods can be employed. Tables for the three-parameters solution have been prepared recently by Cooley and Cohen (1970) and are discussed later in the text.

2.12.1d Double-sided truncation. This problem has been recently studied by Nelson (1967). Let us assume a lower truncation point x_A and an upper point x_B. Then $x_A \leqslant x \leqslant x_B$. Furthermore, the parameters of the Gaussian distribution are $\bar{x}$ and σ, for the truncated distribution $\bar{x}_t$ and σ_t. Then we can write:

$$\bar{x}_t = \bar{x} + a\sigma \qquad [2.116a]$$

$$\sigma_t = b\sigma \qquad [2.116b]$$

or: $a = (\bar{x}_t - \bar{x})/\sigma$ [2.116c]

and: $b = \sigma_t/\sigma$ [2.116d]

Both a and b are functions of the standardized truncations:

$$A_t = (x_A - \bar{x})/\sigma \qquad [2.116e]$$

$$B_t = (x_B - \bar{x})/\sigma \qquad [2.116f]$$

Nelson has given a graphical solution, which is valid for b when $x_B - x_A \geqslant 0.4\sigma$. When $x_B - x_A \leqslant 0.4\sigma$ then:

$$\sigma \approx (x_B - x_A)/\sqrt{12} = 0.2887\,(x_B - x_A) \qquad [2.117a]$$

The $\bar{x}$ can then be computed from [2.116a] or from:

$$\bar{x}_t = (x_B - x_A)/2 + c\sigma \qquad [2.117b]$$

$$c = [\bar{x}_t - (x_B + x_A)/2]/\sigma \qquad [2.117c]$$

The c can be interpreted as the count (in units of standard deviations) from the center $(x_B + x_A)/2$ to the population mean $\bar{x}_t$.

The sign of A_t and B_t determines the sign for a and c. For $B_t < 0$, the $a < 0$, for $B_t \geqslant 0$, the $a > 0$. The sign of c follows from $B_t \geqslant 0$ and $-B_t \leqslant A_t$, then $c \leqslant 0$. The computation of the parameters a, b and c can be performed from:

$$a = [f(A_t) - f(B_t)]/[F(B_t) - F(A_t)] \qquad [2.118a]$$

$$b = \left[1 - \frac{B_t f(B_t) - A_t f(A_t)}{F(B_t) - F(A_t)} - a^2\right]^{1/2} \qquad [2.118b]$$

$$c = a - (A_t + B_t)/2 \qquad [2.118c]$$

Nelson's method is mainly designed to estimate the truncated mean and variance when $\bar{x}$ and σ are known. In most problems in atmospheric science the problem is reversed, i.e. $\bar{x}_t$ and σ_t are known. Then an iterative process assuming $\bar{x}$ and σ could be based on the set of equations [2.116].

Another solution was given by Gupta (1952) and later modified by Prescott (1970). See Section 2.12.1g.

2.12.1e Maximum-likelihood method for one-sided truncation. The problem has been solved by Cohen (1959). As it is customary we denote with x_m, s the parameters of the empirical (truncated) and with $\bar{x}$ and σ the ones of the regular Gaussian distribution; the truncation point is x_0. Then we compute for type A the function $\theta(\hat{\gamma})$, with:

$$\hat{\gamma} = s^2/(x_m - x_0)^2 \qquad [2.119a]$$

Cohen (1959 and 1961b) has given a table for the determination of θ. Then we find the estimators:

$$\hat{\bar{x}} = x_m - \hat{\theta}(x_m - x_0) \qquad [2.119b]$$

$$\hat{\sigma} = s^2 + \hat{\theta}(x_m - x_0)^2 \qquad [2.119c]$$

The type B, censored samples with known truncation point x_0, requires:

$$\hat{\bar{x}} = x_m - \hat{\lambda}(x_m - x_0) \qquad [2.119d]$$

$$\hat{\sigma}^2 = s^2 + \hat{\lambda}(x_m - x_0)^2 \qquad [2.119e]$$

where $\hat{\lambda} = \lambda(\hat{\gamma}, h)$, and h stands again for $\phi(x_0)$.
The type C, censored samples with unknown terminus, requires:

$$\hat{\bar{x}} = x_m - \hat{\lambda}(x_m - x_n) \qquad [2.119f]$$

$$\hat{\sigma}^2 = s^2 + \hat{\lambda}(x_m - x_n)^2 \qquad [2.119g]$$

again, $\hat{\lambda} = \lambda(\hat{\gamma}, h)$, where now:

$$\hat{\gamma}' = s^2/(x_m - x_n)^2 \qquad [2.119h]$$

for a left-side truncation $\bar{x} < x_m$, and reversed for the right side $\bar{x} > x_m$. The $\theta(\hat{\gamma})$ can be computed from:

$$\theta(\xi) = Z(\xi)/[(Z(\xi) - \xi)] \qquad [2.119i]$$

where: $Z(\xi) = f(\xi)/[1 - F(\xi)]$ for truncated samples $\qquad [2.119j]$

and: $Z(\xi) = [\Psi/(1 - \Psi)] f(\xi)/F(\xi)$ for censored samples $\qquad [2.119k]$

and $\xi = (x_0 - x_m)/s$; $\Psi = \emptyset(x_0)$.
One can express, however, for computational simplification $\theta(\hat{\gamma})$, by:

$$\theta(\hat{\gamma}) = [1 - Z(Z - \xi)]/(Z - \xi)^2; \qquad [Z \equiv Z(\xi)]$$

The λ can be obtained by a similar procedure.

Cooley and Cohen (1970) have computed tables to facilitate the process. We calculate $\hat{\gamma}$ or $\hat{\gamma}'$ from the truncated frequency distribution, enter the tables, and find the estimators $\hat{\bar{x}}$ and $\hat{\sigma}^2$ from the respective equation. In the type-B and -C case, the $\Psi = \emptyset(x_0)$ as the degree of truncation must be known. Cohen's (1959) earlier version contains tables for θ only; they have been revised later (1961b). The expanded and improved version by Cooley and Cohen (1970) is available in report form.

2.12.1f Least-square solution. The author (1957) has attacked the problem as a curve-fitting task to fit a fractional Gaussian frequency density through the given empirical frequency. Let us denote with Y_i the empirical class frequency for the variate x_i, with y_i the complete Gaussian distribution. Then the given fraction of a truncated curve can be fitted by least-square methods by matching the quasi-pair S_q^2 and $\bar{X}_q$ with σ^2 and $\bar{X}$. Hence:

$$S_q^2 = (\Sigma Y_i^2 \cdot x_i^2)/\Sigma Y_i^2 = (\Sigma y_i^2 \cdot x_i^2)/\Sigma y_i^2 \qquad [2.120a]$$

$$\bar{X}_q = (\Sigma Y_i^2 \cdot x_i)/\Sigma y_i^2 = (\Sigma y_i^2 \cdot x_i)/\Sigma y_i^2 \qquad [2.120b]$$

Since we can compute the S_q^2 and $\bar{X}_q$ for the Gaussian distribution from the right side of [2.120], we are able to determine $\bar{x}$ and σ. The author (1957) has established tables with entry S_q^2 and $\bar{X}_q$ and exit σ^2 and ΔX, with $\Delta X = \bar{x} - X_r$. The X_r is the reference point $X_r = (X_{max} - X_{min})/2$ for the computation of the x_i-values in [2.120a] (e.g. for three classes with $X_1 = 0$, $X_2 = 1$, $X_3 = 2$ the reference point is $X_r = 1.0$ with $x_1 = -1.0$, $x_2 = 0$, $x_3 = 1.0$, etc.).

Equations [2.120a and b] are approximations. The exact equations by least-square solutions are:

$$\Sigma Y_i y_i = N \Sigma y_i^2 \qquad [2.121a]$$

$$\Sigma Y_i y_i x_i = N \Sigma y_i^2 x_i \qquad [2.121b]$$

$$\Sigma Y_i y_i x_i^2 = N \Sigma y_i^2 x_i^2 \qquad [2.121c]$$

The Y_i stands for the class frequency in the empirical (truncated) distribution and the y_i is the respective ordinate of the normal distribution:

$$y_i = [1/(\sigma\sqrt{2\pi})] \exp[-(x_i - \bar{x})^2/(2\sigma^2)] \qquad [2.121d]$$

Since N is not known, we must first determine an ordinate ratio from [2.121a]. Assume, the Y_i is part of a normal distribution, $N \cdot f'(x_i)$, and $y_i = f(x_i)$, where $f'(x_i)$ and $f(x_i)$ are supposed to be identical. Since the $f'(x_i)$ is fitted from the empirical data, we would assume:

$$\Sigma[Y_i - Nf'(x_i)] = 0 \qquad [2.121e]$$

then the ratio: $\Sigma Y_i^2/\Sigma y_i^2 = N^2 \qquad [2.121f]$

or: $\Sigma Y_i y_i/\Sigma y_i^2 = N \qquad [2.121g]$

because: $\Sigma y_i^2 = 1.0$, $\Sigma Y_i^2 = N^2$ and $\Sigma Y_i y_i = N$

This procedure leads to [2.120a] or the modified version:

$$\Sigma Y_i y_i x_i / \Sigma Y_i y_i = \Sigma y_i^2 x_i / \Sigma y_i^2 \qquad [2.122a]$$

$$\Sigma Y_i y_i x_i^2 / \Sigma Y_i y_i = \Sigma y_i^2 x_i^2 / \Sigma y_i^2 \qquad [2.122b]$$

The two equations can be solved by electronic computers, although the author (1957) has prepared tables for 3, 4, 5, 6, 7 and 9 classes. The solution necessitates a simultaneous solving of the two equations [2.122a and b]. One would condense the equations to one single ratio:

$$\Sigma y_i^2 x_i / \Sigma y_i^2 x_i^2 = a(\sigma, \xi) \qquad [2.122c]$$

where: $\xi = X_r - \bar{x}$ \qquad [2.122d]

and solve for: $\Sigma Y_i^2 x_i / \Sigma Y_i^2 x_i^2 = a_t$ \qquad [2.122e]

from the empirical distribution. There are several solutions of $a(\sigma, \xi)$ to match a_t depending on the number of classes and σ and ξ. The ratio can be employed only in conjunction with a parameter checking by either [2.122a] or [2.122b] in order to pick the appropriate solution.

One may even gain some benefit by the availability of several solutions with respect to judgment for decomposing mixed Gaussian populations. This will be discussed in the later section in more detail. The solution to the set of equations [2.121] or [2.122] are relatively easy to calculate with the aid of electronic computers compared with most of the other methods of truncations. The other methods are based upon moments with reference to the truncation point, while in our case the reference is the center of the given x-scale:

$$x_m = (\Sigma x_i)/i_{max}, \quad i = 1, 2, \ldots i_{max}$$

2.12.1g Gupta's method for censored samples. This maximum-likelihood method developed by Gupta (1952) is very simple. We first consider the x_0 as the truncation point (greatest observation), $\emptyset(x_0)$ is known and x_m and s are the parameters (estimators) from the truncated distribution. We set:

$$d = (x_0 - x_m) \qquad [2.123a]$$

$$p = n/N \qquad [2.123b]$$

where n is the observed sample size, N the total number of observations, and:

$$N = n + n_s \qquad [2.123c]$$

where n_s stands for the number censored (above x_0).

We need maximum-likelihood estimators for $\bar{x}$ and σ of the total population, and have available x_m and s of the censored sample. The first equation for the estimator $\hat{\bar{x}}$ requires:

TRUNCATED DISTRIBUTIONS

$$\hat{\bar{x}} = x_m + (\sigma^2 - s^2)/d \qquad [2.123d]$$

In this equation the only unknown is σ^2. If we find an estimator for σ, the $\hat{\bar{x}}$ can be calculated. In order to find an estimator for σ we employ the following steps. We define a parameter z:

$$z = \eta + A(1-p)/p \qquad [2.123e]$$

with: $\eta = (x_0 - \bar{x})/\sigma \qquad [2.123f]$

$$A = f(\eta)/\phi(\eta) \qquad [2.123g]$$

$$f(\eta) = (1/\sqrt{2\pi}) \exp(-\eta^2/2) \qquad [2.123h]$$

$$\phi(\eta) = (1/\sqrt{2\pi}) \int_\eta^\infty \exp(-t^2/2)\,dt \qquad [2.123i]$$

We find: $\sigma = d/z \qquad [2.123j]$

Since z is defined by η, and η is a function of σ, a relationship between the sample quantities s, d and the unknown η, z is necessary:

$$\Psi = s^2/(s^2 + d^2) = (1 + \eta z - z^2)/(1 + \eta z) \qquad [2.123k]$$

Substitution of [2.123e] will eliminate η:

$$\Psi = [1 - A \cdot z(1-p)/p]/[1 + z^2 - Az(1-p)/p] \qquad [2.133l]$$

For $p = 1.0$ with $z = \eta$, the solution is:

$$z^2 = (1-\Psi)/\Psi \qquad [2.123m]$$

Gupta has tabulated Ψ versus p.

2.12.2 Truncated bivariate and multivariate distributions

The truncations in the multidimensional case have been studied by various authors. Rosenbaum (1961) has developed a method to estimate the parameters of a truncated bivariate normal distribution with the aid of five moments. Khatri and Jaiswal (1963) have modified his results for a singly truncated bivariate normal distribution and obtain direct estimation by eight (or nine) moments. They define:

$$m_{r,s} = E(x-h)^r (y-k)^s \qquad [2.124]$$

where h and k are the respective (lower) truncation points and x and y the variables of the truncated distribution. Then the nine moments are computed:

$$\begin{bmatrix} 1.0 & m_{01} & m_{02} & m_{03} \\ m_{10} & m_{11} & m_{12} & \cdot \\ m_{20} & m_{21} & \cdot & \cdot \\ m_{30} & \cdot & \cdot & \cdot \end{bmatrix} = M_{r,s} \qquad [2.124a]$$

These are the first three moments of each variable and the cross-product xy, x^2y and yx^2 (ρ = correlation n_{xy}).

We may estimate parameters $h_1, h_2, h_3, \eta_1, \eta_2$ (definition [2.124g–h]) from:

$$rm_{r-1,s} = h_1 m_{r+1,s} + h_3 m_{r,s+1} + \eta_1 m_{r,s} \quad (r \geqslant 1, s \geqslant 0) \qquad [2.124b]$$

and: $sm_{r,s-1} = h_2 m_{r,s+1} + h_3 m_{r+1,s} + \eta_2 m_{r,s} \quad (r \geqslant 0, s \geqslant 1) \qquad [2.124c]$

The higher moments than determined in the matrix [2.124a] can be obtained from:

$$m_{r,s+1} = \sigma_2(s\sigma_2 m_{r,s-1} + r\rho\sigma_1 m_{r-1,s} - \xi_2 m_{r,s}) \, r \geqslant 1, s \geqslant 1 \qquad [2.124d]$$

and: $m_{r+1,s} = \sigma_1(r\sigma_1 m_{r-1,s} + s\rho\sigma_2 m_{r,s-1} - \xi_1 m_{r,s}) \, r \geqslant 1, s \geqslant 1$

[2.124e]

with: $\xi_1 = (h - \bar{x})/\sigma_1; \, \xi_2 = (k - \bar{y})/\sigma_2^2$ [2.124f]

$h_1 = 1/[\sigma_1^2(1-\rho^2)]; \, h_2 = 1/[\sigma_2^2(1-\rho^2)]; \, h_3 = -\rho/[\sigma_1\sigma_2(1-\rho^2)]$

[2.124g]

$\eta_1 = (\xi_1 - \rho\xi_2)/[\sigma_1(1-\rho^2)]$ and $\eta_2 = (\xi_2 - \rho\xi_1)/[\sigma_2(1-\rho^2)]$

[2.124h]

Setting the combinations $r, s = 1,0; 2,0; 1,1$ in [2.124b] and $r, s = 0,1; 0,2; 1,1$ in [2.124c] six equations are determined from which the estimators can be calculated. Khatri and Jaiswal (1963) give several ways to compute the estimators. Maximum-likelihood estimators determined by Khatri and Jaiswal are as follows:

$$\hat{\sigma}_1 = (m_{2,0} - m_{1,0}^2)/\psi_1 \qquad [2.125a]$$

$$\hat{\sigma}_2^2 = (m_{0,2} - m_{0,1}^2)/\psi_2 \qquad [2.125b]$$

$$\hat{\xi}_1 = (m_{1,0}/\hat{\sigma}_1) - \phi_{1,0} \qquad [2.125c]$$

$$\hat{\xi}_2 = (m_{0,1}/\hat{\sigma}_2) - \phi_{0,1} \qquad [2.125d]$$

$$\hat{\rho} = [(\psi_1 - \psi_2)(b_3 - b_1 b_2) + \psi_3 \{b_1(b_1 - \xi_1) - b_2(b_2 - \xi_2)\}] /$$
$$[\psi_1 b_1 (b_1 - \xi_1) - \psi_2 b_2 (b_2 - \xi_2)] \qquad [2.125e]$$

with: $\psi_1 = \phi_{2,0} - \phi_{1,0}^2$ [2.126a]

$$\psi_2 = \phi_{0,2} - \phi_{0,1}^2 \tag{2.126b}$$

$$\psi_3 = (m_{1,1} - m_{0,1} m_{1,0})/(\sigma_1 \sigma_2) \tag{2.126c}$$

$$\phi_{1,0} = b_1 + \rho b_2 \tag{2.126d}$$

$$\phi_{0,1} = b_2 - \rho b_1 \tag{2.126e}$$

$$\phi_{2,0} = 1 + \xi_1 b_1 + \rho^2 \xi_2 b_2 + \rho b_3 \tag{2.126f}$$

$$\phi_{0,2} = 1 + \rho^2 \xi_1 b_1 + \rho^2 \xi_2 b_2 + \rho b_3 \tag{2.126g}$$

and: $b_1 = z(\xi_1) a(\alpha_2)/D$ [2.127a]

$$b_2 = z(\xi_2) a(\alpha_1)/D \tag{2.127b}$$

$$b_3 = [(1-\rho^2)/2\pi]^{1/2} z(\alpha_3)/D \tag{2.127c}$$

$$D = \frac{1}{2\pi\sqrt{1-\rho^2}} \int_{\xi_1}^{\infty} \int_{\xi_2}^{\infty} \exp\left[-\frac{x^2 - 2\rho xy + y^2}{2(1-\rho^2)}\right] dx\, dy \tag{2.127d}$$

$$z(x) = (1/\sqrt{2\pi}) \exp(-x^2/2) \tag{2.127e}$$

$$a(x) = \int_x^{\infty} z(t)\, dt \tag{2.127f}$$

$$\alpha_1 = (\xi_1 - \rho \xi_2)/\sqrt{1-\rho^2} \tag{2.127g}$$

$$\alpha_2 = (\xi_2 - \rho \xi_1)/\sqrt{1-\rho^2} \tag{2.127h}$$

$$\alpha_3 = [(\xi_1^2 - 2\rho\xi_1\xi_2 + \xi_2^2)/(1-\rho^2)]^{1/2} \tag{2.127i}$$

The equations can be solved only by iterative procedures. Khatri and Jaiswal have performed the task in the iterative sequence of [2.125]. They state that the moments estimators are not very efficient.

Rosenbaum's (1961) method looks simple at first glance, namely:

$$\hat{\sigma}_1 = s_x/(\hat{m}_{20} - \hat{m}_{10}^2)^{1/2} \tag{2.128a}$$

$$\hat{\sigma}_2 = s_y/(\hat{m}_{02} - \hat{m}_{01}^2)^{1/2} \tag{2.128b}$$

$$\hat{\mu}_1 = x_m - \hat{m}_{10} \hat{\sigma}_1 \tag{2.128c}$$

$$\hat{\mu}_2 = y_m - \hat{m}_{01} \hat{\sigma}_2 \tag{2.128d}$$

$$\hat{m}_{11} = [(\Sigma xy)/n - \hat{\mu}_1 y_m - \hat{\mu}_2 x_m + \hat{\mu}_1 \hat{\mu}_2)]/(\hat{\sigma}_1 \hat{\sigma}_2) \tag{2.128e}$$

with the definitions: $\hat{m}_{10} = (\bar{x}_m - \bar{x})/\sigma_1$ [2.129a]

$$\hat{m}_{20} = [\Sigma(x_i - \bar{x})^2]/(n\sigma_1^2) \tag{2.129b}$$

$$\hat{m}_{01} = (y_m - \bar{y})/\sigma_2 \tag{2.129c}$$

$$\hat{m}_{20} = [\Sigma(y_i - \bar{y})^2]/(n\sigma_2^2) \tag{2.129d}$$

$$\hat{m}_{11} = [\Sigma(x_i - \bar{x})(y_i - \bar{y})]/(n\sigma_1\sigma_2) \qquad [2.129e]$$

The symbols $\bar{x}$, σ_1, $\bar{y}$, σ_2 stand for the complete population, while x_m, y_m, s_x, s_y denote the quantities from the truncated distribution and x_0, y_0 the truncation points. These equations must also be solved by iteration. In order to perform the iteration, the following must be calculated:

$$F(x_0, y_0, \rho)m_{10} = Z(x_0)T(y') + \rho Z(y_0)T(x') \qquad [2.129f]$$

and:
$$F(x_0, y_0, \rho)m_{01} = \rho Z(x_0)T(y') + Z(y_0)T(x') \qquad [2.129g]$$

where ρ is the correlation coefficient between x and y, $F(x_0, y_0, \rho)$ is the total probability in the truncated distribution, $Z(\xi)$ is the frequency density and $T(\xi)$ the cummulative distribution. Furthermore:

$$x' = (y_0 - \rho x_0)/\sqrt{1-\rho^2} \qquad [2.129h]$$

$$y' = (x_0 - \rho y_0)/\sqrt{1-\rho^2} \qquad [2.129i]$$

$$F(x_0, y_0, \rho)m_{20} = F(x_0, y_0, \rho) + x_0 Z(x_0)T(y') + \rho^2 y_0 Z(y_0)T(x')$$
$$+ (\rho\sqrt{1-\rho^2}/\sqrt{2\pi}) \cdot Z(x'') \qquad [2.129j]$$

$$x'' = [(x_0^2 - 2\rho x_0 y_0 + y_0^2)/(1-\rho^2)]^{1/2} \qquad [2.129k]$$

$$F(x_0, y_0, \rho)m_{11} = \rho F(x_0 y_0, \rho) + \rho x_0 Z(x_0)T(y') + \rho y_0 Z(y_0)T(x')$$
$$+ [(1-\rho^2)/2\pi]^{1/2} Z(x'') \qquad [2.129l]$$

The estimates must be computed by an iterative process from [2.129f] through [2.129l] and [2.128a] through [2.128e].

First we can assume appropriate initial values of x_0, y_0 and ρ for [2.129f] through [2.129k] to find Z. Subsequently, the T can be obtained from standard univariate tables, $F(x_0, y_0, \rho)$ from tables of the bivariate normal probability or graphs such as Zelen and Severo (1960). The graphs require $x'\sqrt{1-\rho^2}$, $y'\sqrt{1-\rho^2}$, $x''\sqrt{1-\rho^2}$. The values for $m_{10}, m_{01}, m_{20}, m_{02}$, are then placed into [2.128a–d]. Then new values for x_0 and y_0' can be calculated.

A quadratic equation emerges for ρ, which is circumvented by Khatri's and Jaiswal's (1963) solution:

$$(\hat{x}_0 + \hat{y}_0)\rho^2 - [(\hat{x}_0 + \hat{y}_0)\hat{m}_{11} - \hat{x}_0\hat{y}_0(\hat{m}_{10} + \hat{m}_{01})]\hat{\rho} - (\hat{x}_0 + \hat{y}_0)$$
$$- \hat{x}_0\hat{y}_0(\hat{m}_{10} + \hat{m}_{01}) + y_0 m_{20} + \hat{x}_0\hat{m}_{02} = 0 \qquad [2.129m]$$

The value for $\hat{m}_{11}$ is taken from [2.129e].

Other methods have been derived by Des Raj (1953), based on Birnbaum's (1950) technique for multiple normal population. For a cutting point $Y \geqslant t$ in the bivariate case, Birnbaum states the following moments for the X-variate:

$$E(X) = \rho\lambda(t) \qquad [2.130a]$$

$E(X^2) = 1 + \rho^2 t \lambda(t)$ [2.130b]

$\sigma^2(X) = 1 + E(X)[\rho t - E(X)]$ [2.130c]

$t > [E^2(X) - 1]/[\rho E(X)]$ [2.130d]

$E(X^{3\cdot}) = E(X)[3 - \rho^2(1 - t^2)]$ [2.130e]

and: $E[X - E(X)]^3 = E(X)\rho^2[\{\lambda(t) - t\}\{2\lambda(t) - t\} - 1]$ [2.130f]

where: $\lambda(t) = f(t)/\phi(t)$ [2.130g]

and: $f(t) = (1/\sqrt{2\pi}) \cdot \exp(-t^2/2)$ [2.130h]

$\phi(t) = (1/\sqrt{2\pi}) \int_t^\infty \exp(-t^2/2) dt$ [2.130i]

When only one variate is truncated, Cohen's (1955) method for the truncation or censoring of the bivariate normal distribution would also be available. For singly truncated samples $x \geqslant x_0$, the parameters of the population can be estimated by:

$\hat{\sigma}_x^2 = v_1/(\hat{Z} - \hat{\xi})$ [2.131a]

$\bar{x} = x_0 - \hat{\sigma}_x \hat{\xi}$ [2.131b]

$\hat{\xi} = (x_m - \bar{x})/\hat{\sigma}_x$ [2.131c]

$Z(\xi) = f(\xi)/\phi(\xi)$ [2.131d]

with: $f(\xi) = (1/\sqrt{2\pi}) \exp(-\xi^2/2)$ [2.131e]

$\phi(\xi) = \int_\xi^\infty f(\xi) dx$ [2.131f]

$v_k = (1/n) \Sigma (x_i - x_0)^k$ [2.131g]

and $\hat{\xi}$ as the solution from:

$[1 - \xi(Z - \xi)]/(Z - \xi)^2 - v_2/v_1^2 = 0$ [2.131h]

In singly censored samples, the $Z(\xi)$ is replaced by $Y(\xi)$, i.e.:

$Y(\xi) = (N/n)[f(\xi)/\{1 - \phi(\xi)\}] = (N/n)Z(-\xi)$ [2.131i]

N is the total number of observations, n_t the number of truncated, and n the remaining number of observations, $N = n_t + n$. Other details may be taken from the quoted literature.

2.12.3 Truncated binomial and negative binomial

(1) *Binomial.* This problem was considered by Finney (1949) and later by Rider (1955). We assume the notation:

$$f(x) = \binom{n}{k} p^x (1-p)^{n-k} \qquad [2.132a]$$

We must estimate p from a truncated distribution, where n_k classes are missing at the lower end. Then the estimate for p is:

$$p = (m_2 - n_k m_1)/[(n-1)m_1 - (n_k - 1)nm_0] \qquad [2.132b]$$

where: $m_0 = \sum_{n_k}^{n} n_x = N - n_t \qquad [2.132c]$

$$m_1 = \sum_{n_k}^{n} x n_x = X_m \qquad [2.132d]$$

and: $m_2 = \sum_{n_k}^{n} x^2 n_x = s^2 + x_m^2 \qquad [2.132e]$

are the moments with reference to zero and $n_x = Nf(x)$.
The estimate for N follows from the equation:

$$N - m_0 = N \sum_{0}^{k-1} \binom{n}{x} p^x q^{n-x} = N \sum_{0}^{n-1} f(x) \qquad [2.132f]$$

Shah (1966) has established solutions for doubly truncated binomial distributions. The reader may refer to the literature.

(2) *Negative binomial.* A solution for the truncated negative binomial distribution has been suggested by Rider (1955). The negative binomial frequency can be stated as:

$$f(x) = \binom{-k}{x} (-1)^x \frac{\gamma^x}{(1+\gamma)^{k+x}} \qquad [2.133a]$$

Only the class $x = 0$ is truncated. Then:

$$\gamma = (m_3 m_1 - m_2 m_1 + m_1^2 - m_2^2)/[m_1(m_2 - m_1)] \qquad [2.133b]$$
$$1 + \gamma = (m_3 m_1 - m_2^2)/[m_1(m_2 - m_1)] \qquad [2.133c]$$
$$k = (2m_2^2 - m_2 m_1 - m_3 m_1)/(m_3 m_1 - m_2 m_1 + m_1^2 - m_2^2) \qquad [2.133d]$$

$$m_3 = \sum_{n_k}^{n} x^3 n_x \qquad [2.133e]$$

$$N - m_0 = N(1 + \gamma)^{-k} \qquad [2.133f]$$

The m_i denotes again the moments estimators reference zero (see [2.132c—e]). Some of the estimates from the moments are not very efficient (see later Sampford's (1955) approach).

TRUNCATED DISTRIBUTIONS

(3) *Maximum likelihood*

(a) *Binomial.* The maximum-likelihood estimation for p must be obtained (Rider, 1955) from the first moment v_1 of the truncated sample:

$$v_{1,n_k'} = \frac{np - \sum_{x=0}^{n_k-1} \binom{n}{x} xp^x(1-p)^{n-x}}{1 - \sum_{x=0}^{n_k-1} \binom{n}{x} p^x(1-p)^{n-x}} \quad [2.134a]$$

where n_k denotes the number of missing classes.
For the values $n_k = 1$ and $n_k = 2$ (one or two missing classes):

$$v_{1,1} = np/[1-(1-p)^n] \quad [2.134b]$$

$$v_{1,2} = [np - np(1-p)^{n-1}][1-(1-p)^n - np(1-p)^{n-1}] \quad [2.134c]$$

(b) *Negative binomial.* The equations for maximum-likelihood estimations of the truncated negative binomial ([2.133a]) are difficult to solve. Sampford (1955) has given an alternative method. Let us first look at his moments solution (zero class truncated):

$$ky/[w(1-w^k)] = x_m \quad [2.135a]$$

$$ky(1+ky)/[w^2(1-w^k)] = s^2 + x_m^2 \quad [2.135b]$$

where x_m and s denote the sample values, and:

$$w = 1/(1+\gamma) \quad [2.135c]$$

$$y = 1 - w = \gamma/(1+\gamma) \quad [2.135d]$$

This can be transformed to:

$$1 + k = (x_m - 1 + s^2/x_m)w/(1-w) \quad [2.135e]$$

$$\text{and: } 0 = x_m w^{k+1} + s^2 w/x_m - 1 \quad [2.135f]$$

with solutions for k and w, i.e. the parameters k and γ in [2.133a]; With the definition:

$$\phi(x) = (-x \ln x)/(1-x) \quad [2.135g]$$

we can rewrite [2.135f] as follows:

$$x_m \cdot \exp[-(x_m - 1 + s^2/x_m)\phi(w)] + s^2 w/x_m = 1 \quad [2.135h]$$

If we define:

$$w_0 = 1/(x_m + s^2/x_m) < 1.0 \quad [2.135i]$$

then the solution for w must be within the boundaries:

$$w_0 < w < 1.0 \quad [2.135j]$$

This can be expressed in terms of $\bar{x}$ and σ as:

$$1 - \exp[-(x_m - 1 + s^2/x_m)] < (x_m - 1 + s^2/x_m)/x_m < \ln(x_m + s^2/x_m)$$
[2.135k]

Sampford recommends an iterative procedure for the solution of [2.135h], namely:

$$w_{i+1} = \frac{x_m}{s^2}[1 - x_m \exp\{-(x_m - 1 + s^2/x_m)\phi(w_i)\}]$$
[2.135l]

We can solve [2.135h] also by trial and error.

The maximum-likelihood equations are given by:

$$nk/[w(1-w^k)] - n\bar{x}/(1-w) = 0$$
[2.136a]

$$(n \ln w)/[(1-w^k)] + \sum_{x=1}^{\infty} n_x \sum_{j=1}^{x} 1/(k+j-1) = 0$$
[2.136b]

Haldane (1941) has rewritten this as:

$$(n \ln w)/(1+w^k) + \sum_{j=1}^{R} (k+j-1)^{-1} \sum_{i=j}^{R} n_i = 0$$
[2.136c]

where R is the highest observed value of x.

Sampford's alternate method is:

$$\phi(w) = (-\hat{w} \ln \hat{w})/(1 - \hat{w}) = (\hat{k}/nx_m) \sum_{j=1}^{R} (\hat{k} + j - 1)^{-1} \sum_{x=j}^{R} n_x$$
[2.136d]

Furthermore: $\psi(\hat{w}, \hat{k}) = \hat{w}(1 - \hat{w}^k)x_m/[(1-\hat{w})k] = 1$ [2.136e]

$\phi(x)$ was defined above ([2.135g]) from which the value of w can be obtained. Then the k must be determined from [2.136e]. Sampford points out that cases may exist in which no maximum-likelihood solution may be found with $k > 0$.

Equation [2.136e] can be brought into different form to give an iterative solution:

$$k_{i+1} = [k_i \psi(w_i k_i)]$$
[2.136f]

This process converges rather slowly according to Sampford.

2.12.4 Truncated Poisson distributions

The problem of the truncation of a Poisson distribution has been extensively studied by Cohen (1954) and others. Let us assume a Poisson distribution:

$$f(x) = e^{-h} h^x/x!$$
[2.137]

TRUNCATED DISTRIBUTIONS

with $h = \bar{x}$ and a truncation point x_c to the left, one of x_d to the right. Then the estimation equation by maximum likelihood is (Cohen, 1954):

$$(x_m/\bar{x}) - 1 - [f(x_c - 1) - f(x_d)]/[F(x_c) - F(x_{d+1})] = 0 \qquad [2.137a]$$

where $f(x)$ denotes the usual probability density and:

$$F(x_c) = \sum_{x=x_c}^{\infty} f(x), \text{ or } F(x_{d+1}) = \sum_{x=x_{d+1}}^{\infty} f(x) \qquad [2.137b]$$

We may distinguish several special cases (x_m denotes the sample mean):

(1) Single truncation to the left, for $x_d \to \infty$ the $f(x_d) = 0$, $F(x_{d+1}) = 0$:

$$(x_m/\bar{x}) - 1 - f(x_c - 1)/F(x_c) = 0 \qquad [2.137c]$$

(2) Single truncation to the right, $x_c = 0$ and $f(x_c - 1) = 0$, $F(x_c) = 1.0$:

$$x_m/\bar{x} - 1 + f(x_d)/(1 - F(x_{d+1})) = 0 \qquad [2.137d]$$

(3) Doubly censored samples, n_1 and n_2 denote the unmeasured observations to the left and right, respectively ($x_c < x < x_d$), total $N = n_1 + n_2 + n_t$. The estimation function based on maximum likelihood is:

$$(x_m/\bar{x}) - 1 - (n_1/n_t)[f(x_c - 1)(1 - F(x_c))] + (n_2/n_t)[f(x_d)/(F(x_d + 1))]$$
$$= 0 \qquad [2.137e]$$

(4) Singly censored to the left, $n_2 = 0$
(5) Singly censored to the right, $n_1 = 0$

Cohen has also given the variances of the estimators $\hat{\bar{x}}$. Subsequently, the formulae for the doubly truncated or censored sample is listed. The reduction can readily be performed by the reader. Generally we find the relationship:

$$\text{Var}(\hat{\bar{x}}) = -\left[\frac{d^2 L}{d\bar{x}^2}\right]^{-1} \qquad [2.137f]$$

The second derivatives for the computation of the variance are as follows:

(a) Doubly truncated Poisson:

$$\frac{1}{n}\frac{d^2 L}{d\bar{x}^2} = -\frac{x_m}{\bar{x}^2} - \left[\frac{f(x_c - 2) - f(x_c - 1) - f(x_d - 1) + f(x_d)}{F(x_c) - F(x_d + 1)}\right]$$

$$\left[\frac{f(x_c - 1) - f(x_d)}{F(x_c) - F(x_d + 1)}\right]^2 = \text{Var}(\hat{\bar{x}}) \qquad [2.137g]$$

(b) Doubly censored Poisson:

$$\frac{1}{n}\frac{d^2L}{d\bar{x}^2} = \frac{x_m}{\bar{x}^2} - \frac{n_1}{n}\left[\frac{f(x_c-2)-f(x_c-1)}{1-F(x_c)} + \left\{\frac{f(x_c-1)}{1-F(x_c)}\right\}^2\right]$$

$$+ \frac{n_2}{n}\left[\frac{f(x_d-1)-f(x_d)}{F(x_d+1)} - \left\{\frac{f(x_d)}{F(x_d+1)}\right\}^2\right] = \text{Var}(\hat{\bar{x}}) \qquad [2.137\text{h}]$$

The estimators derived by Cohen are biased but an unbiased estimator for truncation does not exist for truncation on the right. Tate and Goen (1958) have developed an unbiased minimum variance estimator for truncation on the left. Their equation for the final generalized case (for $x_c = c \geqslant 1$) is:

$$h_c(t) = tG^c_{n,t-1}/G^c_{n,t} \qquad [2.138]$$

where $G^c_{n,t}$ is the generalized Stirling number:

$$G^c_{n,t} = \frac{(-1)^n t!}{n!}\sum\frac{n!}{k_1!\ldots k_{c+2}!}\frac{(-1)^{k_1}k_1^{(t-s)}}{(t-s)!P_s} \qquad [2.138\text{a}]$$

Furthermore:

$$s = \sum_{j=0}^{c}jk_{j+2},\ P_s = \prod_{j=0}^{c}(j!)^{k_{j+2}},\ \text{and}\ c = x_c.$$

The summation terms have been arranged that $k_i = 0, 1, \ldots, n$ for $i = 1, 2, \ldots, x_c + 2; t - x_{c+1}, +1 \ldots$ The summation is taken over k_i that $\sum_{i=1}^{c+2} k_i = n$. (The t replaces the x in [2.137].) More details may be found in the cited literature, where also tables for a solution with $x_0 = 0$ and $x_c = 1$ are published.

Cohen (1961a) developed later a different version than in 1954 for a Poisson distribution truncated on the right with truncation point x_d. The estimation equation is now:

$$x_m = hF(x_d - 1)/F(x_d) \qquad [2.139]$$

This can be simplified by the polynomial equation:

$$\sum_{i=0}^{x_d} h^i[x_d!/(i-1)! - x_m \cdot x_d!/i!] = 0 \qquad [2.139\text{a}]$$

where h denotes the parameter of the Poisson distribution, the estimator h is the positive root of [2.139a].

It should be noted that $a!/b! = 0$ for $b < 0$ and $0! = 1$. The solution becomes quite simple when only the zero class is truncated. As Finney and and Varley (1955) have demonstrated, the estimation equation is:

$$x_m = h(1 - e^{-h}) \qquad [2.139\text{b}]$$

with variance:

TRUNCATED DISTRIBUTIONS

$$\text{Var}(\hat{h}) = h^2/[Nx_m(h+1-x_m)] \quad [2.139c]$$

It may be added that Selvin (1971) has attempted to combine all binomial type distributions by stating a generalized estimating equation, but the method is valid for one parameter only.

2.12.5 Truncated gamma distributions

The problem of estimating the parameters of a truncated incomplete gamma function has been studied by Cohen (1950), Des Raj (1953), Chapman (1956) and others. Des Raj has developed a method, when the third moment ν_3 is known. Let x_0 be the left and x'_0 the right truncation point, n_1 and n_2 the respective unmeasured observations and n_t the observations in the truncated sample, $n_1 + n_2 + n_t = N$. We distinguish three cases: (1) n_1 and n_2 are not known (truncated sample); (2) n_1 and n_2 are known (censored sample); and (3) The sum of $n_1 + n_2$ is known.

Des Raj has derived the following solutions:

(1) *Estimation by moments* (ν_3 is known):

Let
$$f(x) = [\beta^\alpha \Gamma(\alpha)]^{-1} x^{\alpha-1} e^{-x/\beta} \quad [2.140]$$

be the p.d.f. with the notation as earlier introduced. We further denote with:

$$m_r = \sum_{i=1}^{n_t} (x_i - x_0)^r / n_t \quad [2.141]$$

the moments given by the truncated sample with reference to the terminus x_0. The estimating equations can be written:

$$m_1 = \sigma[-\xi_1 + \theta(Z_1 - Z_2) - \nu_3 R Z_2/(2\sigma)] \quad [2.141a]$$

$$m_2 = \sigma^2[1 + \xi_1^2 + \theta(\alpha_3/2 - \xi_1)(Z_1 - Z_2) - (R Z_2/\sigma)\{1 + (\nu_3/2)(\nu_3/2 + R/\sigma)\}] \quad [2.141b]$$

The following definitions have been employed:

$$\xi_1 = (x_0 - \bar{x})/\sigma, \quad \xi_2 = (x'_0 - \bar{x})/\sigma \quad [2.141c]$$

$$R = x'_0 - x_0 \quad [2.141d]$$

$$\theta = 1 + \nu_3 \xi_1/2 \quad [2.141e]$$

$$Z_1 = f(\xi_1) \left[\int_{\xi_1}^{\xi_2} f(t)\,dt \right]^{-1} \quad [2.141f]$$

$$Z_2 = f(\xi_2) \left[\int_{\xi_1}^{\xi_2} f(t)\,dt \right]^{-1} \quad [2.141g]$$

and ν_3 as customary stands for the third central moment. The following density function (origin at mode):

$$f(t) = c[1 + \nu_3 t/2]^{A^2-1} e^{-At} \qquad [2.141h]$$

with $t = (x - \bar{x})/\sigma$ and $1/c = (A^2)^{A^2-0.5}[\exp(-A^2)]\,\Gamma(A^2)$ is another writing of the incomplete gamma function with $x = X - c$, $c \geqslant 0$, and $A = 2/\nu_3$. The equations [2.141a and b] for ξ^* and σ^* must be solved by iteration, e.g. the Newton-Raphson method.* Z_1 and Z_2 can be extracted from tables of the incomplete gamma function, with $f(\xi)$, the probability density for ξ_1 or ξ_2. The $\bar{x}^*$ can then be added from:

$$(x_0 - \bar{x}^*)/\sigma^* = \xi_1^* \qquad [2.141i]$$

The solution as described above is specified for case (1). For case (2) the Y_1 and Y_2 are substituted for Z_1 and Z_2, respectively, as follows:

$$Y_1 = (n_1/n_t)\,f(\xi_1)\left[\int_{-A}^{\xi_1} f(t)\,dt\right]^{-1} \qquad [2.141j]$$

$$Y_2 = (n_2/n_t)\,f(\xi_2)\left[\int_{\xi_2}^{\infty} f(t)\,dt\right]^{-1} \qquad [2.141k]$$

The Z_1 and Z_2 is replaced by X_1 and X_2, respectively for case (3):

$$X_1 = (N - n_t)\,n_t^{-1}\,f(\xi_1)\left[\int_{-A}^{\xi_1} f(t)\,dt + \int_{\xi_2}^{\infty} f(t)\,dt\right]^{-1} \qquad [2.141l]$$

$$X_2 = (N - n_t)\,n_t^{-1}\,f(\xi_2)\left[\int_{-A}^{\xi_1} f(t)\,dt + \int_{\xi_2}^{\infty} f(t)\,dt\right]^{-1} \qquad [2.141m]$$

(2) *Maximum-likelihood estimators* (ν_3 is known). The estimation equations are:

$$\sigma[-\xi_1 + \theta(Z_1 - Z_2) - \nu_3 R Z_2/(2\sigma)] - m_1 = 0 \qquad [2.142a]$$

and: $(1/n_t)(2/\nu_3 - \nu_3/2)\sum_{1}^{n_t}[\theta + \nu_3 x_1/(2\sigma)]^{-1} - 2\nu_3 + (Z_1 + Z_2) = 0$

$$[2.142b]$$

with $x_1 = x - x_0$. Equation [2.142a] is the same form as [2.141a]. Again, we substitute Y_1 and Y_2 in case (2) and X_1 and X_2 in case (3) for Z_1 and Z_2, respectively.

The equations must be solved by iteration.

The equations simplify, when the sample is singly truncated to the left, $x_0 \to \infty$. We have:

---

* The star is used to indicate the moments estimator in our case.

TRUNCATED DISTRIBUTIONS

$$m_1 = \sigma[\theta t(\xi_1) - \xi_1] \qquad [2.143a]$$

$$m_2 = \sigma^2[\theta + (\nu_3/2 - \xi_1)(\theta t(\xi_1) - \xi_1)] \qquad [2.143b]$$

θ has been defined by [2.141e], m_1 and m_2 by [2.141a and b], and:

$$t(\xi_1) = (n_1/n_t)\lambda(\xi) \qquad [2.143c]$$

$$\lambda(\xi_1) = [1 + (\nu_3/2)\xi_1]^{A^2-1} \cdot \exp(-A\xi_1)\left[\int_{-A}^{\xi_1}(1+\nu_3 t/2)^{A^2-1} e^{-At} dt\right] \qquad [2.143d]$$

In these equations the σ can be eliminated and we write:

$$m_2/m_1^2 = [1/(\theta t(\xi_1) - \xi_1)][\theta/(\theta t(\xi_1) - \xi_1) - \nu_3/2 - \xi_1] \qquad [2.143e]$$

Then: $\sigma^* = m_1[\theta^* t(\xi_1^*) - \xi_1^*]^{-1} \qquad [2.143f]$

$$\bar{x} = x_0 - \xi_1^* \sigma_1^* \qquad [2.143g]$$

The solutions for case (2) and (3) are the same this time; only [2.143e] changes to [2.143h] while [2.143f and g] remain:

$$R(\xi_1) = (1/n_t)(2\nu_3 - \nu_3/2)\Sigma[\theta + (x_1/m_1 A)(\theta t(\xi_1) - \xi_1)]^{-1}$$
$$- 2/\nu_3 + t(\xi_1) = 0 \qquad [2.143h]$$

(3) *Estimation, when ν_3 is unknown.* We estimate σ and obtain two equations for ξ_1 and ν_3, where the left side denotes the moments from the truncated observations and n_1 is known:

$$m_2/m_1^2 = [1/(\theta t(\xi_1) - \xi_1)][\theta/(\theta t(\xi_1) - \xi_1) + \nu_3/2 - \xi_1] \qquad [2.144a]$$

$$m_3/m_1^3 = [1/(\theta t(\xi_1) - \xi_1)^2][\nu_3 \xi_1 + 2 + (\nu_3 - \xi_1)\{\theta/(\theta t(\xi_1) - \xi_1) + \nu_3/2 - \xi_1\}] \qquad [2.144b]$$

When the truncation is on the right, the estimation equations are:

$$m_1 = \sigma[BT(\xi_1) - \xi_1] \qquad [2.145a]$$

$$m_2 = \sigma^2[B + (\nu_3/2 - \xi_1)\{BT(\xi_1) - \xi_1\}] \qquad [2.145b]$$

$$m_3 = \sigma^3[B(\nu_3 - \xi_1) + \{(\nu_3/2)(\nu_3 - \xi_1) + \xi_1^2 + 2\}\{BT(\xi_1) - \xi_1\}] \qquad [2.145c]$$

$$B = -(1 + \nu_3 \xi_1/2)$$

$$T(\xi_1) = (n_1/n_t)(-B)^{A^2-1} e^{-A\xi}\left[\int_{\xi_1}^{\infty}(1+\nu_3 t/2)^{A^2-1} e^{-At} dt\right]^{-1} \qquad [2.145d]$$

(4) *Estimation by Chapman.* Chapman (1956) has developed a solution of an asymptotic variance-covariance matrix for the estimations, since Des Raj's (1953) solution requires sometimes elaborate iteration procedures. His

equations are:

$$\hat{a} = (1/\Delta)[(v^T M_0^{-1} v)(u^T M_0^{-1} w) - (u^T M_0^{-1} v)(v^T M_0^{-1} w)] = 1/\hat{\beta} \quad [2.146a]$$

$$\hat{b}' = (1/\Delta)[(u^T M_0^{-1} u)(v^T M_0^{-1} w) - (u^T M_0^{-1} v)(u^T M_0^{-1} w)] = \hat{\alpha} - 1 \quad [2.146b]$$

The covariance matrix for a, b' is:

$$\begin{bmatrix} (1/\Delta)(v^T M_0^{-1} v) & (-1/\Delta)(u^T M_0^{-1} v) \\ (-1/\Delta)(u^T M_0^{-1} v) & (1/\Delta)(u^T M_0^{-1} u) \end{bmatrix} \quad [2.146c]$$

and: $\Delta = (u^T M_0^{-1} u)(v^T M_0^{-1} v) - (u^T M_0^{-1} v)^2$ [2.146d]

The M_0^{-1} is the inverse of M_0 ([2.146e]), $u, v, w; u^T, v^T, w^T$ are vectors, the T indicating the transpose. The elements of the vectors are:

$$u_i = \xi_{i+1} - \xi_i \quad [2.147a]$$

$$v_i = \ln \xi_i - \ln \xi_{i+1} \quad [2.147b]$$

$$w_i = y_i - \ln h_i + \ln h_{i+1} \quad [2.147c]$$

with: $y_i = \ln(n_i/n) - \ln(n_{i+1}/n)$ [2.147d]

The h_i is half the class width for a grouping of the observations into i number of classes, with n_i being the number of observations in the class interval $(\xi_i - h_i, \xi_i + h_i)$. Furthermore:

$$\xi_1 - h_1 = 0, \; \xi_r + h_r = T_r \; \text{and} \; \xi_i + h_i = \xi_{i+1} - h_{i+1}$$

with a truncation point T_r, i.e. $0 \leq x \leq T_r$. Let us define $q_i = n_i/n$ then the matrix (moments matrix) becomes:

$$M_0 = \begin{bmatrix} \frac{1}{n}\left(\frac{1}{q_1} + \frac{1}{q_2}\right) & -\frac{1}{n}\left(\frac{1}{q_2}\right) & 0 & \cdots & 0 \\ -\frac{1}{n}\left(\frac{1}{q_2}\right) & \frac{1}{n}\left(\frac{1}{q_2} + \frac{1}{q_3}\right) & -\frac{1}{n}\left(\frac{1}{q_3}\right) & & 0 \\ 0 & -\frac{1}{n}\left(\frac{1}{q_3}\right) & \frac{1}{n}\left(\frac{1}{q_3} + \frac{1}{q_4}\right) & \cdots & 0 \\ \cdot & \cdot & \cdot & \cdots & \cdot \\ \cdot & \cdot & \cdot & \cdots & \cdot \\ \cdot & \cdot & \cdot & \cdots & \cdot \\ 0 & 0 & 0 & \cdots & \frac{1}{n}\left(\frac{1}{q_{r-1}} + \frac{1}{q_r}\right) \end{bmatrix}$$

[2.146e]

According to Chapman, the solution is simple except for the determination of M_0^{-1} from M_0.

2.13 MIXED DISTRIBUTIONS

In the previous sections on frequency distributions we have silently assumed that all observations come from one and the same population, and all observations would be available on a random selection basis. Unfortunately, these assumptions are not always valid in atmospheric data collections. We may think only of the concept of air masses. It is immaterial whether one accepts or disagrees with the various classification schemes of air masses; the fact still remains that atmospheric conditions in the tropical and polar regions are different, representing separate individual physical conditions and hence non-identical populations of observed elements. Locations with a mixture of air masses must, therefore, display mixtures of populations. The separation of observations by a fixed calendar date such as the first of January, etc. does not automatically guarantee that data now come from a homogeneous population; only the homogeneity of the physical background assures this. This was outlined previously in the discussion and definition of homogeneity (Section 1.3). Although most of the meteorlogical observations are mixtures, many statistical problems can be treated by ignoring this mixture. In fact, there are numerous occasions where the individual population cannot be separated because of close similarity. Furthermore, random fluctuations may amount to more than the differences between the distributions of the mixtures. The conditions under which different individual populations can be separated will be considered later in detail.

As is known, the statistical background for the generation of an individual single frequency distribution sample is among others the condition of random sampling from a homogeneous population. This postulate is not always fulfilled, especially not in physical science and in atmospheric physics in particular. The problem also arises in biology, empirical Bayes procedures, social science or other problems of statistical analysis where a heterogeneous background makes sampling from one population impossible. Consequently the consideration of a mixture of distributions is an important topic in statistical analysis of atmospheric data.

It is obvious that knowledge of the individual populations from which any observation has been drawn could easily lead to the separation of the mixtures. We would merely divide the data into the individual collectives and then establish the frequency distribution for any of the parted samples, which now have a homogeneous background, coming from one individual population. Although successful in many cases, it fails in others. In fact, many times meteorological observations are given without individual background and a-priori grouping by populations is not possible. Thus, this method of separation cannot be employed in every problem. Indeed, observations are often recorded or already made, and the goal of the statistical analysis is now to find out whether there is more than one population.

To begin with the task of decomposition, we need to accept the fact that the job is very difficult, may lead to ambiguity and may not have a solution with physical meaning; although formally it can be mathematically solved. The problem has already been considered by Pearson (1894), who dealt with the basic principle of estimation for a mixture of two Gaussian normal distributions. In atmospheric science, the mixture of Gaussian frequencies is the most important case of all mixture processes. It will therefore be presented in more detail. We treat first the univariate case.

By and large, three conditions, but not necessarily simultaneously, must be used for successful separation of two and more populations. We must know: (1) the number of populations; (2) parts of the frequency distribution, sufficiently un-influenced by the other population(s); and (3) some parameters of the populations.

Since the second condition is difficult to meet for other than Gaussian distributions, provision (1) or (3) must then be fulfilled for all other distribution types.

In addition, chances for separation are enhanced when the variances of the populations are unequal for coinciding means or equal for differing means. The larger the distance between two means of partial collectives the better are the chances of separation. Bimodal or sometimes multimodal distributions can be separated relatively well, while unimodal distributions with superposition of components may evade any attempts. The distance between means and the chances of separation were recently correlated by Choi and Bulgren (1968). The success of separation decreases with diminishing distance. Although Pearson had discussed the problems already as early as 1894, the topic attracted little attention among statisticians until very recently.

This writer (1955a) has pointed out, however, that some of the procedures for decomposing the heterogeneous distribution into homogeneous partial collectives may be mathematical methodology without physical background. In this case the result would lack physical reality.

2.13.1 *Gaussian univariate mixed distributions*

The separation into individual populations can be performed largely by five methods. Again, it should be stressed that we assume the individual collectives can be associated with different physical processes and their background confirms the reality of their existence. We can apply the following techniques: (1) moments method; (2) truncation method; (3) Fourier transform; (4) maximum-likelihood solutions; and (5) graphical methods.

The five methods vary in complexity. Furthermore, one method may be successful in one case, and may fail in the other. Subsequently more details on these different solutions will be given.

MIXED DISTRIBUTIONS

2.13.1a *Separation by moments method.* This problem was introduced by Pearson (1894). He computed five moments for two superimposed collectives and finally arrived at some results. Rao (1952) simplified the method by employing the cumulants of the distribution. The moments solution was also discussed by Doetsch (1928) and later studied by Robertson and Fryer (1970), who added the determination of bias and accuracy of the moments estimators to the solution. This is a very important contribution. As the writer has earlier pointed out (1954a) the moments method has two disadvantages. First, the number of components must be known, and second, every Gaussian distribution requires two parameters plus the knowledge of the fractional proportion of the observations which is one more unknown. This system already leads to at least eight estimators for three populations and necessitates the calculation of as many moments. The number can be reduced by assumptions like equal variance or means, etc. for the individual populations. These postulations allow simplifications but may not reflect the true physical reality. Furthermore, the errors of moments become larger with increasing order, as can be readily concluded from the fact that empirical frequencies become unreliable towards the extremes. Since the extremes, however, enter into the computations of higher-order moments with high weight, especially the even-order moments may have a larger error attached. The computation of the parameters by moments may therefore encounter large errors. Not until very recently has such an error analysis been made. Robertson and Fryer's (1970) error terms already necessitate elaborate matrices for five moments with permutations of derivatives up to the fifth order. Simultaneously bias, covariance and variances of the moments estimators are then provided, and since many of the matrix entries are zero some simplification of the computations result.

Let us assume we have a mixture of Gaussian distributions and let us call them $g_k(x)$. The frequency density of the total distribution is denoted by $f(x)$. We find then:

$$Nf(x) = \sum_1^n n_k g_k(x) \qquad [2.148]$$

As customary: $g(x) = A \exp[-(x - \bar{x})^2/(2\sigma^2)]$ [2.148a]

and: $g_k = a_k \exp[-(x - \bar{x}_k)^2/(2\sigma_k^2)]$ [2.148b]

$\qquad a_k = 1/(\sigma_k \sqrt{2\pi})$ [2.148c]

A is defined analogously. We can now write:

$$f(x) = N^{-1} \sum_1^n n_k a_k \exp[-(x - \bar{x}_k)^2/(2\sigma_k^2)] \qquad [2.148d]$$

The n_k/N assumes the role of a weighting factor of the total contribution of the partial population, with:

$$\sum n_k = N \qquad [2.148e]$$

The moments of the distribution follow as usual, namely:

$$M_r = \sum_{x_1}^{x_N} x_i^r f(x) \quad [2.149a]$$

and:
$$m_{rk} = \sum_{x_{ik}}^{x_{nk}} x_i^r g_k(x) \quad [2.149b]$$

The latter expresses moments with reference to the respective mean $\bar{x}_k$ of the partial collective.

In order to equate the moments of the total distributions by the moments of the partial components, we must replace the m_{rk} by the respective M_{rk} which denotes the moments with respect to the general mean x. Then:

$$M_r = \sum n_k M_{rk} \quad [2.149c]$$

Let the origin of the X-scale (abscissa) now be the total observation mean $\bar{x}$. Then we define:

$$\bar{X}_k = \bar{x}_k - \bar{x} \quad [2.150]$$

We derive the following moments equations:

$$\sum_1^k n_k = N \quad [2.150a]$$

$$\sum n_k \bar{X}_k = 0 \quad [2.150b]$$

$$N^{-1} \sum n_k (\bar{X}_k^2 + \sigma_k^2) = M_2 = \sigma^2 \quad [2.150c]$$

$$N^{-1} \sum n_k (\bar{X}_k^3 + 3\bar{X}_k \sigma_k^2) = M_3 \quad [2.150d]$$

$$N^{-1} \sum n_k (\bar{X}_k^4 + 6\bar{X}_k^2 \sigma_k^2 + 3\sigma_k^4) = M_4 \quad [2.150e]$$

$$N^{-1} \sum n_k (\bar{X}_k^5 + 10\bar{X}_k^3 \sigma_k^2 + 15\bar{X}_k^4 \sigma_k^4) = M_5, \text{etc} \quad [2.150f]$$

It is obvious that for two components alone no easy solution to determine $n_1, n_2, \bar{x}_1, \bar{x}_2, \sigma_1$ and σ_2 can be found.

Pearson (1894) has carried out the elaborate procedure for two mixed populations. He sets:

$$p_1 = \bar{x}_1 + \bar{x}_2 \quad [2.151a]$$

$$p_2 = \bar{x}_1 \cdot \bar{x}_2 \quad [2.151b]$$

$$p_3 = p_1 p_2 \quad [2.151c]$$

$$\lambda_4 = 9M_2^2 - 3M_4 \quad [2.151d]$$

$$\lambda_5 = 30M_2 M_3 - 3M_5 \quad [2.151e]$$

and obtains the three equations:

$$M_3^2 - 4M_3 p_3 - 2p_3^2 - \lambda_4 p_2 + 6p_2^3 = 0 \quad [2.152a]$$

$$5M_3^2 p_3 - 2p_2^3 + 4p_3 p_2^3 - 20M_3 p_2^3 - \lambda_5 p_2^2 = 0 \qquad [2.152b]$$

$$p_3 = (2M_3^3 - 2M_3\lambda_4 p_2 - \lambda_5 p_2^2 - 8M_3 p_2^3)/(4M_3^2 - \lambda_4 p_2 + 2p_2^3) \qquad [2.152c]$$

This leads to a 9th-order equation in p_2 by substituting [2.152c] into [2.152a]. When p_2 is known, p_3 can be determined from [2.152c], then $p_1 = p_3/p_2$. Finally:

$$c^2 - p_1 c + p_2 = 0 \qquad [2.152d]$$

gives the roots c_1 and c_2 which are the solutions $\bar{x}_1$ and $\bar{x}_2$. Then σ_1 and σ_2 must be determined from [2.150c and d]. This is a tedious way and becomes even more complex for more than two components.

Before going into more than two components, let us consider three special cases of the two-component solution.

(1) The k means $\bar{x}_k$ are known, then the $\bar{X}_k$ are known (see [2.150]).
(2) The mean $\bar{x}_k \equiv \bar{x}$ for all k, with $\bar{X}_k = 0$.
(3) The variances are alike, $\sigma_k = \sigma_c$ for all k.

The case (1) simplifies the computation considerably, as $\bar{x}_k$ and $\bar{X}_k$ is known. Then we compute n_1 or n_2 from the two equations, which are sufficient to determine n_1 and n_2; namely:

$$n_1 + n_2 = N \qquad [2.153a]$$

$$n_1 \bar{x}_1 + n_2 \bar{x}_2 = 0 \qquad [2.153b]$$

This leads to:

$$n_1 = N\bar{X}_2/(\bar{X}_2 - \bar{X}_1) \qquad [2.153c]$$

$$n_2 = N\bar{X}_1/(\bar{X}_2 - \bar{X}_1) \qquad [2.153d]$$

The next two moments determine σ_1 and σ_2:

$$n_1(\bar{X}_1^2 + \sigma_1^2) + n_2(\bar{X}_2^2 + \sigma_2^2) = N\sigma^2 \qquad [2.153e]$$

$$n_1(\bar{X}_1^3 + 3\bar{X}_1 \sigma_1^2) + n_2(\bar{X}_2^3 + 3\bar{X}_2 \sigma_2^2) = NM_3 \qquad [2.153f]$$

This provides the following solution, which also has been described by Court (1949):

$$\sigma_1^2 = \sigma^2 + 2\bar{X}_1 \bar{X}_2/3 - \bar{X}_1^2/3 - M_3/(3\bar{X}_2) \qquad [2.153g]$$

$$\sigma_2^2 = \sigma^2 + 2\bar{X}_1 \bar{X}_2/3 - \bar{X}_2^2/3 - M_3/(3\bar{X}_1) \qquad [2.153h]$$

When mean $\bar{x}$ and $\bar{x}_k$ coincide (case 2) only two equations remain:

$$n_1 + n_2 = N \qquad [2.154a]$$

$$n_1 \sigma_1^2 + n_2 \sigma_2^2 = N\sigma^2 \qquad [2.154b]$$

All odd moments become zero, and the higher moments are functions of σ multiplied by a factor. No independent new equation is therefore added to determine the parameters. We must look for restoring the number of equations. This can be done by calculating the moments around a different point x_0. Then $\bar{X}_k$ in [2.150a through f] becomes $\bar{X}_k = \bar{x} - x_0$. This changes the right side, too. Instead of σ^2 we have $M_2' = \sigma^2 + \bar{X}_k$, and $M_3' = M_3 + 3\bar{X}_k\sigma^2 + \bar{X}^3$. It is obvious, M_2' and M_3' are known, and we can compute σ_1 and σ_2 from:

$$\sigma_1^2 = M_2' + 2\bar{X}_1\bar{X}_2/3 - \bar{X}_1^2/3 - M_3'/(3\bar{X}_2) \qquad [2.155a]$$

$$\sigma_2^2 = M_2' + 2\bar{X}_1\bar{X}_2/3 - \bar{X}_2^2/3 - M_3'/(3\bar{X}_1) \qquad [2.155b]$$

We need one additional equation to determine n_1 and n_2. We can add the frequency at point $\bar{x}$, namely:

$$n_1/\sigma_1 + n_2/\sigma_2 = N/\sigma \quad (\bar{x} \equiv \bar{x}_1 \equiv \bar{x}_2) \qquad [2.155c]$$

Note that the σ is the standard deviation of the mixture. Since σ_1 and σ_2 are known from the solutions above the determination of the proportion of the mixture is left. The two equations for this task are:

$$n_2 = N\sigma_2(1 - \sigma\sigma_1)/(\sigma_2 - \sigma_1) \qquad [2.155d]$$

$$n_1 = N\sigma_1(\sigma\sigma_2 - 1)/(\sigma_2 - \sigma_1) \qquad [2.155e]$$

The problem of equal variances (case 3) has been treated by Doetsch (1928). We can rewrite the moments by combining $\sigma_k = \sigma_c$ for all k, namely:

$$M_r = \sum_1^{n_k} a_k \sum_{r=0}^{n_r} x^r \exp[-(x-\bar{x}_k)^2/(2\sigma_c^2)] = \sum_{r=0}^{n_r} \binom{n_r}{r} s^{r+1} \mu_r \sum_{k=1}^{n_k} a_k \bar{x}_k^{(n_r-r)} \qquad [2.156]$$

where: $a_k = f_k/(\sigma_c\sqrt{2\pi})$; and $\sum f_k = 1$ \qquad [2.156a]

The system leads (with the notation $s^2 = 2\sigma_c^2$) to the following:

$$M_0 = s\mu_0 \sum_1^{n_k} a_k \qquad [2.157a]$$

$$M_1 = s\mu_0 \sum a_k \bar{x}_k + s^2\mu_1 \sum a_k \qquad [2.157b]$$

$$M_2 = s\mu_0 \sum a_k \bar{x}_k^2 + 2s^2\mu_1 \sum a_k \bar{x}_k + s^3\mu_2 \sum a_k \qquad [2.157c]$$

$$M_3 = s\mu_0 \sum a_k \bar{x}_k^3 + 3s^2\mu_1 \sum a_k \bar{x}_k^2 + 3s^3\mu_2 \sum a_k \bar{x}_k + s^4\mu_3 \sum a_k \qquad [2.157d]$$

$$M_4 = s\mu_0 \sum a_k \bar{x}_k^4 + 4s^2\mu_1 \sum a_k \bar{x}_k^3 + 6s^3\mu_2 \sum a_k \bar{x}_k^2 + 4s^4\mu_3 \sum a_k \bar{x}_k$$
$$+ s^5\mu_4 \sum a_k, \text{ etc.} \qquad [2.157e]$$

with definitions:

$$\mu_k = \int_{-\infty}^{\infty} \exp(-z^2)z^k \, dz \qquad [2.157f]$$

MIXED DISTRIBUTIONS

and: $z = (x - \bar{x}_k)/s$ [2.157g]

In general: $M_r = \sum\limits_{1}^{n_k} a_k \sum\limits_{m=0}^{r} \binom{r}{m} s^{m+1} \bar{x}_k^{r-m} \mu_m$ [2.157h]

We finally arrive at the following system ($x_k \equiv \bar{x}_k$):

$a_1 + \ldots a_n = \varphi_0(s)$ [2.158a]

$a_1 x_1 + \ldots a_n x_n = \varphi_1(s)$ [2.158b]

$a_1 x_1^2 + \ldots a_n x_n^2 = \varphi_2(s)$ [2;158c]

$a_1 x_1^3 + \ldots a_n x_n^3 = \varphi_3(s)$ [2.158d]

$a_1 x_1^4 + \ldots a_n x_n^4 = \varphi_4(s)$, etc. [2.158e]

The following φ-values have been used:

$\varphi_0(s) = M_0/s\sqrt{\pi}$ [2.159a]

$\varphi_1(s) = M_1/s\sqrt{\pi}$ [2.159b]

$\varphi_2(s) = M_2/s\sqrt{\pi} - M_0 s/2\sqrt{\pi}$ [2.159c]

$\varphi_3(s) = M_3/s\sqrt{\pi} - 3M_1 s/2\sqrt{\pi}$ [2.159d]

$\varphi_4(s) = M_4/s\sqrt{\pi} - 3M_2 s/\sqrt{\pi} + 3M_0 s^3/4\sqrt{\pi}$, etc. [2.159e]

For the solution we need the $2n + 1$ equations to determine a_k, $\bar{x}_k$ and σ_c. Note, the s^2 stands for $2\sigma_c^2$, twice the variance.

The problem has been solved by Rao (1952) for two components and the ratio $n_1/N = p$; The ultimate equations can be written as:

$1 = p + q$ [2.160a]

$\bar{x} = p\bar{x}_1 + q\bar{x}_2$ or $p\bar{X}_1 + q\bar{X}_2 = 0$ [2.160b]

$\sigma^2 = \sigma_c^2 + p\bar{X}_1^2 + q\bar{X}_2^2$ [2.160c]

$M_3 = p\bar{X}_1^3 + q\bar{X}_2^3 + 3\sigma_c^2$ [2.160d]

$k_4 = p\bar{X}_1^4 + q\bar{X}_2^4 - 3(p\bar{X}_1^2 + q\bar{X}_2^2)^2$ [2.160e]

with the notation $\bar{X}_k = \bar{x}_k - \bar{x}$, as previously defined. When N is small, say <50, we should multiply σ^2 by $N/(N-1)$ and substitute $k_3 = N^2/[(N-1)(N-2)]$ for M_3.

Furthermore: $k_4 = [N^2/\{(N-1)(N-2)(N-3)\}]$

$\times [(N+1)M_4 - 3(N-1)\sigma^4]$ [2.160f]

We define $t = \bar{X}_1 \bar{X}_2$, and calculate the t as the negative root from the cubic equation:

$t^3 + 0.5 k_4 t + 0.5 M_3^2 = 0$ [2.160g]

(Again, we could introduce k_3^2 instead of M_3^2). Assume we have determined t, then $\bar{X}_1$ is the negative root of:

$$\bar{X}_1^2 + M_3 \bar{X}_1 / t + t = 0 \qquad [2.160\text{h}]$$

Continuation shows:

$$\bar{X}_2 = -(M_3/t) - \bar{X}_1 \qquad [2.161\text{a}]$$
$$\bar{x}_1 = \bar{x} + \bar{X}_1 \qquad [2.161\text{b}]$$
$$\bar{x}_2 = \bar{x} + \bar{X}_2 \qquad [2.161\text{c}]$$
$$p = n_1/N = \bar{X}_2/(\bar{X}_2 - \bar{X}_1) \qquad [2.161\text{d}]$$
$$\sigma_c^2 = \sigma^2 + t \qquad [2.161\text{e}]$$

Robertson and Fryer (1970) have analyzed the accuracy and bias of the moment estimators.

2.13.1b *Truncation method.* It is obvious that the moments method is elaborate and tedious, however one looks upon it. Other methods have been sought for practical application, which lead to a solution faster especially when more than two partial collectives are involved. One method which has also been prepared for a graphical procedure (Daeves, 1952, or Daeves-Beckel, 1948) is based on truncated Gaussian distributions.

It must be assumed that one partial Gaussian component at the left or right side of the given frequency distribution is sufficiently uninfluenced by the adjacent components. Then the techniques of truncations discussed previously can be applied, i.e., we postulate that we know a fraction from $x_{\min}$ to x_t, the truncation point. We can employ any of the previously discussed methods, either Fisher-Hald, Cohen or Essenwanger, or the graphical method by Daeves-Beckel, outlined in Section 2.13.1d.

The other three techniques must employ a kind of trial-and-error method, i.e. one would assume a certain truncation point x_{t1}, then compute σ_1 and $\bar{x}_1$, expand the truncation point to $x_t' = x_t + \Delta x$. Now we compute σ_1' and x_1' and compare. When we still have the same (uninfluenced) partial component, we will find $\bar{x}_1 \approx \bar{x}_1'$ and $\sigma_1 \sim \sigma_1'$. This could also be tested by statistical significance tests (see t- or F-test, Essenwanger, 1974a, section 6.4 and 6.5).

The next step is the computation of the partial components $n_1 g(x)$ and subtraction from $N f(x)$. The remainder is treated by going back to step one above. The computational procedure has been described earlier.

Ideally we would work from both ends of the frequency distributions; and by successive elimination of components the last remainder is a Gaussian component.

We must now add one final balancing step, which is already initiated with the writers method. The remaining differences from the mixture distribution:

MIXED DISTRIBUTIONS

$$\Delta y = \Sigma\, n_k g_k(x) - N f(x) \qquad [2.162]$$

should be a minimum. Hence:

$$\frac{d(\Delta y^2)}{dx} = 0 \text{ (minimum)} \qquad [2.162a]$$

In practice the solution can be found by establishing:

$$\Sigma\, n_k g_k(x) + \Delta y_k(x) = 0 \qquad [2.162b]$$

separately for every individual k, where:

$$\Sigma\, \Delta y_k(x) = \Delta y(x)$$

The individual Δy_k follow a ratio in accordance with their frequency components:

$$\Delta y_1(x): \Delta y_2(x): \ldots \Delta y_k(x) = n_1 g_1(x): n_2 g_2(x): \ldots n_k g_k(x) \qquad [2.162c]$$

In other words, the Δy is split into Δy_k sections, which have the same ratio as the ordinates (frequency densities) of the partial components $g_k(x)$ at every individual x. Since after this process every ordinate value (frequency density), i.e. $n_1 g_1(x) + \Delta y_1(x)$ ([2.162b]), or the (balanced) frequency density of the partial Gaussian distribution is known, the n_k, $\bar{x}_k$ and σ_k can be calculated (estimated) by the usual moments estimator procedures. Thus the separation is only a first step.

This "balancing" process should be continued as long as significant changes in the composition of the partial collectives appear.

In practical applications it is less important to determine $\bar{x}_1$ and σ_1 completely accurately the first time as long as the number of components and their approximate locations results. One may interject that then the method is not objective but rather very subjective. Let us analyze whether this is true.

As pointed out earlier, there are cases in which the mixed frequency distribution cannot be decomposed. When little difference exists among the x_k-values, and further the σ_k are close, only the moments method could separate the components. But again, in the moments method we must postulate the number of Gaussian components, and with more than two components the estimators for the partial collectives are tedious and elaborate to resolve from the system of equations. In fact, the writer has not seen any solution by the moments method with more than two components, unless limiting postulations have been made such as all variances are alike, or assumptions for the location of mean values, etc. While some subjectivity may be introduced in the process of separation by the truncation method, the subjectivity in the moments method is found in the assumptions. Consequently, little difference exists between the two methods.

Separation of two components is usually not difficult by any method, although elaborate for the moments method under non-restrictive conditions. Much computational effort can be saved, however, by the truncation method, even when the process of balancing is included and must be repeated several times. The recognition of the components is the primary work. The balancing is then routine.

It may be further stressed that the physical reality of the partial components is not guaranteed by any method. It is very likely that different populations exist in a distinct bimodal (or multimodal) distribution. These cases are relatively easy to decompose. The intricate problem arises when more than two components are mixed, and success of decomposition cannot always be assured.

The physical reality must be studied in addition to it. By and large, the physical processes which may lead to mixture, should in general be known. One may also suspect when observations should come from a single population. Examples of mixtures have been given by the writer in various articles (e.g. Essenwanger, 1954a, 1955a and b; Schneider-Carius and Essenwanger, 1955, and Brandtner and Essenwanger, 1957). Other examples from different disciplines have been referenced at the beginning of this section on mixture. The Example 2.14 will demonstrate more details.

2.13.1c *Fourier transform.* This technique was developed by Doetsch (1928, 1936), originally planned to eliminate the Doppler effect in spectroscopic fine structure, but it has also potential to retract a certain temperature structure to a temperature T_0.

The method permits a very detailed separation of partial components. The number of collectives need not be known, but we must make tedious calculations. Today they can be facilitated by electronic computers. It is cautioned only that too many partial components may be generated, as almost every peak in the observed frequency is converted into a partial component. This assumption may hold true in physics for the processes mentioned above, and hence the method would be suitable. In atmospheric science some caution should be exercised. Reservations against the method have been summarized by the writer (1955c) who has compared solutions by the truncation method and the Fourier transform. The method has recently also been described by Gregor (1969).

Doetsch takes advantage of the Fourier transform of the Gaussian distribution. For one Gaussian component we may write $f(x)$ as:

$$f(x) = A \exp[-(x - \bar{x})^2/(2\sigma^2)] = \Sigma\, a_\nu \sin(\nu\alpha_x + \beta_\nu) \qquad [2.163]$$

where the right side is merely the expression of the Gaussian distribution by a series of Fourier terms (Essenwanger, 1974a, section 4.2). The following abbreviations have been taken:

$$A = 1/\sigma\sqrt{2\pi} \qquad [2.163a]$$

MIXED DISTRIBUTIONS

$$\alpha_x = 2x\pi/p \qquad [2.163b]$$

The α_x, β_ν are given in radians, but both α_x and β_ν could also be expressed in angular values, replacing the 2π by 360. Should we desire to utilize the n_x instead of the relative frequency $f(x)$, we would multiply A by N. The symbols $\bar{x}$ and σ are self-explanatory and a_ν and β_ν can be computed as given by the writer (1974a, section 4.2, Fourier analysis); ν is an integer indicating the harmonic number of the component, and p stands for the period length. Since the Gaussian curve ranges from $-\infty$ to ∞, p would be infinity. It should be replaced, however, by $p = x_{\max} - x_{\min}$, the observed range, where $f(x_{\max})$ or $f(x_{\min}) \sim 0$. The p represents the basic period of the Fourier analysis, in our case a somewhat arbitrary constant, but it is of little consequence here that p cannot be determined with accuracy.

We assume now that $f(x)$ comprises m Gaussian populations. Hence we summarize:

$$\sum_{k=1}^{m} a_k \exp[-(x-\bar{x}_k)^2/(2\sigma_k^2)] = \sum_{1}^{p} B_\nu \sin(\nu\alpha_x + \beta_\nu) \qquad [2.163c]$$

where now the left side is summed over the collectives k and the right side over the total period p (range). The a_k denotes $n_k/(\sigma_k \sqrt{2\pi})$. Since an individual collective can be written in the form:

$$\sum_{1}^{p} a_{\nu k} \sin(\nu\alpha_x + \beta_{\nu k}) = g_k(x) \qquad [2.163d]$$

the $\beta_{\nu k}$ and $a_{\nu k}$ must be synchronized to β_ν, since for the total period of length p only one phase angle per Fourier term is permitted, and we obtain:

$$B_\nu = \sum_{k=1}^{n_k} a_{\nu,k} \qquad [2.163e]$$

As demonstrated by Doetsch (1936), the Gaussian distribution transforms into a narrow line, when we replace:

$$B_\nu = a_\nu \exp(\nu^2 \pi^2 t/p^2) \qquad [2.163f]$$

and increase t from 0 to $t_{\max}$. It is evident that for $t = 0$ we can determine the $B_\nu = a_\nu$, which leads to the set of a_ν coefficients.

If we thus compute the frequency distribution of:

$$\emptyset(x, t) = \sum_{1}^{p} a_\nu [\exp(\nu^2 \pi^2 t/p^2)] \sin(\nu\alpha_x + \beta_\nu) \qquad [2.164a]$$

which is the Fourier transform of the Gaussian distributions, we find the subsequently generated "waves" with more "impressive" peaks than the original function $f(x)$. The smaller σ_k^2 the faster should this maximum emerge and indicate the location of $\bar{x}_k$. (See Tables 2.12, 2.13, and Example 2.13). The next step is the determination of n_k, which assumes the percentage proportionality ratio when $N = \Sigma f(x) = 1.0$. In general, the N signifies Σn_i. We can derive the n_k-value from:

$$\sum_{x_1}^{x_2} \emptyset(x, t) = n_k \qquad [2.164b]$$

when t is sufficiently large. The partial component of the distribution $\emptyset(x, t)$ can then be recognized as a separate unit between x_1 and x_2. The x_1 and x_2 are the boundaries of the left and right side, respectively, next to $\bar{x}_k$, where the first time $\emptyset(x, t) \leq 0$, indicating that now the component is separated from the other parts.

It is evident that by this method $\bar{x}_k$ can be determined only in units of the given progression of the class interval unless one takes into account the asymmetry of the peaks. Then the spacing of the $\bar{x}_k$ is a vital factor in the success of separation by this method. When $\bar{x}_i = \bar{x}_{i+1}$ or $\bar{x}_i \sim \bar{x}_{i+1}$ the two components cannot be separated by this method and will appear as one single collective. The component with the smallest σ_k^2 will emerge first. Since by that time it may be possible that at least the adjacent $\bar{x}_k$ to the first $\bar{x}_1$ would be known (otherwise the t should be increased and $\emptyset(x, t)$ recomputed), we must make a judgment whether the ordinate value $f(\bar{x}_1)$ can be considered the maximum of a single collective, namely the first component for $x = \bar{x}_1$. Furthermore:

$$f(\bar{x}_1) = a_1 = n_1/(\sigma_1\sqrt{2\pi}) \qquad [2.164c]$$

If this assumption can be made, the σ_1 can easily be determined. Otherwise we would have:

$$f(\bar{x}_1) = a_1 + \Sigma a_i \exp[-(x - \bar{x}_i)]/(2\sigma_i^2) \quad (i = 2, 3, \ldots) \qquad [2.164d]$$

Then we would need additional equations, such as $f(\bar{x}_1 + 1), f(\bar{x}_1 - 1)$, etc. The number of equations depends on how many components contribute significantly to the ordinate $f(x)$. Sometimes $f(\bar{x}_1) \approx a_1$ and we can obtain a tentative σ_1. Since the σ_i appears in the exponent, by and large no simple solution will be found. If we assume two components to be of major influence at $\bar{x}_1$ and $\bar{x}_2$ and the other components could be neglected, we could find:

$$f(\bar{x}_1) = a_1 + a_2 \exp(-\delta_2^2) \qquad [2.164e]$$

$$f(\bar{x}_2) = a_1 \exp(-\delta_1^2) + a_2 \qquad [2.164f]$$

with $\delta_1^2 = (\bar{x}_1 - \bar{x}_2)^2/(2\sigma_1^2)$ and $\delta_2^2 = (\bar{x}_2 - \bar{x}_1)^2/(2\sigma_2^2)$. Setting $c_1 = \exp(-\delta_1^2)$ and $c_2 = \exp(\delta_2^2), f(\bar{x}_1) = f_1, f(\bar{x}_2) = f_2$, we cast:

$$f_2 - f_1 c_1 = a_2(1 - c_1 c_2) \qquad [2.164g]$$

MIXED DISTRIBUTIONS

In this form an iterative procedure by trial and error for σ_1 and σ_2 combinations is not too difficult.

Another approximation has been suggested by Gregor (1969). It is based on area integrals of the Gaussian distribution.

He assumes two strips of the width of 2δ and 4δ, with g_0 the frequency of the component under consideration at $\bar{x}_k$. Further frequencies are $g_i = g(\bar{x}_k + i\delta)$ with $i = \pm 1, \pm 2, \ldots$. Then:

$$\sigma_k^2 = \delta^2 \frac{7(g_2 + g_{-2}) + 26(g_1 - g_{-1}) + 6g_0}{12(g_2 + g_{-2}) + 24(g_1 - g_{-1}) - 6g_0} \quad [2.165a]$$

Gregor also gives a formula for the computation of the frequency n_k. Since the location of $\bar{x}_k$ must be obtained from [2.164a] the n_k may just as well be determined by [2.164b]. Gregor's method is, however, based on the previously defined g_i-values:

$$n_k' = \frac{g_1 + 4g_0 + g_{-1}}{6} \cdot \frac{60\sigma_k^2 - 9\delta^2}{60\sigma_k^2 - \delta^2} \cdot \sigma_k \sqrt{2\pi} \quad [2.165b]$$

or: $$n_k'' = \frac{g_2 - g_{-2} + 4(g_1 + g_{-1}) + 2g_0}{12} \cdot \frac{60\sigma_k^2 + 36\delta^2}{60\sigma_k^2 - 4\delta^2} \cdot \sigma_k \sqrt{2\pi} \quad [2.165c]$$

Subsequently, $f_k = [n_k/(\sigma_k \sqrt{2\pi})] \exp[-(x - \bar{x}_k)/(2\sigma_k^2)]$ is calculated and the difference $|g(x) - f_k|$ is formed, providing the solutions n_k' and n_k'' for n_k. The parameter rendering the smallest maximum deviation $|g(x) - f_k|$ is then adopted. Again, the limitations of the decomposing method by Doetsch are emphasized. The x_k must be sufficiently spaced to permit separation by the Fourier transform. Some smaller partial collective may arise due to random fluctuations in the observed frequency density. The number of components, however, need not be assumed, which is an advantage.

The partial components can either be found simultaneously, when several peaks show up with sufficient discrimination or the first component $\bar{x}_1, \sigma_1, n_1$ can be determined, then subtracted by the procedure:

$$f'(x) = f(x) - f_1(x) \quad [2.164h]$$

The $f'(x)$ is subjected to a new Fourier analysis and transform, repeating the computation of $\emptyset(x, t)$ for the new frequency $f'(x)$. This successive subtraction should lead to finally one last component.

As pointed out by the writer, the method by Doetsch tends to render too many partial components, but is otherwise equivalent to the truncation method. Although Doetsch's method (with modification by Gregor) can be completely computerized the route via the Fourier transform makes the method for most cases much costlier than the truncation methods. If funding is no object, Doetsch's method is very skillful and relatively schematic to apply. (See also Graphical methods, Section 2.13.1e).

TABLE 2.12

Narrowing of the Gaussian distribution ($\emptyset(x, t)$) for $\sigma = 1.0$ and 15 points, $\sigma = 2.0$ and 19 points, and $\sigma = 3.0$ and 23 points)

x	Gauss $f(x) = \emptyset(x, 0)$	$\emptyset(x, t)$ $t = 1$	$t = 2$	$t = 4$	Gauss $f(x) = \emptyset(x, 0)$	$\emptyset(x, t)$ $t = 1$	$t = 2$	$t = 4$	Gauss $f(x) = \emptyset(x, 0)$	$\emptyset(x, t)$ $t = 1$	$t = 2$	$t = 4$
1	0.0%	−0.01%	−0.3%	−68.8%	0.0%	0.0%	0.0%	0.01	0.02%	0.01%	0.01%	0.00%
2	0.0	0.03	0.9	211.5	0.01	0.00	0.0	−0.01	0.05	0.04	0.03	0.02
3	0.0	−0.04	−1.5	−370.3	0.04	0.02	0.0	0.02	0.15	0.12	0.09	0.03
4	0.01	0.03	2.2	559.6	0.22	0.12	0.06	−0.01	0.38	0.32	0.26	0.18
5	0.44	0.06	−2.8	−799.6	0.88	0.60	0.36	0.06	0.87	0.77	0.66	0.43
6	5.40	0.74	3.3	1113.6	2.70	2.17	1.60	0.52	1.80	1.65	1.49	1.18
7	24.20	21.47	−3.7	−1500.0	6.48	5.90	5.14	2.95	3.32	3.14	2.96	2.49
8	39.89	55.44	103.9	1808.0	12.10	12.04	11.83	10.41	5.47	5.34	5.19	4.85
9	24.20	21.47	−3.7	−1500.0	17.60	18.49	19.50	21.93	8.07	8.06	8.04	7.88
10	5.40	0.74	3.3	1113.6	19.95	21.32	23.03	28.26	10.65	10.81	10.99	11.39
11	0.44	−0.06	−2.8	−799.6	17.60	18.49	19.50	21.93	12.58	12.90	13.25	14.00
12	0.01	0.03	2.2	559.6	12.10	12.04	11.83	10.41	13.30	13.68	14.10	15.15
13	0.0	−0.04	1.5	−370.3	etc.	etc.	etc.	etc.	12.58	12.90	13.25	14.00
14	0.0	0.03	0.9	211.5					etc.	etc.	etc.	etc.
15	0.0	−0.01	−0.3	−68.8								

MIXED DISTRIBUTIONS

Example 2.13. Separation by Doetsch method. The technique was described in detail in the preceding section, and from the theoretical aspect no problem would exist. The method can be readily converted for utilization by electronic data processing and has a definite advantage over the moments method because a solution can be readily obtained for a mixture of more than two collectives. Furthermore, the number of collectives need not be known a priori.

Table 2.12 demonstrates the narrowing of the Gaussian distribution for increasing t. Three frequencies have been depicted, $\sigma = 1, 2$ and 3. Since the Gaussian distribution has a range from $-\infty$ to $+\infty$, an arbitrary empirical cut-off had to be selected. For this example the cut-off was set at $f(x) = 0.0001$ or 0.01%. This truncation leads to 15, 19 and 23 classes. As one can readily deduce from the exhibited result in Table 2.12, the peak has increased in value from $t = 0$ to $t = 4$ for all three frequencies, but as expected the most rapid growth is shown for $\sigma = 1$.

The length of the basic period or the position of the median does not influence the augmentation. Thus it is immaterial where the collective would be located within the range of an observed frequency. One problem should be pointed out, however. When the Fourier transformation for the Gaussian frequency $\sigma = 3$ and 27 classes (i.e. $p = 27$) was performed, the tabulation displayed suddenly a decrease of the peak class $x = 14$ for $t = 6$ (see Table 2.13a), which further decreased for $t = 8$. The tabulated values give the impression that the single peak breaks into two (unequal) side lobes. This result cannot be supported by theory.

An examination of the amplitudes of the Fourier series reveals (see Table 2.13b) that the Gaussian frequency for $\sigma = 3.0$ and $p = 27$ can be approximated by six terms of the series, i.e. the percentage, reduction (Essenwanger, 1974a or later, Section 3.1.4) $Z_6^2 \sim 100\%$. Every term beyond it is "noise". In the reconstruction of the frequency the "noise" does not contribute since the amplitudes of the noise terms are below the threshold of truncated digits. Since these components are amplified with increasing t the phase angle of the noise amplitudes becomes very important, however.

Based on theoretical background the expected phase angles $E(\psi_j)$ for a symmetric Gaussian distribution were determined. They are listed in the last column of Table 2.13b. The phase angles ψ_j deviating from the expected value were marked with an asterisk.

A rerun of the Fourier analysis with double and triple precision did not change the phase angles. The frequency density for $\sigma = 3.0$ was, therefore, reconstructed with 16 digits accuracy. The Fourier representation for this more precise frequency density is given in Table 2.13b under A'_j and ψ'_j. Although the deviation of the phase angle ψ'_j from the expected value $E(\psi_j)$ has taken an orderly pattern, this step has not eliminated every discrepancy.

It can be concluded that expansion of the frequency at both sides to include the fuller range of the Gaussian distribution and increase of the

TABLE 2.13

Effect of accuracy (phase angle for small amplitudes) a. $\phi(x, t)$ for $\sigma = 3.0$ and $p = 27$ points

x	$t = 0$	$t = 2$	$t = 4$	$t = 6$	$t = 8$
1	0.001%	0.0%	−0.01%	−0.8%	−82.5%
2	0.004	0.001	0.00	0.2	31.7
3	0.02	0.007	0.01	0.4	18.0
4	0.05	0.03	−0.00	−1.0	−65.1
5	0.15	0.09	0.07	1.5	108.3
6	0.38	0.26	0.13	−1.8	−146.2
7	0.87	0.66	0.49	2.4	178.3
8	1.80	1.49	1.12	−1.5	−203.9
9	3.32	2.96	2.55	4.4	227.1
10	5.47	5.19	4.79	1.9	−239.7
11	8.07	8.04	7.95	10.1	266.2
12	10.65	10.98	11.31	9.2	−261.7
13	12.58	13.25	14.06	17.6	304.2
14	13.30	14.10	15.05	13.5	−284.4
15	12.58	13.25	14.06	17.9	332.2
	etc.	etc.	etc.	etc.	etc.

b. Fourier terms ($\sigma = 3.0, p = 27$ points)

j	A_j	ψ_j	Z_j^2	A_j'	ψ_j'	$E(\psi_j)$
1	5.80%	277	0.798	5.80	277	277
2	2.80	103	0.983	2.79	103	103
3	0.83	290	0.999	0.83	290	290
4	0.15	117	0.9999	0.15	117	117
5	0.017	303	0.9999	0.017	303	303
6	0.001	130	1.0	0.001	130	130
7	$0.67 \cdot 10^{-4}$	317	1.0	$0.67 \cdot 10^{-4}$	317	317
8	$0.23 \cdot 10^{-4}$	323*	1.0	$0.14 \cdot 10^{-4}$	323*	143
9	$0.22 \cdot 10^{-4}$	150*	1.0	$0.12 \cdot 10^{-4}$	330	330
10	$0.23 \cdot 10^{-4}$	337*	1.0	$0.9 \cdot 10^{-5}$	337*	156
11	$0.34 \cdot 10^{-5}$	343	1.0	$0.6 \cdot 10^{-5}$	343	343
12	$0.51 \cdot 10^{-5}$	170	1.0	$0.4 \cdot 10^{-5}$	350*	170
13	$0.54 \cdot 10^{-5}$	177*	1.0	$0.1 \cdot 10^{-5}$	357	356

* Deviations from expected value, $E(\psi_j)$ = expected phase angle.

precision would eventually bring the calculated phase angles in line with expectation. Although this procedure may be satisfactory from the theoretical point of view, it cannot be recommended in practical work because it would only lead to unnecessary computer expenses without adding real information. The rising of the peak as generally expected can be accomplished by a truncation of the noise, in our example the limitation to six

MIXED DISTRIBUTIONS

terms. Although this truncation slows down the rise of the peak it suffices in practical work. Consequently, it is essential in practical work to examine the amplitudes of the Fourier components for noise or to realize that a rise and subsequent decrease of the peak is caused by this noise.

An example of the Doetsch method with observed data is presented in Example 2.15 later.

2.13.1d *Maximum-likelihood method and order statistics.* In the past sections several methods of separating mixtures have been presented, but they are not based on the maximum-likelihood principles when the entire mixture is considered. One could ask, why the maximum-likelihood method has not been applied. This question has been discussed by Choi and Bulgren (1968). They state that already for a finite mixture and moderate sample size computations are very complicated. The likelihood function is:

$$L(x) = \prod_{i=1}^{N} \left[\sum_{j=1}^{k} n_j f_j(x_i) \right] \qquad [2.166]$$

with $\sum_{j=1}^{k} n_j f_j(x)$ the density function, N the number of observations, k the number of mixtures. Then n_j is a polynomial of kth degree.

Choi and Bulgren have derived computable estimators, however, based on order statistics. They assume in their procedure that the parameters θ_j of:

$$P_{G(x)} = \sum_{j=1}^{k} n_j F(x, \theta_j) \qquad [2.166a]$$

are known and that the number of components is known. Hence their procedure estimates only n_j (the mixtures) by the maximum-likelihood method. This technique is only one part of the general solution, and the reader may refer to the literature.

Behboodian (1970) attempts to solve the $\bar{x}_k$ and σ_k^2 of the samples as a weighted sample mean of variance, respectively. Then:

$$\bar{x} = \sum_{j=1}^{k} n_j \bar{x}_j \qquad [2.167a]$$

$$\sigma^2 = \sum_{j=1}^{k} n_j (\sigma_j^2 + \bar{x}_j^2) - \left(\sum_{j=1}^{k} n_j \bar{x}_j \right)^2 \qquad [2.167b]$$

a formula which we encounter in similar form in the correlation ratio (Essenwanger, 1974a, section 3.4). Behboodian proves the $\bar{x}$ and σ^2 are the maximum-likelihood estimators. This procedure can be used only to compute the mean when the partial components are known; or to determine one missing component when the others are given.

Fryer and Robertson (1972) have compared several methods for estimation of mixed normal distributions. They have studied moments estimation, multinomial maximum-likelihood and minimum χ^2-estimates as obtained by grouping of the underlying variable. They conclude that the methods do not differ essentially so far as the bias is concerned, but for the mean squared error the moments method is less accurate. They advise, however, that the moments method seems preferable for distributions whose estimators are difficult to calculate.

The problem of maximum-likelihood estimation for mixtures has recently again been studied by Dick and Bowden (1973) for mixtures of only two normal distributions. Their solution is limited, however, to cases where independent information is available on one of the samples. They conclude that the sample variance of the estimates can be as much as three times as large as an estimated asymptotic variance for a small number of observations and means $\bar{x}_k$ which are not well separated. This result is in agreement with the reservations as expressed in our discussions.

2.13.1e *Graphical methods.* Various schemes have been developed to decompose mixtures of Gaussian frequencies by graphical methods. Among the leading procedures is the one developed by Daeves-Beckel (1948). This method is based on the determination of a partial collective by identifying the components from the margins as described in the truncation method.

The crucial point is the decision, which location should the truncation point x_t be given. In the procedure introduced by Daeves-Beckel either the total $f(x)$ or $F(x)$ is plotted on probability paper. The partial components appear then as fractional straight lines, either in logarithmic ordinate (parabola for Gaussian c.f.d.) or in an ordinate reflecting the Gaussian frequency such as the probability paper. Then $f(x)$ is a hyperbola. From the slope and by scaling, the parameters n_j, $\bar{x}_j$ and σ_j of the components can be determined. In the following a similar technique by Bhattacharya (1967) is described, which is very simple.

We plot the ratio of the frequency density in the class interval x (with class width w) at the center point of the class (i.e. x_c) on semi-logarithmic paper (linear abscissa x, logarithmic scale for ordinate y). This process can be mathematically expressed, namely we draft the ratio:

$$y_{i+1}/y_i = n(x_i + w)/[n(x_i)] \qquad [2.168]$$

at the location of x_i in progressing order of x_i. The n denotes the observed frequency in the class interval x_i with center x_i (i.e. $y_i = n(x_i)$, etc.).

On graph paper with linear ordinate we can then plot:

$$\Delta(\ln y) = \ln y_{i+1} - \ln y_i \qquad [2.169]$$

Thus no special graph paper is necessary although the semi-log paper would save the conversion to logarithms.

MIXED DISTRIBUTIONS

Partial components are now reflected in this diagram as straight lines with negative slope. The number of straight lines (with negative slope) indicates how many partial components are present. This fact can readily be verified, as proven by Bhattacharya. The condition for $\Delta(\ln y)$ can be derived from [2.168] as follows:

$$\Delta(\ln y) \sim -w(\sigma_j^2 - w^2/12)(x - \bar{x}_j + w/2)/\sigma_j^4 \qquad [2.169a]$$

This formula is an approximation under the assumption that the jth component is dominant. Since the variate x on the right side appears as a first-order term, the $\Delta(\ln y)$ gives a straight line with negative slope $-w(\sigma_j^2 - w^2/12)/\sigma_j^4$.

The variance σ_j^2 can be determined from the equation:

$$c\sigma_j^4 - \sigma_j^2 + w^2/12 = 0 \qquad [2.169b]$$

which is a quadratic equation with $c = (b \tan \theta)/aw$ and with scaling parameters a and b as discussed later. This leads to the equation for mean and variance of the partial components:

$$\hat{x}_{mj} = \hat{\lambda}_j + w/2 \qquad [2.170a]$$

$$\hat{s}_j^2 = (aw/b) \cot \theta_j - w^2/12 \qquad [2.170b]$$

The θ is the slope angle of the respective straight line, counted from the abscissa in clockwise rotation (see Fig. 2.9). The $\hat{\lambda}_j$ designates the x-coordinate of the intercept of the straight line with $\Delta(\ln y) = 0$ (i.e. the zero ordinate, identical with the x-axis).

It should be noted that lines with negative slopes not reaching the abscissa and as such missing the intercept indicate partial collectives with overlap. They cannot readily be determined in this first round. If such conditions exist, the obvious and readily distinguishable partial components should be determined first. These should then be subtracted from the y_i and the remainder can be treated again by [2.169] with subsequent steps.

The a and b are scaling factors. They must be measured as follows: The b represents the length of $\Delta x = 1.0$, e.g. $l_x = 1$ cm. The a corresponds to the length of $\Delta(\ln y) = 1.0$, e.g. $l_y = 10$ cm (measured from the abscissa to the ordinate 1.0). In our case the scaling factor would render:

$$a/b = l_y/l_x = 10 \qquad [2.170c]$$

If $\Delta(\log y) = 1.0$ and $l_y = 10$ cm (for common logarithm) then $a = l_y \cdot \ln e$

or: $a/b = (l_y \ln e)/l_x = 10 \cdot (0.43\ldots)/1 = 4.3\ldots \qquad [2.170d]$

This smaller scaling factor is compensated by a different slope and consequently a different $\cot \theta$, which is smaller in the $\Delta(\ln y)$ plot than in the $\Delta(\log y)$ graph. Examination of the different scales will show that the slope of the straight line has a smaller angle in the $\Delta(\log y)$ diagram for the same difference between y_{i+1} and y_i. A smaller angle means a larger cotangent value and $(wa/b)\cot\theta$ provides the same numerical value.

The last unknown, the n_j fraction of the partial collective, probably causes the most tedious work in this graphical solution. Several methods are available. Three basic principles are presented by Bhattacharya:

(1) $f(x_i) = \sum\limits_{j=1}^{k} n_j P_j$ [2.171a]

(2) $f(x_i) = n_j P_j + n_{j+1} P_{j+1}$ over the rth line [2.171b]

(3) $f(x_i) \sim n_j P_j$ over the rth line [2.171c]

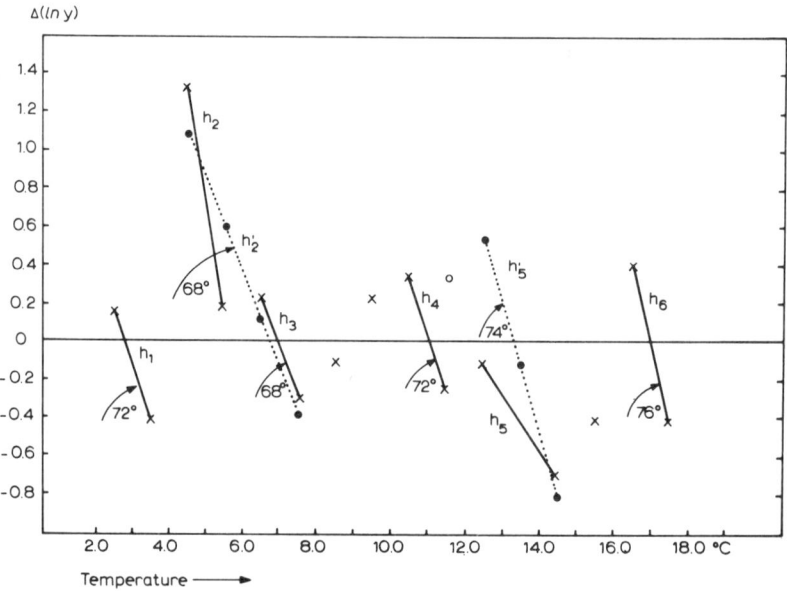

Fig. 2.9. Separation of mixed distribution into individual (Gaussian) components by graphical method (Bhattacharya) for the average daily temperature data at Karlsruhe (Germany) in October (1890—1950) with Grosswetterlage "HM". × = first plot, ● = second plot.

The three principles are based on various types of assumptions which will be analyzed now. It is evident that neither the true $f(x_i)$ nor the count P_j, i.e. the expected frequency of the partial component, is available, The $f(x_i)$ must therefore be replaced by the empirical observation $f(x_i) \sim y_i$ or $n(x_i)/N$. The $n(x_i)$ can also directly be substituted. Then $\Sigma n(x_i) = N$ instead of $\Sigma f(x_i) = 1.0$. With $P_j(x_i)$, abbreviated P_j, we denote a frequency density of the partial collective j at the point x_i for a standardized Gaussian distribution (n_j has been set unity). Hence, we may mathematically formulate:

$P_j(x_i) = F(t_L) - F(t_U)$ [2.172]

MIXED DISTRIBUTIONS

where: $t_L = (x_i - \bar{x}_j - w/2)/\sigma_j$ [2.172a]

and: $t_U = (x_i - \bar{x}_j + w/2)/\sigma_j$ [2.172b]

are the lower- and upper-class boundaries of the cumulative Gaussian frequency for component j. The $\bar{x}_j$ and σ_j will be replaced by the estimators $\hat{x}_{mj}$ and $\hat{s}_j^2$ from the graphical solution. Since the true mean and variance are not known, the P_j, although calculated, must be considered as an estimate.

It is self-evident that k equations of either conditions (1) through (3) need to be available for the determination of the k estimates of n_j. Some principles of balancing the errors of the observations must be found.

Under (1) we could formulate:

$$\sum_{j=1}^{k} \hat{n}_j \sum P_j P_h = \sum y P_h \qquad [2.173]$$

with h running from $1 \ldots k$. The summation Σ would be performed over the total range from x_{min} to x_{max}. This equation is equivalent to the following matrix notation:

$$\begin{bmatrix} \sum P_1^2 & \sum P_2 P_1 & \ldots & \sum P_k P_1 \\ \sum P_1 P_2 & \sum P_2^2 & \ldots & \sum P_k P_2 \\ \sum P_1 P_3 & \sum P_2 P_3 & \ldots & \sum P_k P_3 \\ \cdot & \cdot & & \cdot \\ \cdot & \cdot & & \cdot \\ \cdot & \cdot & & \cdot \\ \sum P_1 P_k & \sum P_2 P_k & \ldots & \sum P_k P_k \end{bmatrix} \cdot \begin{bmatrix} n_1 \\ n_2 \\ n_3 \\ \cdot \\ \cdot \\ \cdot \\ n_k \end{bmatrix} = \begin{bmatrix} \sum y P_1 \\ \sum y P_2 \\ \sum y P_3 \\ \cdot \\ \cdot \\ \cdot \\ \sum y P_k \end{bmatrix} \qquad [2.174a]$$

This is a symmetric matrix, and the solution for n_k follows the scheme described in Chapter 4 for the solutions of linear equations (Section 4.14). Although the terms are summed over the entire field of the x-scale, in practice the partial components do not necessarily extend over the entire range (see Example 2.14).

A simplified system, which is also very trivial and quick to calculate, is the following:

$$\begin{bmatrix} \sum_{h_1} P_1 & \sum_{h_1} P_2 & \ldots & \sum_{h_1} P_k \\ \sum_{h_2} P_1 & \sum_{h_2} P_2 & \ldots & \sum_{h_2} P_k \\ \sum_{h_3} P_1 & \sum_{h_3} P_2 & \ldots & \sum_{h_3} P_k \\ \cdot & \cdot & & \cdot \\ \cdot & \cdot & & \cdot \\ \cdot & \cdot & & \cdot \\ \sum_{h_k} P_1 & \sum_{h_k} P_2 & \ldots & \sum_{h_k} P_k \end{bmatrix} \cdot \begin{bmatrix} n_1 \\ n_2 \\ n_3 \\ \cdot \\ \cdot \\ \cdot \\ n_k \end{bmatrix} = \begin{bmatrix} \sum_{h_1} y \\ \sum_{h_2} y \\ \sum_{h_3} y \\ \cdot \\ \cdot \\ \cdot \\ \sum_{h_k} y \end{bmatrix} \qquad [2.174b]$$

where the summation Σ_{h_k} indicates that it will only be carried out for the x-classes of the h_k th line, which can be seen from the diagram $\Delta(\ln y)$ versus x. Under ideal conditions of no overlap (or no significant overlap) only the diagonal elements remain in the P matrix. Then the solution is trivial (see later [2.175a or b], case (1)).

While the simplification of the set of equations to [2.174] was limited to one line h_k, the system (2) accepts the summation over two neighboring lines h_k and h_{k+1} but assumes that all other influences can be neglected. The matrix scheme is, therefore:

$$\begin{bmatrix} \sum_{h_k} P_k^2 & \sum_{h_k} P_k P_{k+1} \\ \sum_{h_{k+1}} P_{k+1} P_k & \sum_{h_{k+1}} P_{k+1}^2 \end{bmatrix} \cdot \begin{bmatrix} n_k \\ n_{kh} \end{bmatrix} = \begin{bmatrix} \Sigma y P_k \\ \Sigma y P_{k+1} \end{bmatrix} \quad [2.174c]$$

In the final procedure (3) only the h_kth line is considered and it is assumed that all non-diagonal components are zero. This postulation may be approximately fulfilled for the center of the h_k th line if no overlapping occurs. Then an explicit expression can be provided, namely:

$$n_k = \Sigma y P_k / \Sigma P_k^2 \quad [2.175a]$$

$$\text{or: } n_k = \Sigma y / \Sigma P_k \quad [2.175b]$$

A further estimate is derived from $\Delta(\ln y)$. This technique is useful when no tables of the Gaussian distribution are available or it may serve as a check.

$$\ln(w\, n_j/\sigma_j) = (\Sigma \ln y)/n + \Sigma\, t^2/n + w^2/(24\sigma_j^2) + \ln\sqrt{2\pi} \quad [2.176a]$$

$$\text{where: } t^2 = (x - \bar{x}_j)/[2(\sigma_j^2 + w^2/12)] \quad [2.176b]$$

Again, summation is restricted over the classes of the h_k th line for the component j. The n denotes the number of terms (classes) which have been summed. The difference in the symbols k and j is intentional because it may occur that not all h_k lines with negative slope will represent an initial partial component. E.g. we may have six lines, only four of these lines may be acceptable in the first round (see Example 2.14). It should be indicated that the respective kth line corresponding to the jth partial component must be selected.

Example 2.14. Decomposing of mixed frequency distribution into Gaussian components. The data sample of the daily mean temperatures at Karlsruhe (Germany) in October (1890–1950) for the "Grosswetterlage HM" has been chosen to illustrate the procedure for the graphical method of decomposing data samples into partial Gaussian components. This data sample was selected because the author previously (1955b) had demonstrated the separation into partial components by the truncation method. Utilization of the graphical technique by Bhattacharya (1967) provides an opportunity to compare the results from the truncation method and the graphical scheme. To preclude the outcome depicted in Fig. 2.10, the two solutions are very similar and differ only in small details which could be attributed to data deficiencies and error tolerance. Now the detailed procedure will be discussed.

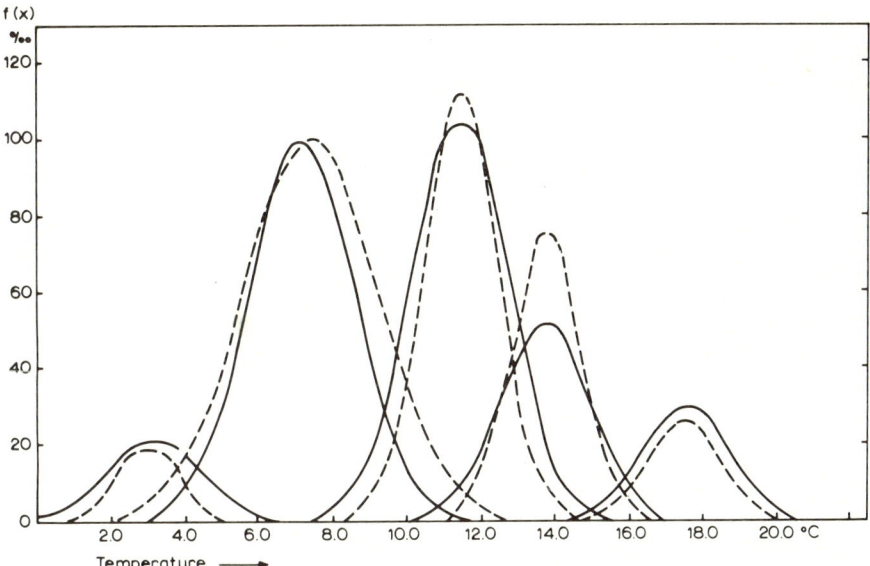

Fig. 2.10. Comparison of partial components from mixed distributions as obtained by truncation method (Essenwanger) and graphical method (Bhattacharya) for the average daily temperature data at Karlsruhe (Germany) (see Fig. 2.9).
——— = Bhattacharya, ———— = Essenwanger.

Essenwanger			Bhattacharya		
x_{mi}	S_i	n_i	x_{mi}	S_i	n_i
3.0	0.8	42⁰/₀₀	3.2	1.24	69⁰/₀₀
7.4	1.8	450	7.2	1.39	356
11.5	1.0	277	11.5	1.24	338
13.8	0.9	140	13.8	1.16	155
17.5	0.9	90	17.5	1.08	84
		999			1000

As presented in Section 2.13.1e, the first step is the formation of ln y and calculation of $\Delta(\ln y)$. These steps can be examined from the respective columns of Table 2.14, where the original frequency is given in per mill for classes of 1°C. The column T_{Li} lists the lower-class boundary and the x, the central class value. The $\Delta(\ln y)$ was plotted versus the x as exhibited by Fig. 2.9. Six lines with negative slope h_1 through h_6 are displayed, but h_2 and h_5 do not reach the abscissa and are disregarded for the moment.

TABLE 2.14

Frequency distribution of daily mean temperature in October at Karlsruhe, Germany (1890–1950) for Grosswetterlage HM (graphical method)

T_{Li} (°C)	x (°C)	y_i (°/₀₀)	$\ln y$	$\Delta \ln y$	First $P(x)$	$y_{2,i}$	$\ln y_2$	$\Delta(\ln y_2)$	Second $P(x)$ (final)	Δ_B	Δ_E
2.0	2.5	23	3.13	0.17	23	0			13	−13	−3
3.0	3.5	27	3.30	−0.41	28	−1			22	+1	3
4.0	4.5	18	2.89	1.33	27	−9			28	−2	0
5.0	5.5	68	4.22	0.19	43	25	3.22		31	−13	−12
6.0	6.5	82	4.41	0.23	81	1	0.00		54	14	11
7.0	7.5	104	4.64	−0.30	105	2	0.69		89	−8	−6
8.0	8.5	77	4.34	−0.12	88	−11			97	6	4
9.0	9.5	68	4.22	0.23	71	−3			73	3	−7
10.0	10.5	86	4.45	0.35	95	−9			60	9	4
11.0	11.5	122	4.80	−0.25	114	8	2.08	0.32	92	−4	−2
12.0	12.5	95	4.55	−0.10	84	11	2.40	0.57	120	+5	0
13.0	13.5	86	4.45	−0.46	33	53	3.97	−0.14	109	−14	1
14.0	14.5	54	3.99	−0.69	8	46	3.83	−0.79	77	7	0
15.0	15.5	27	3.30	−0.41	6	21	3.04		47	5	−4
16.0	16.5	18	2.89	0.41	18	0			23	3	11
17.0	17.5	27	3.30	−0.41	27	0			20	−1	2
18.0	18.5	18	2.89		18	0			24	4	0
19.0	19.5								16	3	+3
Σ		1000							5	−5	−3

T_{Li} = lower class boundary of temperature class.
x = class center.
$y_i = n_i/N$ where n_i is observed class frequency.
$P(x) = \sum_j P_j(x)$

$y_{2,i} = y_i - P_i(x)$.
$\Delta_B = y_i - P_i(x)$ after second recomputation by method Bhattacharya.
Δ_E = the same difference by the author's (1954a) method.

MIXED DISTRIBUTIONS

TABLE 2.15

Mean and variance as derived from the graphical solution plot, Fig. 2.9

$\hat{x}_{mj}$ $\lambda_j + w/2$	$\hat{\theta}_j$	$\cotg \theta$	$wa/b = 5$ $(wa/b) \cotg \theta$	$\hat{s}_j^2$	$\hat{s}_j$	$1/\hat{s}_j$
a. First extraction						
2.7 + 0.5	72	0.3249	1.624	1.541	1.241	0.806
7.0 + 0.5	68	0.4040	2.020	1.937	1.392	0.718
11.0 + 0.5	72	0.3249	1.624	1.541	1.241	0.806
17.0 + 0.5	76	0.2493	1.246	1.163	1.078	0.928
b. Second extraction						
6.7 + 0.5	68	0.4040	2.020	1.937	1.392	0.718
13.3 + 0.5	74	0.2867	1.433	1.350	1.161	0.861

$\hat{s}_j^2 = (wa/b) \cotg \theta - 0.083$ (see [2.170b]).

Table 2.15 discloses the determination of $\hat{x}_{mj}$ and $\hat{s}_j$ for four components $j = 1, 2, 3, 4$. The columns should be self-explanatory. The calculation of the n_j is demonstrated in Table 2.16. We build the difference $x - x_{mj}$ and compute $t = (x - x_{mj})/s_j$. From tables of the Gaussian distribution (e.g. Owen, 1962; Beyer 1966; Hald 1952; or Appendix in this book) the $F(t)$ is extracted. Finally, $P_j(x) = F(t, x + 1) - F(t, x)$.

A simple procedure for gaining n_j is based on [2.174b]. The summations are carried out over the four lines h_1, h_3, h_4, h_6. The part b of Table 2.16 shows the four equations, from which the n_j are estimated as given on the right side. E.g. ΣP_1 over classes 2 and 3 provides 0.573 and Σy for the same classes is 0.050. As required by the dominance of h_1 the ΣP_2 is 0.008 over the same classes and can be neglected compared with ΣP_1. The P_3 and P_4 have no values in the classes of h_1. Consequently $n = 0.050/0.573 - 0.087$. The n_4-value can also be found by simple division. Thus n_2 can be calculated from h_3, which makes n_3 the only unknown from h_4.

The multiplication $n_j \cdot P_j(x)$ and the summation $\Sigma x \, n_j P_j(x)$ over the x provides the first estimate of $P(x)$. The subtraction $y_i(x) - P(x)$ (see Table 2.14) leads to the new $y_{2,i}$. The steps $\ln y_2$ and $\Delta (\ln y_2)$ are self-explanatory (the logarithm can only be taken for positive $y_{2,i}$!). This leaves only the section around h_5', and h_5' can now be plotted replacing h_5. The line h_5' is fitted through the three points. Furthermore, the lines h_2 and h_3 must be combined, because no second line reaching the zero-abscissa can be found for h_2. The fitting by a first-order curve through the points 1.33, 019, 0.23, and −0.30 was performed according to the rules as later outlined in Section 3.1 (see Fig. 2.9).

The extraction of the $\hat{x}_{mj}$ and θ with derivation of $\hat{s}_j$ for these two components is contained in Table 2.15b and should be self-explantory.

TABLE 2.16

Auxiliary table, demonstrating calculation of P_j and n_j

a. First extraction of four components (h_1, h_3, h_4, h_6)

		$j=1$ $x_{m1}=3.2;$ $s_1=1.24$			$j=2$ $x_{m2}=7.5;$ $s_2=1.39$				$j=3$ $x_{m3}=11.5;$ $\hat{s}_3=1.24$				$j=4$ $x_{m4}=17.5;$ $s_4=1.08$				y $(^o/_{oo})$
	$x-w/2$	$x-x_{m1}$	t	$F(t)$ $(^o/_{oo})$	$P_j(x)$ $(^o/_{oo})$	$x-x_{m2}$	t	$F(t)$ $(^o/_{oo})$	$P_j(x)$ $(^o/_{oo})$	$x-x_{m3}$	t	$F(t)$ $(^o/_{oo})$	$P_j(x)$ $(^o/_{oo})$	$x-x_{m4}$	t	$F(t)$ $(^o/_{oo})$	$P_j(x)$ $(^o/_{oo})$	
h_1 {	2.0	−1.2	−1.87	31	135													23
	3.0	−0.2	−0.97	166	270													27
	4.0	0.8	−0.16	436	303													18
	5.0	1.8	0.64	739	188													68
h_3 {	6.0	2.8	1.45	927	61													82
	7.0		2.26	988	11													104
	8.0		3.06	999	1													77
	9.0					−4.5	3.23	1	5									68
h_4 {	10.0					−3.5	2.51	6	30									86
	11.0					−2.5	1.80	36	104									122
	12.0					−1.5	1.08	140	219									95
	13.0					−0.5	0.36	359	282									86
	14.0					0.5	−0.36	641	219	−4.5	−3.63	2	2					54
	15.0					1.5	etc.	860	104	−3.5	−2.82	22	20					27
h_6 {	16.0					2.5		964	30	−2.5	−2.01	113	91	−3.5	−3.25	1	9	18
	17.0					3.5		994	5	−1.5	−1.21	345	232	−2.5	−2.32	10	72	27
	18.0					4.5		999	1	−0.5	−0.40	655	310	−1.5	−1.39	82	241	18
										0.5	etc	887	232	−0.5	−0.46	323	354	
										1.5		978	91	0.5	0.46	677	241	
										2.5		998	20	1.5	etc.	918	72	
										3.5			2	2.5		990	9	
										4.5						999	1	

b. Determination of n_j

$h_1: n_1 \cdot 0.573 = 0.050$ $n_1 = 0.087$
$h_3: n_1 \cdot 0.012 + n_2 \cdot 0.501 = 0.186$ $n_2 = 0.185/0.501 = 0.369$
$h_4: n_2 \cdot 0.035 + n_3 \cdot 0.542 = 0.208$ $n_3 = 0.195/0.542 = 0.360$
$h_6: \quad n_4 \cdot 0.591 = 0.045$ $n_4 = 0.076$

$\Sigma n_j = 0.892$

c. Extraction of two components (h_2' and h_5')

$x_{m_2} = 7.2; s = 1.39$ $x_{m_5} = 13.8; s = 1.16$

$x - w/2$	$x - x_{m_2}$	t	$F(t)$ (⁰/₀₀)	$P_j(x)$ (⁰/₀₀)	$x - x_{m_5}$	t	$F(t)$ (⁰/₀₀)	$P_j(x)$ (⁰/₀₀)
2.0								
3.0	−4.2	−3.02	1	10				
4.0	−3.2	−2.30	11	46				
5.0	−2.2	−1.58	57	138				
6.0	−1.2	−0.86	195	249				
7.0	−0.2	−0.14	444	272				
8.0	0.8	0.57	716	186				
9.0	1.8	1.29	902	76				
10.0	2.8	2.01	978	19	−3.8	−3.27	1	7
11.0	3.8	2.73	997	3	−2.8	−2.41	8	53
12.0					−1.8	−1.55	61	184
13.0					−0.8	−0.69	245	323
14.0					0.2	0.17	568	281
15.0					1.2	1.03	849	122
16.0					2.2	1.89	971	26
17.0					3.2	2.75	997	3
18.0								

TABLE 2.17

Determination of the n_j (after second round, five components)

a. Table of $P_j(x)$ and y_i

T_{L_i}		P_1 (⁰/₀₀)	P_2 (⁰/₀₀)	P_3 (⁰/₀₀)	P_4 (⁰/₀₀)	P_5 (⁰/₀₀)	y (⁰/₀₀)
h_1	2.0	270					23
	3	303	10				27
	4	188	46				18
h_2'	5	61	138				68
	6	11	249				82
	7	1	272	002			104
	8		186	020			77
	9		76	091	1		68
h_4	10		19	232	7		86
	11		3	310	53		122
	12			232	184		95
h_5'	13			091	323	001	86
	14			020	281	009	59
	15			002	122	072	27
	16				26	241	18
h_6	17				3	354	27
	18					241	18

b. Set of equations

h_1: $0.573\, n_1 + 0.010\, n_2$ = 0.050
h_2': $0.261\, n_1 + 0.705\, n_2 + 0.002\, n_3$ = 0.272
h_4: $\quad\quad\quad\quad 0.022\, n_2 + 0.542\, n_3 + 0.060\, n_4$ = 0.208
h_5': $\quad\quad\quad\quad\quad\quad\quad\quad + 0.343\, n_3 + 0.788\, n_4 + 0.009\, n_5$ = 0.235
$\quad\quad\quad\quad\quad\quad\quad\quad\quad\quad\quad\quad + 0.029\, n_4 + 0.595\, n_5$ = 0.045

c. Results

$n_1 = 0.046/0.573 = 0.081$ adjusted 0.080
$n_2 = 0.248/0.700 = 0.355\quad\quad\quad\quad 0.352$
$n_3 = 0.199/0.542 = 0.367\quad\quad\quad\quad 0.363$
$n_4 = 0.104/0.750 = 0.138\quad\quad\quad\quad 0.137$
$n_5 = 0.041/0.595 = 0.069\quad\quad\quad\quad 0.068$
$\quad\quad\quad\quad\quad\quad\quad\quad\quad\; 1.010 \quad\quad\quad\quad 1.000$

We turn to Table 2.16c, where the calculation of $F(x)$ and $P_j(x)$ is shown in detail following the method as explained for Table 2.16a.

Finally, the $P_j(x)$ for the five components after this second round (now renamed P_1 through P_5) are listed in Table 2.17a. The summation of the respective lines as indicated by the class combination leads to the set of

MIXED DISTRIBUTIONS

equations of Table 2.17b. The n_1 through n_5 may be gained by successive elimination, replacing the unknowns by an estimate from the first round. The n_3 and n_4 are provided by the answers from two equations h_4 and h_5'.

Table 2.18 exhibits the utilization of the full matrix scheme [2.174a]. The given calculations, summations, and the establishment of the matrix Table 2.17b should be trivial and explanation will be skipped. The answer for n_1 through n_5 follows the procedure as outlined for the solution of a system of linear equations (see Section 4.14).

Finally, Table 2.19 can be established. For this particular example the tentative values from Table 2.17. have been utilized. Table 2.18 discerns that the two answers for the set of n_j do not differ significantly.

The final $P(x)$ from Table 2.18 has been placed back into Table 2.14 and $\Delta_B = y_i = P(x)$ has been computed. The five partial components derived by the author and the graphical method are portrayed in Fig. 2.10, and the difference $\Delta_E = y_i - P(x)$ for the result by the author's method is listed in the last column of Table 2.14. Although in general the differences for the Δ_E appear somewhat lower, a statistical significance between the two composite frequencies cannot be detected by application of the χ^2-test. Since the Δ_B shows three large values at $x = 4.5, 5.5$, and 12.5, the $\chi_B^2 \sim 1.6 \chi_E^2$ (χ_B^2 and χ_E^2 have been calculated from Δ_B and Δ_E, respectively, Table 2.14).

If we were to test the significance of the approximation of the analytical five-component frequency to the observed, we must convert the $f(x)$ into actual counts n_i for $N = 221$. Then $\chi_B^2 \sim 6.5$ and $\chi_E^2 \sim 4.0$ (including the margins $x = 1.5$ and 19.5). The degrees of freedom are 19—1 classes minus $n_p = 5 \times 3$, i.e. the number of parameters estimated from the samples. This would leave only 3, or for $17 - 1 - n_p$ only 1 degree. Both, the $\chi^2 \sim 4.0$ and 6.5 would not be significant at the 95-% level. For 17 classes (excluding the margins) the $\chi_E^2 \sim 3.8$ and $\chi_B^2 \sim 4.4$. While χ_E^2 corresponds with the 95-% value for 1 degree of freedom, the 4.4 would reach the 96-% level. We may consider this as a border case, because the n_j from the full matrix solution has not been used (see below).

For a larger N the same χ^2 would exceed the significant threshold. Increase of N may also smooth some of the peaks in the histogram and the deviations Δ_B or Δ_E may decrease, especially some maxima. Then the χ^2-value would decrease, too, and again significance may not be reached.

The utilization of the n_j from the full matrix solution reduces the χ_B^2 from 6.5 to 5.2. This implies a better fit as one should expect. The χ_B^2 of 4.4 for 17 classes would reduce to 3.7, which is in line with χ_E^2. The detailed tabulations have been omitted because the primary goal was a demonstration of how well the short method for the calculation of the n_j would work. Figure 2.10 exhibits the components from the full matrix solution, however.

Significance may also be checked by other test procedures, but tne question of significance is of secondary interest in this example. Furthermore, the author (1955b) has elaborated on some reasons for the inhomogeneity of the physical conditions leading to the mixture of frequencies.

TABLE 2.18
Full matrix solution
a. Summation of $P_j(x)$

P_1^2 (⁰/₀₀)	P_1P_2 (⁰/₀₀)	P_1P_3	P_1P_4	P_1P_5		P_2^2 (⁰/₀₀)	P_2P_3 (⁰/₀₀)	P_2P_4	P_2P_5
73	3	0.000	0.000	—			1		—
92	9					2	4		
35	9					19	7		
4	3					62	4	0.00012	
						74	1	0.00016	
						36			
						6			
Σ = 0.204	0.025	0	0	0		0.199	0.017	0.0003	0

P_3^2 (⁰/₀₀)	P_3P_4 (⁰/₀₀)	P_3P_5		P_4^2 (⁰/₀₀)	P_4P_5 (⁰/₀₀)	P_5^2 (⁰/₀₀)		yP_1 (⁰/₀₀)	yP_2 (⁰/₀₀)	yP_3 (⁰/₀₀)	yP_4 (⁰/₀₀)	yP_5
8	1	0.0002										1
54	17	0.0001			3	5		6	1	2		2
96	43			3	9	58		8	8	6	6	4
54	30			34	6	125		4	20	20	18	10
8	6			104	1	58		0.004	28	38	29	4
				79				0.001	14	22	15	
				15					5	8	3	
				1					2	1	1	
0.220	0.097	0.0003		0.236	0.019	0.246		0.023	0.079	0.097	0.072	0.021

172   FREQUENCY DISTRIBUTIONS

MIXED DISTRIBUTIONS

b. Matrix for [2.174a]

$$\begin{bmatrix} 0.204 & 0.024 & 0 & 0 & 0 \\ 0.024 & 0.199 & 0.017 & 0.0003 & 0 \\ 0 & 0.017 & 0.220 & 0.097 & 0.0003 \\ 0 & 0.0003 & 0.097 & 0.236 & 0.019 \\ 0 & 0 & 0.0003 & 0.019 & 0.246 \end{bmatrix} \cdot \begin{bmatrix} n_1 \\ n_2 \\ n_3 \\ n_4 \\ n_5 \end{bmatrix} = \begin{bmatrix} 0.023 \\ 0.080 \\ 0.097 \\ 0.077 \\ 0.021 \end{bmatrix}$$

Solution   from lines, Table 2.17

$n_1 = 0.071$ 0.081
$n_2 = 0.359$ 0.355
$n_3 = 0.341$ 0.367
$n_4 = 0.158$ 0.138
$n_5 = 0.086$ 0.069

$\Sigma = 1.015$ 1.010

TABLE 2.19

Final breakdown of components n_i from Table 2.17 (unit: $^0/_{00}$)

x	n_1P_1	n_2P_2	n_3P_3	n_4P_4	n_5P_5	$P(x) = \Sigma n_j P_j$	y_i
(0.5)	2					2	0
(1.5)	11					11	0
2.5	22					22	23
3.5	24	4				28	27
4.5	15	16				31	18
5.5	5	49				54	68
6.5	1	88				89	82
7.5		96	1			97	104
8.5		66	7			73	77
9.5		27	33			60	68
10.5		7	84	1		92	86
11.5		1	112	7		127	122
12.5			84	25		109	95
13.5			33	44		77	86
14.5			7	39	1	47	54
15.5			1	17	5	23	27
16.5				4	16	20	18
17.5					24	24	27
18.5					16	16	18
19.5					5	5	0
	80	354	362	137	67	1000	

Finally, it should be stated that the graphical method is a very quick and relatively easy scheme to decompose mixed frequencies, if only a few samples are to be analyzed. Numerous data samples may be better treated by computerized procedures. In a modified version this graphical method by Bhattacharya could be converted for computer application, too.

It should be observed that after determination from the graph the σ_j^2 and x_{mj} are not altered in the future course of action. The adjustment rests in the n_j for Bhattacharya's method. The principle of balancing developed by the author (see truncation method, [2.162a]) allows for adjustment of every individual parameter (estimator). It is possible that these differences in the principle viewpoints play an insignificant role in practical work.

Example 2.15. Decomposition of the data of Example 2.14 by Doetsch's method. The frequency density for the temperature at Karlsruhe (Grosswetterlage HM) has been decomposed into Gaussian components by the graphical method (Example 2.14). It was concluded that the results are in close resemblance with the outcome by the author's truncation method. The moments method would not be applicable because the mixture contains

MIXED DISTRIBUTIONS 175

more than two components and thus the mathematical and numerical procedure is too elaborate. A comparison with the Fourier transform method by Doetsch follows now.

Table 2.20a lists the function $\emptyset(x, t)$ for three different period lengths p, namely 20, 30 and 40 points in order to evaluate the influence of noise caused by a particular length of periodicity. In this case no influence can be discovered, although later this subject is taken up again.

Table 2.20b provides the ratio of the rising peaks, $\emptyset(x, t)/\emptyset(x, 0)$ (for positive $\emptyset$-values only). We discover eight classes with increase. Although the component with the maximum growth as the one with the smallest sigma should have been determined and subtracted, and the remaining data should have been re-analyzed it can be concluded from Table 2.12 that the distribution for $\sigma = 1$ generates very little side-lobes until $t = 4$. Thus the procedure to subtract the partial components one by one would not change the outcome with respect to the number of components or the variances. The indication of eight components compared with five by the graphical or the truncation method appears to be a genuine feature of the Fourier transform method. This fact was already pointed out in Section 2.13.1c and earlier in an article by the author (1955c).

Part c of Table 2.20 delineates an interesting point. It is known that the component with the smallest sigma would rise most. A ranking of the ratio which was given in Table 2.20b was performed and is reflected in Table 2.20c. Again, we notice the close agreement between the results for the three different periods of length p. While little or no change from $t = 1$ to $t = 3$ can be observed for most rankings, a drastic shift can be noticed from component rank 7 to 2 for $t = 1$ to $t = 2$. This modification is independent of the length of p. Another major change in ranks takes place from 8 to 5. These large jumps can be produced by some influence of side lobes. Since the σ-values of the individual components are not significantly different, even for the five-component solution with some broader dispersion of the frequency density of the collectives, this shift in ranking may be immaterial, and could merely be a display of the sensitivity of the Fourier transform.

The last column of Table 2.20c indicates the ranking of the variances by the graphical method. Some of the components ostensibly have been combined which are split in the Fourier transform.

In order to enlighten the functioning of the Doetsch method, as recommended by him, the collectives with the shortest σ were extracted. In our case these are the margin components. The remaining frequency was re-analyzed. Table 2.20d exhibits the results. The recalculation for $t = 1$ discerns again a rise for every second class. Gaussian distributions with small dispersion appear, therefore, as a characteristic property of the Doetsch method.

A modification is demonstrated in Table 2.20e. The observed frequency $f(x)$ was smoothed by overlapping three classes averages. This new frequency

TABLE 2.20

Description of the observations for Karlsruhe by Doetsch's method

a. Listing of $\emptyset(x, t)$ for three selected lengths of p

x	$\emptyset(x, 0)$	$p = 20$			$p = 30$			$p = 40$		
		$t = 1$	$t = 2$	$t = 3$	$t = 1$	$t = 2$	$t = 3$	$t = 1$	$t = 2$	$t = 3$
1	23	17.7	−223.0	−3799	18.7	−188.1	3094	19.3	−173.8	−2846
2	27	61.2	436.5	4743	59.9	400.9	4075	59.3	387.1	3835
3	18	−46.1	−546.8	−5538	−44.7	−511.3	−4897	−44.1	−498.0	−4664
4	68	130.1	676.7	6096	128.6	641.6	5474	128.1	628.5	5247
5	82	29.4	−518.2	−6125	30.8	−483.6	−5517	31.4	−470.7	−5294
6	104	164.9	711.5	6351	163.4	677.1	5752	162.9	664.4	5531
7	77	32.9	−468.5	−5999	34.4	−434.1	−5403	34.9	−421.5	−5183
8	68	93.9	558.0	5980	92.5	523.4	5381	91.9	510.8	5161
9	86	40.1	−459.4	−5887	41.6	−424.2	−5279	42.1	−411.4	−5057
10	122	188.5	722.8	6112	187.0	686.6	5488	186.4	673.7	5263
11	95	44.1	−449.0	−5580	45.7	−411.4	−4933	46.2	−398.2	−4703
12	86	127.5	554.7	5268	125.8	515.3	4590	125.2	501.8	4355
13	54	28.6	−318.1	−4543	30.4	−276.4	−3825	31.0	−262.4	−3583
14	27	39.7	322.7	4105	37.8	278.1	3335	37.2	263.5	3083
15	18	−4.2	−264.9	−3708	−2.2	−217.0	−2874	−1.5	−201.8	−2610
16	27	46.5	271.2	3403	44.5	219.4	2484	43.8	203.4	2206
17	18	14.3	−153.0	−2995	16.1	−97.2	−1967	16.9	−80.1	−1674

b. Listing of ratio $\emptyset(x, t)/\emptyset(x, 0)$ (positive values only)

x	t = 1	t = 2	t = 3	t = 1	t = 2	t = 3	t = 1	t = 2	t = 3
1	0.77	—	—	0.81	—	—	0.84	—	—
2	2.26	16.17	175.	2.22	14.85	151	2.20	14.33	142
3	—	—	—	—	—	—	—	—	—
4	1.91	9.96	90	1.89	9.44	80	1.88	9.24	77
5	0.36	—	—	0.38	—	—	0.38	—	—
6	1.59	6.88	61	1.57	6.51	55	1.57	6.39	53
7	0.43	—	—	0.45	—	—	0.45	—	—
8	1.38	8.21	88	1.36	7.70	79	1.36	7.51	76
9	0.47	—	—	0.48	—	—	0.49	—	—
10	1.55	5.92	50	1.53	5.62	45	1.53	5.52	43
11	0.46	—	—	0.48	—	—	0.49	—	—
12	1.48	6.45	61	1.46	5.99	53	1.46	5.83	51
13	0.53	—	—	0.56	—	—	0.57	—	—
14	1.47	11.95	152	1.40	10.30	124	1.38	9.76	114
15	—	—	—	—	—	—	—	—	—
16	1.72	10.04	126	1.65	8.13	92	1.62	7.53	82
17	0.79	—	—	0.89	—	—	0.94	—	—

TABLE 2.20 (Continued)

c. Ranking of peaks

x	$t=1$	$t=2$	$t=3$	$t=1$	$t=2$	$t=3$	$t=1$	$t=2$	$t=3$	Graphical
1										
2	1	1	1	1	1	1	1	1	1	
3										3–4
4	2	4	4	2	3	4	2	3	4	
5										
6	4	6	6	4	6	6	4	6	6	
7										5–6
8	8	5	5	8	5	5	8	5	5	
9										
10	5	8	8	5	8	8	5	8	8	2
11										
12	6	7	7	6	7	7	6	7	7	
13										7–8
14	7	2	2	7	2	2	7	2	2	
15										
16	3	3	3	3	4	3	3	4	3	1

d. Extraction of marginal collective and recalculation of $\emptyset(x, t)$ with $p = 20$

x	$\emptyset(x, 0)$	$t = 1$	$\emptyset(x, 1)/\emptyset(x, 0)$
3	9	−48.0	
4	66	127.3	1.93
5	82	30.7	
6	104	164.4	1.58
7	77	33.1	
8	68	93.9	1.38
9	86	40.1	
10	122	188.4	1.54
11	95	44.3	
12	86	126.8	1.51
13	54	30.1	
14	24	36.6	1.54
15	6	−10.1	

MIXED DISTRIBUTIONS

e. Doetsch's method applied to smoothed data for Karlsruhe

x	$f'(x)$	$\emptyset(x, 0)$	$t=1$	$t=2$
0	7	7.2	8.0	14.8
1	17	16.5	17.9	21.9*
2	23	23.9	21.4	12.7
3	38	36.7	35.7	43.4*
4	56	57.6	53.5	37.1
5	85	83.1	90.2	105.1*
6	88	90.2	96.4	107.8*
7	83	80.7	76.0	61.0
8	77	79.3	76.6	81.7
9	92	89.8	87.6	76.9
10	100	102.9	108.9	120.7*
11	101	99.4	104.7	113.1*
12	78	79.2	78.5	69.9
13	56	55.2	56.3	66.9*
14	33	33.5	28.4	14.8
15	24	23.8	22.3	23.2
16	21	20.9	22.5	26.2*
17	15	15.2	15.9	12.7
18	6	5.8	6.3	14.9
19	0	0.0	−6.0	−23.0

* Rising peaks.

$f'(x)$ is given in the column next to x in Table 2.20e. A subsequent analysis by the Doetsch method resulted still in the same number of peaks, indicating a partial component every second class as displayed in the set of Tables 2.20a through d. The Fourier series for twenty classes, resulting in ten terms was now truncated at the eighth term. The regenerated frequency $\emptyset(x, 0)$ with these lesser terms is listed in the column $\emptyset(x, 0)$. As illustrated, this frequency is not exactly the $f'(x)$ but is close. The loss is 0.2% of the variance. Now $\emptyset(x, t)$ was pursued for increasing t. The results for $t = 1$ and $t = 2$ are contained in Table 2.20e.

The rising peaks are marked next to the frequency in the column $t = 2$, indicating now only six independent components. (The rising in adjacent classes must be considered as one peak.) This outcome would be more in line with the graphical or truncation method. Consequently, smoothing and truncation together are recommended if a reduction in the originally indicated number of partial components is desired for the Doetsch method.

2.13.1f *Multivariate case.* The previous methods treat only the univariate (or one-dimensional) case. Day (1969) has recently introduced the k-variate case; but has given only solutions for the mixture of two collectives. The postulation is made that the covariance matrices are the same for these two partial components.

Day considers the moments equations but with $k > 1$ the estimators are poor. The maximum-likelihood equations increase in difficulty as k rises, but computation of estimators is feasible for $k \leq 10$. The minimum χ^2 equations appear unmanageable for $k > 1$ and the computations for a Bayes estimation seem out of proportion. Thus only moment and maximum-likelihood solution is given below.

We assume that a sample of $x_{1,k} \ldots x_{N,k}$ has been obtained from a k-dimensional population, i.e. the observations x_i are k-dimensional vectors. Then the density function* can be written as:

$$f(x) = (2\pi)^{-k/2} \|\mathbf{M}_c\|^{-0.5} [n_1 \exp\{-0.5(x_i - \bar{x}_1)\mathbf{M}_c^{-1}(x_i - \bar{x}_1)'\}$$
$$+ (1 - n_1) \exp\{-0.5(x_i - \bar{x}_2)\mathbf{M}_c^{-1}(x_i - \bar{x}_2)'\}] \qquad [2.177a]$$

where $\bar{x}_1$ and $\bar{x}_2$ stand for the mean vectors (in dimension k) of the components, n_1 is the proportional share of the first component, and $\mathbf{M}_c$ is the common covariance matrix of the components (see [2.76a]), in our case the two partial collectives. $\mathbf{M}_c^{-1}$ is the inverse, and $\|\mathbf{M}_c\|$ the determinant. The prime indicates that it is not the identical x of the preceding term.

In this set of frequency density equations we can find $k^2/2 + 5k/2 + 1$ unknown parameters, for $k = 1$ we would have four parameters ($n_1, \bar{x}_1, \bar{x}_2$ and covariance). For $k = 2$ we must estimate eight parameters.

In the k-variate case we have k first, $k(k + 1)/2$ second, $k + k(k - 1) + k(k - 1)(k - 2)/6$ third and $k + 3k(k - 1)/2 + k(k - 1)(k - 2)/2 + k(k - 1)(k - 2)(k - 3)/24$ fourth moments. For $k > 1$ we must select the proper third and fourth moments or functions thereof. For $k = 2$ we have 2 first, 3 second, 4 third and 5 fourth moments, i.e. 14 moments, and we need 8 estimators. Let us denote the overall mean by $\bar{X} = f(\bar{x}_1, \bar{x}_2, \ldots, \bar{x}_k)$, then:

$$\bar{X} = n_1 \bar{x}_1 + (1 - n_1) \bar{x}_2 \qquad [2.177b]$$

We define $d = \bar{x}_1 - \bar{X}$, with $d = (d_1, \ldots d_k)$. Then the moments (reference mean) are given by:

$$v_{ij} = \sigma_{ij} + n_1(1 - n_1)^{-1} d_i d_j \qquad [2.177c]$$
$$v_{ijl} = n_1(1 - 2n_1)(1 - n)^{-2} d_i d_j d_l \qquad [2.177d]$$
$$v_{ijlm} = v_{ij} v_{lm} + v_{il} v_{jm} + v_{jm} v_{jl} + n_1(1 - 6n_1 + 6n_1^2)(1 - n_1)^{-3} d_i d_j d_l d_m$$
$$\qquad [2.177e]$$

with σ_{ij} symbolizing the (i, j)-th element of $\mathbf{M}_c$, the covariance matrix of the components of the mixture.

We denote the sample moments with S. Then [2.177c—e] can be rewritten with respect to sample moments as stated below:

---

* See also [2.76].

MIXED DISTRIBUTIONS

$$S_i = x_{mi} \quad (i = 1, \ldots k) \tag{2.178a}$$

$$S_{ij} = \sigma_{ij} + n_1(1-n_1)^{-1} d_i d_2 \quad (j = 1, \ldots, k) \tag{2.178b}$$

$$S_{ijl} = \tilde{n}_1(1-2n_1)(1-n_1)^{-2} d_i d_j d_l \quad (l = 1, \ldots, k) \tag{2.178c}$$

$$S_{ijlm} - S_{ij}S_{lm} - S_{ij}S_{jm} - S_{im}S_{jl} = n_1(1-6n_1+6n_1^2)(1-n)^{-3} d_1 d_2 d_3 d_m$$
$$(m = 1, \ldots, k) \tag{2.178d}$$

where x_m, S and $\tilde{n}$ are the moments estimators.

Under rotations of the sample space (see later Section 3.5) we obtain as the third and fourth moments invariant:

$$S_{iii} = \tilde{n}_1(1-2\tilde{n}_1)(1-\tilde{n}_1)^{-2} d_i^3 \tag{2.178e}$$

$$[n_1(1-n_1)(1-2n_1)]^{-1/3}(1-6n_1+6n_1^2)(1-2n_1)^{-1}$$
$$= \left[\frac{1}{N} \sum_{l=1}^{k} D_l^4 - 2 \sum_{i=1}^{k} \lambda_i^2 - \left(\sum_{i=1}^{k} \lambda_i^2 \right) \right] / Q^{2/3} \tag{2.178f}$$

$$Q = \sum_{i=1}^{k} S_{iii}^2 + 3 \left[\sum_{i=1}^{k} \sum_{j<i} (aS_{iij}^2 + bS_{iii}S_{ijj}) \right]$$
$$+ 6 \left[\sum_{i=1}^{k} \sum_{j<i} \sum_{l<j} (cS_{iij}S_{jll} + dS_{ijl}^2) \right] \tag{2.178g}$$

with $a + b = 1$, $c + d = 1$ for $k > 2$, and $c = d = 0$ for $k = 2$.

We have abbreviated D_l for the distance of x_l from the centroid of the sample and λ_i (with $i = 1, \ldots, k$) as the eigenvalue of the sample covariance matrix.

The maximum-likelihood functions for the same conditions as above are as follows:

$$\sum_{i=1}^{N} (e_{1i} - e_{2i})/\text{denom} = 0 \tag{2.179a}$$

$$\sum_{i=1}^{N} (x_i - \hat{\bar{x}}_1) \hat{n}_{1i}/\text{denom} = 0 \tag{2.179b}$$

$$\sum_{i=1}^{N} (x_i - \hat{\bar{x}}_2)(1 - \hat{n}_1) e_{2i}/\text{denom} = 0 \tag{2.179c}$$

with the abbreviations:

$$\text{denom} = \hat{n}_1 e_{1i} + (1 - \hat{n}_1) e_{2i} \tag{2.179d}$$

$$e_{ji} = \exp[-0.5(x_i - \hat{\bar{x}}_j) \Sigma^{-1} (x_i - \hat{\bar{x}}_j)'] \tag{2.179e}$$

FREQUENCY DISTRIBUTIONS

The differentiation with respect to $\mathbf{M}_c$ provides:

$$-n\hat{\mathbf{M}}_c + \sum_{i=1}^{n}[(x_i-\hat{\bar{x}})'(x_i-\hat{\bar{x}}_1)\hat{n}_1 e_{1i} + (x_i-\hat{\bar{x}}_2)'(x_i-\hat{\bar{x}}_2)(1-\hat{n}_1)e_{2i}]/$$
$$(\text{denom}) = 0 \qquad [2.179\text{f}]$$

Mean and covariance matrix are given for the mixture by:

$$\bar{X} = n_1 x_1 + (1-n_1)x_2 \qquad [2.180\text{a}]$$
$$\mathbf{CV} = \mathbf{M}_c + n_1(1-n_1)(x_1-\bar{x}_2)'(x_1-\bar{x}_2) \qquad [2.180\text{b}]$$

where **CV** stands for the covariance matrix of the mixture. The maximum-likelihood solutions are:

$$\hat{\bar{X}} = \bar{x} = (1/N)\sum_{1}^{N} x_i \qquad [2.180\text{c}]$$

$$\hat{\mathbf{CV}} = (1/N)\sum_{1}^{N}(x_i-\bar{x})'(x_i-\bar{x}) \qquad [2.180\text{d}]$$

This elaborate system can be reduced by introducing:

$$\hat{P}(1|x_j) = \hat{n}_1 e_{1j}/[\hat{n}_1 e_{1j} + (1-\hat{n}_1)e_{2j}] \qquad [2.180\text{e}]$$
and: $\hat{P}(2|x_j) = 1 - \hat{P}(1|x_j) \qquad [2.180\text{f}]$

Then $P(k|x_j)$ is the probability that the observation x_j arises from the component k.
Furthermore we rewrite:

$$(1/\hat{n}_1)\sum_{j=1}^{N}\hat{P}(1|x_j) = (1-n_1)^{-1}\sum_{j=1}^{N}\hat{P}(2|x_j) \qquad [2.181\text{a}]$$

$$\hat{\bar{x}}_1 = [\sum_j x_j \hat{P}(1|x_j)]/\sum_j \hat{P}(1|x_j) \qquad [2.181\text{b}]$$

$$\hat{\bar{x}}_2 = [\sum_j x_i \hat{P}(2|x_j)]/\sum_j \hat{P}(2|x_j) \qquad [2.181\text{c}]$$

$$\hat{\mathbf{M}}_c = N^{-1}\sum_j[(x_j-\hat{\bar{x}}_1)'(x_j-\hat{\bar{x}}_1)\hat{P}(1|x_j) + (x_j-\hat{\bar{x}}_2)'(x_j-\hat{\bar{x}}_2)\hat{P}(2|x_j)]$$
$$[2.181\text{d}]$$

Now we reformulate:

$$P(1|x) = [1+\exp(ax'+b)]^{-1} \qquad [2.181\text{e}]$$

with:

$$\hat{a} = [\hat{\mathbf{CV}}^{-1}(\hat{\bar{x}}-\hat{\bar{x}}_2)']/[1-\hat{n}_1(1-\hat{n}_1)(\hat{\bar{x}}-\hat{\bar{x}}_2)\hat{\mathbf{CV}}^{-1}(\hat{\bar{x}}_1-\hat{\bar{x}}_2)'] \qquad [2.181\text{f}]$$
$$\hat{b} = \hat{a}(\hat{\bar{x}}_1+\hat{\bar{x}}_2)'/2 + \ln[(1-n_1)/n_1] \qquad [2.181\text{g}]$$

The replacement of M_c by CV is an important step, since inversion of CV is simpler than M_c (compare also [2.181d] with [2.180d], CV needs to be calculated only once). This is a set of equations of the general nature:

$$\hat{a} = \phi_1(\hat{a}, \hat{b}, x_1, \ldots x_n) \qquad [2.181h]$$

$$\hat{b} = \phi_2(\hat{a}, \hat{b}, x_1, \ldots x_n) \qquad [2.181i]$$

It can be solved by customary mathematical procedures although a solution is tedious. If electronic data processing is available, an iterative procedure is probably the simpler scheme. More than one solution may exist, however. These likelihood equations can be expanded for mixtures of more than two collectives (see Day, 1969).

2.13.2 Mixtures of other than Gaussian distributions

Up to this point only mixtures of Gaussian components were discussed. While mixtures of Gaussian distributions have been described in the meteorological literature, little information can be found on mixtures of other distribution types. This may be attributed to the fact that these other mixtures play a minor role in atmospheric physics. Their treatment will, therefore, be only brief.

In the last decade some articles and reports on various types of mixtures have appeared in the statistical literature. Generally, various types of mixtures have been identified. Some authors call the distribution a mixture when:

$$f(x) = \Pi f_k(x_k) \qquad [2.182a]$$

This is not a mixture in the sense as discussed by this author. The individual functions of f_k are all of the same type, and this form may rather be classified into the category of multivariate distributions or mixture of secondary kind. A mixture as specified in our sense would require the summation of populations or:

$$f(x) = \Sigma f_k(x_k) \qquad [2.182b]$$

Some of these distributions have also been called "compound" distributions. E.g. for precipitation frequencies Friedman and Jones (1957) and Thom (1968) suggest the use of the incomplete gamma function with the zero class added, $G(x) = P + (1 - P) F(x)$, etc. (Essenwanger, 1974a, section 7.7). Chakravarti et al. (1967) call the non-central χ^2-distribution a mixture of a sequence of χ^2 and a Poisson distribution. Many mixtures of frequency densities as specified by [2.182a] can be found in Johnson and Kotz (1969, 1970a, b, 1972a). The reader may refer to the literature. The expression mixture as adopted in this text shall be limited to the summation of frequencies of the same type.

It should be pointed out that a mixture of two Poisson distributions is a Poisson distribution again (Essenwanger, 1974a, section 2.2). It may be a worthwhile analysis task in physics to identify the individual mixture components of the Poisson distribution. Due to the composite property of the Poisson distribution the frequency can be readily represented by statistical description of a single curve, even if a mixture exists. In contrast this simple representation is usually not the case in mixtures of Gaussian distributions. In this latter case the observed frequency and the single analytical model would disagree if no attempt is made to determine the individual collectives and to formulate a description by a mixture model. A typical example is the bimodal distribution arising from superposition of two Gaussian components.

In other distribution types, such as the binomial, Weibull, or gamma, the problem of mathematical approximation of mixtures by one single curve can usually be solved without too much discrepancy between analytical and observed frequency, and is similar to the Poisson model. It is, therefore, not necessary to go into every detail similar to the discussion of the Gaussian case.

Binomial mixture. Mixtures of binomial distributions were studied by Rider (1962), Blischke (1964) and Hald (1968). We may generally state (with the customary symbols, $p = 1 - q$, see Essenwanger, 1974a, section 2.1):

$$f(x) = \sum_{k=1}^{k_c} w_k \binom{n_k}{x} p_k^x q_k^{n_k - x} \qquad [2.183a]$$

and: $\sum w_k = 1$ [2.183b]

with $w_k \geqslant 0$, serving as the proportion (or weight) of the individual mixtures. It is commonly assumed that the components n_k are equal, i.e. $n_1 = n_2 = \ldots = n$. Then $n \geqslant 2 k_c - 1$ provide enough equations for the parameter estimation.

It can be shown that:

$$E(\phi_\nu) = \sum_{k=1}^{k_c} w_k p_k^\nu \sim \varphi_\nu \qquad [2.183c]$$

where ϕ_ν denotes the factorial moments and φ_ν the empirical factorial moments (see Essenwanger, 1974a, section 7.5.7).

$$\phi_\nu = \frac{1}{N} \sum_{i=1}^{N} \frac{x_i(x_i - 1) \ldots (x_i - \nu + 1)}{n(n-1) \ldots (n - \nu + 1)} = N^{-1} \sum_{i=1}^{N} x_i^{[\nu]}/n^{[\nu]} \qquad [2.183d]$$

The center part of [2.183d] reflects the writing for the individual x_i observations, and the right side indicates the corresponding abbreviation in

MIXED DISTRIBUTIONS

mathematical symbols. In most solution procedures for the parameters w_k and p_k the n is assumed to be known for the binomial, $n = x_m^2/(x_m - s^2)$. The n denotes in general the highest non-zero-frequency class number (integer) or in short the number of classes.

We must now establish a set of linear equations from [2.183c] by equating the $\Sigma w_k \cdot p_k^v$ with the observed factorial moments as required:

$$\varphi_0 = 1.0 = \Sigma w_k \qquad [2.183e]$$

$$\varphi_1 = \Sigma w_k p_k \qquad [2.183f]$$

$$\varphi_2 = \Sigma w_k p_k^2 \qquad [2.183g]$$

etc.

We need $2k_c - 1$ equations. We solve the first $k_c - 1$ equation for the w_k terms and utilize the remaining k_c equations for the determination of the p_k. The latter can be reduced to one equation, a polynomial of degree k_c:

$$Z^k + B_{k-1} Z^{k-1} + \ldots + B_1 Z + B_0 = 0 \qquad [2.183h]$$

The B's are functions of the φ_k only, and the p_k are the roots of this polynomial. For two components, $k_c = 2$, we must meet $n > 3$ and need φ_1, φ_2 and φ_3.

From the second-order equation:

$$B_2 Z^2 + B_1 Z - B_0 = 0 \qquad [2.183i]$$

we obtain the solution:

$$p_1, p_2 = (A \pm \sqrt{A^2 - 4A\varphi_1 + 4\varphi_2})/2 \qquad [2.183j]$$

with: $A = (\varphi_3 - \varphi_1 \varphi_2)/(\varphi_2 - \varphi_1^2) \qquad [2.183k]$

Subsequently: $w_1 = (\varphi_1 - p_2)/(p_1 - p_2) \qquad [2.183 l]$

and w_2 is $1 - w_1$ (see [2.183b]).

If $n = 2k_c - 1$, then the estimators have 100% efficiency relative to maximum-likelihood estimators according to Blischke (1964), if $n \geqslant 2k_c - 1$ the estimators are less efficient, but approximate 100% with increasing n.

Mixture of two Poisson distributions. Let us assume, we have the mixture:

$$f(x) = w_1 \exp(-h_1)h_1/x! + w_2 \exp(-h_2)h_2/x! \qquad [2.184a]$$

then according to Cohen (1964), Cohen and Fall (1967) we obtain the solution with the same set of equations as stated above. We need to replace only p_1, p_2 by h_1 and h_2, respectively. The solution can be modified (see Cohen, 1964). This modified version improves the efficiency by utilizing the first two moments and the zero class. We find:

$$h_1 = (\varphi_2 - \varphi_1 h_2)/(\varphi_1 - h_2) \qquad [2.184b]$$

Unfortunately the h_2 must be obtained by an iterative process from:

$$(\varphi_1 - h_2)/(G - h_2) = [N_0/N - \exp(-h_2)]/[e^{-G} - \exp(-h_2)] \quad [2.184c]$$

The abbreviations denote:

$$G = (\varphi_2 - \varphi_1 h_2)/(\varphi_1 - h_2) \quad [2.184d]$$

and N_0 is the number of observations in the zero class, N the total number of observations.

In a short note Gross (1970) has derived the convolution of Poisson distributions with the zero class missing for the cases that the parameters of the distributions are identical or vary. This problem does not deal, however, with the computation of estimators for the parameters of the individual components.

Mixture of two exponential distributions. Let us assume, we have the mixture:

$$f(x) = w_1 a_1 \exp(-a_1 x) + w_2 a_2 \exp(-a_2 x) \quad [2.185]$$

for $x \geq 0$, $a_2 > a_1 > 0$ and $0 \leq w_1 \leq 1.0$ with the usual notation for w_1 as defined above. This postulation leads to the estimating equations:

$$\bar{x} - a_2^{-1} = w_1(a_1^{-1} - a_2^{-1}) \quad [2.185a]$$

$$\mu_2/2 - a_2^{-2} = w_1(a_1^{-2} - a_2^{-2}) \quad [2.185b]$$

$$\mu_3/6 - a_2^{-3} = w_1(a_1^{-3} - a_2^{-3}) \quad [2.185c]$$

where μ_i stands for the non-central moment as defined earlier. We abbreviate:

$$A = a_1^{-1} + a_2^{-1} \quad [2.185d]$$

and: $B = (a_1 \cdot a_2)^{-1} \quad [2.185e]$

We compute the estimators for A and B as:

$$A = [\mu_3/6 - \bar{x} \cdot \mu_2/2]/[\mu_2/2 - \bar{x}^2] \quad [2.185f]$$

$$B = \bar{x} \cdot A - \mu_2/2 \quad [2.185g]$$

The estimators for a_1 and a_2 can then be calculated from [2.185d and e]. Furthermore:

$$w_1 = (\bar{x} - a_2^{-1})/(a_1^{-1} - a_2^{-1}) \quad [2.185h]$$

with w_2 as customary (see [2.183b]).

Mixture of two Weibull distributions. We postulate:

$$f(x) = w_1 f_1(x) + w_2 f_2(x) \quad [2.186]$$

with $w_i > 0$, $w_2 = 1 - w_1$, and $0 < w_i < 1.0$.

MIXED DISTRIBUTIONS

The two frequency density distributions $f_1(x)$ and $f_2(x)$ are Weibull distributions with parameters $\delta_1, \gamma_1, \theta_1$ and $\delta_2, \gamma_2, \theta_2$. First we assume $\gamma_1 = \gamma_2 = 0$ in order to simplify the solution. We arrive after some arithmetic at the following terms:

$$\theta_2 = \frac{\bar{x} w_2 c_1 b_2 \pm [-\bar{x}^2 w_1 w_2 b_2 b_1 c_2 + \mu_2 w_1 w_2^2 b_2 b_1^2 c_1^2 + \mu_2 w_1^2 w_2 b_1^4 c_2]^{1/2}}{w_2^2 c_1^2 b_2 + w_1 w_2 b_1^2 c_2} \quad [2.186a]$$

and: $\theta_1 = \dfrac{\bar{x} - w_2 c_1}{w_1 b_1} \theta_2$ [2.186b]

Furthermore, the basic (non-central) moments equations can be stated

$\bar{x} = w_1 \theta_1 b_1 + w_2 \theta_2 c_1$ [2.186c]

$\mu_2 = w_1 \theta_1^2 b_2 + w_2 \theta_2^2 c_2$ [2.186d]

$\mu_3 = w_1 \theta_1^3 b_3 + w_2 \theta_2^3 c_3$ [2.186e]

$\mu_4 = w_1 \theta_1^4 b_4 + w_2 \theta_2^4 c_4$ [2.186f]

$\mu_5 = w_1 \theta_1^5 b_5 + w_2 \theta_2^5 c_5$ [2.186g]

The following abbreviations have been used:

$b_i = \Gamma(i/\delta_1 + 1)$ [2.186h]

$c_i = \Gamma(i/\delta_2 + 1)$ [2.186i]

It is obvious that no explicit solution can be found for δ_1, δ_2 and w_1. Cohen and Fall (1967) recommend a graphical solution for w_1. We plot the cumulative (mixed) distribution on Weibull plot paper. The mixture will be a curved line, while a true Weibull distribution would be a straight line. Then we draw a tangent to the lower position. At the abscissa value x where the cumulative frequency ends (at 99.9%) we read off the percentage value of the second curve and have the proportionality w_1 and w_2 (see Fig. 2.11). This may introduce only a relatively small error according to Cohen and Fall (1967), especially since only the first four moments are needed (Fall, 1970).

We can then solve [2.186e] for δ_1 and δ_2 by an iterative procedure. Assume a value of δ_1 and δ_2, substitute into [2.186a and b] and obtain a first value of θ_1 and θ_2, then check the solution by [2.186f]. Attention should be given that the solution may lead to more than one set of estimates, as [2.186a and b] have two solutions. The total procedure is elaborate and can run into considerable computer costs when a larger number of samples must be estimated. The procedure simplifies when some of the parameters are known.

It should be further called to the attention that a single Weibull distribution whose γ-value is not zero (as assumed) may also produce a curved line in the

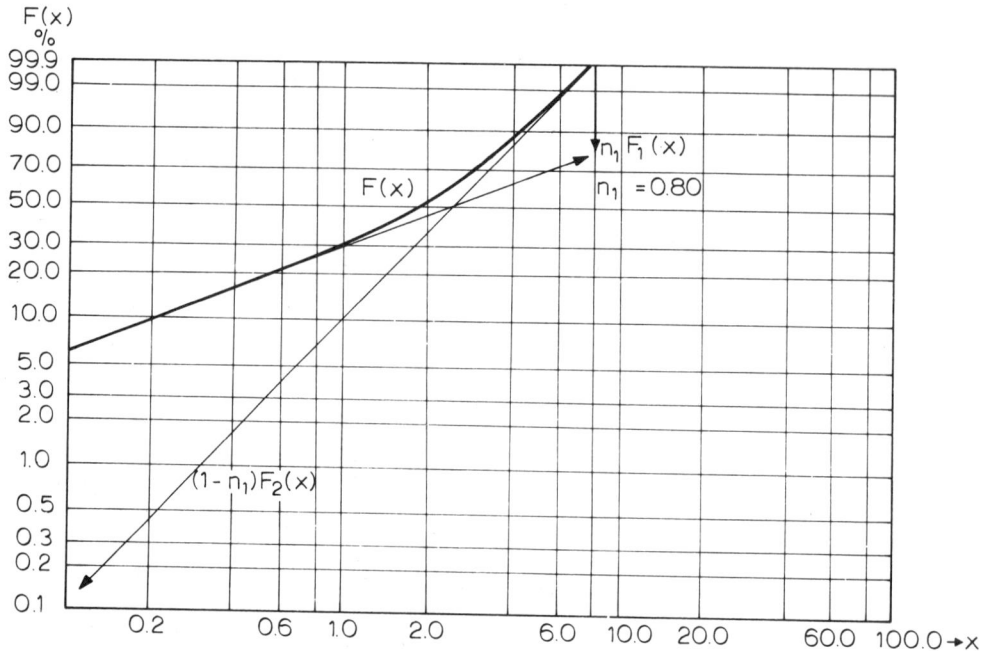

Fig. 2.11. Separation of two Weibull distributions. (Reproduction by permission from Mr. Fall and the American Statistical Association.)

Weibull diagram (see Fig. 2.12). The determination of γ (e.g. by procedures as discussed for a single Weibull distribution, see Section 2.11) and replotting of a three-parameter distribution model against the observed values will quickly show whether a mixture would be a necessary concept. The three-parameter model for a single distribution would not suffice in the case of a true mixture with clearly separable components, but may sometimes be a tolerable approximation. The tolerance of deviations should also take into account the computer expenses arising for the mixed distribution when the number of samples is large. Since the assumption of γ_1 and $\gamma_2 = 0$ has been made, the four-parameter Weibull distribution (see Section 2.11) may be another alternative to the three-parameter Weibull or mixture of two Weibull densities.

MIXED DISTRIBUTIONS

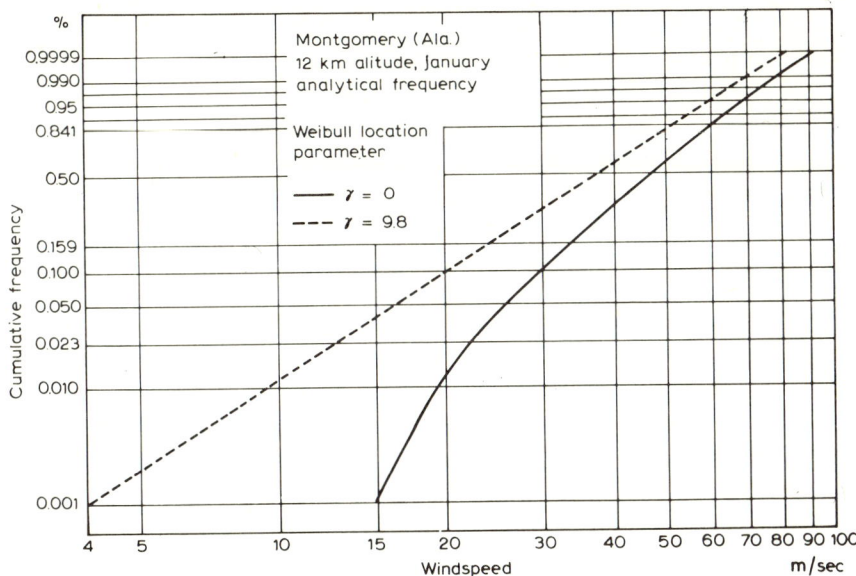

Fig. 2.12. Cumulative frequency distribution of the wind speed for Montgomery (Alabama) at 12 km altitude in January (1956—1964) on the Weibull plotting paper under the assumption of $\gamma = 0$ and $\gamma = 9.8$ m/sec.

2.14 FOLDED (GAUSSIAN) NORMAL DISTRIBUTION

Although a thorough treatment of this type has been given by Leone et al. (1961), applications in the general statistical literature are seldom. This type of distribution is suitable, however, for frequency distributions of measurements without the algebraic sign if the observations with sign follow the Gaussian law. Leone et al. (1961) have quoted typical examples in industrial practical work, and Essenwanger et al. (1961) have investigated the suitability for quality control procedures of wind shear values. Since parameters in atmospheric science can be found where the sign is of secondary importance this topic may be briefly presented. Application in atmospheric science could exist for other than distributions of the absolute (zonal) wind speed or wind shear.

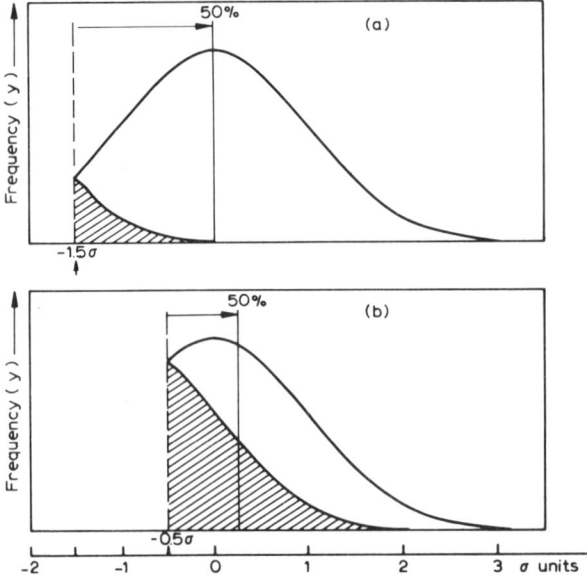

Fig. 2.13. Folding of the Gaussian distribution (a) and relationship to the 50% (median) threshold (b).

Let us assume that $x = |z|$, i.e. $x \geqslant 0$, folded at $z = 0$. We further postulate that $Z = z - \bar{z}$ has a Gaussian distribution. Then x is typically a folded normal distribution. Although the folding point $x_F = \bar{z}$ is known, the $\bar{z}$ and the folded portion of the frequency is unknown. Leone et al. (1961) have shown that the frequency density for x can be expressed as the superposition of two normal components, namely:

$$f(x) = (\sigma_z \sqrt{2\pi})^{-1} \cdot [\exp\{-\tfrac{1}{2}(Z_1/\sigma_z)^2\} + \exp\{-\tfrac{1}{2}(Z_2/\sigma_z)^2\}] \text{ for } x \geqslant 0$$
[2.187]

FOLDED (GAUSSIAN) NORMAL DISTRIBUTION

(see Fig. 2.13). Now $Z_1 = z - \bar{z}$ and $Z_2 = z + \bar{z}$.
This frequency can be transformed (see Johnson, 1962) to:

$$f(x) = (\sigma_z^{-1}\sqrt{2/\pi}) \exp[-(x^2 + \bar{z}^2)/(2\sigma_z)] \cdot \cosh(\bar{z}x/\sigma_z^2) \quad [2.187a]$$

where cosh is the hyperbolic cosine function. With the substitution:

$$\theta_z = \bar{z}/\sigma_z \quad [2.187b]$$

this density function assumes the form:

$$f(x) = (\sigma_z^{-1}\sqrt{2/\pi}) \exp(-\theta_z^2/2) \cdot \exp[-x^2/(2\sigma_z^2)] \cosh(\theta_z x/\sigma_z) \quad [2.187c]$$

Direct integration leads to the moments of x.

$$\mu_1(x) = \bar{x} = (\sigma_z \sqrt{2/\pi}) \exp[-\tfrac{1}{2}(\bar{z}/\sigma_z)^2] + \bar{z}[1 - 2F(-\bar{z}/\sigma_z)] \quad [2.188a]$$

with the definition:

$$F(a) = (1/\sqrt{2\pi}) \int_{-\infty}^{a} \exp(-t^2/2)\, dt \quad [2.188b]$$

The variance follows as:

$$\sigma_x^2 = \bar{z}^2 + \sigma_z^2 - [\sigma_z\sqrt{2/\pi} \exp\{-0.5(\bar{z}/\sigma_z)^2\} + \bar{z}\{1 - 2F(-\bar{z}/\sigma_z)\}]^2 \quad [2.188c]$$

Elandt (1961) has defined an incomplete normal moment for $\sigma_z = 1$:

$$I_r(a) = (1/\sqrt{2\pi}) \int_{a}^{\infty} z^r \exp(-z^2/2)\, dz \quad [2.189]$$

for $r \geq 1$. For $r = 0$ this moment can be expressed as:

$$I_0(a) = 1 - F(a) \quad [2.189a]$$

Consequently, for $r > 1$ we can now write (see Elandt, 1961):

$$I_r(a) = (a^{r-1}/\sqrt{2\pi}) \exp(-a^2/2) + (r-1)I_{r-2}(a) \quad [2.189b]$$

The general formula for the non-central moments is then given by:

$$\mu_r(x) = \sigma_z^r \sum_{j=0}^{r} \binom{r}{j} \theta_z^{r-j}[I_j(-\theta_z) + (-1)^{r-j} I_j(\theta_z)] \quad [2.190]$$

with θ_z of [2.187b]. This leads to the following moments:

$$\bar{x} = \sigma_z\sqrt{2\pi} \exp(-\theta_z^2/2) - \bar{z}[1 - 2I_0(-\theta_z)] \quad [2.190a]$$

(Note the difference in the sign for the second term between [2.188a] and [2.190a]).

$$\sigma_x^2 = \bar{z}^2 + \sigma_z^2 - \bar{x}^2 \quad [2.190b]$$

$$\nu_3(x) = 2[\bar{x}^3 - \bar{z}^2\bar{x} - (\sigma_z^3/\sqrt{2\pi})\exp(-\theta_z^2/2)] \qquad [2.190c]$$

$$\nu_4(x) = (\bar{z}^4 + 6\bar{z}^2\sigma_z^2 + 3\sigma_z^4) + (8\bar{x}\sigma_z^3/\sqrt{2\pi})\exp(-\theta^2/2) +$$
$$+ 2(\bar{z}^2 - 3\sigma_z^2)\bar{x}^2 - 3\bar{x}^4$$

$$\mu_2(x) = \bar{z}^2 + \sigma_z^2 \qquad [2.190d]$$

$$\mu_3(x) = (\bar{z}^2 + 2\sigma_z^2)\bar{x} - \bar{z}\sigma_z^2[1 - 2F(\theta_z)] \qquad [2.190e]$$

$$\mu_4(x) = \bar{z}^4 + 6\bar{z}^2\sigma_z^2 + 3\sigma_z^4 \qquad [2.190f]$$

For the half normal distribution where the normal distribution is folded at the median, the density function simplifies to:

$$f(x) = \sigma_z^{-1}\sqrt{2/\pi}\exp[-x^2/(2\sigma_z^2)] \qquad [2.191]$$

and the moments (with $\theta_z = 0$) can be stated as:

$$\bar{x}_h = \sigma_z\sqrt{2/\pi} = 0.7979\,\sigma_z \qquad [2.191a]$$

$$\sigma_h^2 = \sigma_z^2(1 - 2/\pi) = 0.3634\,\sigma_z^2 \qquad [2.191b]$$

$$\nu_{3,h} = \sigma_z^3(4/\pi - 1)(\sqrt{2/\pi}) = 0.2180\,\sigma_z^3 \qquad [2.191c]$$

$$\nu_{4,h} = \sigma_z^4(3 - 4/\pi - 12/\pi^2) = 0.5109\,\sigma_z^4 \qquad [2.191d]$$

$$\bar{x}/\sigma_x = 1.3237 \qquad [2.191e]$$

Inspection of [2.187] and [2.191] reveals that estimators are needed for $\bar{z}$ and σ_z^2. Leone et al. (1961) have suggested a moments fit, based on the customary estimators x_m and s^2 for $\bar{x}$ and σ_x^2. As readily can be checked we derive from [2.188a and c] the following:

$$\sigma_z^2 = (\bar{x}^2 + \sigma_x^2)/(1 + \theta_z^2) \qquad [2.192a]$$

with θ_z as defined (see [2.187b]).
Division of [188c] by the squared value of [188a] provides:

$$A_x = \theta_x^2/(1 + \theta_x^2) \qquad [2.192b]$$

and: $G(\theta_z) = A_x = (C_1 + \theta_z C_2)^2/(1 + \theta_z^2) \qquad [2.192c]$

where: $C_1 = \sqrt{2\pi} \cdot \exp(-\theta_z^2/2) \qquad [2.192d]$

$C_2 = 1 - 2F(-\theta_z)$ and $\theta_x = \bar{x}/\sigma_x \qquad [2.192e]$

Equation [2.192c] relates θ_x with θ_z, or for $\sigma_z = 1$, θ_x with $\bar{z}$. It can be solved for $\theta_z = \bar{z}/\sigma_z$, in units of σ_z. These two ratios have been tabulated by Leone et al. (1961). E.g. $\bar{z}/\sigma_z = c_1$, $\sigma_x/\sigma_z = c_2$. Then $\sigma_z = \sigma_x/c_2$ and $\bar{z} = c_1 \cdot \sigma_z$.

Elandt (1961) has defined $G(\theta_z)$ in slightly different terms. It becomes:

$$G(\theta_z) = (c_1 + \theta_z c_3)^2/(1 + \theta_z^2) \qquad [2.192f]$$

with $c_3 = 1 - 2I_0(-\theta)$.

A simpler formula has been derived by Elandt employing the fourth moment, namely:

$$B_x = \mu_4(x)/\mu_2^2(x) = (3 + 6\theta_z^2 + \theta_z^4)/(1 + \theta_z^2)^2 = h(\theta_z) \qquad [2.193a]$$

A direct estimate can, therefore, be obtained for θ_z from:

$$(1 + \hat{\theta}_z^2)^2 B_x - (3 + 6\theta_z^2 + \theta_z^4) = 0 \qquad [2.193b]$$

This is a quadratic equation in $\theta_z^2 = t$, with explicit solution:

$$t = [(3 - B_x) \pm \sqrt{2(3 - B_x)}]/(B_x - 1) = \theta_z^2 \qquad [2.193c]$$

Immediately the limiting condition $B_x < 3$ can be recognized. Equation [2.192a] can now be utilized to find an estimator for σ_z^2, and $\bar{z} = \theta_z \cdot \sigma_z$ provides the estimator for $\bar{z}$.

A maximum-likelihood solution has been described by Leone et al. (1961). The following version has been derived by Johnson (1962):

$$\partial L/\partial \theta = -N\hat{\theta}_z + \hat{\sigma}_z^{-1} \sum_{i=1}^{N} x_i \tanh(\hat{\theta}_z x_i/\sigma_z) = 0 \qquad [2.194a]$$

and: $\partial L/\partial \sigma = -N/\hat{\sigma}_z + \sigma_z^{-3} \Sigma x_i^2 - \hat{\theta}_z \sigma_z^{-2} \sum_{1}^{N} \tanh(\hat{\theta}_z x_i/\sigma_z) = 0$ [2.194b]

where tanh is the hyperbolic tangent. The combination of both equations leads to:

$$\hat{\sigma}_z^2 (1 + \hat{\theta}_z^2) = \sum_{1}^{N} x_i^2/N = m_2 \qquad [2.194c]$$

The calculations for finding the maximum-likelihood estimators are very complex and in most cases too tedious to be of practical value. Hence Johnson has recommended an initial approximation by replacing:

$$\tanh(\hat{\theta}_z x_i/\hat{\sigma}_z) = \hat{\theta}_z x_i/\sigma_z - (\hat{\theta}_z x_i/\sigma_z)^3/3 \qquad [2.194d]$$

After some arithmetic the solution for $\hat{\theta}^2$ is:

$$\hat{\theta}_z^2 \sim m_2 \sqrt{3/m_4} - 1 \sim \sqrt{3/B_x} - 1 \qquad [2.194e]$$

with the substitutions $\mu_2(x) = m_2$ and $\mu_4(x) = m_4$. This equation can be compared with the solution from the moments fit [2.194c]. Johnson (1962) has elaborated on the error variance for θ_z and σ_z^2. Unfortunately, the error variance is only reliable for $\theta_z < 0.5$ and $\theta_z > 3$. In the first case $\bar{z} < 0.5\,\sigma_z$ (table by Leone et al., 1961) which means the folding occurs close to the mean. The second case is of little practical interest because folding at the 3-σ point produces virtually a Gaussian distribution with some minute changes at one margin. The error variances are therefore omitted here.

CHAPTER 3

CURVE FITTING

3.1 GENERAL

3.1.1 Introduction

Basic methods of curve fitting have been introduced by Essenwanger (1974a, section 4). It may be restated that in the process of curve fitting a mathematical expression (other than a frequency distribution) is sought for a series of observations. The two major tools are the polynomial representation and the Fourier analysis. The basic background described by Essenwanger (1974a) will not be repeated here. However, some additional remarks are in order.

Although polynomial representation and Fourier analysis are widespread in the literature, these functions are not the only forms of a mathematical representation. Any suitable mathematical expression can be employed, and the constants for the function can be obtained by the various techniques of curve fitting. The least-square fit is usually the simplest one. The polynomial representations have some advantages because the theoretical background has been given extensive treatment in the mathematical and statistical literature.

In many applications of polynomial representation the first decision is to answer the question whether an orthogonal system or a non-orthogonal system should be selected. Let us write the observations y_{ij} as:

$$y_{ij} = a_0 + \sum_{j=1}^{n} a_i x_{ij} \qquad [3.1a]$$

where the x_{ij} is a polynomial (e.g. $x_{i1} = X_i$, $x_{i2} = X_i^2$ ...), and i stands for the number of the observation, j for the number of the polynomial term, $j < i$. The a_0 is a general constant. If the x_{ij} are not independent the constants $a_j (j = 0, 1 \ldots n)$ must be recomputed for every chosen j-value. The orthogonal system:

$$y_{ij} = b_0 + \sum_{1}^{n} b_j \phi_{ij} \qquad [3.1b]$$

has the advantage that the coefficients must be calculated only once, because the independence is the purpose of the orthogonal system.

Another benefit of the orthogonal system is the direct relationship with the "left variance" or "percentage reduction". The left variance ϵ_L^2 must be found from:

$$\epsilon_L^2 = \Sigma(y_{ij} - Y_i)^2/N \qquad [3.2]$$

where the Y_i denotes the observation, y_{ij} the analytical curve from the mathematical expressions on the right-hand side of either [3.1a] or [3.1b]. This left variance can be expressed in explicit form of any individual orthogonal polynomial term, because the coefficients do not change (see above). If we define:

$$\sigma_{\phi_j}^2 = \sum_{i=1}^{N} \phi_{ji}^2/N \qquad [3.2a]$$

then the left variance is:

$$\epsilon_L^2 = \sigma^2 - \sum_{j=1}^{n} b_j^2 \sigma_{\phi_j}^2 \qquad [3.2b]$$

or:

$$PR = (1 - \epsilon_L^2/\sigma^2) \cdot 100 \qquad [3.2c]$$

is the percentage reduction. For more details see Essenwanger (1974a, section 4).

One disadvantage of the orthogonal system lies in the finding of functions which fulfill:

$$\sum_{i=1}^{N} \phi_{ik}\phi_{ij} \begin{cases} = 0 \text{ for } k \neq j \\ \neq 0 \text{ for } k = j \end{cases} \qquad [3.3]$$

This fact is not really a handicap because various functions are readily available in the literature such as the Tchebycheff or the Legendre polynomials (see e.g. Pearson and Hartley, 1958; Beyer, 1966; Abramowitz and Stegun, 1964, etc.).

A further drawback of some orthogonal polynomials is the fact that interpolation between given points or extrapolation often cannot readily be performed, while the x variate in the series of [3.1a] can be replaced by any arbitrary numerical value, even extrapolations can be quickly made. The basic problem of inter- and extrapolation has nothing to do with the question whether inter- or extrapolation is a valid procedure in a particular case. This decision of significance or legality is an additional problem which must be answered, too.

It can be summarized that the utilization of [3.1a] or [3.1b] (non-orthogonal or orthogonal system) is entirely a matter of choice depending on analysis goals, the functional relationship of data, even the availability of computational tools, etc., and no a-priori recommendation can be given before the purpose of polynomial representation is known.

One more element may be added. In general, it is silently assumed that data are observed at equal space or time intervals. In this case many tools of curve fitting are applicable. Non-equally spaced observations always create problems unless their spacing follows some systematic pattern, e.g. a logarithmic or a quadratic spacing.

Two avenues for fitting non-equally spaced observations can be taken, either the data are converted or the functions are adjusted. The latter method is often very cumbersome although adjustment can be made for continuous variates. Many times appropriate weighting functions must be introduced to compensate for clustering, and this leads to elaborate and complex fitting tools.

More economically, although usually connected with some loss in accuracy, is the procedure of data interpolation. In many instances a linear interpolation may suffice but techniques based on higher-order terms are available; some of them are treated by Essenwanger (1974b).

The topic of the aliasing effects on power spectrum, etc. had been discussed by Essenwanger (1974a), where equal spacing was assumed (see also Section 3.2.3). It is self-evident that aliasing does exist for unequal spacing (e.g. see Jones, 1970, 1971 and 1972 or Shapiro and Silverman, 1960).

3.1.2 Tchebycheff polynomials

Individual types of polynomials have certain properties peculiar to them, and they are usually selected because of their particular behavior. It cannot be the purpose of this volume to present all individual features of various polynomial systems. However, a few comments on some of the systems in statistical analysis appear appropriate.

It is well known that Taylor's formula in $(x-c)^n$ gives a good approximation to $f(x)$ for x close to the selected value c. The approximation by the Tchebycheff series is uniformly good for any x.

The Tchebycheff (orthogonal) system is based on the goal to minimize the maximum error rather than the sum of the squared deviations as in a least-square fitting. When the task requires that a maximum error never exceeds a certain threshold, the Tchebycheff approximation is the solution. Although tables for Tchebycheff polynomials have been referenced earlier, the basic definition is:

$\phi_0(x) = 1$ [3.4a]

$\phi_n(x) = \cos(n \arccos x)$ with $n = 1, 2 \ldots$ [3.4b]

and:

$\phi_{-n}(x) = \phi_n(x)$ [3.4c]

The interval of x is given by $-1 \leq x \leq 1$. With the notation:

$\theta = \arccos x$ [3.4d]

we find the inverse:

$$x = \cos \theta \qquad [3.4e]$$

A recurrence formula can be developed, namely:

$$\phi_{n+1}(x) - 2x\phi_n(x) + \phi_{n-1}(x) = 0 \qquad [3.4f]$$

for $n \geq 1$, e.g. we could determine ϕ_2 from $\phi_1 = x$ and $\phi_0 = 1$:

$$\phi_2(x) = 2x^2 - 1 \qquad [3.4g]$$

$$\phi_3(x) = 2x(x^2 - 1) - x = 2x^3 - 3x, \text{ etc.} \qquad [3.4h]$$

The polynomials are orthogonal. We find:

$$\sum_{-1}^{1} \omega(x)\phi_j(x)\phi_k(x) = C_\phi = \begin{cases} 0 \text{ for } j \neq k \\ \pi/2 \text{ for } j = k \neq 0 \\ \pi \text{ for } j = k = 0 \end{cases} \qquad [3.4i]$$

The "weight function" is $\omega(x) = 1/(1-x)^{\frac{1}{2}}$.

The Tchebycheff polynomials are also orthogonal for a discrete set of points. By the substitution $x_p = \cos(\pi p/N)$ with $p = 0, 1, ..., N-1$ we can write:

$$\sum_{p=0}^{N-1} \phi_j(x_p)\phi_k(x_p) = \begin{cases} 0 \text{ for } j \neq k \\ N/2 \text{ for } j = k \neq 0 \\ N \text{ for } j = k = 0 \end{cases} \qquad [3.4j]$$

In order to obtain whole numbers for the set of $\phi_j(x_p)$ polynomials ($X_p = 1 ... n$), a formula by Yates can be employed (see Fisher and Yates, 1963). It is customary to replace $\phi(x)$ by $\xi(x)$, namely:

$$\xi_{r+1} = \lambda_r \left[\xi \xi_r - \frac{r^2(n^2 - r^2)}{4(4r^2 - 1)} \xi_{r-1} \right] \qquad [3.5]$$

with:

$$\xi = \tfrac{1}{2}(n - x) \qquad [3.5a]$$

and $x = 1, 3, 5 ... (2n - 1)$. The n denotes the number of points. The λ_r is any multiplier, but specifically the factor which is necessary to reduce the set of ξ_{n+1} to the set of smallest whole numbers. In details, we find:

$$\xi_0 = 1 \qquad [3.5b]$$

$$\xi_1 = \xi \qquad [3.5c]$$

$$\xi_2 = \lambda_2 \left[\xi^2 - \frac{n^2 - 1}{4 \cdot 3} \right] \qquad [3.5d]$$

$$\xi_3 = \lambda_3 \left[\xi^3 - \frac{3n^2 - 7}{20} \right] \tag{3.5e}$$

$$\xi_4 = \lambda_4 \left[\xi^4 - \frac{6n^2 - 26}{4 \cdot 7} \xi^2 + \frac{3(n^2 - 9)(n^2 - 1)}{4 \cdot 140} \right] \tag{3.5f}$$

$$\xi_5 = \lambda_5 \left[\xi^5 - \frac{5(n^2 - 7)\xi^3}{18} + \frac{15n^4 - 230n^2 + 407}{1008} \xi \right] \tag{3.5g}$$

$$\xi_6 = \lambda_6 \left[\xi^6 - \frac{5(3n^2 - 31)}{4 \cdot 11} \xi^4 + \frac{5n^4 - 110n^2 + 329}{4 \cdot 44} \xi^2 - \frac{10(n^2 - 1)(n^2 - 9)(n^2 - 25)}{4 \cdot 7437} \right] \tag{3.5h}$$

The coefficients of the system:

$$z_{ij} = b_0 + \sum_{j=1}^{n} b_j \phi_{ij} \tag{3.1c}$$

would then be computed by:

$$b_j = \sum_{i=1}^{n} Y_i \phi_{ij} / \phi_j^2 \tag{3.1d}$$

where $\phi_j^2 = \sum_{i=1}^{n} \phi_{ij}^2$ and Y_i stands for the observation. In the ξ-notation this would read:

$$b_j = \Sigma Y_i \xi_{ij} / \xi_j^2 \tag{3.1e}$$

and:

$$\xi_j^2 = \sum_i \xi_{ij}^2$$

3.1.3 Legendre polynomials

Another important series of orthogonal polynomials over the interval $-1 \le x \le 1$ is the Legendre system. It can be shown that the series of Legendre polynomials follows Rodrigues' formula:

$$P_n(x) = \frac{1}{2^n n!} \frac{d^n}{dx^n} (x^2 - 1)^n \tag{3.6}$$

This leads to:

$$P_0(x) = 1 \tag{3.7a}$$

$$P_1(x) = x \qquad [3.7b]$$
$$P_2(x) = (3x^2 - 1)/2 \qquad [3.7c]$$
$$P_3(x) = (5x^3 - 3x)/2 \qquad [3.7d]$$
$$P_4(x) = (35x^4 - 30x^2 + 3)/8 \qquad [3.7e]$$
$$P_5(x) = (63x^5 - 70x^3 + 15x)/8 \qquad [3.7f]$$

It is evident that it becomes more and more difficult to derive the higher-order polynomials from Rodrigues' formula. A recursion law is therefore used:

$$nP_n(x) = (2n-1)xP_{n-1}(x) - (n-1)P_{n-2}(x) \quad \text{for} \quad n \geq 2 \qquad [3.6a]$$

The Legendre polynomials are orthogonal, i.e.

$$\int_{-1}^{+1} P_h P_k \, dx = \begin{cases} 2/(2n+1) & \text{for} \quad h = k = n \\ 0 & \text{for} \quad h \neq k \end{cases} \qquad [3.6b]$$

Any function $Y(z)$ would be represented by Legendre polynomials with the transformation $y(x) = Y(z)$. Then:

$$y(x) = \sum_{n=0}^{\infty} a_n P_n(x) \qquad [3.8]$$

The coefficients must be determined from:

$$a_n = [(2n+1)/2] \int_{-1}^{+1} y(x) P_n(x) \, dx \qquad [3.8a]$$

and here begins the difficulty in practical work with discrete points. If $Y(z)$ is a function which can be expressed in analytical terms, and the integral can be solved explicitly, the representation of any function by Legendre polynomials is trivial. Such examples can be found in almost any text on mathematics or numerical analysis where polynomials are covered. In the atmospheric sciences we are mostly interested, however, to express a discrete function $Y(z)$ by polynomials. While the coefficients for the Tchebycheff series are simple to calculate even in this case, the usual procedure of replacing the integral by the summation sign is insufficient for a small number of points, i.e. we cannot merely state:

$$a_n = [(2n+1)/2] \sum_{x=-1}^{1} y(x) P_n(x) \Delta x \qquad [3.8b]$$

This replacement would be a permissible approximation for a large number of points, say probably for about 30 or more and the number of terms $n \ll 30$. For a small number, i.e. seven, this formula generally does not provide the coefficients a_n accurate enough to be of value.

GENERAL

We may evaluate the success of engaging [3.8b] by calculating two polynomial characteristics, the variance Var_{Pn} and an integral, which we may call S_{Ln}. The two parameters are defined by:

$$\text{Var}_{Pn} = \int_{-1}^{+1} P_n^2(x)\,dx = 2/(2n+1) \qquad [3.9a]$$

and:

$$S_{Ln} = \int_0^{+1} P_n(x)\,dx = \sum_{v=0}^{n/2} (-1)^v \frac{1 \cdot 3 \cdot 5 \ldots (2n-2v-1)}{2^v v!(n-2v)!} \frac{1}{(n-2v+1)}$$

$$= 0 \text{ for even } n \neq 0 \qquad [3.9b]$$

e.g. the true value for S_{L5} would be $S_{L5} = \frac{21}{24} - \frac{35}{24} + \frac{15}{24} = \frac{1}{16}$. It is easy to prove that $S_{L1} = \frac{1}{2}, S_{L2} = 0, S_{L3} = \frac{1}{8}, S_{L4} = 0$, etc. Against these expected values the empirical counterparts can be obtained:

$$\text{Var}'_{Pn} = \sum_{i=1}^{s} P_n^2(x_i)\,\Delta x \to 2/(2n+1) \qquad [3.9c]$$
$$s \to \infty$$

The summation:

$$S'_{Ln} = \sum_{x_i=0}^{x_i=1} P_n(x_i)\,\Delta x \qquad [3.9d]$$

is somewhat more difficult to calculate due to considerations in the marginal class intervals. If the two border points $x_i = 0$ and $x_i = 1$ are utilized, the $P_n(x_i = 0)$ and $P_n(x_i = 1)$ must be multiplied by $\Delta x/2$. Otherwise, $\sum_{x_i=0}^{x_i=1} \Delta x = 1$ is not fulfilled.

Before examples for $\text{Var}_{P'_n}$ and $S_{L'_n}$ can be given, the transformation procedure from z to x must be discussed. It is self-explanatory that the two ranges must equal, i.e. $x_r = z_r$. Therefore, the transformation equation provides:

$$x/x_r = (z-z_0)/z_r \qquad [3.10]$$

The z_0 is a reference parameter corresponding to $x = 0$. Since most of the observed discrete variables can be arranged in steps of class intervals, two versions of the transformation must be accommodated. Let us assume that seven points $Y(z_i)$ are given. We number the variate z from $z_1 = 1$ through $z_7 = 7$ (with unity steps). If other scales are given, they can be reduced to this basic form (see later Table 3.1). The transformation in this case can be written (with $x_r = 2$ and $z_r = z_7 - z_1 = 6$):

$$x/2 = (z-4)/6 \qquad [3.11a]$$

or: $3x + 4 = z \qquad [3.11b]$

We shall call this Version 1.

If we consider $z_1 = 1$, with a lower-class boundary of $z_{11} = 0.5$ and the upper boundary of z_u as $z_{ru} = 7.5$, the $z_r = z_{7u} - z_{11} = 7$, and:

$$x/2 = (z-4)/7 \quad [3.11c]$$

or: $3.5x + 4 = z \quad [3.11d]$

This may be called Version 2. The resulting Legendre polynomials for these two interpretations are given in Table 3.1. The respective Var'_P and S'_L parameters are listed in Table 3.2 for four different number of points.

TABLE 3.1

Legendre polynomial terms for seven discrete points

z	P_0	Version 1, $\Delta x = 2/6$				
		$x = P_1$	P_2	P_3	P_4	P_5
1	1	-1.0	1.0	-1.0	1.0	-1.0
2	1	-0.667	0.167	0.259	-0.427	0.306
3	1	-0.333	-0.333	0.407	0.012	-0.333
4	1	0	-0.500	0	0.375	0
5	1	0.333	-0.333	-0.407	0.012	0.333
6	1	0.667	0.167	-0.259	-0.427	-0.306
7	1	1.0	1.0	1.0	1.0	1.0

z	P_0	Version 2, $\Delta x = 2/7$ (x at midpoint of class)				
		$x = P_1$	P_2	P_3	P_4	P_5
1	1	-0.857	0.602	-0.289	-0.019	0.260
2	1	-0.571	-0.010	0.391	-0.383	0.081
3	1	-0.286	-0.378	0.370	0.098	-0.347
4	1	0	-0.500	0	0.375	0
5	1	0.286	-0.378	-0.370	0.098	0.347
6	1	0.571	-0.010	-0.391	-0.383	-0.081
7	1	0.857	0.602	0.289	-0.019	-0.260

It is self-evident that the expected Var_P and S_L are best approximated for the largest subdivision, namely 31 points. Version 2 renders a slightly better approximation than Version 1. The deviation increases with ascending polynomial order. These findings can be applied directly to the utilization of [3.8b]. In other words, at least about thirty points are needed to calculate the coefficients accurately enough by mere summation. The determination of the integral by numerical methods will be treated below. Some basic differences between Legendre and Tchebycheff polynomials will be pointed out next.

The Legendre polynomials provide the least-square solution to curve fitting while the Tchebycheff polynomials produce the minimum deviation of the

GENERAL

TABLE 3.2

Summation of [3.9c] and [3.9d]. See text

	V'_{Pn}					S'_L				
	P_1	P_2	P_3	P_4	P_5	P_1	P_2	P_3	P_4	P_5
True value for integral	2/3 0.667	2/5 0.400	2/7 0.286	2/9 0.222	2/11 0.182	1/2 0.500	0 0	−1/8 −0.125	0 0	1/16 0.0625
Version 2										
7 pionts	0.653	0.360	0.213	0.130	0.111	0.490	−0.010	−0.135	−0.033	0.0016
11 points	0.661	0.384	0.254	0.175	0.124	0.496	−0.004	−0.129	−0.013	0.0371
21 points	0.665	0.395	0.277	0.208	0.161	0.499	−0.001	−0.126	−0.007	0.0554
31 points	0.666	0.398	0.282	0.215	0.172	0.499	−0.0005	−0.126	−0.001	0.0593
Version 1										
7 points	0.704	0.509	0.489	0.502	0.470	0.500	0.028	−0.056	0.091	0.176
11 points	0.680	0.440	0.363	0.343	0.340	0.500	0.01	−0.100	0.033	0.1052
21 points	0.670	0.410	0.306	0.254	0.229	0.500	0.003	−0.119	0.008	0.0734
31 points	0.668	0.404	0.295	0.237	0.203	0.500	0.001	−0.112	0.004	0.0673

maximum departure. This can be rephrased. After Legendre and Tchebycheff polynomials of the same order have been fitted, the maximum difference between the analytical and observed curve would be equal or less for the reconstructed curve from the Tchebycheff series. If we sum the squared deviation of the reconstructed curve from the observed $y(x)$ the Legendre series would generally show a smaller value but in no case more than the same value. This result is to be expected from the theoretical point of view.

In practical work the curve fitting procedure for the Legendre polynomial series may not always provide the least squared sum of deviations because the calculation of the coefficients is an approximation of an integral for a discrete function, and the approximation error is reflected in the sum of the squares. Furthermore, the discrete points of the Legendre polynomials are not fully orthogonal.

The coefficients of the Tchebycheff series are procured by a straight forward solution, namely

$$A_{Tch} = [\Sigma y(x) \cdot \phi_n(x)] / \Sigma \phi_n^2(x) \qquad [3.12]$$

As pointed out, the determination of the coefficients of the Legendre series depends on the answer of the question how accurate can the integral $\int_{-1}^{+1} y(x) P_n(x) \, dx$ be obtained in practical work. This problem is treated now.

3.1.3a *Determination of the constants.* For discrete observations $Y(z)$ the integral in [3.8a] does not have an analytical solution, and the trivial conversion to the summation as is common practice in statistical analysis is not accurate enough. Some numerical methods must be engaged.

A class of approximation formulae can be stated as outlined by Hamming and Pinkham (1966), namely:

$$\int_{x_1=-1}^{x_s=1} f(x_i)\,dx = \Sigma\, w_i f(x_i)\Delta x + C\Delta x \qquad [3.13]$$

where w_i denotes a specific weighting factor and C designates a correction term. We abbreviate $f(x_1 = -1)$ by f_1, $f(x_i)$ by f_i, and $f(x_s = 1)$ by f_s. Then the approximation (with symmetric weights) can be written:

$$\Sigma\, w_i f(x_i) = af_1 + bf_2 + cf_3 + bf_4 + \ldots + cf_{s-2} + bf_{s-1} + af_s \qquad [3.13a]$$

The weights are independent of s. As shown by Hamming and Pinkham (1966):

$$b + c = 2 \qquad [3.13b]$$

$$2a - c = 0 \qquad [3.13c]$$

Only two equations are given, which leaves the constant a open for arbitrary definition. If $a = 1/2$, then $b = c = 1$, and we recognize Gregory's formula. For $a = 1/3$, the $c = 2$ and $b = 4$, and Simpson's composite rule is found.

The correction term is based on the derivatives of the $f(x_i)$ or in short the $\Delta^r f(x)$, where the r stands for the order of the differences. Thus $\Delta^3 f(x)$ means the third-order differences (e.g. $\Delta_1 = x_1 - x_2$, $\Delta_1^2 = \Delta_1 - \Delta_2$, $\Delta_1^3 = \Delta_1^2 - \Delta_2^2$ etc.; the exponent signifies the order and not the power).

The correction term can be expressed as:

$$C = \sum_{r=1}^{\infty} q_r [\Delta^r f_{s-r} + (-1)^r \Delta^r f_s] \qquad [3.13d]$$

In practical work the summation is limited to s terms or less, since s is the number of points for which the observations of $Y(z)$ are given. Hence, the coefficient is only good up to the order $s - 1$ or s (Simpson). This restriction limits the series to $s - 1$ or s terms. Since in our problem:

$$f(x) = y(x) P_n(x) \qquad [3.13e]$$

the a_s for a third-order curve $y(x)$ and 7 given points could not be expected to render the proper numerical value (see Examples 3.1 and 3.2). The q_r in [3.13d] are coefficients of t in a power series expansion:

$$\sum_{r=1}^{\infty} q_r t^r = -\left[\frac{1}{\ln(1-t)} + a + \frac{b(1-t)}{t(2-t)} + \frac{c(1-t)^2}{t(2-t)}\right] \qquad [3.13f]^*$$

As developed by Hamming (1971), after some lengthy arithmetic one finds:

---

* The power series for $\ln(1-t) = -(t + t^2/2 + t^3/3, \ldots)$.

GENERAL 205

$$q_r = g_r + (\tfrac{1}{2} - a)\frac{1}{2^r} \qquad [3.13g]$$

where: $\sum_{r=1}^{\infty} g_r t^r = -\dfrac{t}{12} - \dfrac{t^2}{24} - \dfrac{19}{720}t^3 - \dfrac{9}{480}t^4 - \dfrac{863}{60480}t^5$

$$-\frac{275}{24192}t^6 - \frac{33953}{3628800}t^7, \quad \text{etc.} \qquad [3.13h]$$

These are the coefficients of the Gregory correction (setting $t = 1$). Simpson's formula is based on $a = 1/3$. Consequently, the first two terms (of t and t^2) disappear and the q_r series becomes:

0, 0, $-4/720$, $-4/480$, $-548/60480$, $-212/24192$, $-29228/24192$

This means, the correction starts with the third-order differences, and $C_{\text{Simp}} < C_{\text{Greg}}$.

For $a = 0$ the correction terms provide the series of q_r:

1/6, 1/12, 13/360, 1/80, 41/30240, $-43/12096$, $-9889/1814400$ etc.

More details for practical work follow in Example 3.1.

It is self-evident that more sophisticated numerical methods can be employed. The reader is referred to the literature.

3.1.3b *An Iterative Method.* Although the previously introduced methods by Gregory or Simpson are very simple to put into practice, the truncation (e.g. after the seventh term such as in [3.13h]) and the limitation of the order of $y(x)$ as dictated by the number of points are sources of error for any approximation of empirical data. It should be repeated that the order of $f(x)$ is not identical with $y(x)$ or $P_n(x)$ but is the composition of the order of $y(x)$ plus $P_n(x)$. Thus for an assumed sixth order $y(x)$ the a_5 requires at least eleven points for a sensible solution. The Gaussian quadrature, although accurate to $2s - 1$ degrees, requires special tables (e.g. Abramowitz and Stegun, 1964) and special spacings of the $y(X_i)$. Since the observations if not specially designed are hardly available at these X_i-values, the $y(x_i)$ must be interpolated at $y(X_i)$. This introduces interpolation errors, and the coefficients may not be any better than those found by other methods.

A trivial iteration method may be successful, since in our case we deal with an orthogonal system, for which:

$$\int_{-1}^{1} P_k P_h \, dx = 0 \quad \text{for } k \neq h \qquad [3.6b]$$

This orthogonality condition is rigorously fulfilled for the discrete points of the Tchebycheff polynomials. The orthogonality condition is completely fulfilled for the *continuous* Legendre polynomials. This means that if we were to find:

$$\sum_{s=1}^{7} P_k(x_s) P_h(x_s) = c \qquad [3.6c]$$

as given in Table 3.1, the c should be zero for $h \neq k$ and V'_{Pn} for $k = h$. It is illustrated in Table 3.2 that this condition is not exactly met. (E.g. the reader can readily check that the c is not exactly zero for the seven-point polynomials given in Table 3.1.) Consequently, the a_n cannot be precisely determined by plain summation, and the subtraction of the term $a_k P_k(x_i)$ from $y(x)$ leaves a remainder of the kth order in $y(x_i)$ which can be utilized to iterate the coefficient.

Generally we would not need an iteration for a_0 and a_1. We should remember, however, that:

$$a_0 = \tfrac{1}{2} \int y(x) \, dx \qquad [3.14a]$$

$$a_1 = \tfrac{3}{2} \int y(x) x \, dx \qquad [3.14b]$$

The substitution of the integral by the summation sign does not always give the correct value of the coefficient. In this case Simpson's method may be superior to Gregory's procedure. Since the order of $f(x)$ is identical with $y(x)$, these coefficients can generally be determined well enough for higher powers of $y(x)$.

Now let us assume that:

$$a_{k1} = a_k \pm a_{e1} \qquad [3.15a]$$

The representation of $y(x)$ in Legendre polynomials can be written:

$$y(x) = a_0 P_0 + a_1 P_1 + \ldots a_k P_k + \ldots a_n P_n \qquad [3.15b]$$

and:

$$y(x) - a_k P_k = y_k(x) \qquad [3.15c]$$

Thus we have removed only $a_{k1} P_k$, which leaves:

$$y_k(x) = y(x) - a_{k1} P_k = a_0 P_0 + \ldots a_{e1} P_k + \ldots a_n P_n \qquad [3.15d]$$

This shows that in $y(x)$ the term $a'_k P_k$ has been replaced by $a_{e1} P_k$ instead of being completely removed. Now:

$$\sum_{x=-1}^{x=1} y_k(x) \cdot P_k(x) \neq 0 \qquad [3.15e]$$

Had the term been completely removed, [3.15e] would be zero, since all other terms of $y_k(x)$ would not furnish any contribution because $\int P_k P_h = 0$ for $k \neq h$. This result is independent from the a_j coefficients!

The presently remaining term would provide:

$$a_{e1} \cdot \int P_k P_k \, dx \neq 0$$

or:

GENERAL

$$\Sigma\, y(x) P_k(x)\, \Delta x = [2/(2k+1)](a_{\epsilon 1} \pm a_{\epsilon 2}) \qquad [3.15\mathrm{f}]$$

It is, therefore, possible to determine the coefficient by an iterative procedure provided the series $a_{\epsilon 1}, a_{\epsilon 2}, a_{\epsilon 3} \ldots$ converges. We calculate first:

$$a_{k1} = [(2k+1)/2]\, \Sigma\, y(x) P_k(x)\, \Delta x \qquad [3.16]$$

The next step is the subtraction of $a_{k1} P_k(x)$, i.e.:

$$y_{k1}(x) = y(x) - a_{k1} P_k(x) \qquad [3.16\mathrm{a}]$$

Then we find:

$$a_{k2} = a_{\epsilon 1} = [(2k+1)/2]\, \Sigma\, y_{k1}(x) P_k(x)\, \Delta x \qquad [3.16\mathrm{b}]$$

etc.

Finally: $a_k = \sum_{i=1}^{n_j} a_{ki} \qquad [3.16\mathrm{c}]$

This iteration method can be combined with Gregory's or Simpson's procedure for the first coefficients, e.g. a_0 and a_1 are determined by either one of the methods and $a_2 \ldots a_s$ are found by iteration, because the higher-order terms may not be given accurately enough by Gregory's or Simpson's technique.

The combination would then take the following form. We start with:

$$y_1(x) = y(x) - (a_0 P_0 + a_1 P_1) \qquad [3.17\mathrm{a}]$$

and:

$$a_{21} = (s/2) \cdot \Sigma\, y_1(x) P_2(x)\, \Delta x \qquad [3.17\mathrm{b}]$$

It is advantageous to introduce some weights analogously to the formula [3.13]. This modification can be written:

$$a_{kj} = [(2k+1)/2] \cdot \Sigma\, y_j(x_i) w_i P_k(x_i)\, \Delta x \qquad [3.17\mathrm{c}]$$

This weighting function with $\Sigma\, w_i \Delta x = 1.0$ serves largely to dampen the influence of the marginal values $y_j(x = 1)$ and $y_j(x = -1)$. Similar to Gregory's rule we start with $w_{j1} = 0.5$ and $w_{js} = 0.5$ in version one, while all other $P_k(|x| \neq 1)$ receive the weight $w_j = 1$. For progressive iterations the w_{j1} and w_{js} would be changed to < 0.5. We may find suitable dampening with:

$$w_{j1} = w_{js} = 1/[2 + (n/2 - c)(j-1)] \quad \text{for} \quad n > 2c \qquad [3.17\mathrm{d}]$$

This reduces the influence of the margins. Otherwise the coefficients by iteration may not converge. Divergence in the iteration process can be generated by small remainders a_ϵ which lead to larger deviations at the margins. These in turn enter with unbalanced weights after an a_{k1} has been removed, and the series of a_ϵ diverges rather than converges.

The parameter c of [3.17d] can be taken as a constant, e.g. $c = 1.0$. It can be utilized, however, to derive a set of coefficients a_k with the iteration method which produces the smallest sum of squared deviations from the given $y(x)$. In other words, the parameter c can be varied, and the set of coefficients is adopted which displays the minimum of the summed squares. This solution may not be identical with the expected least-square solution but may come close to it (see Example 3.3). The reason for the dependence of the set of coefficients a_k on c is trivial. The parameter c controls the weights. In Simpson's or Gregory's formula the weights have been exactly determined, while the weights of the margins cannot readily be determined, but can be found by variation of c. Since the Legendre polynomials provide the least-square solution, the adoption of the answer with the minimum sum of squares is in agreement with the properties of the Legendre polynomials. Finally, [3.17d] is not the only way to weight the margins. The reader may prefer different formulae, but [3.17d] is one recommendation which the author has tried.

3.1.3c *Matrix solution and orthogonalization* It was pointed out that the discrete Legendre polynomials comprise a non-orthogonal system. The coefficients of this system should then properly be calculated by procedures for a non-orthogonal system (see Essenwanger, 1974a). This is equivalent to converting the "covariance matrix" (left) into the "coefficient matrix" (right):

$$\begin{bmatrix} \Sigma P_0^2 & \Sigma P_1 P_0 & \Sigma P_2 P_0 & \cdots & \Sigma P_n P_0, & \Sigma P_0 y \\ \Sigma P_0 P_1 & \Sigma P_1^2 & \Sigma P_2 P_1 & \cdots & \Sigma P_n P_1, & \Sigma P_1 y \\ \cdot & \cdot & \cdot & & \cdot & \cdot \\ \cdot & \cdot & \cdot & & \cdot & \cdot \\ \cdot & \cdot & \cdot & & \cdot & \cdot \\ \Sigma P_0 P_n & \Sigma P_1 P_n & \Sigma P_2 P_n & \cdots & \Sigma P_n^2, & \Sigma P_n y \end{bmatrix} \rightarrow \begin{bmatrix} 1.0 & 0 & 0 & \cdots & 0, & A_0 \\ 0 & 1.0 & 0 & \cdots & 0, & A_1 \\ \cdot & \cdot & \cdot & & \cdot & \cdot \\ \cdot & \cdot & \cdot & & \cdot & \cdot \\ \cdot & \cdot & \cdot & & \cdot & \cdot \\ 0 & 0 & 0 & \cdots & 1.0, & A_n \end{bmatrix}$$

[3.8c]

This conversion has been treated by the author (1974a, section 3) and is equivalent with the diagonalization of a matrix (see Section 4.13 and Example 3.10).

This technique does not provide "Legendre coefficients" unless the matrix contains a sufficient number of terms (i.e. orders of P_i), e.g. the following coefficients are obtained for an approximation of $y(x)$ being a third plus fourth-order Tchebycheff polynomial of seven points (see Table 3.8 later). The last column in each version of the subsequent short table is identical with the Legendre coefficients:

GENERAL

Version 1					Version 2				
a_0	a_1	a_2	a_3	a_4	a_0	a_1	a_2	a_3	a_4
0	0	0	–	–	0	0	0	–	–
0	−0.8	0	1.8	–	0	0.20	0	2.86	–
−1.3	−0.8	−6.5	1.8	10.8	0.71	0.20	4.42	2.86	20.01

We learn from this table that the Legendre coefficients are not the most advantageous coefficients for an incomplete system, but the solutions converge with the inclusion of a sufficient number of (or all possible) terms (see also Example 3.4). Some readers may prefer this method of calculating coefficients since it is exact and it certainly proves advantageous once the number of terms in the series has been decided upon. As in any non-orthogonal system, the addition of terms requires a recalculation of coefficients, however, while Legendre coefficients determined as outlined would not necessitate recomputation.

The reader may ask whether the discrete Legendre polynomials could not be orthogonalized. Without doubt, orthogonalization is technically feasible, and the author has produced an orthogonalized set of polynomials for the seven-point Legendre polynomials which were given in Table 3.1. This orthogonalized set is exhibited in Table 3.1a. It must be reported first that Version 1 and Version 2 merged to only one set after this orthogonalization procedure.

TABLE 3.1a

Orthogonalized set of discrete Legendre polynomials of Table 3.1

P_1	P_2	P_3	P_4	P_5
−0.5669	0.5455	−0.4083	0.2417	−0.1092
−0.3780	0	0.4083	−0.5641	0.4363
−0.1890	−0.3273	0.4083	0.0806	−0.5457
0	−0.4365	0	0.5641	0
0.1890	−0.3273	−0.4083	0.0806	0.5457
0.3780	0	−0.4083	−0.5641	−0.4363
0.5669	0.5455	0.4083	0.2417	0.1092

A closer perusal of the orthogonalized set reveals that the columns of Table 3.1a are now identical with the Tchebycheff seven-point polynomials of Table 3.3 except for rounding and a multiplication factor. This conversion to the Tchebycheff polynomials by orthogonalization has been observed for $N > 7$ (calculation by the author went up to $N = 20$). Identity with this Tchebycheff system implies, however, that this orthogonalized set has also assumed the properties of the Tchebycheff polynomials. Consequently, there would be no reason why the Tchebycheff polynomials could not be employed a priori, since the original purpose of utilizing the Legendre series is defeated with the change of properties. It is reiterated that for a small n the discrete Legendre series would not be very advantageous while its application for a larger N (e.g. $N > 30$) should prove useful.

In conclusion, the Legendre series is different from the Tchebycheff series in its theoretical approach to curve fitting. Some difficulty arises when the Legendre series is applied to a discrete function $y(x)$. The procedures introduced to calculate the coefficients by numerical methods are certainly not exhaustive. Some readers may prefer more sophisticated techniques. Nevertheless, the techniques are restricted by the form in which the original observations — i.e. the function $y(x)$ — are given and the methods presented can be readily adapted for electronic computers without complex programming.

Finally, it should be added that Legendre polynomials could also be fitted to unequally spaced observations. Adjustments of weights for this case would be necessary as outlined by [3.13].

As the Legendre polynomials for discrete points become simpler to handle with an increasing number of points, they should prove to be a useful replacement for the Tchebycheff polynomials in problem solutions where the practical application of the Tchebycheff polynomial method apparently shows a weakness.

Example 3.1 Curve-fitting procedures for Legendre polynomials In order to illustrate some of the difficulties in the fitting process of a discrete function by Legendre polynomials the seven-point Tchebycheff fourth-order curve has been selected because the mathematical expression $y(x)$ is known, while for empirical data the very task is to find a mathematical expression. We can, however, determine the correct numerical value of the Legendre coefficients for this ϕ_4-curve by solving the integral and compare the numerical answer for the discrete case. The correct solution is presented first.

Let us write the fourth-order Tchebycheff term ϕ_4 (see [3.5f]) as:

$$\xi_4 = (7\xi^4 - 67\xi^2 + 72)/12$$

where $\xi = (7-z)/2$ with $z = 1, 3, ..., 13$.

The transformation equation from ξ to the x-variate of the Legendre series is (see Table 3.1):

Version 1: $z = 6x + 7$, consequently $\xi = -3x$

Version 2: $z = 7x + 7$, with $\xi = -7x/2$

The coefficients of the polynomial series are according to [3.7a]:

$$a_n = [(2n+1)/2] \int_{-1}^{1} \xi_4 P_n(x)\, dx \qquad [3.7a']$$

GENERAL

The integration shall be illustrated for $P_0 = 1$, where we substitute ξ into [3.7a'] and obtain:

$$a_0 = (1/2) \int_{-1}^{1} \tfrac{1}{2}[7 \cdot (-3)^4 x^4 - 67(-3)^2 x^2 + 72] \, dx$$

The integration renders:

$$a_0 = \tfrac{1}{24}[7 \cdot 81 \cdot x^5/5 - 67.9 x^3/3 + 72x]_{-1}^{1} = -13/10$$

Version 2 furnishes for [3.7a'] the expression:

$$a_0 = \frac{1}{24}\left[7 \cdot \frac{49^2}{4^2} \frac{x^5}{5} - 67 \cdot \frac{49}{4} \frac{x^3}{3} + 72x\right]_{-1}^{1} = 0.709\ldots$$

For convenience to the reader the Tchebycheff polynomials for seven points are listed in Table 3.3. The Legendre polynomials were given in Table 3.1.

TABLE 3.3

Tchebycheff's ϕ_{ji} for seven points

i	ϕ_{0i}	ϕ_{1i}	ϕ_{2i}	ϕ_{3i}	ϕ_{4i}	ϕ_{5i}
1	1	−3	5	−1	3	−1
2	1	−2	0	1	−7	4
3	1	−1	−3	1	1	−5
4	1	0	−4	0	6	0
5	1	1	−3	−1	1	5
6	1	2	0	−1	−7	−4
7	1	3	5	1	3	1

From the respective column ϕ_{4i} of Table 3.3 it can be seen immediately, that the algebraic answer for a_0 by [3.8b] in Version 2 renders:

$$a_0 = \tfrac{1}{2} \sum_{i=1}^{7} P_{ni} \phi_{4i} \Delta x = 0 \quad (\text{for } \Delta x = 2/7)$$

With the substitution of $\Delta x/2 = 1/6$ instead of $\Delta x = 1/3$ for $P_0(x = 1)$ and $P_0(x = -1)$ in Version 1 the coefficient a_0 amounts to:

$$a_0 = \tfrac{1}{2} \sum_{i=1}^{7} P_{0i} \phi_{4i} \Delta x = \tfrac{1}{2}(-3\Delta x) = \tfrac{1}{2}(-3 \cdot \tfrac{1}{3}) = -1/2$$

The first discrepancy is obvious, namely:

$a_0 = 0$ (Version 2) compared with a correct value of 0.71

and: $a_0 = -0.5$ (Version 1) compared with $-13/10$

Since the Tchebycheff polynomials are constructed such that:

$$\lambda \cdot \sum_{i=1}^{n_i} \phi_{ji} = 0$$

no summation for any number of discrete points ϕ_{ji} or ξ_{ji} ([3.5], i.e. without an interpolation) will provide the proper value of the coefficient a_0. Here Simpson's or Gregory's formula excell and will lead to the correct Legendre coefficient.

TABLE 3.4

Coefficients of the Legendre polynomial series for a fourth-order Tchebycheff term with seven points

	True	Version 1					Version 2		
		G	S	iteration only	with a_0	a_0 and a_2 known	true	w/o it	it
a_0	-1.30	-1.30	-1.30	-0.50	-1.30	-1.30	0.71	0	0
a_1	0	0	0	0	0	0	0	0	0
a_2	-6.50	-6.50	-6.53	-7.96	-6.90	-6.50	4.42	0	0
a_3	0	0	0	0	0	0	0	0	0
a_4	10.80	23.5	23.1	10.70	10.68	10.80	20.01	9.90	16.86
a_5	0	0	0	0	0	0	0	0	0

G = Gregory, S = Simpson, it = iteration.

TABLE 3.5

Recomputed fourth-order Tchebycheff polynomial term for seven points for the coefficients as given by Table 3.4

i	y(x)	Version 1					Version 2	
		G	S	only	a_0	$a_0 + a_2$	w/o it	it
1	3	15.2	15.2	2.2	2.5	3.0	-0.2	-0.3
2	-7	-12.2	-12.3	-6.4	-7.0	-7.0	-3.8	-6.6
3	1	1.2	1.2	2.3	1.1	1.0	1.0	1.7
7	6	10.6	10.6	7.5	6.2	6.0	3.7	6.4
5	1	1.2	1.2	2.3	1.1	1.0	1.0	1.7
6	-7	-12.2	-12.3	-6.4	-7.0	-7.0	-3.8	-6.6
7	3	15.2	15.2	2.2	2.5	3.0	-0.2	-0.3

The iteration is based on $c = 1.0$ in [3.17d].

Table 3.4 displays the results for the determination of the first six Legendre polynomial coefficients for the $\phi_4 \equiv \xi_4$ Tchebycheff term by seven different methods. The reconstructed ϕ_{4i} provided by these coefficients is contained in Table 3.5. The table of coefficients (Table 3.4) shows that the iteration procedure comes closer to the correct value for the higher-order terms, while Simpson's or Gregory's technique excells for the lower-order terms. Three ways of iteration are displayed, the pure iteration alone, taking a_0 from either Simpson's or Gregory's scheme, and finally starting with a_0 and a_2 from Simpson or Gregory. The latter proves best. In our case due to the fourth order of the Tchebycheff term the iteration procedure alone would have been better than taking the series of coefficients by Gregory's or Simpson's rule.

The values of the coefficients a_0 through a_5 for Version 2 are listed in the right part of Table 3.4. It is obvious that without iteration the coefficient a_4 is not close enough.

Again, as previously observed, the deviations of the empirical from the expected coefficient for a_0 and a_2 are caused by the structure of the Tchebycheff polynomials for seven points. While interpolation (and extrapolation of the margin points) for the given ϕ_4 would provide a better basis to come closer to the expected value for a_0 and a_2, the seven points do not furnish sufficient information to enable a closer approximation. This conclusion confirms what has been pointed out previously to the reader. It may be repeated that the seven point ϕ_4 and the techniques for the determination of the coefficients were chosen for demonstration purposes only. Had the author been confronted only with the problem of finding a Legendre polynomial representation for the seven discrete points ϕ_4, his recommendation would have been either the matrix method or a calculation by Version 1 with the a_0 through a_2 coefficients by Gregory's or Simpson's rule and an iteration of a_3 through a_5, or the (linear) inter- and extrapolation of ϕ_4 and utilization of fourteen-point Legendre polynomials in Version 2. These two procedures lead to a very close representation of the given seven-points.

The reader may ask the question whether the calculated approximate coefficients would provide a good representation of the originally given seven points despite the deviations from the expected values. In other words we ask whether the closed system of Legendre polynomials leads to a tolerable answer.

The limitations imposed on Gregory's or Simpson's system can be readily recognized by inspection of the results as exhibited in Table 3.5. The obvious large deviations by the Gregory or Simpson method are produced by the overestimate of a_4, while evidently the deviation of a_2 from the true value is less critical in this case. While Simpson's procedure in this example is virtually identical with Gregory's method, this fact is not true in general. Although Simpson's formula requires a smaller correction, particular cases exist where Gregory's scheme is better.

Again, the main purpose of this example is the demonstration of the calculation of the coefficients and the delineation of the difficulties involved with the computation of coefficients of the Legendre series from a small number of discrete points. The system coefficients by the matrix method would lead to a better fit than the Legendre coefficients.

Example 3.2 Iteration procedure. An iterative procedure for the determination of the higher-order coefficients has been introduced. This method will now be elucidated more than it was in the previous text or by Example 3.1. The third- and fourth-order terms of the Tchebycheff polynomials as displayed in Table 3.3 have been added together and form the given data for this example. Since an even- and odd-order term has been combined the Legendre polynomial series would consist of non-zero coefficients for a_0 through a_4.

The following Tables 3.6 and 3.7 are self-explanatory. Part a of Table 3.6 exhibits the calculated coefficients by the individual methods, while part b discloses the iteration steps for the individual coefficients.

Part c of Table 3.6 gives the coefficients for Version 2, where the weighting for calculation of the coefficients has been employed as outlined by Gregory or Simpson with corrections by differences. Surprisingly, Version 2 emerges with the closer approximation for Gregory's or Simpson's rule than Version 1 (see Table 3.7). Again, it should be cautioned that Version 2 is not necessarily always the better procedure.

As expected, Table 3.7 discloses that the combination of methods renders the best approximation. The largest deviations exist at the margins.

3.1.4 *Percentage reduction and left variance*

It was previously pointed out that the goal in curve-fitting can also be classified as an attempt to describe the variance of the function y by a mathematical expression. If the match is perfect, the variance σ_y^2 of the given data and of σ_{ya}^2, the analytical counterpart, are identical. We can, therefore, mathematically formulate a criterion of the success in curve-fitting by defining a left variance:

$$\epsilon_L^2 = \Sigma\,(y_i - y_{ai})^2/N. \qquad [3.18]$$

The explained variance is then:

$$\epsilon_E^2 = \sigma_y^2 - \epsilon_L^2 \qquad [3.18a]$$

The measure: $Z_R^2 = \epsilon_E^2/\sigma_y^2 = 1 - \epsilon_L^2/\sigma_y^2$ [3.19a]

can be called reduction, and:

$$Z_{PR}^2 = Z_R^2 \cdot 100\% \qquad [3.19b]$$

is then the percentage reduction.

It was previously mentioned that the left variance can be readily determined in an orthogonal system with the aid of the coefficients and the

TABLE 3.6

Legendre coefficients for a third- and fourth-order Tchebycheff term (seven points) combined

	Version 1						Version 2			
	a. coefficients				b. individual steps in iteration			c. coefficients		
	true	G	S	it 1	it 2	step		true	G	S	it
							a_2 a_3 a_4				
a_0	−1.30	−1.30	−1.30	−1.30	−1.30	1	−2.32 3.08 24.38	0.71	−0.69	−0.69	0.0
a_1	−0.80	−0.80	−0.80	−0.80	−0.80	2	−1.01 −1.47 −18.72	0.20	−0.22	−0.22	0.0
a_2	−6.50	−6.50	−6.53	−5.20	−6.50	3	−1.00 0.18 5.91	4.42	−2.16	−2.18	0.0
a_3	1.80	1.80	1.79	1.80	1.81	4	−0.84 0.01 −0.80	2.86	1.73	1.73	2.8
a_4	10.80	23.5	23.1	11.20	10.80	5	−0.71 — 0.03	20.01	14.73	14.77	16.87
a_5	0	0	0	−0.000	0.003	6	−0.61 — 0.00	0	−0.43	−0.43	−0.19

For G, S, it 1, it 2 see Table 3.7.

TABLE 3.7

Recomputed function from coefficients of Table 3.6 part a

y(x)	Version 1				Version 2		
	G	S	it 1	it 2	G	S	it
2.0	14.2	14.2	3.7	2.0	−2.68	−2.69	−1.17
−6.0	−11.2	−11.3	−6.0	−6.0	−5.53	−5.56	−5.38
2.0	2.2	2.2	1.6	2.0	2.43	2.44	2.76
6.0	10.6	10.6	5.5	6.0	5.91	5.95	6.33
0	0.2	0.2	−0.4	0	0.72	0.74	0.55
−8.0	−13.2	−13.3	−8.0	−8.0	−7.07	−7.08	−7.54
4.0	16.2	16.2	5.7	4.0	−1.84	−1.85	0.55

G = Gregory method, S = Simpson's rule.
it 1 = a_0 and a_1 by Gregory, a_2 through a_5 by iteration.
it 2 = a_0 through a_2 by Gregory, a_3 through a_5 by iteration.

respective variance of the orthogonal terms. Thus, for the Tchebycheff system the variance of the individual term is $\sigma^2_{\phi k}$, hence:

$$z_k^2 = a_k^2 \sigma^2_{\phi k} / \sigma_y^2 \qquad [3.20a]$$

and:
$$Z_R^2 = \sum_1^{n_k} z_k^2 \qquad [3.20b]$$

In the Fourier series:

$$z_k^2 = A_k^2 / (2\sigma_y^2) \qquad [3.20c]$$

and analogously we would expect [3.20a] to hold for the Legendre polynomials.

In order to gain some insight and to understand the difficulties with the Legendre system a digression into some details of the left variance appears beneficial.

Since we are not dealing with the estimation problem the question of dividing by N or $N-1$ for the variance is immaterial, and we can write:

$$N\sigma_y^2 = \Sigma(y - \bar{y})^2 = \Sigma(y - y_a + y_a - \bar{y})^2 \qquad [3.21]$$

or: $\Sigma(y - \bar{y})^2 = \Sigma(y - y_a)^2 + 2\Sigma(y - y_a)(y_a - \bar{y}) + \Sigma(y_a - \bar{y})^2$

$$[3.21a]$$

Again, the y represents the given or observed data, the y_a its mathematical expression. The reader will recognize that the first term on the right-hand side is the left variance.

The last two terms can be combined, and:

$$\Sigma(y_a - \bar{y})(2y - y_a - \bar{y}) = -\Sigma(y_a - \bar{y})^2 + 2\Sigma(y - \bar{y})(y_a - \bar{y}) \quad [3.21b]$$

GENERAL

Since the reference for y_a is a_0, it may be introduced into the above equation. The first term on the right-hand side becomes:

$$-\Sigma(y_a - \bar{y})^2 = -[\Sigma(y_a - a_0)^2 + 2\Sigma(y_a - a_0)(a_0 - \bar{y}) + \Sigma(a_0 - \bar{y})^2] \quad [3.21c]$$

We can now combine the middle term with the last term of [3.21b]:

$$-2\Sigma(y_a - a_0)(a_0 - \bar{y}) + 2\Sigma(y - \bar{y})(y_a - a_0 + a_0 - \bar{y}) =$$
$$2\Sigma(y_a - a_0)(-a_0 + \bar{y} + y - \bar{y}) + 2\Sigma(y - \bar{y})(a_0 - \bar{y}) \quad [3.21d]$$

The last term is zero by definition of $\bar{y}$. Dividing by N, introducing σ_y^2 and ϵ_L^2, we find with the two remaining terms from [3.21c] and the one from [3.21d]:

$$\epsilon_L^2 = \sigma_y^2 + (a_0 - \bar{y})^2 + \Sigma(y_a - a_0)^2 - 2\Sigma(y - a_0)(y_a - a_0) \quad [3.21e]$$

We may further substitute:

$$y_a - a_0 = a_1 P_1 + a_2 P_2 + \ldots a_n P_n \quad [\text{see } 3.8]$$

where P_j denotes the Legendre polynomials. It is self-evident that [3.21e] has validity for any analytical function y_a, and P can be replaced by any variate. We now introduce [3.8] and find:

$$\epsilon_L^2 = \sigma_y^2 + (a_0 - \bar{y})^2 + \Sigma(a_1 P_1 + a_2 P_2 + \ldots a_n P_n)^2 -$$
$$2\Sigma(y - a_0)(a_1 P_1 + \ldots a_n P_n) \quad [3.21f]$$

$$\epsilon_L^2 = \sigma_y^2 + (a_0 - \bar{y})^2 + M_z \quad [3.21g]$$

The last two terms can be expressed in matrix form, where the last term has been split and is given in the first two lines*:

$$M_z = \begin{bmatrix} -2a_1 \Sigma yP_1 & -2a_2 \Sigma yP_2 & \ldots & -2a_n \Sigma yP_n \\ 2a_0 a_1 \Sigma P_1 & 2a_0 a_2 \Sigma P_2 & \ldots & 2a_0 a_2 \Sigma P_2 \\ \hline a_1^2 \Sigma P_1^2 & a_1 a_2 \Sigma P_1 P_2 & \ldots & a_1 a_n \Sigma P_1 P_n \\ a_1 a_2 \Sigma P_1 P_2 & a_2^2 \Sigma P_2^2 & \ldots & a_2 a_n \Sigma P_2 P_n \\ \cdot & \cdot & & \\ \cdot & \cdot & & \\ \cdot & \cdot & & \\ a_1 a_n \Sigma P_1 P_n & a_2 a_n \Sigma P_2 P_n & \ldots & a_n^2 \Sigma P_n^2 \end{bmatrix} \quad [3.22]$$

---

* The first two lines could have been written as to make M_z a complete symmetric matrix. For practical purposes (see Example 3.3) the terms were combined.

It is evident that the second line comprises all terms whose value is zero if the summation sign for the Legendre polynomials is replaced by the integral. For discrete points of Legendre polynomials however, the even-order terms are not zero, i.e. $\int_{-1}^{1} P_2 = 0$ but $\Sigma P_2 \neq 0$ (or $= 0$ for $N \to \infty$). In addition, the $\int_{-1}^{1} P_i P_j = 0$ for $i \neq j$, but $\Sigma P_i P_j$ is not necessarily zero. We have $\Sigma P_i P_j \neq 0$ for $i = 2k$ and $j = 2h$, or $i = 2k - 1$ and $j = 2h - 1$ where $k = 1, ..., h = 1, ...$

Further, by substituting the polynomial expression y, we find:

$$\int y P_j = a_j \int P_j^2 \qquad [3.22a]$$

The orthogonality condition requires that only the $a_j P_j$ term in y would remain. Consequently, for a true orthogonal system we define:

$$M_{z0} = (a_1^2 \Sigma P_1^2 + a_2^2 \Sigma P_2^2 + ... a_n^2 \Sigma P_n^2) \qquad [3.22b]$$

and: $\epsilon_L^2 = \sigma_y^2 + (a_0 - \bar{y})^2 - M_{z0} \qquad [3.22c]$

turns into the well known formula with $M_{z0} \equiv \epsilon_E^2$. The reference term $(a_0 - \bar{y})^2$ becomes zero for $a_0 \equiv \bar{y}$. If M_{z0} did not include this reference term, and $\sigma_y^2 \equiv M_{z0}$ or $\sigma_y^2 \equiv M_z$, then this reference term would remain as a left variance (see [3.22c]).

Consequently M_z or M_{z0} will contain this item whenever a_0 deviates from $\bar{y}$. It would be immaterial to redefine the reduction since we could always subtract the reference term by stating:

$$\epsilon_{L2}^2 = \epsilon_L^2 - (a_0 - \bar{y})^2 \qquad [3.23a]$$

The ϵ_L^2 is the actual left variance, however, and therefore an adjustment could be made, namely:

$$S_y^2 = \sigma_y^2 + (a_0 - \bar{y})^2 \qquad [3.23b]$$

Furthermore: $Z_R^2 = \epsilon_{E2}^2 / S_y^2 \qquad [3.23c]$

with: $\epsilon_{E2}^2 = S_y^2 - \epsilon_L^2 \qquad [3.23d]$

It is self-evident that:

$$\epsilon_{L2}^2 = \sigma_y^2 - M_z \quad \text{and} \quad \epsilon_{L2}^2 < \epsilon_L^2 \qquad [3.23e]$$

Equation [3.22b] implies that the left variance is reduced with every added term (unless $a_j = 0$). This trivial result for an orthogonal system is not applicable to a non- or semi-orthogonal system without qualification.

The outcome of the summation of the elements in M_z cannot be predicted a priori without favorable circumstances such as most terms becoming zero nor can it be expressed as functions of other terms and only a limited number remains such as in the orthogonal case. It is possible that M_z is positive, which increases the left variance (see Example 3.3 or 3.4). Although at the first instance this may appear as a paradox, analysis of the particular case will

GENERAL

confirm the correctness (i.e. the coefficients are not independent in a non-orthogonal system).

In any case where [3.20a] loses its validity, the (percentage) reduction of the individual term must then be calculated from:

$$z_k^2 = Z_{k+1}^2 - Z_k^2 \qquad [3.20d]$$

In the case that M_z is positive, the z_k^2 would become negative. This fact implies for practical work that the series could not be truncated at this point and more terms must be carried until a reduction of the variance takes place. Fortunately, these cases are not too frequent in practical work. For more details see Example 3.3.

Example 3.3 Left variance for Legendre polynomial. Let us assume that the given data are the two terms:

$$y = \phi_3 + \phi_4$$

for the Tchebycheff polynomials for seven points. This example has been selected because the coefficients of the Legendre polynomials series can be calculated by integration. Table 3.8 results.

TABLE 3.8

Data for Example 3.3

i	ϕ_3	ϕ_4	y	Version 1 a_{i-1}	Version 2 a_{i-1}
1	−1	3	2	− 1.3	0.709
2	1	−7	−6	− 0.8	− 0.204
3	1	1	2	− 6.5	4.424
4	0	6	6	+ 1.8	2.858
5	−1	1	0	10.8	20.008
6	−1	−7	−8	0	0
7	1	3	4		

The $y_a = a_0 + a_1 P_1 + a_2 P_2 + a_3 P_3 + a_4 P_4$ with the coefficients of Table 3.8, and is identical with the data. Thus we have a perfect match. Now $\bar{y} = 0$ and:

$$\sigma_y^2 = 160/7 - 22.857$$

where $\sigma_{\phi_3}^2 = 6/7$ and $\sigma_{\phi_4}^2 = 22.0$. Hence $z_3^2 = 3.8\%$ and $z_4^2 = 96.2\%$ for the Tchebycheff series.

The matrix M_z (Version 1) becomes for the six coefficients a_0 through a_5:

$$M_z = \begin{bmatrix} 0 & 0 & -1.71 & -44.00 & 0 \\ 0 & 2.82 & 0 & -6.20 & 0 \\ 0.28 & 0.0 & -0.28 & 0.0 & 0 \\ 0.0 & 15.26 & 0 & -16.67 & 0 \\ -0.28 & 0.0 & 1.14 & 0 & 0 \\ 0.0 & -16.67 & 0 & 41.76 & 0 \\ 0.0 & 0.0 & 0 & 0 & 0 \end{bmatrix} \begin{matrix} \text{line sum} \\ -45.71 \\ -3.38 \\ 0.0 \\ -1.41 \\ 0.86 \\ 25.00 \\ \underline{0.0} \\ -24.55 \end{matrix}$$

The summation of all elements of $M_z = -24.55$. Consequently:

$$\epsilon_L^2 = 22.86 + 1.69 - 24.55 = 0$$

is confirmed.

The left variance of the individual term sequence is:

$$\epsilon_{0L}^2 = \Sigma(y-a_0)^2/N = 24.55 = 22.86 + 1.69$$
$$\epsilon_{1L}^2 = \Sigma(y-a_0-a_1P_1)^2/N = 24.83 = 24.55 + 0.28$$
$$\epsilon_{2L}^2 = \Sigma(y-a_0-a_1P_1-a_2P_2)^2 = 42.91 = 24.83 + 15.26 + 2.82$$
$$\epsilon_{3L}^2 = \epsilon_L^2 = 0$$

The sequence of left variance is therefore:

Term	0	1	2	3	4	5
Version 1	24.55	24.83	42.91	41.76	0	0
Version 2	23.36	23.37	26.83	25.96	0	0

In Version 2 (matrix not given here) we find:

$$\sigma_y^2 + (a_0 - \bar{y})^2 = 22.86 + 0.50 = 23.36$$

It should be added that in our example a third- and fourth-order term comprise the data of y. It is, therefore, no surprise that the first terms of the Legendre series do not contribute to the reduction of the left variance (see Table 3.9).

The positive reduction begins in both versions with the third-order term. Although the actual percentage contributions of the third- and fourth-order terms are not completely identical with the numbers from the Tchebycheff system, the important features run parallel; namely a small contribution from a third-order term and a considerable dominance of the fourth-order term. It

GENERAL

TABLE 3.9

Percentage reduction

a. Cumulative value

		Term					
		0	1	2	3	4	5
Version 1	with σ_y^2	−7.4	−8.6	−87.7	−82.7	100%	100%
	with S_y^2	0.0	−1.2	−74.8	−70.1	100%	100%
Version 2	with σ_y^2	−2.2	−2.3	−17.4	−13.6	100%	100%
	with S_y^2	0.0	−0.1	−14.9	−11.1	100%	100%

b. Individual terms

		0	1	2	3	4	5
Version 1	with σ_y^2	−7.4	−1.2	−79.1	5.0	182.7	0
	with S_y^2	0.0	−1.2	−73.6	4.7	170.1	0
Version 2	with σ_y^2	−2.2	−0.1	−15.1	3.8	113.6	0
	with S_y^2	0.0	−0.1	−14.8	3.8	111.1	0

may be further concluded that the representation by the coefficients a_0 through a_2 is inadequate. In fact, the assumption of zero for these three coefficients above would leave a smaller left variance than the actual value. In practical work this would imply that we could not truncate the series after the second term. Even the third term would not contribute a sufficient amount to bring the reduction out from the negative, i.e. we still have added to the variance instead of reducing it.

In practical work one may very seldom encounter a hypothetical case with no first- and second-order contribution (e.g. see Example 3.4). It should be kept in mind, however, that the seven-point Legendre polynominal system is not truly orthogonal in the discrete case. Consequently, we find paradoxes such as the increase of the left variance, which may be puzzling in the first instance, but proves to be legal under the total system integration.

Although the negative contribution to the reduction may be an unfamiliar thought, it has its legal place in the data analysis and interpretation. Assume as given in Example 3.3 the fourth-order term contributes 183% of the variance. It would imply, however, that this fourth-order term alone would not be a good representation. Since the second-order term provides −79%, the combination of the two terms should render a good representation and a close approximation. This can be confirmed by the little table below. It is obvious that neither coefficient alone would accomplish a good fit and a 183% reduction alone is as bad as the −79%. The final left variance can be calculated from the last column as $17.83/7 = 2.55$. The same answer is provided by the matrix $\mathbf{M}_z$, where we consider $a_0 = a_1 = a_3 = 0$. Then:

a_2P_2	a_4P_4	y_a	y	$(y_a-y)^2$
−6.5	10.8	4.3	2.0	5.29
−1.1	−4.6	−5.7	−6.0	0.09
2.2	0.1	2.3	2.0	0.09
3.2	4.1	7.3	6.0	1.69
2.2	0.1	2.3	0.0	5.29
−1.1	−4.6	−5.7	−8.0	5.29
−6.5	10.8	4.3	4.0	0.09

$$\begin{bmatrix} 0 & 0 & 0 & -44.0 \\ 0 & 0 & 0 & 0 \\ 0 & 0 & 0 & 0 \\ 0 & 15.26 & 0 & -16.67 \\ 0 & 0 & 0 & 0 \\ 0 & -16.67 & 0 & 41.67 \end{bmatrix} = M_z = 20.32$$

$\epsilon_L^2 = 22.86 + 0 - 20.32 = 2.54$

Note that $a_0 = \bar{y}$.

Example 3.4 Tchebycheff and Legendre polynomials. Examples 3.1 through 3.3 dealt mostly with theoretical aspects of curve fitting. It is now time to demonstrate the approximation of empirical data by Legendre and Tchebycheff polynomials. Two wind profiles at 2 km altitude level intervals were arbitrarily selected, January 1, 1957 and 1958 at Montgomery. The following Table 3.10 exhibits the empirical data and the approximation by polynomials up to the fifth-order. Since the correct coefficients for Legendre polynomials cannot be determined a priori, the effect of the approximation cannot be directly shown. It may be inferred, however, that the reconstructed curve from the Legendre polynomials should have a smaller sum of the squared deviations from the analytical data than for the Tchebycheff approximation. As can be readily checked, however, both sums are about the same. This may be seen as a confirmation of an earlier conclusion that for less than about twenty points the advantage of the Legendre series over the Tchebycheff series may not show up in practical work.

As an added feature, the percentage reduction is displayed. No problems are apparent for the Tchebycheff series, whereas the third-order term in Version 1 is negative which demonstrates a slight increase of the left variance for both dates. The percentage reductions for the individual terms have been calculated by [3.20d]. While S_y^2 is the basis for the reduction in Version 1, the $a_0 \equiv \bar{y}$ in Version 2 and $S_y^2 = \sigma_y^2$.

Although differences in the percentage reduction between the three systems

TABLE 3.10

Comparison of Tchebycheff and Legendre polynomials representation of a wind profile

	1 Jan. 57							1 Jan. 58					
$y(x)$	Tche	Legendre		matrix			$y(x)$	Tche	Legendre		matrix		
m/sec		V1	V2	V1	V2		m/sec		V1	V2	V1	V2	

a. Recomputed wind profile (Montgomery)

6	5.7	3.1	6.1	5.8	5.8		5	3.7	0.7	6.1	3.7	3.7
12	13.2	12.5	12.7	13.2	13.2		14	18.3	18.9	17.8	18.3	18.3
23	20.8	20.3	20.4	20.8	20.8		23	21.0	21.9	20.2	21.0	21.0
28	30.1	29.5	30.1	30.1	30.1		31	24.7	24.6	24.3	24.7	24.7
41	39.5	38.9	39.8	39.5	39.5		27	32.7	31.8	32.7	32.7	32.7
44	45.7	45.5	46.2	45.7	45.7		40	42.5	41.3	42.7	42.5	42.5
47	45.7	45.8	46.2	45.7	45.7		51	48.4	47.6	48.9	48.4	48.4
38	37.8	38.2	38.4	37.8	37.8		46	45.5	45.4	46.2	45.5	45.5
23	23.9	24.1	24.4	23.9	23.9		34	32.2	32.8	33.4	32.3	32.3
11	10.5	10.4	10.4	10.5	10.5		12	14.8	14.8	15.2	14.8	14.8
11	11.1	11.0	8.5	11.1	11.1		10	9.1	6.7	6.1	9.1	9.1

b. Coefficients

a_0	25.82	27.21	25.82	27.52	25.90			26.64	28.51	26.64	28.82	26.57
a_1	0.43	+ 2.13	2.33	1.44	2.90			0.95	4.35	5.20	4.52	6.14
a_2	− 1.49	− 29.56	− 30.10	− 29.01	− 38.81			− 1.45	− 25.22	− 29.14	− 25.25	− 28.78
a_3	− 0.04	4.21	− 2.04	− 4.18	− 0.28			− 0.18	− 12.69	− 9.81	− 12.75	− 6.27
a_4	0.83	9.40	14.23	9.91	14.50			0.23	0.37	4.01	2.79	4.08
a_5	0.54	6.00	7.78	5.41	8.71			1.10	11.35	15.85	10.94	17.61

c. Percentage reduction for S_y^2 (see [3.23])

1st	0.9	0.9	0.9	0.8	0.9			4.2	4.1	4.2	4.1	4.0
2nd	86.9	82.7	86.9	84.0	86.6			76.2	76.2	76.1	76.5	76.2
3rd	0.3	− 0.8	0.2	− 1.1	0.1			5.7	− 0.1	5.8	0.1	4.9
4th	9.0	13.0	8.9	11.9	9.2			0.7	0.2	0.6	0.8	0.7
5th	2.1	3.0	1.9	3.5	2.3			8.1	13.7	7.3	13.3	9.0
Total	99.2	98.8	98.8	99.1	99.1			94.9	94.1	94.0	94.8	94.8

V1 = Version 1, V2 = Version 2.

exist, the numbers are equivalent and imply the same integrated effect. The second-order term dominates considerably. Besides this second-order term, a fourth-order term contributes to the 1957 data and a fifth-order term for 1 January 1958. The other components may be considered to have minor influence.

3.1.5 Miscellaneous polynomial techniques

Polynomials can be classified either as the discrete type (e.g. Tchebycheff) or continuous type (Legendre). Therefore, the reader should be able with the aid of the presented material to solve problems with other types of polynomials. e.g. Laguerre polynomials have been used by Francis (1972) for numerical models.

One other problem, the aliasing process with equally and unequally spaced observations in curve fitting has been called to the attention of the reader. Consideration of aliasing is especially important in spectral analysis (e.g. see Jones, 1972 or Section 3.2.4).

One more principle in polynomial analysis may be added. Recently Dixon (1969) and Dixon et al. (1972) have elaborated on a global analysis of meteorological data. Dixon (1969) has recommended orthogonal functions to be fitted to multidimensional data rather than non-orthogonal systems. As pointed out by the author in the preceding section, an orthogonal system may sometimes be very advantageous. The Legendre polynomials served as a typical example of deficiences of a non-orthogonal system. Although orthogonal for the integral form, this property is not rigidly fulfilled when discrete steps are taken.

The idea of employing orthogonal polynomials for multi-dimensional problems is not new. Malone (1956) has earlier attempted to represent climatological maps by polynomials, and Bryson and Kuhn (1956) and Essenwanger et al. (1958) have dealt with half-hemispheric 500 mbar map description. Cehak (1962) has later given a survey of data representation by orthogonal polynomials and Dixon et al. (1972) have recently expanded on these discussions. Dixon's scheme goes back to [3.1b], which is expanded for vectors. Let us assume that we have the problem of finding z in a two-dimensional system with the variates x and y up to the second-order. Then all possible combinations are:

$$z = a_0 + a_1 x + a_2 y + a_3 x^2 + a_4 xy + a_5 y^2 \qquad [3.24a]$$

In orthogonal terms this would be converted to:

$$z = a_0 \phi_0 + a_1 \phi_1 + a_2 \phi_2 + a_3 \phi_3 + a_4 \phi_4 + a_5 \phi_5 \qquad [3.24b]$$

The ϕ_j denotes orthogonal polynomials which are functions of the variables x and y. The coefficients of the orthogonal system are then simply:

GENERAL

$$a_j = \Sigma f(z) \phi_{jz}/\phi_j^2 \qquad [3.24c]$$

where $f(z)$ stands for the element such as the pressure or the geopotential height, temperature, etc. The ϕ_j is then the summation of $\Sigma \phi_{ji}$ and the ϕ_j^2 is symbolically written for $\Sigma \phi_{ji}^2$ (see Section 3.9.2).

The determination of the ϕ_j-function is not trivial but the problem can be solved by various techniques. Two-dimensional Tchebycheff polynomials have been utilized by Bryson and Kuhn (1956) and Essenwanger et al. (1958). Furthermore, it is possible to establish a set of orthogonal vectors by the Gram–Schmidt process which is described later in Chapter 4 (matrix-algebra).

Let us assume we have three variates x, y, z up to the second order. The necessary vector terms are then:

$1, x, y, z, x^2, y^2, z^2, xy, xz, yz$

The orthogonal set could then be stated as:

$$\phi_0 = 1 \qquad [3.25a]$$
$$\phi_1 = x - (\phi_0 x/\phi_0^2)\phi_0 \qquad [3.25b]$$
$$\phi_2 = y - (\phi_1 y/\phi_1^2)\phi_1 - (\phi_0 y/\phi_0^2)\phi_0 \qquad [3.25c]$$
$$\phi_3 = z - (\phi_2 z/\phi_2^2)\phi_2 - (\phi_1 z/\phi_1^2)\phi_1 - (\phi_0 z/\phi_0^2)\phi_0 \qquad [3.25d]$$

etc.

The solution of this system is outlined in Chapter 4. An example of a description of climatological fields by orthogonal polynomials (Tchebycheff) has also been given by Shih Yung-nien (1965). Another method by Kutzbach and Wahl (1965) for the representation of the 500 mbar data (Northern Hemisphere) is later delineated in Section 3.3.5.

3.2 SPECTRAL ANALYSIS

The basic techniques in power-spectrum analysis, Fourier series and periodogram analysis have been presented in Essenwanger (1974a, sections 4.2 and 4.3). The knowledge of these fundamentals is assumed. Neither the equations nor the examples shall be repeated here, and the reader is referred to the standard statistical literature or Essenwanger (1974a). Some additional topics deserve treatment, however, as modifications of the ordinary concepts of harmonic waves and waves with constant amplitudes may be necessary in atmospheric science. Thus, the first section familiarizes the reader with the case where the amplitude is modified by a trend. Later, the case of non-harmonic waves is introduced. Some problems in estimation of some spectral characteristics round out the discussion with some reference to power-spectrum analysis in the literature.

3.2.1 Power-spectrum of $x \cdot \sin \alpha$

Previously we have discussed the power-spectrum and Fourier analysis when the data comprise pure cycles. We learned that the power-spectrum is able to reflect a cycle in the line spectrum with less computational efforts than the harmonic analysis. It was further pointed out that the treatment of the trend (whether to remove it or not) may influence the result, but a decision whether to eliminate a trend or not cannot always be made a priori. Certain cycles with long period lengths may be lost in the power-spectrum or be combined in the zero-class of the line spectrum depending on the chosen maximum lag of analysis. Conversely, the harmonic analysis may require too many unnecessary components to be computed. Both methods have in common that they are somewhat sensitive to the length of the basic period, be it the number of (observed) data or period of the first cycle in the harmonic analysis or the number of lags in the power-spectrum. These are not the only problems which make the power-spectrum (and harmonic analysis) a tool which should be considered with caution.

It is well known that the harmonic analysis and power-spectrum can be computed for any function, whether the observed frequency constitutes a cyclic function, pure sine wave or any other type of function. The first case (Example 3.5) treats an analysis of the function $x \cdot \sin \alpha$.

We assume:

$$f(x) = xA \sin(\alpha_x + \alpha_0) \qquad [3.26]$$

with $A = 5.0$, $p = 30$, $x = (X_i - 15)/10$ and $X_i = 0, 1 ..., p$ and $\alpha_x = X_i \cdot 2\pi/p$. The α_0 has been set zero for simplification. It would not alter the result. In order to assure continuity with the previous examples (Essenwanger, 1974a), the total length of the specified data by [3.26] was extended to 240 points (i.e. $8p$). This leads to a proper computation of the autocorrelation function

SPECTRAL ANALYSIS

TABLE 3.11

Results of Fourier analysis and power-spectrum for two given conditions (see text, Examples 3.5 and 3.6)

j	Fourier				Power-spectrum			
	length	A	Z_A^2	ΣZ_A^2	L_j	ΣL_j	Ls_j	given
a. Single amplitude								
0					−0.014	0.141	0.141	
1	30	1.176	0.302	0.302	0.324	0.310	0.303	100%
2	15	1.609	0.565	0.367	0.570	0.880	0.398	
3	10	0.615	0.082	0.949	0.067	0.947	0.174	
4	7.5	0.336	0.025	0.973	0.027	0.974	0.032	
5	6.0	0.217	0.010	0.984	0.008	0.982	0.012	
6	5.0	0.155	0.005	0.989	0.008	0.990	0.007	
7	4.28	0.119	0.003	0.992	0.002	0.992	0.003	
8	3.75	0.096	0.002	0.994	0.003	0.995	0.002	
9	3.33	0.081	0.001	0.995	0.000	0.996	0.001	
10	3.00	0.070	0.001	0.996	0.002	0.996	0.001	
11	2.73	0.063	0.001	0.997	0.000	0.997	0.001	
12	2.50	0.058	0.001	0.998	0.001	0.998	0.001	
13	2.31	0.055	0.001	0.999	0.000	0.999	0.001	
14	2.14	0.053	0.000	0.999	0.001	1.000	0.001	
15	2.00	0.026	0.000	1.000	0.000	1.000	0.001	
b. Two amplitudes, short x								
0					−0.013	−0.013	0.295	
1	30	4.324	0.646	0.646	0.656	0.064	0.364	50%
2	15	1.047	0.038	0.684	0.056	0.800	0.232	50%
3	10	2.560	0.227	0.911	0.219	0.918	0.140	
4	7.5	1.169	0.047	0.958	0.042	0.960	0.077	
5	6.0	0.709	0.018	0.976	0.016	0.976	0.020	
6	5.0	0.492	0.008	0.984	0.008	0.984	0.009	
7	4.28	0.370	0.005	0.989	0.005	0.988	0.005	
8	3.75	0.296	0.003	0.992	0.003	0.992	0.003	
9	3.33	0.247	0.002	0.994	0.002	0.994	0.002	
10	3.00	0.214	0.002	0.996	0.002	0.996	0.002	
11	2.73	0.191	0.001	0.997	0.001	0.997	0.001	
12	2.50	0.176	0.001	0.998	0.001	0.998	0.001	
13	2.31	0.166	0.001	0.999	0.001	0.999	0.001	
14	2.14	1.160	0.000	1.000	0.001	1.000	0.001	
15	2.00	0.079	0.000	1.000	0.000	1.000	0.001	

up to lag 30. The line spectrum for the data was then computed up to lag 15 as exhibited in Table 3.11. The Fourier analysis as seen in Table 3.11 was processed with a basic period 30 to make it compatible with the power spectrum.*

---

* All examples 3.5–3.7 have no trend.

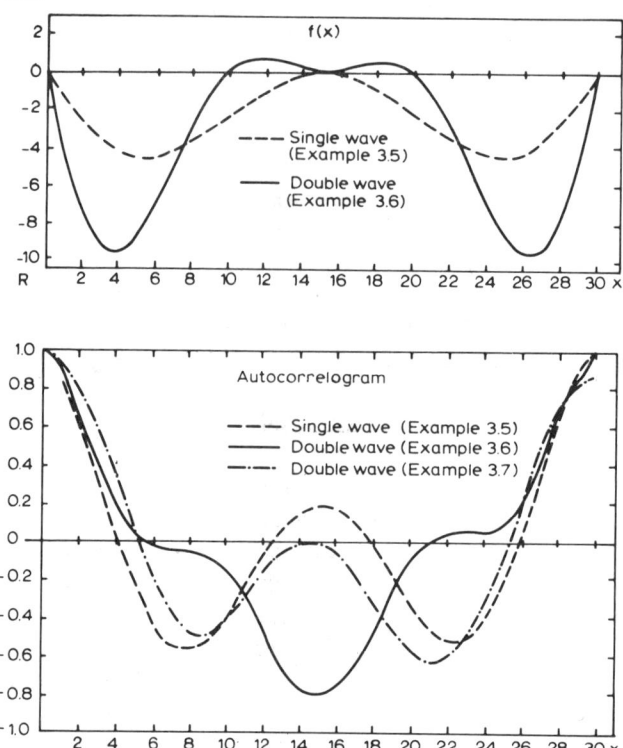

Fig. 3.1. Correlogram for $x \cdot \sin \alpha$ data.

We learn from Table 3.11a and Fig. 3.1 that the results of the Fourier analysis and the power spectrum agree in principle. Both display a peak at a period of 15 points, while the actual 30-point period which is the postulation for the data emerges with only 30% contribution. Evidently the distortion has been caused by the multiplication factor x.

In this particular example we know the assumptions which produced the line spectrum. Thus, an interpretation of the effect was easy. We should keep in mind, however, that in most cases the original conditions are not known. Interpretations of the power spectrum (and the Fourier analysis) must therefore be made with caution and awareness that composition of functions leads also to a power spectrum though the power spectrum is not a true image of the original waves and the real functions involved.

It can be readily explained that a 15-point cycle emerges. Visual inspection of Fig. 3.1 reveals that the function $f(x)$ is negative for all x. The power spectrum is based on the deviations from the mean value A_0. This creates a double wave for representation, also visible in the autocorrelation function. Therefore a 15-point cycle with the maximum amplitude in the line spectrum is generated. The effect cannot be compensated in the spectrum (see column

L_j). Should we try a smoothing of the original line spectrum the peak would shift to even shorter wave lengths.

Examples 3.6 and 3.7 cover the power spectrum and Fourier analysis for two waves with factor multiplications. For simplification we have assumed that the two waves are harmonic waves:

$$f(x) = x[A_1 \sin(\alpha_x + \beta_1) + A_2 \sin(2\alpha_x + \beta_2)] \qquad [3.26a]$$

The phase angles β_1 and β_2 have again been set zero for simplification, also $A_1 = A_2 = 5.0$. The basic period was chosen as $p = 30$ to permit comparison with the previous case. The first multiplier was the same as previously, $x = (X_i - 15)/10$; $X_i = 0, 1, \ldots p - 1$; $\alpha_x = X_i \cdot 2\pi/p$. The total length of the data was once more 8 cycles. The result is exhibited in Fig. 3.1 and Table 3.11b as the second curve (*Example 3.6*). We recognize the double wave pattern in the top part and in the autocorrelation function. The harmonic analysis and the power-spectrum for the basic period of 30 (maximum lag 15) again agree well, although both outcomes do not conform with the assumptions. In this particular case the spectra display a cycle of 30 and 10 points with predominance of the 30-point period length. It may be reiterated that we can interprete the harmonic analysis or power-spectrum as a mere attempt to approximate a given frequency curve in terms of the Fourier series. Since the Fourier series in its fundamental form has no room for a variable amplitude, the findings of Table 3.11b must be anticipated.

The last demonstration (*Example 3.7*) is based on the same $f(x)$ as in the second case except that now the x varies by a long-term periodicity, $x_i = (X_i - 120)/10$, with $X_i = 0, 1 \ldots 239$ and α_i still the short term $\alpha_i = X_i' 2\pi/p$ with $X_i' = 0, 1 \ldots 29$. The full cycle would then be $8p = 240$ points. The wave has not been drawn in the top part of Fig. 3.1 since we would need to extend the graph to 240 points. We would have a similar pattern as in the first case, superimposed by a second harmonic, the amplitude decreasing up to 120 points length, a symmetry point at 120 and increasing to the end (x becomes positive).

The result of the analysis is disclosed in Table 3.12a. We consider the power spectrum first. As displayed, with lag 15 of the autocorrelation function and a basic period of 30 points the 30- and 15-point cycles become visible as the major components. The approximation resembles almost the postulated pattern. It can therefore be concluded that the true length of a cycle could be discovered in the power-spectrum when the superimposed modification (in our case the multiplication by x) possesses a proportionate larger dimension than the cycles involved. In Example 3.6 the x-factor had the same basic length as the longest cycle (30), while in this Example 3.7 the x-factor is assumed to be eight times longer than the basic period for the power-spectrum.

Table 3.12b exhibits the answers from the Fourier analysis. Several lengths of the basic period were engaged, e.g. the original series of 240 observations were shortened to $p = 120, 80$, etc. by:

TABLE 3.12

Power-spectrum and Fourier analysis for two harmonic waves, long period factor x

a. Power-spectrum

j	length	L_j	ΣL_j	Ls_j	Given
0		0.059	0.059	0.251	
1	30	0.470	0.535	0.378	50%
2	15	0.465	1.000	0.362	50%
3	10	0.007	etc.	0.107	
4	7.5	−0.017		−0.006	
5	6.0	0.007		0.000	
6	5.0	−0.001		etc.	
7	etc.	0.004			
8		−0.002			
9		−0.002			
10		−0.001			
11		0.002			
12		−0.001			
13		0.001			
14		−0.001			
15		0.001			

b. Fourier analysis

j	240 points		120 points		80 points		60 points		30 points	
	length	Z_j^2	length	Z_j^2	length	Z_j^2	length	Z_j^2	length	Z_j^2
1	240	0.022	120	0.094	80	0.069	60	0.556	30.0	0.646
2	120	0.023	60	0.133	40	0.203	30	0.133	15.0	0.038
3	80	0.027	40	0.315	26.7	0.230	20	0.018	10.0	0.227
4	60	0.033	30	0.032	20.0	0.003	15	0.008	7.5	0.047
5	48	0.045	24	0.036	16.0	0.344	12	0.191	6.0	0.017
6	40	0.078	20	0.004	13.3	0.114	10	0.046	5.0	0.008
7	34.3	0.226	17.1	0.103	11.4	0.019	8.6	0.019	4.3	0.005
8	30.0	0.008	15.0	0.002	10.0	0.007	7.5	0.010	3.8	0.003
9	26.7	0.088	13.3	0.176	8.9	0.003	6.7	0.001	3.3	0.002
10	24.0	0.009	12.0	0.046	8.0	0.002	6.0	0.003	3.0	0.002
11	21.8	0.000	10.9	0.020	7.3	0.001	5.5	0.002	2.7	0.001
12	20.0	0.001	10.0	0.011	6.7	0.001	5.0	0.002	2.5	0.001
13	18.5	0.007	9.2	0.007	6.2	0.001	4.6	0.001	2.3	0.001
14	17.1	0.026	8.6	0.005	5.7	0.000	4.3	0.001	2.1	0.001
15	16.0	0.132	8.0	0.003	5.3	0.000	4.0	0.001	2.0	0.000
16	15.0	0.000	7.5	0.002						
17	14.1	0.166								
18	13.3	0.044								
19	12.6	0.020								
20	12.0	0.011								

SPECTRAL ANALYSIS

$$X_i' = (x_i + x_{i+p})/a$$

with $a = 2, 3, 4, ..., 8$.

We learn from Table 3.12b that the Fourier analysis reflects a multiple wave pattern, although we would never be able to pin the cycles down as it has been achieved with the power-spectrum. This can be readily explained. In the Fourier analysis the phase angle of a cycle plays an important role. Had we analyzed the first 120 and second 120 points separately, etc. leaving the symmetry point at 120 as a separation boundary, we could have obtained a different answer. The 120th point is the location where the phase switch occurs (from $-X$ to $+X$). Had we employed the periodogram analysis with a plot of the phase diagram (i.e. plotting the harmonic terms as vectors with components of the individual subdivision, see later Example 3.8) it would have led to the discovery of this phase shift. The power-spectrum is insensitive in this case to the phase shift which occurs at the center of the observations, because this shift comprises only a fraction of the total data. Therefore, the two cycles are correctly disclosed by the power-spectrum analysis.

In summary we learn from the examples that a variety of tools need to be employed to enable the proper interpretation of the observational data which have been made. If power-spectrum and Fourier analysis are both calculated for the same data, one will not always find agreement in their spectra. The dissimilarity is then an effect of a varying phase angle in the data caused, e.g., by a multiplication factor. The changing phase angle can be readily discovered in a periodogram analysis. Assumptions of a multiplier can be verified by trial and error.

Example 3.8 Periodogram analysis of the data of Example 3.7

As previously mentioned the Fourier analysis would not enable us to discover the true cycles and the possible function for $x \sin \alpha$ (Example 3.7). How would the periodogram analysis fare? This question is answered by the illustration in Table 3.13, which provides the calculations for the periodogram.

In this case the total material of $N = 240$ points was divided into subsections of length p (from $p = 10$ to $p = 40$) and the amplitude and phase angle for every individual subsection was computed. Then the scalar mean of the amplitude was obtained by:

$$As_p = \left(\sum_1^n A_i \right) / n$$

with n to be determined from $n = N/p$ (integer).

The vector mean was computed as summed over the components of the Fourier series:

$$Av_p = \left[\left(\sum_1^n a_i \right)^2 + \left(\sum_1^n b_i \right)^2 \right]^{-0.5} / n$$

The percentage contribution of the amplitude to the periodogram is:

$$Z_p^2 = (Av_p)^2 \bigg/ \sum_{10}^{40} (Av_p)^2$$

This would amount to the percentage reduction for an infinite number of components and proper spacing of the periods (harmonics). The reader may refer for details to Stumpff (1937) and other texts. For our purpose it is unessential that Z_p^2 is not identical with the percentage reduction.

TABLE 3.13a

Periodogram analysis of [3.26a], long term x

p	As_p	Av_p	Z_p^2	q	p	As_p	Av_p	Z_p^2	q
10	19.6	2.56	0.001	0.130	25	29.4	3.89	0.003	0.132
11	26.8	0.84	0.000	0.031	26	26.6	9.35	0.018	0.351
12	30.6	5.21	0.006	0.170	27	28.0	21.05	0.090	0.753
13	29.8	3.06	0.002	0.102	28	28.1	23.44	0.$\overline{112}$	0.833
14	32.4	14.67	0.$\overline{044}$	0.452	29	26.7	14.97	0.$\overline{046}$	0.559
15	32.7	1.04	0.$\overline{000}$	0.032	30	30.5	4.32	0.004	0.141
16	34.5	17.76	0.065	0.514	31	29.3	13.54	0.037	0.462
17	33.2	7.81	0.$\overline{012}$	0.234	32	28.9	22.71	0.105	0.786
18	29.7	1.10	0.000	0.037	33	27.4	25.88	0.$\overline{126}$	0.907
19	31.6	4.89	0.005	0.154	34	30.4	24.67	0.$\overline{124}$	0.812
20	34.7	1.58	0.000	0.045	35	27.5	20.87	0.089	0.759
21	30.9	3.51	0.003	0.113	36	28.1	10.83	0.024	0.385
22	30.6	4.98	0.005	0.162	37	26.9	3.38	0.002	0.125
23	28.6	3.88	0.003	0.135	38	25.7	6.17	0.008	0.239
24	31.4	4.62	0.004	0.147	39	24.4	10.04	0.020	0.411
					40			0.038	0.489

p = period length, As_p = scalar mean amplitude, Av_p = vector mean amplitude, $Z_p^2 = (Av_p)^2 \sum_{10}^{40} (Av_p)^2$; $q = Av_p/As_p$

We learn from inspecting the vector amplitudes in Table 3.13a that the result of the periodogram analysis is in agreement with the Fourier analysis of 240 points (Table 3.12b) with the only exception that the amplitudes are entered in sequence of p instead of harmonic waves. The length of the basic period p was defined to increase by one point. The true cycles 30 and 15 are again indicated only by two peak amplitudes (14 and 16 for 15, and 28 and 33 for 30) equivalent to the Fourier analysis. Since in the Fourier series a 33- or 28-point wave is not a harmonic of the basic period the maxima occur at 34.3 and 26.7 cycle length which is as close to the periodogram wave length as could be expected. The double maxima at 14 and 16 points appear as cycles of similar length in both analysis tools.

If we now study the average scalar amplitudes As_p of the periodogram we observe a crest at $p = 16$ and 30 with another maximum at $p = 20$ and other

SPECTRAL ANALYSIS

TABLE 3.13b

Vector diagrams

	15 points		28 points		30 points		33 points	
	a_i	b_i	a_i	b_i	a_i	b_i	a_i	b_i
1	−57.6	23.6	−48.8	− 0.2	−52.5	−4.3	−50.4	−22.8
2	−47.4	−21.5	−32.6	12.4	−37.5	−4.3	−19.2	−36.8
3	−42.6	17.1	−15.3	18.5	−22.5	−4.3	− 0.1	−18.4
4	−32.4	−15.0	− 2.1	14.1	− 7.5	−4.3	− 1.1	− 0.7
5	−27.6	10.7	1.1	0.3	7.5	−4.3	−12.7	1.3
6	−17.4	− 8.6	− 9.0	−14.2	22.5	−4.3	−23.9	− 7.8
7	−12.6	4.3	−29.5	−17.3	37.5	−4.3	−25.1	−28.0
8	− 2.4	− 2.2	−50.9	− 1.2	52.5	−4.3		
9	2.4	− 2.2						
10	12.6	4.3						
11	17.4	− 8.6						
12	27.6	10.7						
13	32.4	−15.0						
14	42.6	17.1						
15	47.4	−21.5						
16	57.6	23.6						

fluctuations. A closer perusal of the periodogram analysis reveals the true reason for the split of the maximum in the vector components, namely an extinction of the waves by a phase shift. This phase shift can be clearly discovered from Fig. 3.2 where the vector diagrams consist of the summation of the individual amplitudes of the individual vectors, with length of amplitude A and phase angle β. True periodicities progress along a straight line. Quasi-periodicity should render a quasi-progression along a line and random motions lead to irregular patterns. The significance of the total vector amplitude can be tested by the criteria discussed by Essenwanger (1974a, section 4.3.2).

Figure 3.2 displays the periodogram vectors for 15, 28, 30 and 33 points. We can see that the almost zero vector amplitude for 15 points results from an abrupt phase shift of 180° after 8 cycles corresponding with the x-factor. A similar turn is noticed for 30 points. The vector diagrams for 28 and 33 points exhibit a pattern of an almost opposite winding, indicating that the true cycle lies between. Since in both the 28 and 33 vector summation diagrams the endpoint of the vector moves away from zero, the vector amplitude remains at a relatively high level after summation, while the amplitude for the 30-point period almost disappears. Similar patterns could be shown for the 14 and 16 points, but will be omitted here.

We learn from this example that the detailed analysis with the aid of a periodogram can reveal more details than the vector summary of the Fourier analysis. Admittedly, the computer time procuring the information for the periodogram analysis is about equal or more than for the Fourier analysis and

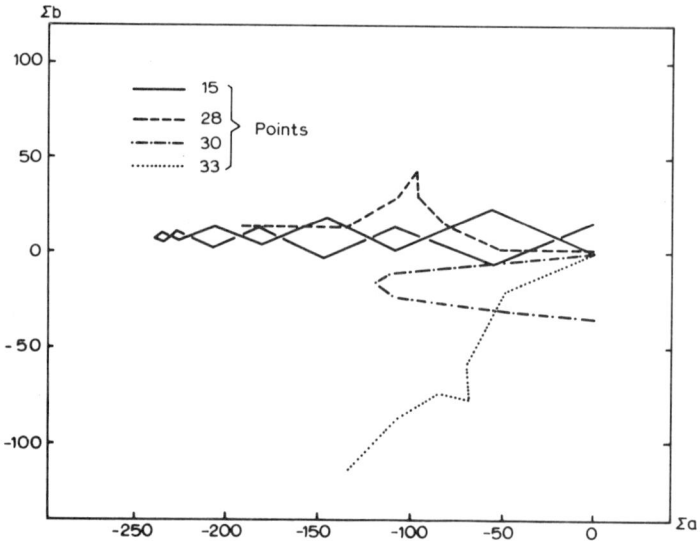

Fig. 3.2. Vector diagram for $x \cdot \sin \alpha$ data.

considerably exceeds the time (and cost) for the power-spectrum analysis. It may sometimes be necessary, however, to investigate certain observational series by periodogram analysis to learn more details than Fourier or power-spectrum analysis can provide.

Patterns with a split of the peak between two adjacent periods and a dip of the vector amplitude between them are indicative of linearly varying amplitudes of a Fourier term.

3.2.2 Power-spectrum and periodogram of non-harmonic waves

Previously all examples have been presented where the dominant cycle and the cycle length of the waves were based on harmonic numbers of the basic period (i.e. p/i; $i = 1, 2 \ldots n$).

We shall now discuss Example 3.9 where this is not the case.

It may be remarked that in reality one should not call this the analysis of "non-harmonic" components. We should precisely talk about an analysis where the basic period and the factual maximum lag length or the cycles of the data are non-harmonic. In other words, assume we have $p_1 = 11$, $p_2 = 28$, $p_3 = 37$. Then the basic period is $11 \cdot 28 \cdot 37 = 11396$, where now the three waves result in harmonic numbers. If our record goes to 500 or 1000 observations only we obtain a fraction of the real basic period and non-harmonic cycles, i.e. the three different waves cannot be expressed by whole numbers of the postulated basic period. It is evident that power-spectrum and Fourier analysis disclose only what has been represented by the short record of 500

SPECTRAL ANALYSIS

TABLE 3.14
Power-spectrum from non-harmonic waves of period length 11, 28 and 37

j	p_j	L_j	ΣL_j	Ls_j	j	L_j	ΣL_j	Ls_j	j	L_j	ΣL_j	Ls_j
0		0.162	0.162	0.310		0.043	0.043	0.307		−0.058	−0.058	0.294
1	28	0.485	0.647	0.337	32	0.616	0.659	0.350	36	0.708	0.650	0.366
2	14	0.165	0.813	0.254	16	0.034	0.693	0.234	18	−0.013	0.638	0.219
3	9.3	0.230	1.043	0.146	10.7	0.322	1.015	0.176	12	0.276	0.914	0.172
4	7.0	−0.068	0.975	0.025	8.0	−0.236	0.991	0.063	9	0.115	1.029	0.114
5	5.6	0.038	1.013	−0.000	6.4	0.011	1.002	−0.000	7.2	−0.049	0.980	0.007
6	etc.	etc.	etc.	etc.	5.3	−0.005	0.996	0.001	6.0	etc.	etc.	etc.
	2.0		1.000		2.0		etc.	etc.	2.0			1.000
							1.000	1.000				

20 points

j	p_j	L_j	ΣL_j	Ls_j
0		−0.103	−0.103	0.260
1	40	0.687	0.584	0.365
2	20	0.078	0.661	0.225
3	13.3	0.107	0.768	0.140
4	10.0	0.282	1.050	0.159
5	8.0	−0.079	0.971	0.032
6	6.7	0.043	1.015	−0.001
	etc.	etc.	etc.	etc.
	2.0		1.0000	

22 points

j	L_j	ΣL_j	Ls_j
	−0.099	−0.099	0.210
44	0.572	0.473	0.343
22	0.249	0.722	0.248
14.7	−0.079	0.644	0.099
11	0.367	1.011	0.175
8.8	−0.020	0.991	0.076
7.3	0.012	1.002	−0.000
etc.	etc.	etc.	etc.
2.0		1.000	

or 1000 values. The analysis would not reflect the true conditions and would depend on the chosen maximum lag. Table 3.14 exhibits the calculations for the composition of three waves with equal amplitude 5.0 and periodicity of 11, 28 and 37. Five cases have been completed for lags from 14, 16, 18, 20 and 22 point length.

It is evident that for the short total length of 240 points and the defined lags the fluctuation of side lobes of the periodicities influences the power-spectrum, and several negative values appear in the line spectrum L_j. These are the uncompensated undesirable side maxima typical for power-spectrum analysis and should be interpreted as such. The smooth spectrum Ls_j has virtually suppressed that effect (see also Blackman and Tukey, 1958).

The true cycles cannot be pinpointed in the spectrum, but are at least found in the vicinity of the correct length. Thus, the analysis for the 14-point lag indicates a longer cycle than 28 and an additional one at the shorter range. In the 18-point lag analysis the 37- and 11-point periodicity is well reflected but the 28-point cycle is virtually lost. The best pattern is delivered by the 22-point analysis, although the true length of 37 cannot be recognized.

It should be reiterated that the examples were selected to give the reader some practical background for the interpretation of power spectra. We have seen that it is difficult to pinpoint the correct cycle even under simplified assumptions. We should keep in mind that in dealing with empirical data random fluctuations of the observations are superimposed on the cycle, and the power-spectrum becomes an "estimate". We must, therefore, test the statistical significance of the peaks in the spectrum to assure that these crests are not created by random fluctuations. Nevertheless, statistical significance alone does not guarantee that the true cycle has now the very same length for the periodicity as the answer in the power-spectrum displays, nor does this mean we have now determined its exact proportion of the total contribution (see also Section 3.2.3). It merely indicates the approximate location of significant waves. This is exactly the same information as generally given by the Fourier series, too. Several spectra of different maximum lags should therefore be considered. This important point is generally neglected although it also has been suggested by Blackman and Tukey (1958). The computer costs are not high once the autocorrelation is established. Further reduction in computations can be gained by the fast Fourier transform, an algorithm discussed also by Essenwanger (1974a, section 4.3). Furthermore, the power-spectrum is not the only method to study autocorrelation functions.

The periodogram analysis for the non-harmonic data is given in Table 3.15. We notice three peaks in the Av_p, z_p^2 and q columns at $p = 11$, 28 and 38, with a slight shift in the length for the longest wave. This should be expected as the total data set is not sufficient to reflect the longest wave with more accuracy. This deficiency is a common property of the power-spectrum. It should be called to attention, however, that the shorter cycles are correctly placed. A side maximum at $p = 24$ as displayed in the three columns from

TABLE 3.15

Periodogram analysis for non-harmonic waves of period length 11, 28 and 37

p	As_p	Av_p	z_p^2	q	p	As_p	Av_p	z_p^2	q
10	5.52	0.46	0.001	0.082	25	5.96	0.37	0.001	0.061
11	5.28	4.83	0.108	0.915	26	6.29	2.59	0.031	0.412
12	5.79	0.38	0.001	0.066	27	6.00	4.50	0.093	0.749
13	5.51	0.11	0.000	0.020	28	6.25	4.86	0.109	0.777
14	5.40	0.20	0.000	0.037	29	6.29	3.67	0.062	0.584
15	5.95	0.14	0.000	0.023	30	6.55	1.99	0.018	0.303
16	3.50	0.19	0.000	0.053	31	5.68	1.16	0.001	0.205
17	5.51	0.19	0.000	0.033	32	6.06	0.30	0.000	0.049
18	3.61	0.35	0.001	0.097	33	6.19	0.62	0.002	0.099
19	4.83	0.42	0.001	0.086	34	6.38	1.42	0.009	0.221
20	5.12	0.49	0.001	0.095	35	5.89	3.19	0.047	0.541
21	5.40	0.69	0.002	0.126	36	6.16	4.08	0.076	0.661
22	5.50	0.41	0.001	0.075	37	6.01	4.86	0.109	0.807
23	5.57	0.39	0.004	0.159	38	5.54	5.00	0.118	0.911
24	5.85	1.27	0.007	0.217	39	5.69	4.92	0.112	0.865
					40	5.69	4.15	0.080	0.730

Symbols are the same as in Table 3.13a.

Av_p to q is produced by the presence of three waves and the cut-off at 240 points and proves not to be statistically significant.

Thus, the periodogram analysis fares better in this case than the Fourier series and the power-spectrum. This may not always be the case as Example 3.7 illustrated where the phase angle of waves leads to extinction of cycles although a detailed plotting of the periodogram waves uncovers the phase shift. The analysis work involved may be tedious, however.

In atmospheric physics many "quasi-periodicities" exist where the beginning of the new cycle at the same wave length is completely independent from the previous event. This creates a phase shift at the same wave. The analysis of the entire data set and the periodogram sequence (e.g. see Table 3.15) may then not reveal the true wave pattern. Studies should then follow the periodogram as outlined in Fig. 3.2. Again, in search for periodicities a variety of methods or tools should be considered to find finally the proper conclusion.

3.2.3 Estimation of spectra, separation of waves and aliasing

Expectancy and significance of spectra have been treated previously by Essenwanger (1974a). Hartley (1949) has also examined the statistical significance in the analysis of harmonic waves. Further elaboration may not be necessary. The separation of waves and the aliasing effect deserve more attention, however.

We have seen in the previous sections that the finite length of observations affects the outcome of the spectrum analysis. In these examples the basic waves were known a priori and the maximum length of lags and the spectrum were adjusted. In most empirical data investigations these facts are not known. The uncertainty in determining unknown frequencies from observations was investigated by various authors.

Stumpff (1937) has already pointed out that two periods of length p_1 and p_2 with wave length angle α_1 and α_2 such as:

$$\alpha_1 = 2\pi/p_1 \quad \text{or} \quad p_1 = 2\pi/\alpha_1 \qquad [3.27a]$$

$$\alpha_2 = 2\pi/p_2 \quad \text{or} \quad p_2 = 2\pi/\alpha_2 \qquad [3.27b]$$

lead to a resulting wave-length variation of:

$$p_\alpha = 2\pi/(\alpha_2 - \alpha_1) = p_1 p_2/(p_1 - p_2) \qquad [3.27c]$$

In order to separate these waves in a spectrum diagram, the total number N of the observations must then be at least $N > p + p_\alpha$, i.e. in the analysis of an interval with length p we must be able to shift the interval at least by p_α. If the two waves p_1 and p_2 are adjacent (neighboring) waves in a harmonic wave pattern, e.g. $p = rp_1$ or $p = (r+1)p_2$, then $p_\alpha = p$ and $N \geq 2p$.

This uncertainty of the estimate was expanded by Priestly (1962) who formulated the variation for the estimated smooth spectrum Ls_j as follows:

$$a_{th}\sigma_j + b_j = cLs_j \qquad [3.28]$$

where σ is the standard deviation of the estimate and b the bias. The constant a_{th} indicates the exceedance threshold of a Gaussian distribution, e.g. the 95-% level requires $a_{th} = 1.96$. The c is the desired fraction of tolerance for the decrease of the amplitude $f(\nu)$, e.g. for 10% of $f(\nu)$ the $c = 0.1$, etc.

Bartlett's (1948, 1950a) estimate requires:

$$\sigma_j^2 \sim \frac{2}{3}\frac{m}{N} \cdot Ls_j^2 \qquad [3.29a]$$

and: $b_j \sim [1/(mB_f)] Ls_j \qquad [3.29b]$

Hence for Bartlett's estimate we obtain:

$$a_{th}\sqrt{\frac{2}{3}\frac{m}{N}} + \frac{1}{mB_f} = c \qquad [3.29c]$$

In this equation the unknowns are m and B_f. With them the N could be determined.

The B_f is defined by the bandwidth where Priestly has assumed:

$$B_W = B_f/2 \qquad [3.28a]$$

For Bartlett: $B_W = 2\pi/m \qquad [3.29d]$

SPECTRAL ANALYSIS

which also means that either the bandwidth or the record length must be known. Note that these are the factors comprising the bias. The bandwidth B_f is commonly introduced as the half power point of the narrowest peak of $f(\nu)$. Eliminating B_f, we find:

$$a_{th} \sqrt{\frac{2m}{3N}} = c - \frac{1}{4\pi} \qquad [3.29e]$$

We rearrange the equation for N and can write:

$$N = \frac{32 m \pi^2 a_{th}^2}{3(4\pi c - 1)^2} \qquad [3.29f]$$

This leads usually to a very high requirement for N in Bartlett's estimate

In Tukey's method the bias and variance for the smooth spectrum Ls_j is:

$$b_j \sim \frac{0.23}{B_f^2} \frac{\pi^2}{m^2} Ls_j \qquad [3.30a]$$

and:
$$\sigma_j^2 \sim \frac{4m}{5N} \cdot Ls_j^2 \qquad [3.30b]$$

This leads to:

$$a_{th} \sqrt{\frac{4}{5}\frac{m}{N} + \frac{0.23}{B_f^2}\frac{\pi^2}{m^2}} = c \qquad [3.30c]$$

Again, introducing $B_f \sim 4\pi/m$, we derive:

$$a_{th} \sqrt{\frac{4}{5}\frac{m}{N} + 0.23/16} = c \qquad [3.30d]$$

and finally:
$$N = \frac{0.8 m a_{th}^2}{c - 0.014375} \qquad [3.30e]$$

This gives a far smaller requirement of N and thus the method via autocorrelation and smoothing as discussed by Essenwanger (1974a, section 4.3.3) needs a shorter period of record than the criterion by Bartlett in order to render the same accuracy of estimate for the power-spectrum. In other words, Tukey's criterion is more efficient.

We conclude furthermore that m enters into [3.29f] and [3.30e] as the principle parameter from the spectrum. Priestly (1962) has given optimum values for m. For Bartlett's scheme he derives:

$$m_B = B_f^{-1}\sqrt{3N} \qquad [3.31]$$

For Tukey's assumption the optimum estimate is:

$$m_T = \sqrt{\frac{2116}{7748} \pi^4 \frac{N}{B_f^4}} \qquad [3.31a]$$

or, after some transformations:

$$m_T = 1.28/(B_f' \cdot \Delta x) \qquad [3.31b]$$

where: $B_f' = B_f/(2\pi \Delta x)$ [3.31c]

with Δx standing for the unit of observation, which is generally the frequency:

$f = 1/(2\Delta x)$ (cycles per second or other times or length unit)

Expressed in different notation we have $\Delta x = T/N$ where T is the length of the sample records. The m would represent in this case the maximum lag in the autocorrelation function.

According to Taubenheim (1969) the maximum lag should stay within 5–10% of the total observation, i.e. N is 10–20 times longer than m. This is a minimum requirement. The m_T (Tukey) stays well within this guideline. Since N is larger in Bartlett's model the fraction is even less. More details may be found in the literature (Bartlett, 1948, 1950a; Tukey, 1949, 1961; Daniell, 1946; and Parzen, 1957a, 1957b, 1961).

It can be added that Parzen (1961) recommends smoothing of the autocorrelation (autocovariance) function by the weighting function:

$w_\tau = 1 + 6(\tau/m)^2(\tau/m - 1)$ for $\tau < m/2$

and: $w_\tau = 2(1 - \tau/m)^3$ for $m/2 < \tau < m$

e.g. the original series $R_0, R_1 ..., R_\tau, ..., R_m$ is then corrected to $w_0 R_0, w_1 R_1, w_2 R_2, ... w_\tau R_\tau, ... w_m R_m$. The weights for the minimum lag $m = 10$ would be: $w_0 = 1$, $w_1 = 1 - 0.054 = 0.946$, ... $w_6 = 0.128$, ... $w_{10} = 0$. This leads to equivalent degrees of freedom of $3 \cdot 7N/m$ compared to $8N/(3n)$ for the Blackman-Tukey (1958) estimate.

Another effect leading to a fictitious periodicity is the so-called "aliasing" effect. Assume we have a series of data at equidistant intervals (e.g. time). Since we need two more points besides the initial observation to determine a sine-wave, the minimum length of a periodicity is $2\Delta x$, where Δx is the distance between two observations (either time or space). Thus the minimum cycle has the length $L_N = 2\Delta x$. The frequency is then given by:

$$\nu_N = 1/L_N = 1/(2\Delta x) \text{ (cycles per unit)} \qquad [3.32]$$

This is the so-called Nyquist frequency. In the case $\Delta x = 1$, this frequency would be $\nu_N = 1/2$ cycle per unit.

SPECTRAL ANALYSIS

If the observations were produced by a cycle smaller than ν_N, this cycle would not be discovered, and a different periodicity with a length $p > L_N$ would appear in the spectrum (see Fig. 3.3).

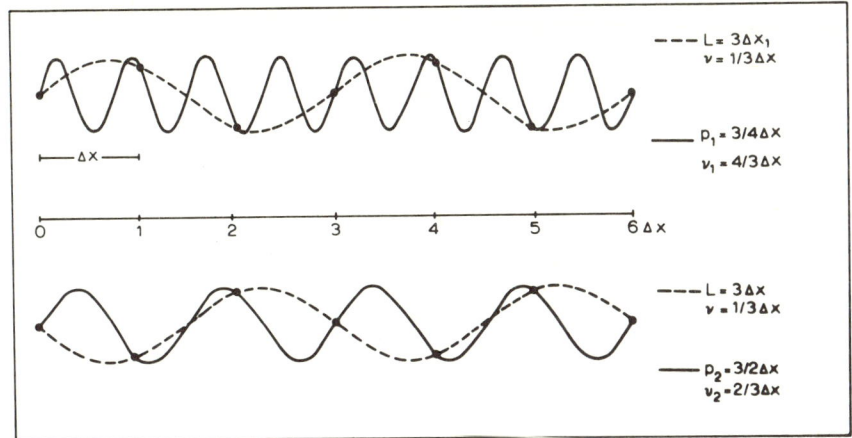

Fig. 3.3. Illustration of the aliasing effect in power-spectrum analysis.

The series of alias frequencies which we cannot distinguish can be listed as:

$$\nu, \quad 2\nu_N \pm \nu, \quad 4\nu_N \pm \nu \text{ etc.} \qquad [3.32a]$$

The first frequency ν must be smaller than ν_N.

Let us assume $\nu = 1/3$ for $\Delta x = 1$. Then the first alias frequency above ν would be $\nu_1 = 4/3$ or $\nu_2 = 2/3$, depending on the sign. Since $p = 1/\nu$, the length of the cycles would be $p_1 = 3/4$ or $p_2 = 3/2$ units. The two waves are smaller than L_N and could create a fictitious period $p > L_N$ (e.g. p_1 or p_2 could appear as $p' = 3$ different only in the phase angle by $180°$).

If we know the highest (significant) cycle ν_l, we can write in turn:

$$\Delta x_{\max} = 1/2\nu_l = p_l/2 \qquad [3.32b]$$

and determine the maximum distance of the observations. (Assume $\nu_l = 2/3$. The maximum length would be $\Delta x = 3/4$ units. Then $p_l = 3/2$). The aliasing effect can be suppressed by appropriate filtering of data, but ν_l or p_l must be known or assumed. Filtering is covered by Essenwanger (1974b).

3.2.4 Spectra of meteorological data

Further examples of meteorological spectra beyond the illustrations given by Essenwanger (1974a) will not be added here because various discussions can be found in the atmospheric-science literature.

Panofsky et al. (1958), Landsberg et al. (1959), and Bolgiano (1959) have probably been among the first authors to make extensive utilization of spectral analysis. While Panofsky et al. and Bolgiano were largely concerned about the prospects in turbulence analysis, Landsberg et al. have dealt with daily, weekly, and monthly cycles of average temperature and precipitation. These articles outline the principal areas of application in meteorology: the field of turbulence analysis and the general climatic fluctuations.

Many subsequent publications on spectral analysis can be found in the literature on turbulence topics such as articles by, e.g. Bolgiano (1962), Monin (1962), Burns (1963), Burns and Rider (1965), Kao and Woods (1964), or Pinus et al. (1967); and on climatic fluctuation papers by Doberlitz (1969), Monin and Vulis (1971), or Julian (1971) exist. An exhaustive list of the literature on power spectra, however, cannot be given here.

The problem of spectral analysis and its relevance to linear prediction of meteorological time series has been investigated by Jones (1964), who later has reappraised the periodogram in spectral analysis (Jones, 1965), and has investigated the problem of spectral estimation with unequally spaced observations (Jones, 1970).

The quoted literature above exemplifies that the availability of electronic data processing has made it possible to incorporate techniques such as the power-spectrum analysis into the standard tools of statistical analysis for investigations and the processing of atmospheric data. The examples and discussions on this topic by Essenwanger (1974a) and here, should be sufficient for the reader to initiate properly the calculation of power spectra and to evaluate and interpret the results.

Information on turbulence spectra is available in various textbooks on this topic, such as Tennekes and Lumley (1972) or Haugen (1973).

It may be of interest to add a vector diagram and vector amplitude distribution for some turbulence data (Fig. 3.4). The observations were taken at a meteorological tower in Huntsville (Alabama) where the wind measurements from the anemometers were recorded once every second at a height of 18 feet for a period of about one hour. A cycle of 40 sec emerged with a peak amplitude in a periodogram from 5 to 100 sec length. The south component of the wind was then selected for the construction of a vector diagram and amplitude distribution as given in Fig. 3.4. The total of 3740 observations recorded led to 94 amplitudes for periods of 40 sec length.

The distribution of the Fourier amplitude $A = (a^2 + b^2)^{0.5}$ with components a and b is displayed in the lower portion of Fig. 3.4. This "scatter diagram" exhibits an elongated distribution. This non-circular shape can be confirmed from the variances of $\sigma_a^2 = 0.42$ and $\sigma_b^2 = 0.21$, whose ratio of 2.04 supports the hypothesis that the variances are significantly different. Consequently, we may assume an elliptical rather than a circular bivariate distribution. Since the correlation coefficient between a and b is not significantly different from zero ($r_{ab} = 0.11$), the transformation angle of 10° becomes unimportant. No transformation to independent coordinates is necessary because a and b are virtually independent.

SPECTRAL ANALYSIS

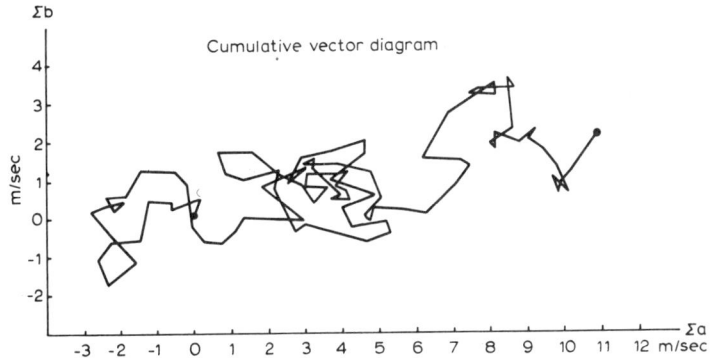

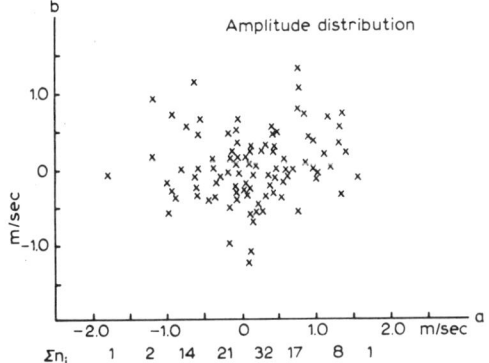

Fig. 3.4. Cumulative vector diagram and distribution of amplitudes for turbulence data (see text).

Although the mean amplitude exhibited a peak in the periodogram, Schuster's criterion (Essenwanger, 1974a, section 4) renders no significance. We calculate $\epsilon^2 = 4\sigma^2/N = 4 \cdot (1.32)/94$. Thus $\epsilon^2 = 0.0563$ or $\epsilon = \pm 0.237$. Thus the resultant vector of $\overline{A} = 0.117$ is not significantly different from zero, i.e. randomly produced. We learn from inspection of the vector summation, illustrated in the top part of Fig. 3.4, that some sequences of the data exist where the phase angles of the vectors are aligned, but irregular (random) sequences dominate. The "quasi-periodicity", e.g. over periods of 400 sec, is not sufficient to outweigh the random influence of the other data. In fact, this quasi-periodicity could itself be random play.

3.3 BESSEL FUNCTIONS

3.3.1 General

Although known to mathematicians and physicists, Bessel functions have not played a great role in statistical analysis. This section will introduce the reader to them, and give a brief survey. Today's availability of electronic computers has removed many restraints which formerly may have contributed to their limited application in atmospheric physics. Kutzbach and Wahl (1965) have exemplified the usefulness of Bessel functions in their mathematical description of scalar fields.

Several avenues are open to derive Bessel functions. One of the most convenient ways to define these functions is as solutions to the differential equation:

$$\frac{d^2u}{dx^2} + \frac{1}{x}\frac{du}{dx} + \left(1 - \frac{n^2}{x^2}\right)u = 0 \qquad [3.33]$$

Bernoulli, Euler, Fourier, Laplace and Bessel have stated problems which led to the above differential equation. Bessel studied the elliptic motion of a planet and found for a quantity called the mean anomaly:

$$\mu = \phi - \epsilon \sin \phi \qquad [3.33a]$$

where ϕ is the eccentric anomaly. If ϵ remains constant, while ϕ and μ vary, we obtain:

$$\phi - \mu = \sum_{1}^{\infty} A_r \sin r\mu \qquad [3.33b]$$

where A_r is a function of the eccentricity ϵ of an ellipse. He formulated this function as:

$$A_r = (2/r\pi) \int_0^\pi \cos r(\phi - \epsilon \sin \phi) \, d\phi \qquad [3.33c]$$

Laplace's equation may be interesting in this connection, as it leads to "cylindrical harmonics", in which capacity Kutzbach and Wahl (1965) have utilized the function. Laplace's version states:

$$\frac{d^2R}{d\rho^2} + \frac{1}{\rho}\frac{dR}{d\rho} + \left(1 - \frac{n^2}{\rho^2}\right)R = 0 \qquad [3.33d]$$

where we use the cylindrical coordinates ρ, ϕ and z, with $x = \rho \cos \phi$ and $y = \rho \sin \phi$. Furthermore, Laplace's equation $\nabla^2 \psi = 0$ must be fulfilled by a potential function ψ such that:

$$\frac{\partial^2 \psi}{\partial \rho^2} + \frac{1}{\rho}\frac{\partial \psi}{\partial \rho} + \frac{1}{\rho^2}\frac{\partial^2 \psi}{\partial \phi^2} + \frac{\partial^2 \psi}{\partial z^2} = 0 \qquad [3.34a]$$

BESSEL FUNCTIONS   245

It is assumed that $\psi = R\Phi Z$, where R, Φ and Z are only functions of just ρ, ϕ and z, respectively. Since $(d^2\Phi/d\phi^2)/\Phi = -n^2$, $\Phi = \exp(\pm in\phi)$, $Z = \exp(\pm kz)$, $d^2Z/dz^2 = k^2$, $v = k\rho$, $R_n(v)$ has the form:

$$V = R_n(k\rho)[\exp(\pm in\phi)] \cdot [\exp(\pm kz)] \qquad [3.34b]$$

as a solution. The complex function $\exp(\pm in\phi)$ can be converted by Eulers formula into trigonometric components with $i = \sqrt{-1}$.

3.3.2 Definition of Bessel functions and recurrence relations

Two solutions were derived by the method of Frobenius:

$$y_1 = c_0 \sum_{s=0}^{\infty} (-1)^s x^{n+2s}/[2 \cdot 4 \dots (2s)(2n+2)(2n+4) \dots (2n+2s)] \qquad [3.35a]$$

and:
$$y_2 = c_0 \sum_{s=0}^{\infty} (-1)^s x^{2s-n}/[2 \cdot 4 \dots (2s)(2-2n)(4-2n) \dots (2s-2n)] \qquad [3.35b]$$

It is evident that n has been set $-n$ for y_2. For integral $n \geq 0$ the series behaves like:

$$1 - x^2/2^2 + x^4/(2^2 \cdot 4^2) - x^6/(2^2 \cdot 4^2 \cdot 6^2) + \dots \qquad [3.36]$$

A general solution is:

$$y = Ay_1 + By_2 \qquad [3.37]$$

If n is a positive integer, the solution y_2 does not exist, and analogously for n being a negative integer y_1 is non-existent.

We give the constant c_0 the value $1/(2^n \cdot n!)$. Then we can write for $n > 0$, but not necessarily an integer:

$$J_n(x) = \sum_{s=0}^{\infty} \frac{(-1)^s}{\Pi(s)\Pi(n+s)} \left(\frac{x}{2}\right)^{n+2s} \qquad [3.38]$$

This solution is also valid when n is not a positive integer. Generally, $\Pi(a) = a!$ and $\Pi(0) = 1$. Furthermore $\Pi(a) = \Gamma(a+1)$, or $\Pi(a) = a\Pi(a-1)$. Equation [3.38] is called the Bessel Function of the first kind, and n is the order of the function. If n is not an integer, $J_n(x)$ and $J_{-n}(x)$ are two linearly independent solutions:

$$J_{-n}(x) = \sum_{s=0}^{\infty} \frac{(-1)^s}{\Pi(s)\Pi(-n+s)} \left(\frac{x}{2}\right)^{-n+2s} \qquad [3.38a]$$

where now n is a positive integer. The factor $1/\Pi(-n+s)$ is zero for $s = 0$, 1, 2 ... $(n-1)$ and is finite for all other values of n. Hence:

$$J_{-n}(x) = (-1)^n J_n(x) \quad [3.38b]$$

We denote as usual $J'_n = [dJ_n(x)]/dx$. Then we find the following recurrence formula:

$$xJ'_n = nJ_n - xJ_{n+1} = -nJ_n + xJ_{n-1} \quad [3.39]$$

We further deduce:

$$2J'_n = J_{n-1} - J_{n+1} \quad [3.39a]$$

$$J'_0 = -J_1 \quad [3.39b]$$

$$2nJ_n/x = J_{n-1} + J_{n+1} \quad [3.39c]$$

where $n = N/2$. When N is odd, then:

$$J_{-0.5} = [x^{-0.5}/\{2^{-0.5}\Pi(-0.5)\}][1 - x^2/2 + x^4/(1 \cdot 2 \cdot 3 \cdot 4) - \ldots \text{etc.}]$$

$$= [2/(\pi x)]^{0.5} \cos x \quad [3.40a]$$

$$J_{0.5} = [2/(\pi x)]^{0.5} \sin x \quad [3.40b]$$

With [3.39c] and [3.40a and b] we can now calculate the expressions $J_{k+0.5}$, where k can be any positive or negative integer, e.g. $k = 0$, $n = 0.5$, $2n = 1$, or:

$$(J_{0.5}/x) = J_{0.5} + J_{1.5} \quad [3.41]$$

$$J_{1.5} \cdot \sqrt{\pi x/2} = \frac{\sin x}{x} - \cos x \quad [3.41a]$$

$$J_{-1.5} \cdot \sqrt{\pi x/2} = -\sin x - (\cos x)/x \quad [3.41b]$$

etc.

3.3.3 Complex Bessel functions

We substitute $x = it$, with the algebraic meaning $i = \sqrt{-1}$. Then:

$$J_n(it) = [(it)^n/\{2^n \Pi(n)\}][1 + t^2/\{2(2n+2)\} + t^4/\{2 \cdot 4(2n+2)(2n+4)\} + \ldots] \quad [3.42]$$

This function is usually expressed as:

$$I_n(t) = i^{-n} J_n(it) = \sum_{s=0}^{\infty} \frac{1}{\Pi(s)\Pi(n+s)} \left(\frac{t}{2}\right)^{n+2s} \quad [3.42a]$$

This is called the modified Bessel function of the first kind. We can deduce:

$$I_{-n}(t) = I_n(t) \quad [3.43]$$

BESSEL FUNCTIONS

$$I'_0 = I_1 \qquad [3.43a]$$

If n is not an integer, $I_{-n}(t)$ is an independent solution of:

$$t^2 \frac{d^2y}{dt^2} + t \frac{dy}{dt} - (n^2 + t^2)y = 0 \qquad [3.44]$$

If n is an integer, we find a second solution $K_n(t)$. Exploitation of the function $K_n(t)$ is widespread. Its definition is:

$$K_n(t) = \frac{\pi}{2 \sin n\pi} [I_{-n}(t) - I_n(t)] \qquad [3.44a]$$

where n is not an integer. This function is known as the modified Bessel function of the second kind.

When $|x|$ is large and n is fixed we can write:

$$I_n(x) \sim (e^x/\sqrt{2\pi x})[1 - (4n^2 - 1)/8x + (4n^2 - 1)(4n^2 - 3^2)/2!(8x)^2 - ...]$$
$$[3.43b]$$

For large n we may approximately state:

$$I_n(x) \sim (1/\sqrt{2\pi x}) e^{ny} (1 - x^2)^{-\frac{1}{4}} \qquad [3.43c]$$

with the abbreviation:

$$y = (1 + x^2)^{\frac{1}{2}} - \ln [x^{-1}\{1 + (1 + x^2)^{\frac{1}{2}}\}] \qquad [3.43d]$$

3.3.4 The zeros of the Bessel functions

It is sometimes important to know the zero places of the Bessel functions. In this short survey no exhaustive treatment can be given. The reader may refer to reference books such as Watson (1958), etc.

Since the most common Bessel functions to be found in atmospheric science appear to be $J_n(x)$, some short discussion on this topic is appropriate.

When n is real, $J_n(x) = 0$ has an infinite number of real roots. Stoke's method for obtaining the zeros of $J_n(x) = 0$ depends upon the asymptotic expression for $J_n(x)$.

For large x and $n = 0$ we transform the equation into:

$$J_0(x) \sim \sqrt{2/\pi x} \cdot M \cos(x - \pi/4 - \psi) \qquad [3.45]$$

which can be made zero for the integer k with:

$$x = (k - 1/4)\pi + \psi \qquad [3.45a]$$

The following abbreviations have been employed:

$$\psi = \arctan \left(\frac{1}{8x} - \frac{33}{512x^3} + \frac{3417}{16384x^5} - ... \right) = \arctan (P/Q) \qquad [3.46a]$$

$$M = 1 - 1/16x^2 + 53/512x^4 - \ldots = \sqrt{P^2 + Q^2} \qquad [3.46b]$$

$$P = 1 - \frac{1^2 3^2}{2!(8x)^2} + \frac{1^2 3^2 5^2 7^2}{4!(8x)^4} - \ldots \qquad [3.46c]$$

$$Q = \frac{1}{8x} - \frac{1^2 3^2 5^2}{3!(8x)^3} + \frac{1^2 3^2 5^2 7^2 9^2}{5!(8x)^5} - \ldots \qquad [3.46d]$$

Another form of [3.45a] is:

$$x/\pi = (k - 1/4) + 1/[2\pi^2(4k - 1)] - 31/[6\pi^4(4k - 1)^3] + \ldots \qquad [3.45b]$$

The corresponding formula for $J_1(x) = 0$ can be derived, namely:

$$x/\pi = k + 0.25 - 0.151982/(4k + 1) + 0.015399/(4k + 1)^3$$
$$- 0.245270/(4k + 1)^5 + \ldots \qquad [3.47]$$

Generally we may state:

$$x = a - (m - 1)/(8a) - 4(m - 1)(7m - 31)/[3(8a)^3]$$
$$- 32(m - 1)(83m^2 - 982m + 3779)/[15(8a)^5]$$
$$- 64(m - 1)(6949m^3 - 153855m^2 + 1585749m$$
$$- 6277237)/105(8a)^7 - \text{etc.} \qquad [3.48]$$

where: $a = (\pi/4)(2n - 1 + 4k)$ \qquad [3.48a]

and: $m = 4n^2$ \qquad [3.48b]

This gives the kth root of the nth order term, $J_n(x) = 0$.

3.3.5 Fourier-Bessel expansion

In the field of heat conduction on a cylinder, we must solve $\nabla^2 V = 0$. This problem was studied as early as 1867 by Neumann. We first set:

$$V = e^{-cz} J_0(cR) \qquad [3.49a]$$

where the boundaries of a cylinder are $R = 1$, and $0 \leq z \leq +\infty$. The parameter c is then any positive root of the equation:

$$ckJ_0'(c) + hJ_0(c) = 0 \qquad [3.49b]$$

As originally introduced, k would represent the conductivity of the material and h would express the external conductivity. The temperature at any point inside the cylinder would then follow from:

$$V = e^{-cz} J_0(cR) \qquad [3.49c]$$

Since [3.49a] has an infinite number of roots c_j, the result can be brought into the generalized form:

$$\phi = \sum_1^\infty A_s e^{-c_s z} J_0(c_s R) \qquad [3.50]$$

For $z = 0$ we derive:

$$\phi_0 = \sum_1^\infty A_s J_0(c_s R) = f(R) \qquad [3.50a]$$

The limitation for R is $0 \leq R \leq 1.0$. The coefficients may be calculated from:

$$A_s = [2k^2 c_s^2/(h^2 + k^2 c_s^2) J_0^2(c_s)] \int_0^1 J_0(c_s R) f(R) R dR \qquad [3.51]$$

A more general form of the potential ϕ can be written as:

$$\phi = \Sigma (A \cos n\alpha + B \sin n\alpha) e^{-cz} J_n(cR) \qquad [3.52]$$

The summation is over n and c independently.

We limit the quantities n to integer values and require that the c-values are the positive roots of $J_n(c) = 0$, the potential becomes periodic in $\alpha + 2\pi$ and vanishes at $z = +\infty$, and also for $R = 1$. Again, for $z = 0$ we find:

$$\phi = \Sigma (A \cos n\alpha + B \sin n\alpha) J_n(cR) \qquad [3.52a]$$

It can be shown that for an arbitrary finite, one-valued function $f(R, \alpha)$ continuous over the circle $R = 1$, with validity of [3.52a], the coefficients can be readily obtained. Assumed we have:

$$f(R, \alpha) = \Sigma\Sigma (A_{n,s} \cos n\alpha + B_{n,s} \sin n\alpha) J_n(c_s R) \qquad [3.52b]$$

$$\text{then: } A_{n,s} = [2/\pi \{J_n'(c_s)\}^2] \int_0^1 \int_0^{2\pi} f(R, \alpha) \sin n\alpha J_n(c_s R) R d\alpha dR \qquad [3.53a]$$

$$\text{and: } B_{n,s} = [2/\pi \{J_n'(c_s)\}^2] \int_0^1 \int_0^{2\pi} f(r, \alpha) \sin n\alpha J_n(c_s R) R d\alpha dR \qquad [3.53b]$$

Based on the above procedure, Kutzbach and Wahl (1965) and Kutzbach (1966) have deduced a system of representation of scalar fields with orthogonal functions. They have adopted the following mathematical representation for the scalar field observation $f(R, \alpha)$:

$$f(R, \alpha) = 0.5 \sum_{k=1}^\infty A_{0k} J_0(h_{0k} R) + \sum_{n=1}^\infty \sum_{k=1}^\infty (A_{nk} \cos n\alpha + B_{nk} \sin n\alpha) J_n(h_{0k} R)$$

$$[3.53c]$$

where R and α are polar coordinates and h_{nk} is the kth positive root of the equation $J_n(r) = 0$ (see Section 3.3.4). The expressions above have been known as cylinder functions. The $A_{nk} \cos n\alpha$ and $B_{nk} \sin n\alpha$ of [3.53c] are orthogonal.* The coefficients are determined from:

---

* Effects on orthogonality of replacing integrals by summations have been covered in Section 3.1, Legendre polynomials.

$$A_{nk} = (1/\pi d_{nk})\int_{-\pi}^{\pi}\int_0^1 f(R,\alpha)J_n(h_{nk}R)\cos n\alpha R\,dR\,d\alpha \qquad [3.54a]$$

$$\text{and: } B_{nk} = (1/\pi d_{nk})\int_{-\pi}^{\pi}\int_0^1 f(R,\alpha)J_n(h_{nk}R)\sin n\alpha R\,dR\,d\alpha \qquad [3.54b]$$

The notation d_{nk} stands for a normalizing factor:

$$d_{nk} = \int_0^1 R[J_n(h_{nk}R)]^2 dR = 0.5[J_{n+1}(h_{nk})]^2 \qquad [3.54c]$$

We may carry out one integration from $-\pi$ to π and rewrite:

$$A_{nk} = (1/d_{nk})\int_0^1 [a_n(R)]J_n(h_{nk}R)R\,dR \qquad [3.54d]$$

$$\text{or: } B_{nk} = (1/d_{nk})\int_0^1 [b_n(R)]J_n(h_{nk}R)R\,dR \qquad [3.54e]$$

The $a_n(R)$ and $b_n(R)$ are the familiar coefficients of the Fourier series as previously introduced (Essenwanger, 1974a, Section 4.2).

The variance σ^2 can be computed from the field or from:

$$\sigma^2 = 0.5\sum_k A_{0k}^2 d_{0k} + \sum_n \sum_k (A_{nk}^2 + B_{nk}^2)d_{nk} \qquad [3.55]$$

Then we can define normalized coefficients (denoted by asterisks) by:

$$A_{nk}^* = (A_{nk}\sqrt{d_{nk}})/\sigma \qquad [3.55a]$$

$$B_{nk}^* = (B_{nk}\sqrt{d_{nk}})/\sigma \qquad [3.55b]$$

$$A_{0k}^* = (A_{0k}\sqrt{d_{0k}/2})/\sigma \qquad [3.55c]$$

$$B_{0k}^* = 0 \qquad [3.55d]$$

Representation of a scalar field requires knowledge of the degree of approximations. The percentage reduction is again sought. We conclude:

$$Z^2 = (\epsilon_E^2/\sigma^2)\cdot 100 \qquad [3.56]$$

where ϵ_E^2 is the explained variance (see [3.18a]) and σ^2 the variance. With the present definition we can write:

$$\epsilon_E^2 = \sum_n \sum_k [(A_{nk}^*)^2 + (B_{nk}^*)^2] \qquad [3.56a]$$

Since the coefficients constitute an orthogonal system, the linear correlation coefficient between two fields can be derived from:

$$r_{12} = \sum_n \sum_k [A_{nk(1)}^* A_{nk(2)}^* + B_{nk(1)}^* B_{nk(2)}^*] \qquad [3.56b]$$

This formula is similar to the equation derived for orthogonal polynomials.

Kutzbach and Wahl (1965) have investigated the mathematical representation of 500 mbar data (Northern Hemisphere) with $n = 0$ to 18, $k = 1$ to 6. From the 222 computed coefficients only 34 may be considered contributing significantly (individually over 1%) to the percentage reduction. This representation by only 34 terms is a considerable curtailment of the original mapped information from $72 \times 14 = 1008$ grid points.

The Fourier-Bessel coefficients prove advantageous over spherical harmonics as they do not require assumptions on data in lower latitudes (e.g. equatorial regions) where observations are scarce in many maps. Instead, boundary conditions can be set, even with variable boundaries. The variable boundary affects the resulting coefficients very little according to Kutzbach (1966), but leaves enough flexibility of adjustment.

More information on Bessel functions can be found in the standard book by Watson (1958). Numerous other good publications on the subject exist, e.g. Gray and Mathews (1966), etc. For tables see British Association for the Advancement of Science (1952).

3.4 EMPIRICAL ORTHOGONAL POLYNOMIALS AND EIGENVALUES

3.4.1 Empirical orthogonal functions

In Section 3.1 discussions centered mainly on established orthogonal polynomials in the literature such as the Tchebycheff or Legendre system. It was stated that the orthogonality permits the addition of a term without recalculation of the prior computed coefficients. But it is never known how much the next term will contribute should it be necessary to expand the system. This problem is being solved by adjustment of the polynomial functions to the given set of data. The first term contributes most and every additional term adds less, but the sequence of terms is now controlled. We always add the maximum information from the remaining unconsidered data. This property seems very desirable for various problems. Since this type of polynomial is derived from a given set of observed data rather than an a-priori postulation, the polynomials are called empirical orthogonal functions or empirical polynomials. It is sometimes necessary, however, to fall back to the classical Tchebycheff or Legendre system, as these functions are standard. They may be better suited for comparability between different sets of data. Assume that we find a set of empirical orthogonal polynomials for station A and one for station B. The functions may prove entirely different and comparison between the two stations may be very complex. In turn, the "classical" polynomials utilize the same function in both stations and comparison may be relatively easy. Usually, the derivation of empirical functions is based on the same principles as predicting a certain set of observations $a(x)$, namely:

$$a(x) = c_0 + \sum_{j=1}^{n} c_j X_j(x) + \epsilon(x) \quad [3.57]$$

where $a(x)$ is the quantity to be predicted, the $X_j(x)$ are the predictors, the c_j denotes a set of constants and $\epsilon(x)$ is the residual error of the prediction scheme which depends on the set of constants.

We may assume that the observations $a(x)$ are now expressed in terms of their mean value, and $A(x) = a(x) - \bar{a}$. Then c_0 will be zero. We further denote $P_j(x) = X_j(x) - \bar{X}_j$, the deviation of the predictors from their mean. We rewrite:

$$A(x) = \sum_{j=1}^{n} c_j P_j(x) + \epsilon(x) \quad [3.57a]$$

The task is now to minimize the residual error $\epsilon^2(x)$ by selecting the constants c_j accordingly.

The set of constants c_j can be obtained as customary from a linear regression system (see Essenwanger, 1974a, section 3 or 4):

$$c_j = (-1)^j \|\mathbf{M}_{cj}\| / \|\mathbf{M}_{11}\| \quad [3.57b]$$

where the $\mathbf{M}_{cj}$ are the cofactors in the coefficient matrices as customary (see

Example 3.10) and $\| \|$ denotes the determinant ($j = 1, \ldots n$). We further assume:

$$N^{-1} \sum_{x=1}^{N} P_k(x) \, \epsilon(x) = 0 \quad \text{for} \quad k = 0, \ldots n \qquad [3.57c]$$

This merely postulates that we have a random error uncorrelated with the predictors. Equations [3.57a and b] constitute a similar prediction scheme as previously expressed under [3.1b]. The $a(x)$ here takes the role of y_{1i} and it should not be difficult to calculate the constants c_j from the technique described in previous sections and Essenwanger (1974a, section 3), once the $P_j(x)$ is known. The solution as an eigenvalue and eigenvector problem is covered in Section 3.4.2.

One more equation may be added:

$$\sigma_\epsilon^2 = \sigma_A^2 - \sum_{j=1}^{n} \sigma_{Pj}^2 \qquad [3.57d]$$

where: $\sigma_{Pj}^2 = (c_j^2/N) \sum_{x=1}^{N} P_j^2(x)$ [3.57e]

It is obvious that the last term in [3.57d] expresses the explained variance, and σ_ϵ^2 is zero when the total variance σ_A^2 is matched (see Section 3.1.4). The original number of predictors P_j could amount to as many as observations are available. This is not the ideal goal in prediction. We aim at a systematic reduction in the number of predictors so that a few of them would be able to contain a maximum of information. This goal is equivalent to the determination of a suitable set of new predictors.

We formulate the problem, substituting $P_j(x)$ for $X_j(x)$, as the mathematical solution of:

$$P_j(x) = \sum_{k=1}^{n} \Phi_{kj} q_k(x) \qquad [3.58]$$

where the Φ_{kj} and $q_k(x)$ may be called empirical polynomials for which the condition of orthogonality requires:

$$\sum_{j=1}^{n} \phi_{kj} \cdot \phi_{mj} = \delta_{km} = \begin{cases} 1 \text{ for } k = m \\ 0 \text{ for } k \neq m \end{cases} \qquad [3.58a]$$

In order to minimize the (squared) residual error we set:

$$\sum_{x=1}^{N} Q_k(x) Q_m(x) = \alpha_k \delta_{km} \qquad [3.58b]$$

where $\alpha_k \geq \alpha_{k+1} \geq 0$, and $Q_i(x) = q_i(x) - \bar{q}_i$. The latter is the mean value of the $q_i(x)$ over x.

The n predictors, observed N times each, add up to the total variance:

$$\sigma_p^2 = \sum_{j=1}^{n} \sum_{x=1}^{N} P_j^2 = \sum_{j=1}^{n} \sigma_{Pj}^2 \qquad [3.58c]$$

Let us express the residual error:

$$\sigma_{\epsilon k}^2 = N^{-1} \sum_{x=1}^{N} [\epsilon_k(x) - \bar{\epsilon}_k]^2 \qquad [3.58d]$$

with $\sigma_{\epsilon n}^2 = 0$. We can then relate σ_P^2, $\sigma_{\epsilon n}^2$ and α_j by:

$$\sigma_P^2 - \sigma_{\epsilon k}^2 = N^{-1} \sum_{j=1}^{k} \alpha_j \qquad [3.58e]$$

The $\phi_j(x)$ as the empirical orthogonal function is associated with the jth largest eigenvalue of a matrix $\mathbf{D}$ of eigenvalues. This condition determines the limitation of the number of terms.

Let us denote the matrix $\mathbf{M}_s$ as a matrix of the elements s. Then $\mathbf{M}_X$, $\mathbf{M}_P$, $\mathbf{M}_q$, $\mathbf{M}_Q$ are matrices of N rows and n columns, whose elements are $X_j(x)$, $P_j(x)$, $q_k(x)$, $Q_k(x)$, respectively. We further need the square matrix $\mathbf{M}_\phi$ with elements ϕ_{km}. We can now formulate (for $\mathbf{M}_P$ replacing $\mathbf{M}_X$):

$$\mathbf{M}_P = \mathbf{M}_q \mathbf{M}_\phi^T \qquad [3.59a]$$

$$\mathbf{M}_\phi^T \mathbf{M}_\phi = \mathbf{I} \qquad [3.59b]$$

$$\mathbf{M}_Q^T \mathbf{M}_Q = \mathbf{D} \qquad [3.59c]$$

where $\mathbf{I}$ stands for the identity matrix and $\mathbf{D}$ for a diagonal matrix, a matrix, whose non-diagonal elements are zero. Equations [3.59a–c] are equivalent in meaning to [3.58, 3.58a and b]. The superscript T is used to indicate the transpose, a matrix, whose rows and columns have been interchanged.

We must now determine $\mathbf{M}_\phi$ and $\mathbf{M}_Q$. Let us define a symmetric matrix:

$$\mathbf{B} = \mathbf{M}_P^T \mathbf{M}_P \qquad [3.59d]$$

or: $b_{ij} = \Sigma\, p_{ik} p_{jk} \qquad [3.59e]$

The $\mathbf{B}$ is a matrix whose elements $N P_m P_k$ are proportional to the covariance of the predictors. If the statements [3.59a–c] hold, we can conclude:

$$\mathbf{M}_q = \mathbf{M}_P \cdot \mathbf{M}_\phi \qquad [3.59f]$$

and: $\mathbf{M}_\phi^T \mathbf{B} \mathbf{M}_\phi = \mathbf{D} \qquad [3.59g]$

If $\mathbf{M}_\phi$ satisfies [3.59b and g], then $\mathbf{M}_q$ defined by [3.59f] satisfies statement [3.59a and c].

As a condition of orthogonality, the transpose must be the inverse, namely:

$$\mathbf{M}_\phi^T = \mathbf{M}_\phi^{-1} \qquad [3.59h]$$

provided the inverse matrix exists.

The problem of finding solutions for [3.59b and f] with given $\mathbf{B}$ is a classical topic in matrix analysis (see Chapter 4). It can be solved by diagonalization of the matrix, but is also known in the literature as the determination of the eigenvalues of a matrix or the characteristic roots. The elements ϕ of the matrix M_ϕ are the eigenvectors of B. Once the M_ϕ is found, it is not difficult to solve for M_q, which is equivalent to calculating:

$$q_k(x) = \sum_{j=1}^{n} \phi_{kj} P_j(x) \qquad [3.60]$$

More details can be found in various articles (e.g. Lorenz, 1956; Gilman, 1957; Grimmer, 1963; Muller and Clodman, 1968; and others). See also Example 3.10.

It should be added that the orthogonality condition is also satisfied by the q_k:

$$\sum_{r=1}^{n} q_{kr} q_{mr} = \begin{cases} 0 \text{ for } k \neq m \\ 1 \text{ for } k = m \end{cases} \qquad [3.60a]$$

The n stands for the number of terms in the matrix. More procedural details can be found in Section 3.4.2 and Chapter 4.

Empirical orthogonal functions may in certain ways be considered as natural functions typical for the particular set of data. They have maximum efficiency of representation. The eigenvalues can be arranged in descending order and each eigenvalue represents a contribution to the total variance (see [3.58e]).

The input of data in the form of $a(x)$ or $A(x)$ is not restricted to a single vector. We could derive a field vector f_{ij} substituting $a(x)$ such as:

$$f_{ij}(x) = M_{aij} \qquad [3.57f]$$

where i and j denote area coordinates, or i stands for the location and j for the time. In this form Grimmer (1963) has investigated the monthly surface temperature pattern.

The computational work connected with the calculation of eigenvectors has today been largely reduced by the existence of electronic computers.

3.4.2 *The eigenvalue and eigenvector problem*

The preceding deliberation centered mainly upon the derivation of empirical orthogonal functions from a set of predictors. The limitation to predictors can be waived. Any field which can be represented by a vector, say f_n with n from 1 to N and m as the number of variables, can be represented by empirical orthogonal functions. Examples have been given by Sellers (1957), Obukhov (1960), White et al. (1958), Holmström (1963, 1970), Yudin (1966), Stidd (1967), Kutzbach (1967, 1970), Glahn (1962) or Craddock and Flood (1969) and others. The problem is to find a matrix M_e which satisfies $M_e^T R M_e = D$ where M_e represents the eigenvectors associated with the symmetric matrix R, again the superscript denoting the transpose.

Assume we have a field or elements which we can represent by a vector f_n. We establish the matrix M_f, with m by n dimensions, e.g. f_{ij} means the ith observation for the jth component (which can be the year, month, or any other vector coordinate). Then we need to maximize $(\mathbf{M}_e^T\mathbf{M}_f)^2 N^{-1}/(\mathbf{M}_e^T\mathbf{M}_e)$. This is equivalent to the maximizing process above and requires:

$$\mathbf{M}_e^T\mathbf{M}_e = \mathbf{I} \qquad [3.61a]$$

Since orthogonality is required, the inverse matrix must exist and be related to the transpose as:

$$\mathbf{M}_e^T = \mathbf{M}_e^{-1} \qquad [3.61b]$$

The data matrix $\mathbf{R}$ satisfies:

$$\mathbf{R} = N^{-1}(\mathbf{M}_f\mathbf{M}_f^T) \qquad [3.61c]$$

and is a symmetric matrix with elements:

$$r_{ij} = N^{-1} \sum_{n=1}^{N} f_{in}f_{jn} \qquad [3.61d]$$

The solution is an equation:

$$\mathbf{R}\mathbf{M}_e = \mu\mathbf{M}_e \qquad [3.61e]$$

where μ is the Lagrange multiplier determined by the eigenvectors. We expand now the matrix $\mathbf{M}_e$ to include all $\mathbf{M}_{ei}$ and denote this by $\mathbf{M}_E$. Then we find:

$$\mathbf{R}\mathbf{M}_E = \mathbf{M}_E \cdot \mathbf{D} \qquad [3.61f]$$

or: $\mathbf{M}_E^{-1}\mathbf{R}\mathbf{M}_E = \mathbf{D} \qquad [3.61g]$

where $\mathbf{D}$ stands for a diagonal matrix of eigenvalues.

We can write μ_i for the ith diagonal elements. Then μ_i is the eigenvalue associated with the eigenvector $\mathbf{M}_{ei}$, which is the ith column of $\mathbf{M}_E$. The μ_i are generally randomly distributed in $\mathbf{D}$, but we can sort them without change of results in the order $\mu_1 > \mu_2 > \mu_3 > ... > \mu_m$ as long as eigenvalue and eigenvector are properly related with one another. Combining all equations we summarize:

$$\mathbf{M}_E^T\mathbf{M}_f\mathbf{M}_f^T\mathbf{M}_E = \mathbf{D} \cdot N \qquad [3.62]$$

We may define: $\mathbf{M}_C = \mathbf{M}_E^T\mathbf{M}_f \qquad [3.62a]$

where now $\mathbf{M}_C$ is an $m \times n$ matrix. Then it follows:

$$\mathbf{M}_f = \mathbf{M}_E \cdot \mathbf{M}_C \qquad [3.62b]$$

and further: $f_n = \sum_{i=1}^{m} C_{in}\mathbf{M}_{ei}(n = 1, 2, ..., N) \qquad [3.62c]$

This shows that the observation vector f_n can be expressed as a linear combination of the m eigenvectors. This condition is equivalent to the postulation in [3.57] or [3.57a].

EMPIRICAL ORTHOGONAL POLYNOMIALS AND EIGENVALUES 257

We further conclude from the theory of eigenvectors* that for **R**:

$$\sigma_k^2 = \sum_{i=1}^{k} \mu_i^2 / \sum_{i=1}^{m} \mu_i^2 \qquad [3.62d]$$

This equation gives the fraction of the total explained variance (see equivalent to [3.58d and e]) due to the kth eigenvector.

In terms of matrix operation, we seek the eigenvector matrix **C** for which we can write:

$$\mathbf{MC} = \mathbf{CD} \qquad [3.63a]$$

or: $\mathbf{C^{-1}MC} = \mathbf{D}$ \qquad [3.63b]

where **M** is the matrix to be diagonalized, **D** the diagonal matrix with eigenvalues along the diagonal. Further $\mathbf{C^{-1}} = \mathbf{C^T} \neq 0$, and $\mathbf{C^{-1}C} = \mathbf{I}$.

In our notation **D** is the diagonal matrix ([3.61f]) that contains the eigenvalues and the $\mathbf{M_E}$ contains the eigenvectors for the matrix $\mathbf{M} = \mathbf{R}$. Since the latter is defined from the observational values f_n ([3.61c]), the eigenvalues and vectors are typical only for the particular observations matrix. The eigenvalues can be determined by the characteristic equation after [3.61e]:

$$(\mathbf{R} - \mu \mathbf{I}) \mathbf{M_e} = 0 \qquad [3.64a]$$

or the determinant is identical zero, namely:

$$\| \mathbf{R} - \mu \mathbf{I} \| = 0 \qquad [3.64b]$$

(double bar denotes the determinant).

It is evident from the theory of eigenvectors that the μ_i-values are the roots of the system:

$$\mathbf{RM_{e1}} = \mu_1 \mathbf{M_{e1}} \qquad [3.64c]$$

$$\mathbf{RM_{e2}} = \mu_2 \mathbf{M_{e2}} \qquad [3.64d]$$

Then:

$$\mathbf{M_E} = \begin{bmatrix} e_{11} & e_{21} & \cdots & e_{n1} \\ e_{12} & e_{22} & \cdots & e_{n2} \\ \cdots & \cdots & \cdots & \cdots \\ e_{1n} & e_{2n} & \cdots & e_{nn} \end{bmatrix} \qquad [3.64e]$$

$$\mathbf{RM_E} = \begin{bmatrix} \mu_1 e_{11} & \mu_2 e_{21} & \cdots & \mu_n e_{n1} \\ \mu_1 e_{12} & \mu_2 e_{22} & \cdots & \mu_n e_{n2} \\ \cdots & \cdots & \cdots & \cdots \\ \mu_1 e_{1n} & \mu_2 e_{2n} & \cdots & \mu_n e_{nn} \end{bmatrix} \qquad [3.64f]$$

---

* For eigenvalues from a non-symmetric matrix $\mathbf{M_p}$ where $\mathbf{C} = \mathbf{M_p^T M_p}$ the μ^2 in [3.62d] must be replaced by μ.

or: $RM_E = M_E \begin{bmatrix} \mu_1 & 0 & \cdots & 0 \\ 0 & \mu_2 & \cdots & 0 \\ \cdots & \cdots & \cdots & \cdots \\ 0 & 0 & \cdots & \mu_n \end{bmatrix}$ [3.64g]

More details on mathematical procedures are found in Section 4.1.

Example 3.10 Calculation of empirical polynomials. The general prediction problem has been formulated:

$$A(x) = \Sigma c_j P_j(x) + \epsilon(x) \qquad \text{(see [3.57a])}$$

where $A(x) = a(x) - \bar{a}(x)$, and the other symbols have been introduced with [3.57a]. We assume that $A(x)$ can be predicted from the P_j with a negligible error $\epsilon_j(x) \sim 0$ and that $\Sigma P_j(x) \epsilon(x) = 0$. As stated, the coefficients can be obtained from:

$$c_j = (-1)^j \| M_{cj} \| / \| M_{11} \| \qquad \text{(see [3.57b])}$$

(More details and the case $M_{11} = 0$ are explained later.) We could now assume that the prediction problem has been solved and we would stop at this point.

It is desirable, however, to eliminate any superfluous contribution by the P_j and simplify the dimensions of the system if possible. As discussed in the previous sections this task requires the conversion of the P_j to a system of empirical polynomials. We can write:

$$P_j(x) = \Sigma \Phi_{kj} Q_k(x) \qquad \text{(see [3.58])}$$

where Φ and Q are called empirical polynomials. We seek the matrix M_ϕ and M_Q.

Let us assume that $X_j(x) = M_x$ and is a given data matrix (e.g. temperature observations at three stations for consecutive days). These data are being used to predict the temperature at a fourth station with predictant $A(x)$. As an excerpt* the following data are treated:

$$M_x = \begin{bmatrix} 11.5 \,(°C) & 7.0 & 10.5 \\ 11.5 & 9.0 & 13.5 \\ 10.0 & 10.0 & 13.0 \\ 7.0 & 10.0 & 12.0 \end{bmatrix} \quad \text{with } M_P = \begin{bmatrix} 1.5 & -2.0 & -1.5 \\ 1.5 & 0 & 0.5 \\ 0 & 1.0 & 1.0 \\ -3.0 & 1.0 & 0 \end{bmatrix}$$

The matrix M_P is not symmetric and we must derive a symmetric matrix C.

---

* The data sample is restricted to four days in order to enable the reader to quickly recalculate all the computations.

EMPIRICAL ORTHOGONAL POLYNOMIALS AND EIGENVALUES

$$C = M_P^T \cdot M_P = \begin{bmatrix} 13.5 & -6 & -1.5 \\ -6 & 6 & 4 \\ -1.5 & 4 & 3.5 \end{bmatrix}$$

It is self-evident that C is a correlation matrix such as:

$$C = N^3 \begin{bmatrix} \sigma_1^2 & r_{12}\sigma_1\sigma_2 & r_{13}\sigma_1\sigma_3 \\ r_{12}\sigma_1\sigma_2 & \sigma_2^2 & r_{23}\sigma_2\sigma_3 \\ r_{13}\sigma_1\sigma_3 & r_{23}\sigma_2\sigma_3 & \sigma_3^2 \end{bmatrix} = N^3 \sigma_1^2 \sigma_2^2 \sigma_3^2 \begin{bmatrix} 1 & r_{12} & r_{13} \\ r_{12} & 1 & r_{23} \\ r_{13} & r_{23} & 1 \end{bmatrix}$$

$$= N^3 \sigma_1^2 \sigma_2^2 \sigma_3^2 C'$$

In other words, any correlation matrix can be directly utilized as the basis of an eigenvalue system.

We could now solve for the eigenvalues λ_j by the Jacobi method (see Section 3.5.4), but the route via the characteristic equation (see Section 4.8) is taken here since the 3×3 matrix is still a simple form and leads to only a third-order equation which has an explicit solution. We find for the given matrix C:

$$\lambda^3 - 23\lambda^2 + 95\lambda + 0 = 0$$

which can be transformed to the system:

$$(\lambda - \lambda_1)(\lambda - \lambda_2) \cdot \lambda_3 = 0$$

It is obvious that $\lambda_3 = 0$ is one solution. This leaves a quadratic equation:

$$\lambda^2 - 23\lambda + 95 = 0$$

with: $\lambda_1 = 11.5 + \sqrt{37.25} = 11.5 + 6.1033 = 17.6033$

and: $\lambda_2 = 11.5 - 6.1033 = 5.3967$

The check $(\lambda - 17.6033)(\lambda - 5.3967) = \lambda^2 - 23.0\lambda + 94.9997$ proves that λ_1 and λ_2 are accurate enough (17.6 and 5.4 give 95.04, but more decimals are needed for proper determination of eigenvectors).

The calculation of eigenvectors (see Section 4.13) follows now:

$\lambda_1 = 17.6033$

$$\begin{bmatrix} -4.1033 & -6 & -1.5 \\ -6 & -11.6033 & 4 \\ -1.5 & 4 & -14.1033 \end{bmatrix} \rightarrow \begin{bmatrix} 1 & 1.4222 & 0.3656 \\ 0 & -2.8299 & 6.1934 \\ 0 & 6.1934 & -13.5550 \end{bmatrix}$$

$$\rightarrow \begin{bmatrix} 1 & 0 & 3.5658 \\ 0 & 1 & -2.1886 \\ 0 & 0 & -1 \end{bmatrix}$$

$(e'_1)^T = 3.5658, -2.1886, -1$ divided by $\sqrt{\sum_1^3 (e'_{1i})^2} = \sqrt{18.505} = 4.3010$.

The normalized $e_1^T = 0.829, -0.509, -0.232$ (the T indicates the transpose).

$\lambda_2 = 5.3967$

$$\begin{bmatrix} 8.1033 & -6 & -1.5 \\ -6 & 0.6033 & 4 \\ -1.5 & 4 & -1.8967 \end{bmatrix} \rightarrow \begin{bmatrix} 1 & -0.7404 & -0.1851 \\ 0 & -3.8393 & 2.8893 \\ 0 & 2.8893 & -2.1744 \end{bmatrix}$$

$$\rightarrow \begin{bmatrix} 1 & 0 & -0.7423 \\ 0 & 1 & -0.7526 \\ 0 & 0 & -1 \end{bmatrix}$$

$(e'_2)^T = -0.7423, -0.7526, -1.0$ divided by $\sqrt{2.117} = 1.455$.

$e_2^T = -0.510, -0.517, -0.687$.

$\lambda_3 = 0$

$$\begin{bmatrix} 13.5 & -6 & -1.5 \\ -6 & 6 & 4 \\ -1.5 & 4 & 3.5 \end{bmatrix} \rightarrow \begin{bmatrix} 1 & -4/9 & -1/9 \\ 0 & 10/3 & 10/3 \\ 0 & 10/3 & 10/3 \end{bmatrix} \rightarrow \begin{bmatrix} 1 & 0 & 1/3 \\ 0 & 1 & 1 \\ 0 & 0 & -1 \end{bmatrix}$$

$(e'_3)^T = 1/3, 1, -1$; divided by $\sqrt{19/9} = 2.111$.

$e_3^T = 0.229, 0.608, -0.688$.

The three eigenvectors form the matrix M_ϕ which is displayed below. It is easy to check that $M_\phi^T = M_\phi^{-1}$ and $M_\phi \cdot M_\phi^T = I$. This last check is demonstrated for the matrix elements with three decimals accuracy in M_ϕ:

$$\begin{bmatrix} 0.829 & -0.510 & 0.229 \\ -0.509 & -0.517 & 0.688 \\ -0.232 & -0.687 & -0.688 \end{bmatrix} \begin{bmatrix} 0.829 & -0.509 & -0.232 \\ -0.510 & -0.517 & -0.687 \\ 0.229 & 0.688 & -0.688 \end{bmatrix}$$

$$= \begin{bmatrix} 0.9998 & -0.0007 & 0.0005 \\ -0.0007 & 0.9997 & 0.0001 \\ 0.0005 & 0.0001 & 0.9991 \end{bmatrix}$$

The inaccuracies in the identity matrix are caused by rounding. One could carry more decimal places if higher accuracy is desired. It can be further checked that $M_\phi^{-1} \cdot C \cdot M_\phi = D_\lambda$, namely:

$$\begin{bmatrix} 0.829 & -0.509 & -0.232 \\ -0.510 & -0.517 & -0.687 \\ 0.229 & 0.688 & -0.688 \end{bmatrix} \begin{bmatrix} 13.5 & -6 & -1.5 \\ -6 & 6 & 4 \\ -1.5 & 4 & 3.5 \end{bmatrix}$$

$$\cdot \begin{bmatrix} 0.829 & -0.510 & 0.229 \\ -0.509 & -0.517 & 0.688 \\ -0.232 & -0.687 & -0.688 \end{bmatrix} = \begin{bmatrix} 17.60 & 0 & 0 \\ 0 & 5.40 & 0 \\ 0 & 0 & 0 \end{bmatrix}$$

M_ϕ and M_Q are the empirical polynomials. The calculation of M_Q follows next with $M_P \cdot M_\phi = M_Q$:

$$\begin{bmatrix} 1.5 & -2.0 & -1.5 \\ 1.5 & 0 & 0.5 \\ 0 & 1.0 & 1.0 \\ -3.0 & 1.0 & 0 \end{bmatrix} \begin{bmatrix} 0.829 & -0.510 & 0.229 \\ -0.509 & -0.517 & 0.688 \\ -0.232 & -0.687 & -0.688 \end{bmatrix}$$

$$= \begin{bmatrix} 2.610 & 1.300 & 0 \\ 1.128 & -1.109 & 0 \\ -0.741 & -1.204 & 0 \\ -2.996 & -1.013 & 0 \end{bmatrix}$$

As is necessary, the squared sums of the three columns correspond to the eigenvalues,* i.e.:

$$\sum_1^4 q_{i1}^2 = 17.60 = \lambda_1; \sum_1^4 q_{i2}^2 = 5.40 = \lambda_2; \sum_1^4 q_{i3}^2 = 0 = \lambda_3$$

The eigenvalues represent, therefore, the explained variance. The zeros in the last column of M_Q are no surprise and could have been predicted after the eigenvalues had been determined. They correspond to $\lambda_3 = 0$. In other words, calculations of eigenvectors for insignificant or zero eigenvalues can be

---
* See footnote on page 257.

deleted. The Jacobi method delivers eigenvalues and M_ϕ at the same time. The route via the characteristic polynomial may sometimes save costs since eigenvalues are calculated first. In turn, other difficulties such as finding the roots of the characteristic polynomials (Chapter 4) are associated with the characteristic equation.

The evaluation of the eigenvalues may lead to a conclusion that even the smallest eigenvalue provides a significant non-negligible contribution. Then in most cases very little is gained by the calculation of M_ϕ and M_Q since the original data set M_p could also be orthogonalized and M_p could not be reduced to a form of lesser dimensions. In our case we can derive reduced matrices M_Q^* and M_ϕ^*, namely:

$$M_Q^{*T} = \begin{bmatrix} 2.610 & 1.128 & -0.741 & -2.996 \\ 1.300 & -1.109 & -1.204 & 1.013 \end{bmatrix}$$

$$M_\phi^{*T} = \begin{bmatrix} 0.829 & -0.509 & -0.232 \\ -0.510 & -0.517 & -0.687 \end{bmatrix}$$

It is easy to check that $M_p = M_Q^* \cdot M_\phi^{*T}$ since in our case $\lambda_3 \equiv 0$. In cases where $\lambda_3 \sim 0$ the corresponding relationship can be formulated as $M_P \sim M_Q^* M_\phi^{*T}$ with little loss of information, depending* on the ratio $\lambda_3/(\sum_{1}^{3} \lambda_i)$. In general M_Q^* would be a matrix of lesser dimension than M_P. This fact can thus be interpreted to mean that in our case one of the three stations contributes no new (independent) information. Independence is one prerequisite in linear regression analysis for full efficiency. The correlation between stations can readily be checked from the matrix of linear correlation coefficients:

$$C = \begin{bmatrix} 1 & -0.667 & -0.219 \\ -0.667 & 1 & 0.873 \\ -0.219 & 0.873 & 1 \end{bmatrix}$$

The system of empirical polynomials and the relationship with regression analysis will be further elucidated by subsequent discussion. The general equation [3.57a] can be written in expanded form for our case as:

$$A(x) = c_1 P_1(x) + c_2 P_2(x) + c_3 P_3(x)$$

where the coefficients must be calculated from the coefficient matrix M_c according to [3.57b]. (For simplicity the reference to the function x has been omitted in M_c.)

---

* See footnote on page 257.

$$M_c = \begin{bmatrix} 1 & \Sigma A & \Sigma AP_1 & \Sigma AP_2 & \Sigma AP_3 \\ c_0 & N & \Sigma P_1 & \Sigma P_2 & \Sigma P_3 \\ c_1 & \Sigma P_1 & \Sigma P_1^2 & \Sigma P_1 P_2 & \Sigma P_1 P_3 \\ c_2 & \Sigma P_2 & \Sigma P_2 P_1 & \Sigma P_2^2 & \Sigma P_2 P_3 \\ c_3 & \Sigma P_3 & \Sigma P_3 P_1 & \Sigma P_3 P_1 & \Sigma P_3^2 \end{bmatrix}$$

Now M_{cj} is the co-factor of c_j and M_{11} is the minor determinant (see Section 4.3b). In our case the $\Sigma P_i = 0$, $\Sigma A = 0$ and $c_0 = 0$. Hence:

$$M_c = \begin{bmatrix} 1 & 0 & \Sigma AP_1 & \Sigma AP_2 & \Sigma AP_3 \\ 0 & N & 0 & 0 & 0 \\ c_1 & 0 & \Sigma P_1^2 & \Sigma P_1 P_2 & \Sigma P_1 P_3 \\ c_2 & 0 & \Sigma P_2 P_1 & \Sigma P_2^2 & \Sigma P_2 P_3 \\ c_3 & 0 & \Sigma P_3 P_1 & \Sigma P_3 P_2 & \Sigma P_3 \end{bmatrix}$$

We learn that $M_{11} = N \cdot C$. The matrix C has an eigenvalue of zero, consequently $\|C\| = \|M_{11}\| = 0$ and we have a zero division in the coefficient equations. This means we could not calculate the coefficients for a three-station system. A zero minor determinant implies that the components of the linear system are not independent. We could now utilize a two-station system, but intuitively one would think that not all the available information given by the data has been utilized by a two-station prediction scheme. Before an answer is given the data sample is further explored.

It may be added that if $\|M_{11}\| \sim 0$, no singularity is generated in the rigid mathematical sense but large values of c_j could arise. Then the prediction scheme comprising several terms could be based on small differences between large terms which opens the door to all kinds of random errors.

We continue the elaboration with the calculation of an actual prediction scheme. We assume that a predictand $A(x)$ is given by $A^T(x) = 2, 2, 1, -5$ (°C as deviation from the mean of 15°C). The correlation coefficients can be found as $r_{AP_1} = 0.98$, $r_{AP_2} = 0.56$, $r_{AP_3} = -0.09$. This leads to the matrix:

$$M_c = \begin{bmatrix} 1 & 0 & 21 & -8 & -1 \\ 0 & 4 & 0 & 0 & 0 \\ c_1 & 0 & 13.5 & -6 & -1.5 \\ c_2 & 0 & -6 & 6 & 4 \\ c_3 & 0 & -1.5 & 4 & 3.5 \end{bmatrix}$$

We first consider a one-station model, for which P_1 is selected for its high correlation:

$$A'(x) = 1.56 P_1(x) \tag{1}$$

The following two-station models comprise all the permutations for two stations:

$$A'(x) = 1.73 P_1(x) + 0.40 P_2(x) \tag{2}$$
$$A'(x) = 1.6 P_1(x) + 0.40 P_3(x) \tag{3}$$
$$A'(x) = 4.8 P_2(x) + 5.2 P_3(x) \tag{4}$$

The three-station model has a singularity $\| M_{11} \| = 0$ and coefficients cannot be obtained. The recomputation of the $A'(x)$ for the individual models and $\Delta_i(x) = A(x) - A'(x)$ is illustrated in the following table:

A	A' for models				Differences		Empirical polynomial	
	one	two stations			Δ_1	Δ_2 to Δ_4	A'	Δ_5
	1	2	3	4				
2	2.34	1.80	1.80	1.80	−0.34	0.20	1.82	0.18
2	2.34	2.60	2.60	2.60	−0.34	−0.60	2.59	−0.59
1	0	0.40	0.40	0.40	1.00	0.60	0.38	0.62
−5	−4.68	−4.80	−4.80	−4.80	−0.34	−0.20	−4.78	−0.22
				$\Sigma \Delta_j^2(x)$	1.35	0.80	—	0.86

It is evident that $\Sigma \Delta_1^2 = r_{AP_1}^2 (N \cdot \sigma_A^2) = (0.98)^2 \cdot (34) = 1.35$. This result is confirmed by the table above. It is further seen that in our case (with $\lambda_3 = 0$) any of the two-station models are equivalent. As expected, the two-station model displays some improvement over the one-station model, but due to the high correlation between A and P_1 this improvement is small.

We study a model with empirical polynomials, namely:

$$A(x) = c_1 q_1(x) + c_2 q_2(x) \tag{5}$$

where the $q_j(x)$ are the columns of the matrix M_Q^*. Since M_Q^* is an orthogonal system, the coefficients are simple to calculate:

$$c_j = [\Sigma A(x) q_j(x)] / \Sigma q_j^2(x)$$
$$c_1 = 21.74/17.6 = 1.235$$
$$c_2 = -5.82/5.4 = -1.078$$

The reconstruction of $A'(x)$ leads to the results as displayed in the last two columns of the above table. It should be emphasized that the differences between the ordinary two-station model and the empirical polynomial scheme are non-significant and can be attributed to rounding.

The reader may now ask the delicate question of whether we have really gained something by the empirical polynomials and by all the work which has been put into calculating them. The answer is as difficult as the problem.

In our specific case we must admit that virtually nothing has been gained which could not have been achieved or interpreted by ordinary regression analysis. We would have noticed the singularity in the coefficient calculation, and the models for two stations provide the same good prediction as the empirical polynomials. It must be pointed out that we should not expect much improvement from a multiple-regression system when one predictor alone provides 96% of the variance ($r^2_{AP_1} = 0.9604$). However, we learn that two predictors can lead to a good prediction even when the individual correlations with the predictand are low. Stations 2 and 3 show low correlation with $A(x)$, but the combination of stations 2 and 3 is as good as the prediction by stations 1 and 3, and considerably better than either station 2 or 3 alone. Although in general the addition of predictors will reduce the left variance, the example illustrates that a saturation level will be reached after which additional predictors contribute little or nothing. This saturation depends on the properties of the predictors and the possible contribution by any newly added predictor.

Had a different set of $A(x)$ been given, the results and the general conclusions would have been the same, i.e. a particular prediction model will furnish a specific left variance, which can be achieved by a minimum number of terms. These can be obtained by empirical polynomials. Since $\lambda_3 = 0$ in our case the number of terms is two and any third station of the three stations contributes nothing. In terms of a regression analysis, again, two stations would suffice. This fact is confirmed by the analysis of the small prediction table in this example.

Had λ_3 been small and its eigenvector in $\mathbf{M}_Q$ negligible, the outcome of the best prediction by two stations would depend somewhat on the station pair, and we would have to check for the best permutation. This result is directly derived for empirical polynomials without search. It is obvious that a check for the best system is tedious and may be costly for a large number of possible predictors.

The question of whether the calculation of empirical polynomials is meritorious can therefore not be answered in a unique way. The answer will depend on the problem at hand. It may be generalized, however, that any possible increase in efficiency of a prediction scheme or equivalently any reduction in the number of predictors can be evaluated by the calculation of eigenvalues, although this procedure may not be the only tool. It is less likely, however, that other methods provide as convenient a tool as the calculation of eigenvalues as the number of predictors increases.

It should be added that [3.57a] comprises a system of linear predictors.

This does not necessarily exclude a non-linear predictand $A(x)$ or non-linear predictors $P_j(x)$ in the variate x. The system needs to be linear between A and P_j.

The calculation of eigenvalues cannot provide an answer to the question of whether a smaller number of non-linear predictors exists which would lead to the same or a better prediction result than given by the empirical polynomials of the linear regression system.

This leaves the question of representing atmospheric mapped parameters. e.g. a description of weather maps by empirical polynomials would result in the representation of an individual map by the minimum number of terms (i.e. eigenvalues and eigenvectors) but the empirical polynomials would vary from day to day. Their variation may disclose an unpredictable pattern. Where such variations are undesirable, a standard polynomial system could be advantageous.

Example 3.11 Empirical polynomials, eigenvalues, eigenvectors. In Example 3.10 the $\lambda_3 = 0$, and the prediction models for two stations were equivalent. In order to avoid any ambiguity arising from this special case a further example is added. The given components are again $A(x)$ and $P_j(x)$, with a new station number 3 in the previous model.

$$A(x) = \begin{bmatrix} 2 \\ 2 \\ 1 \\ -5 \end{bmatrix}; P_j(x) = \begin{bmatrix} 1.5 & -2.0 & -0.5 \\ 1.5 & 0 & -0.5 \\ 0 & 1.0 & 1.0 \\ -3.0 & 1.0 & 0 \end{bmatrix}$$

$$C = \begin{bmatrix} 13.5 & -6.0 & -1.5 \\ -6.0 & 6.0 & 2.0 \\ -1.5 & 2.0 & 1.5 \end{bmatrix}$$

As expected, the characteristic equation is third order:

$$\lambda^3 - 21\lambda^2 + 68\lambda - 36 = 0$$

The three roots (= eigenvalues) of the equation are:

$$(\lambda - 17.158)(\lambda - 3.181)(\lambda - 0.661) = 0$$

Analogously as in Example 3.10 we calculate the eigenvectors and M_Q.

$$\mathbf{M}_\phi = \begin{bmatrix} 0.860 & -0.498 & 0.109 \\ -0.489 & -0.746 & 0.453 \\ -0.144 & -0.443 & -0.885 \end{bmatrix};$$

$$\mathbf{M}_Q = \begin{bmatrix} 2.340 & 0.967 & -0.300 \\ 1.362 & -0.525 & 0.606 \\ -0.633 & -1.189 & -0.432 \\ -3.069 & 0.748 & 0.126 \end{bmatrix}$$

The coefficients matrix for the regression model has the following form:

$$\mathbf{M}_c = \begin{bmatrix} 1 & 0 & 21 & -8 & -1 \\ 0 & 4 & 0 & 0 & 0 \\ c_1 & 0 & 13.5 & -6 & -1.5 \\ c_2 & 0 & -6 & 6 & 2 \\ c_3 & 0 & -1.5 & 2 & 1.5 \end{bmatrix}$$

The single-station model is the same as previously derived:

$$A'(x) = 1.56 P_1(x) \tag{1}$$

The following two-station models can be obtained:

$$A'(x) = 1.73 P_1(x) + 0.40 P_2(x) \tag{2}$$
$$A'(x) = (5/3) P_1(x) + P_3(x) \tag{3}$$
$$A'(x) = -2 P_2(x) + 2 P_3(x) \tag{4}$$

In this case we have a three-station model, although with a zero coefficient for station 2:

$$A'(x) = (5/3) P_1(x) + 0 \cdot P_2(x) + P_3(x) \tag{5}$$

The empirical polynomial model provides:

$$A'(x) = 1.290 q_1(x) - 1.270 q_2(x) - 0.670 q_3(x) \tag{6}$$

The prediction table can be established as follows:

268 CURVE FITTING

$A(x)$	$A'(x)$ for models						Differences Δ for models						q_1+q_2	
	1	2	3	4	5	6	1	2	3	4	5	6	$A'(x)$	Δ
2	2.34	1.80	2.0	3.0	2.0	1.99	−0.34	−0.20	0	—	0	0.01	1.80	0.20
2	2.34	2.60	2.0	−1.0	2.0	2.02	−0.34	−0.60	0	1.0	0	−0.02	2.41	−0.41
1	0	0.40	1.0	0	1.0	0.98	1.00	0.60	0	3.0	0	0.02	0.71	0.29
−5	−4.68	−4.80	−5.0	−2.0	−5.0	−4.99	−0.34	−0.20	0	1.0	0	−0.01	−4.92	−0.08
					$\Sigma\Delta^2$	=	1.35	0.80	0	20.0	0	3.0		0.30

Again, an improvement from the one-station to the multiple-station model can be seen. We learn further that now a "best" and a "worst" two-station model exists. Since a "perfect" prediction is made with two stations, the three-station model shows a zero coefficient for the superfluous station 2.

The empirical polynomial model matches the "best prediction", although the two-component model in this example lags behind the regression model in the explained variance. However, the left variance for the two-term empirical polynomial model with 0.30/34 is smaller than 1% of the variance of $A(x)$. This small residual error should be expected because we have cut about 3% (i.e. $\lambda_3/\Sigma \lambda_i = 0.661/21$) from the total data information, but this is not a significant loss.

3.4.3 *Significance of eigenvectors and eigenvalues*

Statistical significance checks in computations of eigenvectors or eigenvalues are sometimes very healthy, but test criteria are difficult to obtain. The most critical problem is the decision of what hypothesis should be verified. Rao (1948), Bartlett (1951) or Lawley (1956) have dealt with this topic from the point of statistical theory. Problems such as the theoretical distribution of eigenvalues, the probability that eigenvalues arise from random data without distinct pattern, etc. could be checked with standard techniques of comparison of frequency distributions (see Essenwanger, 1974a, section 6 and Essenwanger, 1974b).

The size of the eigenvectors or their contribution to the variance may be another judgment. Although all of the eigenvalues or eigenvectors could be about the same size, it does not necessarily mean that they are now caused by chance and that no physical background is the reason for it. Evaluations must also include the possible number of eigenvectors from a certain set of data. One could check the magnitude of the eigenvectors, or the sum of various combinations. The best criterion would generally be an error ϵ_T. Its maximum would be unity for one eigenvector and would be $\epsilon_T = 1/n$ for n eigenvectors. We could check the individual explained variance $\sigma_e^2 = \mu_i^2/\Sigma \mu_i^2$ and check against ϵ_T^2.

More meaningful would be a check on how stable the system is, i.e. whether we have a stationary time series or not. This question can be answered by and large from the data and need not necessarily be confined to the prior computation of eigenvalues.

Another assumption could be the comparison of the analytical data from the model for the truncated series with the observed data. It is evident that the differences between the truncated series and the observed data should provide the left or unexplained variance. The magnitude of the error can be judged from this left variance. Since it is a subjective decision what magnitude of the error is tolerable and how high the percentage reduction of the truncated series must go to be acceptable (see Section 3.1.4), an absolute value of ϵ_T cannot be given to fit every task.

One of the tests which could also be physically meaningful is a check for the stability of the eigenvalues and the eigenvectors against observational errors. True, the random error cannot be eliminated from observations. We can check, however, how far the derived model of linear representation by empirical polynomials is affected by random fluctuations. If the system is more than pure chance, random fluctuations would certainly display little effect and random shuffling of the data may lead to different results by destroying the significance of the pattern. The latter process could reveal the physical background of the data if true randomization of data could be achieved by rearrangement of the data.

A test for the stability of eigenvectors against random errors has been introduced by Muller and Clodman (1968). We start with a data matrix, let this be called M_A (i.e. identical with M_f in the preceding section). If M_A is not a symmetric matrix, we must generate one by:

$$M_C = M_A^T \cdot M_A \qquad [3.65a]$$

The eigenvalue and eigenvector scheme can then be expressed in matrix form as:

$$V_C^T M_C V_C = D_C \qquad [3.65b]$$

$$\text{or: } V_A^T M_A V_A = D_A \qquad [3.65c]$$

with the same meaning as previously introduced. The matrix D contains the eigenvalues and V the eigenvectors. Let the elements of M_A be a_{ij}. We further assume a set of data independent from M_A, which we call M_B. Both data sets M_A and M_B would be subsets of the total series of observations, which we name M_Q. It can be expected that the eigenvalues are different when data set M_B is completely independent from M_A either in time or location.

We further assume a coefficient matrix M_H for which:

$$D_H = M_H^T \cdot M_H \qquad [3.65d]$$

is a diagonal matrix with elements h_{ij} such that:

$$h_{ij} = \sum_1^n v_{ki} a_{kj} \qquad [3.65e]$$

The v_{ki} are the elements of the eigenvector matrix V. Equation [3.65e] would be equivalent to [3.60] or [3.62c].

Finally we need a matrix of random errors M_ϵ. The random errors ϵ shall be so structured that their standard deviation in one column, say j, can be made proportional to the column mean of M_C or M_A. The $\bar{\epsilon}_j = 0$ within one column, hence:

$$\sum_{i=1}^{n_j} \epsilon_{ij}^2 = n\sigma_{\epsilon j}^2 \qquad [3.66a]$$

The postulated relationship to the column mean $\overline{X}_{Aj}$ or $\overline{X}_{Cj}$ will be:

$$\sigma_{\epsilon j} \sim \beta \overline{X}_{C_j} \qquad [3.66b]$$

where β is a proportionality factor. Combination of [3.66a] and [3.66b] provides:

$$\sum_{i=1}^{n_j} \epsilon_{ij}^2 = n\beta^2 \overline{X}_{C_j} \qquad [3.66c]$$

We now generate a matrix with random perturbations superimposed:

$$\mathbf{W}_\beta = \mathbf{M}_C + \mathbf{M}_\epsilon \qquad [3.67]$$

If $\mathbf{M}_A$ is a suitable symmetric matrix, the $\mathbf{M}_C$ could be replaced by $\mathbf{M}_A$. We calculate now eigenvalues and eigenvectors for $\mathbf{W}_\beta$, where the β can be arbitrarily selected. Muller and Clodman (1968) state that for snowfall the β ranges from 0.1 to 0.2; and in their article they let β vary from 0.1 to 0.4. Similar assumptions could be made for individual studies depending on the characteristics of the data under investigation.

We may construct one more set of eigenvalues, $\mathbf{V}_y$ from $\mathbf{M}_y$. The $\mathbf{M}_y$ is a matrix obtained from the independent data set $\mathbf{M}_B$ by rearranging the elements of an individual column in random fashion.

We establish now a test criterion, the variance of the eigenvalues. We define:

$$\mathbf{D}_{AB} = (\mathbf{V}_A^T \mathbf{M}_B)(\mathbf{M}_B^T \mathbf{V}_A) \qquad [3.68]$$

as the diagonal matrix from the data matrix $\mathbf{M}_B$ and the eigenvectors of $\mathbf{M}_A$. We further define:

$$\mathbf{D}_{HAY} = \mathbf{M}_H^T \mathbf{M}_H \qquad [3.68a]$$

for the matrix $\mathbf{M}_H$ with elements from $\mathbf{M}_A$ and eigenvectors $\mathbf{V}_y$ (see definition of elements by [3.65e]). We denote:

$$n\sigma_{AB}^2 = \Sigma (d_{AB} - \overline{d}_{AB})^2 \qquad [3.69a]$$

where the d_{AB} are the diagonal elements of $\mathbf{D}_{AB}$. Then:

$$n\sigma_{HAY}^2 = \Sigma (d_{HAY} - \overline{d}_{HAY})^2 \qquad [3.69b]$$

are the diagonal elements of the matrix equation [3.68a].

We compare now σ_{AB}^2 with σ_{HAY}^2 where the latter is determined for a significance level α. It is silently assumed that the d_{HAY} are normally (Gaussian) distributed and that the threshold for α can be taken as the regular factor of exceedance for the normal distribution (Essenwanger, 1974a, section 6.2).

The result of the checking procedure would be an indication of whether the eigenvectors obtained from $\mathbf{M}_A$ are significantly different with respect to the data $\mathbf{M}_B$.

We can further check σ_{WB}^2 with σ_{HWY}^2 where the definition is equivalent to [3.69a] and [3.69b], respectively. Since $\mathbf{W}$ is actually $\mathbf{W}(\beta)$ we can plot:

$$\Delta\sigma = \sigma_{WB}^2 - \sigma_{HWY}^2 \qquad [3.69c]$$

as function of β. The excess $\Delta\sigma$ would decrease until a level is found where the eigenvectors for **W** produce no significant results for the expansion of the data $\mathbf{M}_B$. The plot of $\Delta\sigma$ will indicate this very easily. The basis for the testing procedure is analogously the diagonal matrix:

$$\mathbf{D}_{WB} = (\mathbf{V}_W^T \mathbf{M}_B)(\mathbf{M}_B^T \mathbf{V}_W) \qquad [3.68b]$$

and: $\mathbf{D}_{HWY} = \mathbf{M}_H^T \mathbf{M}_H \qquad [3.68c]$

for eigenvectors $\mathbf{V}_y$ and the matrix **W**.

The plot of $\Delta\sigma$ versus β can also be interpreted such that the proportionality factor β now represents the additional relative error required to produce insignificant results at the probability level α for the purpose of expanding $\mathbf{M}_B$. The effect of the observational or random error superimposed upon $\mathbf{M}_A$ can be judged in the following way. We compute the correlation matrix:

$$\mathbf{R}_{AW} = \mathbf{V}_A^T \cdot \mathbf{V}_W \qquad [3.69d]$$

We rearrange the matrix $\mathbf{R}_{AW}$ by squaring the elements r_{ii} of $\mathbf{R}_{AW}$ and placing the highest column value r_{ij}^2 into the diagonal of $\mathbf{R}'_{AW}$. We then compute:

$$\tau_{AW} = 1 - \frac{1}{n}\sum_1^n d_{AW} \qquad [3.69e]$$

where now the d_{AW} stands for the maximum values of the r_{ij}^2 of the column. We can perform this summation over the total number of eigenvalues n or a selected number $k < n$. Since $r_{ij}^2 = d_{AW}$ has the function of a correlation, the τ_{AW} will be unity for uncorrelated data and zero for total correlation. This permits judgement of the similarity between two sets of eigenvectors. Mean values of τ_{AW} could be obtained as a function of the different assumptions for β.

Instead of computing τ_{AW}, we could also calculate τ_{AB}, which is based upon:

$$\mathbf{R}_{AB} = \mathbf{V}_A^T \mathbf{V}_B \qquad [3.69f]$$

Since the distribution of τ_{AB} depends on the sample size, the type of the distribution of the elements in $\mathbf{M}_A$ or $\mathbf{M}_B$, the number of eigenvalues of interest, and whether we have a stationary or non-stationary process, one would have to experiment with the factor $\mathbf{R}_{AB}$ to find a suitable criterion. Some suggestions for this procedure have been given by Muller and Clodman (1968).

Some statisticians may object to the test procedure described by Muller and Clodman. They treat the probability level α as a distribution factor rather than compute the F-ratio developed by Fisher (see Essenwanger, 1974a, section 6.5). This modification can be readily achieved by assuming:

$$F_{AB} = \sigma_{HAY}^2 / \sigma_{AB}^2 \qquad [3.70]$$

We would have to prove, however, that this ratio is a true variance ratio in Fisher's strict sense of the definition. Nevertheless, it would be advisable to check results by [3.70] and compare them with the level α obtained from the above description.

We could also check:

$$\tau'_{AB} = 1 - \tau_{AB} = \frac{1}{n}\sum_{1}^{n} d_{AW} \qquad [3.70a]$$

by some significance criterion for the correlation coefficient and then check the difference of the mean eigenvalues for M_A and M_B or the other established matrices. The latter test can be based on the t-test (see Essenwanger, 1974a, section 6.4).

This transformation to standard test models may be more to the liking of statisticians. The only difficulty arises in the proper assessment of degrees of freedom. If the degrees of freedom are only derived from the number of eigenvalues involved, the background of the evaluation is certainly not based on an over-optimistic view, i.e. if the differences prove to be significant, this conclusion may hold up even under more rigid considerations. Even with some doubt about the validity of the variance tests, the valuable contribution by Muller and Clodman (1968) is the development of the models M_B, M_y and M_ϵ for testing. It is obvious that [3.70] can be modified for other combinations of tests.

It may be helpful to know at least the errors of the eigenvalues. A simple method has recently been disclosed by Stumpf (1973).

Starting with the basic equation [3.61c], he has assumed the existence of an invariant operator:

$$I_G = \sum_{k=1}^{\infty} \mu_k \qquad [3.70b]$$

If the lower bound of eigenvectors is denoted by $m_L \leq \mu_i$ and the upper bound by m_U, then:

$$\mu_i \leq m_U = m_L + I_G - \sum_{k=1}^{n} m_{LK} \qquad [3.70c]$$

The upper and lower bounds can be obtained by methods developed by Weinberger (1960) and Stumpf (1973). The reader is referred to this literature.

3.4.4 Empirical polynomials and time-series analysis

Holmström (1970) has analyzed time series with empirical orthogonal functions. He starts with the formulation:

$$\mathbf{f}_m(\tau) = \sum_{m=1}^{M} \mathbf{q}_{nm} \mathbf{\Phi}_n(\tau) \qquad [3.71]$$

which is equivalent to [3.58].

For any point τ as the center of the time series a residual is then obtained:

$$r(t, \tau) = f(t + \tau) - \boldsymbol{\beta}(t)h(\tau) \qquad [3.71a]$$

The residual can be minimized. This leads to:

$$\int_{-\Delta T}^{\Delta T} K(\tau - \tau') h(\tau') \, d\tau' = \lambda h(\tau) \qquad [3.71b]$$

which resembles the eigenvalue problem. The ΔT comprises the point from the origin to the maximum lag. The kernel K is given by:

$$K(\tau - \tau') = R(\tau - \tau') \int_0^T f^2(t) \, dt \qquad [3.71c]$$

where $R(\tau - \tau')$ is the autocorrelation function of argument $(\tau - \tau')$. The function $h(\tau)$ is normalized with:

$$(1/2\Delta T) \int_{-\Delta T}^{\Delta T} h^2(\tau) \, d\tau = 1 \qquad [3.71d]$$

The constant λ is the eigenvalue:

$$\lambda = 2\Delta T \int_0^T \beta^2(t) \, dt \qquad [3.71e]$$

Since the $f(t)$ is the time series the integral in [3.71c] represents the variance. $R(\tau - \tau')$ is the autocorrelation function. Equation [3.71b] can be solved by the transformation:

$$\int_{-\Delta T}^{\Delta T} R(\tau - \tau') \sigma_T^2 h(\tau') \, d\tau' = \lambda h(\tau') \qquad [3.71f]$$

Holmström also studied the effect of a series expansion in the interval ΔT. The problem is complex and no conclusive answer has been found.

A further interesting application of empirical orthogonal functions to time series has been offered by Wallace and Dickinson (1972), who investigated the cross-spectrum matrix of daily rawinsondes for tropical disturbances by determining its eigenvectors. Later Wallace (1972) expanded this analysis by inclusion of a complex empirical orthogonal function expansion to separate the various wave types of tropical wave disturbances.

From the original time series $x_j(t)$ the transformation:

$$z_i(t) = \sum_j e_{ij} x_j(t) \qquad [3.72a]$$

is performed, where e_{ij} are the elements of the eigenvector matrix M_E. This eigenvector matrix is sought for the covariance matrix M_{cov}, whose elements are the covariances:

$$[x_j(t) x_k^T(t)] = M_{cov\,jk} \qquad [3.72b]$$

The procedure has been outlined in Section 3.4.2 and can be written here as:

$$M_{cov}M_E = \mu M_E \qquad [3.72c]$$

Consequently: $\sum e_{ij}e_{kj} = \delta_{ik}$ \qquad [3.73a]

is the orthogonality condition where δ_{ik} is the Kronecker delta (i.e. $\delta_{ik} = 0$ for $i = k$ and $\neq 0$ for $i \neq k$). The elements:

$$Z_{ik} = \sum_t z_i(t) z_k(t) = \mu_i \delta_{ik} \qquad [3.73b]$$

are uncorrelated, and the equation for the fraction of explained variance (e.g. [3.57e] or [3.62d]) and left variance (e.g. [3.57d], etc.) percentage reduction, etc. are applicable. Statistically significant components:

$$\lambda_{ij} = e_{ij}^2 \mu_i / \sigma_x^2 \qquad [3.72c]$$

are selected and further investigated. A further relation is:

$$\lambda_{ij} = (\sum x_j z_i)^2 / [\sum (x_j)^2 \sum (z_i)^2] \qquad [3.73d]$$

$$= [\text{cov}(x_j z_i)]^2 / (\sigma_x \sigma_y)^2 \qquad [3.73e]$$

which is simply the squared correlation coefficient between x_j and z_i.

The treatment of the complex empirical orthogonal function can be developed from the description given in Section 4.11c.

The filtering of the time series as adopted by Wallace and Dickinson (1972) will be presented later (Essenwanger, 1974b).

Another example of an application of empirical polynomials and factor analysis to multivariate time series and linear prediction has been given by Bauer (1971). In the evaluation of the merits of such a tool he goes as far as to recommend that results from rigorously defined statistical numerical prediction techniques should replace the widely accepted principle of persistence as a verification basis in meteorology.

The development of empirical orthogonal functions for the vertical temperature profile from radiometric measurements of satellites has been illustrated by Alishouse et al. (1967).

3.5 FACTOR ANALYSIS

3.5.1 General concepts and problem formulation

From the computation of eigenvectors and eigenvalues only a small step leads to factor analysis with respect to the related computational techniques. This fact can be readily checked from the mathematical formula below. In behavioral science factor analysis as a tool in statistical analysis has been known for more than fifty years, indeed since 1904 when Spearman (see also 1927) introduced the first model. Other references that can be found are Garnett (1919), Burt (1941), Cattell (1952 and 1965), Harman (1967), and many more. Articles on factor analysis in the meteorological literature, however, stem from the most recent years (e.g. Christensen and Bryson, 1966; Kutzbach, 1967; Buell, 1971; Buell and Bundgaard, 1971; or Hannes, 1974). By and large, most sections of physical science have taken little advantage of factor analysis contrary to the essential function it possesses in behavioral science. This difference in application may be attributed to the reality that in physics generally factor-controlled experiments can be set up, and the effect of one individual factor can be closely examined. Since experimentation with the atmosphere is limited, however, factor analysis may be helpful in meteorology. The complexity of factors, partially also unknown, may still prove to be a major difficulty for widespread application to meterological problems.

No comprehensive treatment of the topic can be given in the framework of this textbook. An attempt will be made, however, to present the necessary fundamentals for some practical initial work.

The goal of factor analysis can probably best be formulated as follows. We try to extract a number of principal factors from a set of observations arising from a multivariable population (generally assumed to be Gaussian). Thereby, a minimum number of factors would ideally contain all the information on the observations. Then we can set up a mathematical model for the observations. The usefulness goes, however, beyond that. The feedback would be the set-up of experimental designs based on the model or utilizing it. Even certain statistical test criteria in statistical analysis could be developed as a consequence of factor analysis.

An application of factor analysis in the estimation of multiple correlation and partial regression weights has only recently been treated by Lawley and Maxwell (1973).

3.5.2 The statistical (mathematical) model

Let us assume, a set of observation vectors X_i (with $i = 1, ..., n$) is produced by a population with n dimensions (in general a Gaussian multivariate distribution). We want to extract the f_j principal factors ($j = 1, ..., m$), where

FACTOR ANALYSIS   277

$m < n$. This goal requires that we obtain a minimum number of m factors f_j comprising a maximum of information from the data. Ideally the limited number of factors represents the full information embodied in the data. The factors should be linear functions of the n observation vectors X_i.

In many cases standardized vectors may be desirable. Let us assume that:

$$X_{ij} \doteq (z_{ij} - z_j)/\sigma_{zj} \qquad [3.74]$$

is the standardized vector from the original z_{ij} vector observation. We denote a column vector with small letters and a matrix by M_s with the subscript indicating the elements. Thus we can formulate our problems in general mathematical terms as:

$$x = M_\Phi = M_A f + M_D \varepsilon \qquad [3.75]$$

where x is a vector of n-dimensions, f and ε are random vectors of the dimension m and n, respectively. M_A is a matrix of dimension $n \times m$ and M_D a diagonal matrix $n \times n$. Since f and ε shall be random vectors, we can write for the expectancy E of the product $f \cdot \varepsilon$:

$$E(f\varepsilon) = 0 = \sum_1^{n_i}\sum_1^{n_j} f_{ij}\varepsilon_{ij} \qquad [3.75a]$$

or in matrix notation:

$$f \cdot \varepsilon^T = 0 \qquad [3.75b]$$

The meaning of the matrix and vector term in [3.75] will be explained below.

We assume now that a data matrix M_x with dimensions $n \times N$ can be established and we define:

$$M_r = (M_x \cdot M_x^T)/(N-1) \qquad [3.76a]$$

The defined matrix M_r is therefore a correlation matrix, whose elements are the correlation coefficients between the (standardized) vector observations. The correlation coefficients from the data are the estimators for the elements of M_r. If we expand the formulation [3.74] for the total set of data N, then x becomes the data matrix M_x, and the f and ε would also be expanded to the dimension $m \times N$ and $N \times N$, respectively. Hence:

$$M_x = M_A M_f + M_D \cdot M_\epsilon \qquad [3.74a]$$

This equation can be called a factor pattern equation. Since M_f is restricted to a limited number of specified factors, it is called the factor matrix. The M_ϵ can be identified as a matrix of unspecified factors plus errors. The matrix M_A can now be called the factor pattern.

In [3.74a] only M_x is known, and the separation into the terms at the right side would be an ambiguous task. Subsequently, it will be further narrowed down. We expand first:

$$M_f \cdot M_\epsilon^T = 0 \qquad [3.75c]$$

This follows from the definition in [3.75b]. We can deduce now:

$$\mathbf{M}_r = \mathbf{M}_x \mathbf{M}_x^T = \mathbf{M} \begin{bmatrix} \mathbf{M}_f \mathbf{M}_f^T & \vdots & \mathbf{M}_f \mathbf{M}_\epsilon^T \\ \cdots & \vdots & \cdots \\ \mathbf{M}_\epsilon \mathbf{M}_\phi^T & \vdots & \mathbf{M}_\epsilon \mathbf{M}_\epsilon^T \end{bmatrix} \mathbf{M}^T \quad [3.76b]$$

where we have omitted the division by $N-1$. Since the right side in parenthesis is a matrix multiplication, we may write the results as follows:

$$\mathbf{M}_x \mathbf{M}_x^T = \mathbf{M}_r = \mathbf{M} \begin{bmatrix} \boldsymbol{\Phi} & \vdots & 0 \\ \cdots & \vdots & \cdots \\ 0 & \vdots & \mathbf{I} \end{bmatrix} \quad [3.76c]$$

where $\mathbf{I}$ is the identity matrix and $\boldsymbol{\Phi}$ is a matrix to be determined.

The model of [3.75] can now be rewritten as:

$$\mathbf{M}_r = \mathbf{M}_A \boldsymbol{\Phi} \mathbf{M}_A^T + \mathbf{M}_D \mathbf{M}_D^T \quad [3.76d]$$

for uncorrelated (orthogonal) factors:

$$\boldsymbol{\Phi} = \delta_{jk} \quad [3.76e]$$

where δ_{jk} is the Kronecker delta (i.e. $\delta_{jk} = 0$ for $j \neq k$ and 1 for $j = k$). Consequently, the final answer is:

$$\mathbf{M}_r = \mathbf{M}_A \mathbf{M}_A^T + \mathbf{M}_D \mathbf{M}_D^T \quad [3.77]$$

In addition, one can readily prove that in this case:

$$\mathbf{M}_x \mathbf{M}_f^T = \mathbf{M}_A \mathbf{M}_f \mathbf{M}_f^T = \mathbf{M}_A \quad [3.77a]$$

Consequently, the columns of $\mathbf{M}_A$ are the correlation coefficients between the original components x (variables) and the factors f. This process is also called factor loading. Equations [3.77] and [3.77a] can be considered as the fundamental equations of factor analysis. The left side of [3.77] represents the correlation matrix between the x vectors, the first term at the right side the correlation between x variables and factors f, and the last term the "loading factor" of the diagonal, explained in more details in Section 3.5.3. Let us first define:

$$\mathbf{M}_R = \mathbf{M}_r - \mathbf{M}_D \mathbf{M}_D^T \quad [3.77b]$$

Then a new factor matrix $\mathbf{M}_R$ has been introduced with:

$$\mathbf{M}_R = \mathbf{M}_A \mathbf{M}_A^T \quad [3.77c]$$

(see later [3.77e]). For the moment let us go back to the problem of eigenvectors.

First, we would state in analogy to [3.61f and g]:

FACTOR ANALYSIS

$$M'_R = M_E D M_E^T \qquad [3.78a]$$

This equation* implies the determination of eigenvectors from a matrix M'_R. If we define the diagonal matrix of eigenvalues:

$$D = D_2 \cdot D_2 \qquad [3.78b]$$

and call D_2 the square root of the diagonal matrix, then:

$$M'_R = M_E D_2 D_2 M_E^T \qquad [3.78c]$$

We could further specify:

$$M_B = M_E D_2 \qquad [3.78d]$$

Then the matrix relation of [3.78c] can be rewritten:

$$M'_R = M_B M_B^T \qquad [3.78e]$$

This is identical however to [3.77c]. Ostensively $M_R = M'_R$, and the determination of the matrix M_A is an eigenvalue problem. We may now return to the analysis of [3.77c].

Equation [3.78e] may not be very illustrative. In the language of statistical terminology the matrix M_R is a correlation matrix between variables with elements r_{pq}. The individual element r_{pq} is the correlation between variables ξ_p and ξ_q, or the set of observation vectors X_p and X_q. The M_B or M_A contains the factors, and the correlation would be based on X_p and f_j. We have earlier employed the notation r_{12} for the linear correlation coefficient between the variables or observation vectors 1 and 2. In terms of a factor relationship, the r_{12} is composed of:

$$r_{12} = \sum_1^n r_{1f_j} r_{2f_j} \qquad [3.79a]$$

Similar equations can be written for the other variables. In general:

$$r_{pq} = \sum_{j=1}^n r_{pf_j} \cdot r_{qf_j} \qquad [3.79b]$$

More details on this correlation coefficient follow next.

3.5.3 The communality problem

It is known from the correlation analysis (e.g. Essen., 1974a, section 3) that the correlation coefficient assumes the value 1.0 when the variable is correlated with itself, i.e. $r_{pp} \equiv 1.0$. Thus we would anticipate unity in the diagonal of the matrix M_r. This assumption is the normal principle under which the correlation matrix has been derived.

Examination of [3.79b] (by writing it for $p = q$) reveals that r_{pp} is the sum of the correlations between the individual factors. This sum is not necessarily unity, i.e.:

---

* $M_E^T M'_R M_E = D$

$$r_{pp} = \sum_{j=1}^{n} r_{pfj}^2 \leq 1.0 \qquad [3.79c]$$

We must therefore make an adjustment in the relationship between M_r and M_R.

Let us introduce:

$$M_D M_D^T = M_I - M_{pp} \qquad [3.77d]$$

where M_I is a diagonal matrix with unity elements and M_{pp} is a diagonal matrix whose elements are r_{pp}. Then:

$$M_R = M_r - M_I + M_{pp} \qquad [3.77e]$$

has the non-diagonal elements of M_r and the r_{pp} is the diagonal. When $r_{pp} = 1.0$, then $M_R \equiv M_r$. Equation [3.77c], i.e. $M_R = M_A M_A^T$, resumes now the role of a true factor matrix.

One may interject that from a mathematical and statistical point of view the diagonal term in M_R should assume unity. Under this concept we would call it a "closed model". This closed model, however, is mostly given the name *component* analysis and is in fact based upon the relationship $M_R \equiv M_r$.

Let us remember, however, that we treat the problem of factor analysis. Under this principle the summation from $j = 1$ to n does not necessarily guarantee that the n terms are *all* the factors. Quite contrary, there may be influences completely unknown and not foreseeable. It was further outlined that we are interested in *principal* factors and not *all* of them. Thus $r_{pp} < 1.0$. The model of [3.77] with $M_R \neq M_r$ is then called an "*open model*" and the diagonal values are called "communalities".

The basic idea behind the "open model" is, therefore, the determination of reasonable "estimators" for the communalities. In simple notation, we can state with [3.77]:

$$c_j^2 = 1 - \alpha_j^2 \qquad [3.79d]$$

(Note, we replace r_{pp} by c_j^2, the square indicating a positive value). In matrix notation we find:

$$c_j^2 = a_j^T a_j \qquad [3.79e]$$

where these communalities are expanded applicable to the terms of the matrix M_A.

In a different notation we formulate:

$$c_j^2 = \sum_{p=1}^{m} a_{jp}^2 \, (j = 1, \ldots n) \qquad [3.79f]$$

It is evident that c_j^2 is a function of m, and c_j^2 is not known. We could also say that in [3.79d] α_j^2 is not known. This uncertainty prevents a unique solution and has been called "the Achilles heel of factor analysis" by Cattell, (1965).

FACTOR ANALYSIS

We must therefore substitute an "estimator" for c_j^2. In the past, four different solutions have been attempted (in addition to $\alpha_j^2 = 0$, i.e. $c_j^2 = 1.0$).

(1) Guttman (1956) recommends substituting for c_j^2 the multiple correlation coefficient (Essenwanger, 1974a, section 3) since this is the minimum condition.

(2) We could also set:

$$c_j^2 = \max(r_{jp}) \text{ for } p \neq j \quad [3.80a]$$

which is the maximum correlation of the row j.

(3) We could further employ:

$$c_j^2 = r_{jp}r_{jg}/r_{pq} \quad [3.80b]$$

where r_{jp} and r_{jq} are the largest correlation coefficients in the jth row of $\mathbf{M}_R$. (Note: $p \cdot q \neq j$).

(4) A fourth conjecture would be:

$$c_j^2 = (\Sigma r_{jp})/(n-1) \text{ for } p \neq j \quad [3.80c]$$

The detailed background for the above postulations can be found in the respective literature (e.g. Cattell, 1952, and other quoted references).

3.5.4 Factor loading or computation of the factors

In the previous section the estimation of the communalities has been established. Whichever method is selected we can now assume that $\mathbf{M}_R$ is given and we must calculate the eigenvalues and eigenvectors. One method is to maximize the total communality. We go back to [3.77c], or equivalently to [3.79f], and determine the first and largest eigenvalue. This leads to an iterative method for the solution of factors.

Let us write for the first communality (see [3.79f]):

$$c_1^2 = \sum_{p=1}^{m} \alpha_{1p}^2 \quad [3.81]$$

With the condition of [3.79b] we formulate:

$$r_{pq} = \sum_{j=1}^{} \alpha_{pj}\alpha_{qj} \quad [3.81a]$$

We know that $r_{pp} = c_p^2$. This leads to the system of n equations:

$$\sum_{q=1}^{n} r_{pq}a_{q1}^T - \mu_1 a_{p1} = 0 \quad [3.81b]$$

where μ_1 is the eigenvalue and a_{p1}, a_{q1} are the elements in $\mathbf{M}_A$, or in other words, we determine eigenvalues and eigenvectors of $\mathbf{M}_R$. The formulation of the problem is similar to that previously discussed, namely:

$$\|\mathbf{M}_R - \mu\| = 0 \quad [3.81c]$$

where the diagonal in M_R contains the communalities. The solution is known to be:

$$M_R e_1 = \mu_1 e_1 \quad [3.81d]$$

It can be shown that $c_1^2 \equiv \mu_1$. The eigenvectors are the coefficients of the first factor. We now subtract:

$$M_{R1} = M_R - e_1 e_1^T \quad [3.81e]$$

The residual matrix M_{R1} is now again treated as outlined under process [3.81b through d] and so on. It can be proven that the resulting series of eigenvalues and eigenvectors in this iterative procedure is identical with a closed solution from the matrix M_R. Mathematically we can formulate that the iteration procedure is based on the recursion formula:

$$M_{Rt} = M_{Rt-1} - e_t e_t^T \quad [3.81f]$$

The determination of the eigenvectors is treated in detail in Sections 4.4 and 4.8–4.13.

The iterative procedure can be replaced by a technique developed by Jacobi (1946) in which the matrix M_R is transformed into a diagonal matrix whose eigenvalues can be easily obtained. This is a technique very suitable for high-speed electronic computers. The diagonalization of M_R is performed by a stepwise orthogonal transformation, each step of the form:

$$M_R^s = B_{jk} M_R^s B_{jk}^T \quad [3.82]$$

The matrix B_{jk} is a diagonal matrix with 1's in the diagonal, and all other elements zero except for b_{jj}, b_{kk}, b_{jk} and b_{kj}:

$$b_{jj} = \cos\theta_{jk}$$
$$b_{jk} = -\sin\theta_{jk}$$
$$b_{kj} = \sin\theta_{jk}$$
$$b_{kk} = \cos\theta_{jk} \quad [3.82a]$$

This rule leads to the following matrix B_{jk}:

$$B_{jk} = \begin{bmatrix} 1 & 0 & 0 & & 0 & & 0 \\ 0 & 1 & \cdots & \cdots & \cdots & \cdots & \cdots \\ \vdots & & \cos\theta_{jk} & \cdots & -\sin\theta_{jk} & \cdots & 0 \\ \vdots & & & 1 & & & \\ \vdots & & \sin\theta_{jk} & & \cos\theta_{jk} & \cdots & 0 \\ \vdots & & \cdots & \cdots & \cdots & 1 & 0 \\ 0 & & \cdots & \cdots & \cdots & \cdots & 1 \end{bmatrix} \quad [3.82b]$$

FACTOR ANALYSIS

The transformation angle θ_{jk} is determined by:

$$\tan 2\theta_{jk} = 2r_{jk}/(r_{jj}^2 - r_{kk}^2) \qquad [3.82c]$$

With every individual step the zero elements derived earlier in M_R^s at step s will reappear as non-zero elements. According to Jacobi, however, the process converges and the non-diagonal elements in the final M_R^S are negligible. The diagonal elements are approximations of the eigenvalues. After solution of the eigenvalues μ_1 have been found, the eigenvectors can be determined as outlined earlier and in Section 4.13. The coefficients of each factor F_j are then simply the eigenvalues multiplied by these eigenvectors:

$$a_j^T = e_j \sqrt{\mu_j} \qquad [3.82d]$$

The significance of eigenvalues in factor analysis was investigated by various authors, among them Bartlett (1951), Burt (1952), and Lawley (1956). The topic was presented in Section 3.4.3.

3.5.5 Summary of factor-analysis procedure

The discussion and proof of some details may have distracted from the stepwise procedure, which will be outlined now. It was established that:

$$M_r = M_A M_A^T + M_D \cdot M_D^T \qquad [3.83]$$

is the fundamental equation where M_r is a correlation matrix whose elements are the (linear) correlations between the (standardized) vectors x. The matrix M_A must be determined.

Now: $M_A M_A^T = M_R$ \hfill [3.83a]

and: $M_R = M_r - M_D M_D^T$ \hfill [3.83b]

with: $M_D M_D^T = M_1 - M_{pp}$ \hfill [3.83c]

where M_D is a diagonal matrix. (Definition of M_1 and M_{pp} see [3.77d and e].) The finding of M_A is an eigenvector problem.

First, M_R is a matrix which includes all non-diagonal elements from M_r. The diagonal elements are either: (1) $r_{pp} = 1.0$ (closed model or component analysis); or (2) $r_{pp} < 1.0$ (open model or factor analysis).

Four methods for suitable estimators $c_j^2 = r_{pp}$ have been recommended (Section 3.5.3).

With M_R established, the solution for [3.83a] is an eigenvalue problem:

$$M_R = (M_E D_2)(D_2 M_E^T) = M_E D M_E^T \qquad [3.83d]$$

or: $(M_R - \mu I) M_E = 0$ \hfill [3.83e]

which can be reduced to:

$$\| M_R - \mu I \| = 0 \qquad [3.83f]$$

(the double bar denotes the determinant).

The eigenvalues are the elements of a diagonal matrix $\mathbf{D}$ ([2.78b]), and $\mathbf{D}_2 \cdot \mathbf{D}_2 = \mathbf{D}$. With $\mathbf{D}_2$ or $\mathbf{D}$ known, the matrix $\mathbf{M}_E$ can be calculated. Then the coefficients of each factor can be obtained from the matrices (see [3.82d]):

$$\mathbf{M}_A = \mathbf{M}_E \mathbf{D}_2 \qquad [3.83\text{g}]$$

Finally: $\mathbf{M}_x \mathbf{M}_f^T = \mathbf{M}_A \qquad [3.83\text{h}]$

or, with the assumptions (see also [3.77a]):

$$\mathbf{M}_f = \mathbf{M}_A^T \cdot \mathbf{M}_x \qquad [3.83\text{i}]$$

The calculation of eigenvalues and eigenvectors has been illustrated in Examples 3.10 and 3.11. Other details can be found in Chapter 4. The factor loading procedure has been outlined in Section 3.5.4.

3.5.6 Decision on the number of factors

Of great importance is the number of factors, which in general is unknown. It is obvious that we should have more observations than factors. Let us assume we have twenty factors generating a windspeed measured at a certain location and time. We could make ten measurements and would probably be able to extract the twenty factors. A restriction to five observations does not mean, however, that only four factors are acting now. This would be the number we could mathematically extract as the upper limit. Logically we must assume that the twenty factors in the above concept are still influencing the windspeed. The number of factors can therefore exceed the "rank" of the correlative matrix $\mathbf{M}_R$ if the number of observations is small.

These principles have been discussed in the literature. The first method for the decision on the number of factors would follow a mathematical concept. The rank of the correlation matrix $\mathbf{M}_R$ limits the factors. We could determine the communalities by the multiple correlation. Since the method minimizes the number of factors, Bartlett (1950b, 1951), Burt (1952) and others have recommended a statistical evaluation. Their model is based on statistical significance and the maximum-likelihood procedure. The sign values of $\mathbf{M}_R$ are tested by the χ^2-test (e.g. see Essenwanger, 1974a, Section 6.7):

$$\chi^2 = -C_0[\ln(\theta_1) - (p-k)\ln(\theta_2)] \qquad [3.84]$$

with: $\theta_1 = \mu_{k+1} \cdot \mu_{k+2} \ldots \mu_p \qquad [3.84\text{a}]$

$\theta_2 = (\mu_{k+1} + \mu_{k+2} + \ldots + \mu_p)/(p-k), \qquad [3.84\text{b}]$

$C_0 = N - [2p + 4k + 7 + 2/(p-k)] - 1 \qquad [3.84\text{c}]$

This system has $(p-k+2)(p-k+1)/2$ degrees of freedom. The number of data pairs in the correlation is N. The p is the number of columns (and

FACTOR ANALYSIS

rows) of the covariance matrix M_R while k is the number of significantly different eigenvalues (in the test). We would test for a specific number of eigenvalues, starting with $k_1 + 1$ significantly different eigenvalues. We continue until we exceed the significance level.

A final approach would be the decision on the number of factors by the factor-structure criterion. In this concept the factors are broken down into real factors and error factors, where the real factors are subdivided into trivial and large true factors. This reflection of factors is made by a process called rotation. The rotation will be discussed in the next section. The method is very suitable for calculating factors by high-speed digital computer.

3.5.7 Rotation in factor analysis

We have up to now established a scheme to relate M_R by [3.78e] with M_B. This procedure would reproduce the given matrix M_R to a high degree, let us say 99% or so. The factors deduced by this process may not reflect the real influence behind the measured data nor represent the true physical background for the generation of the matrix M_R. The solution is, therefore, not unique as was pointed out previously.

We now rotate the system from the original solution M_B to a system M_C, where:

$$M_C = M_B \cdot M_T \tag{3.85a}$$

and M_T is a $k \times k$ transformation matrix. M_C is then the $n \times k$ solution matrix after rotation. The crucial question is, therefore, how we obtain the rotation matrix. With M_T known, the computation of M_C is strictly a process of matrix algebra (see also Cattell and Cattell, 1955).

We look for a rotation that will result in factors which correspond to certain scientific entities whose nature has an explanation either in the form of a physical process or as generators of the data. Two choices are available, the simple structure and the cofactor rotation concept.

In the first technique, a solution for the rotation is found where each factor affects only a few variables. In terms of the correlation matrix we would say that the factors correlate with only a few variables, the smallest possible number.

Cattell (1944, 1965) and Cattell and Cattell (1955) have introduced a second method which may lead to technical difficulties, however. The rotation is based on two sets of data whose correlation matrix M_R has been determined. We derive M_{B1} and M_{B2} from these two data sets. It is then attempted to rotate M_{B1} and M_{B2} so that both render the same M_C.

Rotation is neglected by many researchers although it is an important part of the factor analysis. Since rotation is usually a tedious process, even by computers, the tendency to avoid the rotation process is quite understandable.

The rotation could be made by graphical methods, but for more than two factors the multidimensional schemes are not perceptible. One mathematical model is the technique to maximize the variance of the coefficients in [3.84a]. We can require that M_T is orthogonal, but non-orthogonal transformations are permissible.

We calculate:

$$q_{rs} = c_{rs}^2/r_s^2 \begin{cases} r = 1 \ldots n \\ s = 1 \ldots m \end{cases} \quad [3.85b]$$

where c_s are elements of the matrix M_c and r_s^2 is defined by [3.79d] or [3.80a].

For an orthogonal transformation $M_T \cdot M_T^T = I$ is required. We should obtain a maximum for:

$$\sum_{n=1}^{m} \left(\sum_{s=1}^{n} q_{rs}^2 - \bar{q}_s^2 \right) \quad [3.85c]$$

The $\bar{q}_s$ stands for the mean, namely:

$$\bar{q}_s = n^{-1} \sum_{r=1}^{n} q_{rs} \quad [3.85d]$$

For non-orthogonal transformations we calculate:

$$M_X M_C^T = M_C M_G M_G^T = M_C (M_T M_T^T)^{-1} \quad [3.86]$$

We can assume M_T to be a vector with components $(t_1, t_2, \ldots t_m)$ and determinant $\| M_T \| = 1$ for $r = 1, \ldots m$. Thus we minimize:

$$\sum_{s<S}^{n} \left(\sum_{r=1}^{n} q_{rs} \cdot q_{rS} - \bar{q}_s \bar{q}_S \right) \quad [3.86a]$$

with the same definition of q_s as in [3.85d].

As pointed out by Christensen and Bryson (1966) in many meteorological applications the efficiency of the principal components representation of the meteorological field may be sufficient and a transformation to "factors" may not be necessary. This means that in meteorology rotation may not be as important as in psychometric applications.

3.5.8 *Analysis of covariance*

In the previous sections the factor analysis was based on the correlation matrix. An alternate method could be based on employing the covariance instead of the correlation. Since the covariances for the individual variables are not standardized, the results of a factor analysis on covariances may be quite different from the analysis of the correlation matrix. We would relate the two matrices:

$$M_{cov} = M_\Delta M_R M_\Delta \quad [3.87]$$

FACTOR ANALYSIS

where now M_Δ is a diagonal matrix of raw score sigmas which transforms from standard scores to raw score deviation and which would transform M_R into M_{cov}. Hence:

$$M_{cov} = M_\Delta M_A M_A^T M_\Delta \qquad [3.87a]$$

According to Cattell, however, the raw score units are essentially accidential. Thus, not really anything would be gained by employing covariance factor analysis. Nevertheless, Buell (1971) recommends the covariance method for meteorological application in the form of an integral equation system.

As before, we start with:

$$M_{cov} M_A = M_A D \qquad [3.88]$$

in which M_A is a matrix with elements a_{ij} and the M_E stands for the eigenvector matrix; D is the eigenvalue (diagonal) matrix. Buell transforms this system to:

$$\mu_j \cdot e_j(x_i)\sqrt{A_i} = \sum_k [\sqrt{A_i} C(x_i, x_k)\sqrt{A_k}][\sqrt{A_k} e_j(x_k)] \qquad [3.88a]$$

where A_k is an appropriate multiplier, μ_j the eigenvalues, e_j the eigenvectors constructed at the points P_i, and $C(x_i, x_k)$ the covariance function of the observations x_i, x_k. The point P_i indicates a space coordinate. For the jth columns of M_A we find:

$$a_j(x_i) = e_j(x_i)\sqrt{A_j} \qquad [3.89a]$$

The elements of M_{cov} become:

$$c_{ik} = \sqrt{A_j} C(x_i, x_k)\sqrt{A_k} \qquad [3.89b]$$

The conversion from [3.88] to [3.88a] is not trivial.

The coefficients A_j depend on the solution of the integral equation:

$$\mu_j e_j(x) = \int_a^b C(x, X) e_j(X) \, dX \qquad [3.88b]$$

For a trapezoid integration method (and quadrature formula) we find for equally spaced x_j points:

$$A_j = h \quad j = 2, 3, \ldots n-1 \qquad [3.90a]$$

except: $A_j = h/2 \quad j = 1$ and n \qquad [3.90b]

where $h = x_{i+1} - x_i$. The expanded trapezoidal rule replaces the integral (left side) by:

$$\int_{x_0}^{x_m} f(x)\, dx = h\left(\frac{f_0}{2} + f_1 + \ldots f_{m-1} + \frac{f_m}{2}\right) - \frac{mh^3}{12} + f''(\xi) \qquad [3.90c]$$

The evaluation of the error term can be found in Abramowitz and Stegun

(1964). The matrix M_A is orthonormal and the same can be derived for the a_j-values calculated by [3.89a]:

$$\sum_{i=1}^{n} a_j(x_i) a_k(x_i) = \delta_{jk} \qquad [3.89c]$$

The Kronecker delta δ_{jk} has been defined (see [3.76e]).

For points not included in the original solution scheme we can deduce:

$$\mu_i e_i(x) = \sum_{k=1}^{n} C(x, x_k) e_i(x_k) A_k \qquad [3.88c]$$

This possibility permits quick solutions for an extended number of data points. Equation [3.88a] can be employed for a system of reduced size and then the system can be expanded to include all the points. Since the integral representation of [3.88] depends on the domain, this expansion would be permissible as the expansion integrates over the same domain. This contrasts with the results by Palmieria (1968) whose solution by the original [3.88] renders differences between the third and higher eigenvalues.

The expansion of the system can be outlined as follows. Assume the first solution is based on n data points. Then $C(x, x_k) = C_1$. If we define the expanded (weighted) covariance matrix by C_2, then:

$$C^* = M_{A2} D M_{A2} \qquad [3.88d]$$

is an approximation to C_2 with M_{A2} being the approximate matrix of n eigenfunctions for the expanded data points, or:

$$C_2 = C^* + P$$

where the elements of P are corrections to C^* to arrive at C_2. See also Wilkinson (1965).

3.5.9 Modified correlation input

The original factor analysis was based on a Gaussian distribution of the observations. If this assumption is not the case, the correlation coefficient r_{ik} can be replaced by proper substitution. e.g. we can use Spearman's rank correlation or for a contingency table of four fields a through d the tetrachoric coefficient such as:

$$r_{\text{tet}} = \cos [\pi/(1 + \sqrt{ad/bc})] \qquad [3.91a]$$

or another form:

$$r_\varphi = (bc - ad)/\sqrt{(a+b)(c+d)(a+c)(b+a)} \qquad [3.91b]$$

The impact of the substitution of these values and the consequences in a factor analysis may sometimes be very difficult to evaluate.

3.6 ANALYSIS OF TIME SERIES

One of the primary interests in observing meteorological elements besides their spatial relationship is their behavior in time. Not only are we interested in the past history but also in the prediction of future events. Present, past and future become thus an integral part of any consideration of meteorological elements, and the study of their time sequence is not different from other fields except that it is based on physical rather than social or economic behavior.

Some tools such as autocorrelation or power-spectrum have been introduced in individual sections and others will still follow. Nevertheless, a special section on time series seems appropriate and desirable to familiarize the reader with various techniques and procedures because some of the tools such as the Fourier series have been presented from a general point of view. Later (Essenwanger, 1974b) other related topics such as smoothing will be presented. Here discussions will center on particular treatments of time series. One of its goals is the analysis of the underlying structure and physical mechanism producing the finite series of data observed.

Four major topics can largely be distinguished, namely autoregression, moving average, autocorrelation, and spectral density.

In autoregression we treat the current or future value in terms of linear regression analysis, and we will find familiarity with linear regression models and curve fitting by polynomials.

In the moving-average model the observations are confined to a specified time interval, and the average is calculated. The time variation of this average is then subjected to further investigation. Usually the average, freed from random and short-term fluctuations, can be better represented and interpreted than the individual observation because the mean is more stabilized than the individual observation. This tool is best suitable when a strong time dependence exists such as a trend, etc.

The correlation analysis is based largely on time lags or autocorrelations and the representation of the resulting correlograms by functions and curve-fitting procedures.

Related to, but not identical with the correlogram is the spectral density. Many authors may understand only the power-spectrum under this type of modelling. Other procedures such as a Fourier analysis or description by other periodic functions may be equally important.

Although in analysis of time series the procedures can be traced back to these four basic techniques, a mixture of these groups is not excluded. In fact, in many practical applications the various tools will supplement one another, and insight into the structure of the data is generally enhanced by a combination of models.

The sequence of discussion of the individual procedures is somewhat arbitrary, and no logical argument can be brought forward for one or another succession of topics. The author has chosen the progression which follows subsequently.

Before the time-series analysis is further pursued, one should be aware of some postulations. In the power-spectrum analysis it was silently assumed that the autocorrelation function is "stationary" over the period of observations (comprising N data). This expression "stationary" implies that the autocorrelation function of the subperiods would render the same estimate. In meteorological time-series this property is not always true, as summer and winter may lead to different autocorrelation functions. Then restriction to one season, although limiting the observational data, would bring the data more in line with the assumption of a stationary process. In the following a stationary process is assumed whenever this limitation is not specifically waived.

The meteorologist (climatologist) should stress this point of stationarity in all practical applications. It is especially important to evaluate this restriction for user application in other fields. e.g., the engineer may ask for an annual summary, but a seasonal breakdown may serve much better the needs and purpose of the requester. Since the engineer may not be familiar with stationary processes in atmospheric science, the meteorologist should call this fact to his attention.

Finally, the reader should keep in mind that three specifications are involved in time-series analysis: (1) How do we define it? (2) How do we compute it? (3) How can the answer be interpreted? Any individual solutions to one point does not automatically imply that the other two points can be answered, too.

3.6.1 *General representation*

Before any of the specific models are introduced, it is appropriate to explain some basic principles of time-series data. We may decompose any set of time-series data into two basic components symbolized as:

$$X_t = Y(t) + C_t \qquad [3.92]$$

The X_t denotes the observed data as a time sequence where $1 < t < N$ or $1 < t < n_T$, depending on whether N, the total data, is available or only a selection up to n_T.

The first component $Y(t)$ represents the systematic part, the second component C_t the random fluctuations. The latter can comprise irregularities, random error, small-scale variations, instrumental error or similar short time oscillations, which can be in principle expressed in the form of a statistical expectation, e.g.:

$$E(C_t) = \overline{C_t} \qquad [3.92a]$$

$$E(C_t^2) = \sigma_C^2 \qquad [3.92b]$$

In essence we adopt for the interpretation of C_t a frequency model with a specified type of distribution and variance. Under certain conditions (e.g. Gaussian error law) we may find that $\bar{C}_t = 0$, or the frequency distribution is Gaussian and the variance σ_C^2 is the only parameter of significance. This interpretation may not always be realistic and cannot be assumed automatically.

It is self evident that C_t depends as much on the definition of the systematic component $Y(t)$ as on the actual random fluctuations. As such it can be interpreted as the "left variance" of the model. There is no fixed boundary between $Y(t)$ and C_t, and in this sense the systematic part could be expanded until $\sigma_C^2 \to 0$. Unless the $Y(t)$ is known a priori or is zero the C_t cannot be observed. Although instruments can be manufactured to separate $Y(t)$ and measure the variance of C_t, the definition of $Y(t)$ is arbitrary.

We may find that $Y(t)$ is best fitted by polynomials; in other cases it may be a Fourier series or another model. Details on the individual models follow.

3.6.2 The autoregressive model

In the autoregressive model the current value of an observation is expressed in linear terms of previous values, e.g.:

$$X_t = \alpha_1 X_{t-1} + \alpha_2 X_{t-2} + \dots + \alpha_a X_{t-a} + C_t \qquad [3.93a]$$

This model is called an autoregressive model of order a, because [3.93a] resembles the regression model (e.g. [3.1a]), but is based on its own previously observed values. The coefficients α_i may be determined as outlined for [3.1a] (see also Essenwanger, 1974a, section 3 or 4). Usually the individual X_t is given as a deviation from the mean or can be expressed as such. We shall assume that X_t denotes a deviation from the mean in the following discussion. If the model fits the observed data, we shall call it a process rather than a model.

The autoregressive model can be reformulated:

$$(1 - \alpha_1 B - \alpha_2 B^2 - \dots \alpha_a B^a) X_t = C_t \qquad [3.93b]$$

where B is a backward-shift operator:

$$BX_t = X_{t-1} \qquad [3.93c]$$

or: $B^p X_t = X_{t-p}$ \qquad [3.93d]

A forward-shift operator would then be given by:

$$V X_t = X_{t+1} \qquad [3.93e]$$

or: $V^q X_t = X_{t+q}$ \qquad [3.93f]

also: $V^m = B^{-m}$ \qquad [3.93g]

With this definition we abbreviate [3.93b] to:

$\alpha(B) X_t = C_t$ [3.93h]

or in reciprocal form: $X_t = \alpha^{-1}(B) C_t$ [3.93i]

Since we are able to substitute the X_{t-i} in [3.93a] by X_{t-i-1}, with i from 0 to the maximum of the observations, it is concluded from the theoretical point of view that the autoregressive model can yield eventually an infinite series.

The order of the model is given by the power of B. The first-order autoregressive model (process) is sometimes also called a Markov model (process):

$X_t(1 - \alpha_1 B) = C_t$ [3.93j]

The autoregressive model needs estimation of the α_i, which is covered in Section 3.6.9. Then B would be estimated with the aid of [3.93c] or similar procedures.

Estimation of α_1 by:

$$\hat{\alpha}_1 = \sum_{t=2}^{N} X_t X_{t-1} / \sum_{t=2}^{N} X_{t-1}^2$$ [3.93k]

is considered to provide the maximum-likelihood estimator. The distribution of the maximum-likelihood estimator was recently studied by Reeves (1972) who concludes that the distribution function requires the approximation of a linear combination of χ^2-variates.

3.6.3 *The moving-average model*

A moving-average model represents the observed value by a (finite) number b of linear terms of a previously derived variable Z_t. The model is useful when successive values are highly dependent on previous values. First let us consider an auxiliary model.

We could express the series by independent values c_t. In a linear filter model these c_t are regarded as "shocks", randomly drawn from a fixed distribution with zero mean and variance σ_c^2. This type of sequence of random variables $c_t, c_{t-1}, c_{t-2} \ldots$ etc. is also called a white-noise model by the engineer. The representation would be:

$X_t = c_0 + c_t + \varphi_1 c_{t-1} + \varphi_2 c_{t-2} + \ldots \varphi_n c_{t-n}$ [3.94a]

Again, we can reformulate the model with the backward-shift operator B by setting:

$X_t = c_0 + \varphi(B) c_t$ [3.94b]

We call c_0 the "level of noise" and $\varphi(B)$ the transfer function of a filter:

$\varphi(B) = 1 + \varphi_1 B + \varphi_2 B^2 + \ldots \varphi_n B^n$ [3.94c]

If the sequence is finite or infinite but convergent, the filter is considered stable; the process is stationary. Then c_0 is the mean, otherwise c_0 is merely a reference level.

We now define a moving-average model:

$$X_t = Z_t - \beta_1 Z_{t-1} - \dots - \beta_b Z_{t-b} \qquad [3.94d]$$

where the Z_t-parameter (variable) can be identical with the symbol c_t previously employed. The name "moving average" may sometimes be misleading as the weights 1, $\beta_1, \beta_2 \dots \beta_b$ need neither to sum up to unity nor are they restricted in their signs. Again, the coefficients β_i could be determined with the procedure outlined in previous sections (see further details in Essenwanger, 1974a, section 3).

In accordance with previous postulations we may reformulate:

$$X_t = (1 - \beta_1 B - \beta_2 B^2 - \dots - \beta_b B^b) Z_t \qquad [3.94e]$$

or in abbreviated form:

$$X_t = \beta(B) Z_t \qquad [3.94f]$$

3.6.4 A mixed model

The combination of two models into one increases the flexibility of the model in fitting actual time-series. This model is known as the mixed autoregressive–moving-average model. The combination provides:

$$X_t = \alpha_1 X_{t-1} + \dots + \alpha_a X_{t-a} + c + Z_t - \beta_1 Z_{t-1} - \dots \beta_b Z_{t-b} \qquad [3.95a]$$

This can be written in abbreviated form:

$$X_t = \sum_{j=1}^{a} \alpha_j X_{t-j} + c + Z_t + \sum_{k=1}^{b} -\beta_k Z_{t-k} \qquad [3.95b]$$

(The sign is again determined by the fitting process, the c is constant and often omitted, i.e., $c = 0$.)

The Z_t as defined before represents a series of random (independent) variables, and is called a residual process. Expectant values of Z_t according to the definition by [3.94d] can be set as $E(Z_t) = 0$ and $E(Z_t^2) = \sigma_z^2$. We may abbreviate as previously, for $c = 0$:

$$\alpha(B) X_t = \beta(B) Z_t \qquad [3.95c]$$

where $\alpha(B)$ and $\beta(B)$ are the respective model functions.

This formulation represents a mixed autoregressive–moving-average model of order (a, b). e.g. setting of $a = 1$, $b = 1$ provides a first-order model:

$$X_t - \alpha_1 X_{t-1} = Z_t - \beta_1 Z_{t-1} \qquad [3.95d]$$

which is a linear model on both sides (sign convention in conformity with [3.95a]). The process is stationary for $-1 < \alpha < 1$ and invertible for $-1 < \beta < 1$.

Equation [3.95c] provides the transfer function $\varphi(B)$ for white-noise. We state $X_t = \varphi(B) \cdot Z_t$ and derive:

$$X_t = [\beta(B)/\alpha(B)] Z_t \qquad [3.95e]$$

with: $\varphi(B) = \beta(B)/\alpha(B)$ [3.95f]

Hence the transfer function is the ratio of two polynomials, $\beta(B)$ and $\alpha(B)$. Therefore the process X_t is transformed to Z_t by the linear filter $\varphi(B)$. A stationary process requires for the transformation:

$$X_t = \alpha^{-1}(B) C_t = \Psi(B) \qquad [\text{see } 3.93i]$$

that the weights $\Psi(B)$ form a convergent series (see also text after [3.94c]). A nonstationary homogeneous behavior can be expressed in the form:

$$\varphi(B) X_t = \alpha(B)(1-B)^c X_t = \beta(B) Z_t \qquad [3.95g]$$

where $\alpha(B)$ is a stationary autoregressive operator. We can also write:

$$\alpha(B) w_t = \beta(B) Z_t \qquad [3.95h]$$

where: $w_t = \nabla^c X_t$ [3.95i]

with the backward-difference operator: $\nabla^c X_t = X_t - X_{t-c}$ [3.95j]

Substituting [3.93d] provides the relationship:

$$w_t = \nabla^c X_t = (1 - B^c) X_t \qquad [3.95k]$$

The exponent c indicates the cth difference which would then be stationary. In practice it has been proven that the exponent c is less than 3.

A stationary process also requires that the root of:

$$1 + \alpha_1 B + \alpha_2 B^2 + \ldots \alpha_a B^a = 0 \qquad [3.95l]$$

must lie outside the unity circle. The root of:

$$1 + \beta_1 B + \beta_2 B^2 + \ldots \beta_b B^b = 0 \qquad [3.95m]$$

can be on or outside the unity circle. This condition is also necessary for the uniqueness of the representation (Hannan, 1970). The $\alpha_1, \ldots, \alpha_a$ are called the autoregressive parameters, the $\beta_1, \ldots, \beta_b$ the moving-average parameters.

3.6.5 *Moving average and trend*

As previously disclosed the time-series was decomposed into a systematic and random (irregular) part ([3.92]). Sometimes a separation is studied in terms of short- and long-time fluctuations. A moving-average model can then be utilized to suppress the shorter time variations. This will be achieved even better when the short-time oscillation comprises a cycle. A moving average with the same length or a multiple of the time period of the cycle eliminates this oscillation completely. In the spectrum this procedure would be equivalent to a truncation of the short wave lengths.

The resulting curve from the moving-average model may display a trend. In general, the linear trend is best known and has found widespread application but not all trends are linear.

The fitting of a moving average $z(t)$ by a polynomial curve such as:

$$z(t) = a_0 + a_1 t + a_2 t^2 + a_3 t^3 + \ldots a_n t^n \qquad [3.96]$$

will provide information on the degree of the non-linear trend. The trend is linear when $a_i \sim 0$ for $i \geq 2$. The polynomial equation has been previously presented. It is understood that no periodicity is present in the moving average $z(t)$ of [3.96].

It may be worthwhile to add that the coefficients by least-square solution can be found by matrix solution. In the $\Sigma\, t^n$ terms (with $n = 1, \ldots, p$) the t progresses in consecutive integers from 1 to N or n_t, and mathematical formulae exist relating the sum of the powers of t with N or n_t. We find:

$$\Sigma\, t = N(N+1)/2 \qquad [3.97a]$$

$$\Sigma\, t^2 = [(2N+1)/3] \cdot \Sigma\, t \qquad [3.97b]$$

$$\Sigma\, t^3 = (\Sigma\, t)^2 \qquad [3.97c]$$

$$\Sigma\, t^4 = [(3N^2 + 3N - 1)/5] \cdot \Sigma\, t^2 \qquad [3.97d]$$

In climatology secular, long-term, seasonal or annual trends are very often employed. Special testing of trends is treated later (Essenwanger, 1974b).

Certain general considerations to assist in trend selection may be helpful. If possible, the moving average could be plotted such as: (1) z versus t, (2) ln z versus t, (3) z versus ln t, (4) ln z versus ln t.

If any of these plots render a straight line, the appropriate curve has been found. Further, if t progresses arithmetically, and z forms a geometric series we have:

$$z = ab^t \qquad [3.98a]$$

If both z and t show a geometric series, a parabolic relationship is indicated, namely:

$$z = at^b \qquad [3.98b]$$

Constancy of the first differences of z and an arithmetic progression of t indicates a linear trend, namely:

$$z = a + bt \qquad [3.98c]$$

From calculus it can be inferred that a polynomial of pth degree exists when the pth differences (for arithmetically progressing t) render a constant value.

3.6.6 Distribution of residuals in autoregressive–moving-average models

It has been pointed out that [3.95c] can be considered as a transfer function. Box and Pierce (1970) interpret the autoregressive–moving-average model as a transformation of the data to white-noise. This implies that the sequence of errors is uncorrelated. They designate this error as "residuals" when the parameters of the model are estimators, and regard these "residuals" as "estimators" of the error.

We ask now: How many terms will we need in the autoregressive–moving-average model? By adoption of the above conditions the model is considered appropriately chosen when the autocorrelation in the errors is zero. The residuals serve, therefore, as a test in checking the adequacy of the fit. It was first thought that the following approximate χ^2-test statistics (see Essenwanger, 1974a, section 6.7) can be used for the autoregressive model, namely:

$$R_t = n \sum_{k=1}^{m} r_k^2 \sim \chi_m^2 \qquad [3.99a]$$

which is an approximation of the variance of r_k. The unabridged term is:

$$R_t' = n(n+2) \sum_{k=1}^{m} (n-k)^{-1} r_k^2 \sim \chi_m^2 \qquad [3.99b]$$

The m denotes the degrees of freedom. Although the above expressions can be verified for the true error, Durbin (1970) has pointed out that this approximation would not be valid for the residual autocorrelation.

Box and Pierce (1970) and Pierce (1971) have amended the formula. First they have demonstrated that the test statistics are not restricted to the autoregressive model but can be applied to the moving-average and mixed model, too. The degree of freedom for the χ^2 is $m-p$ rather than the pure m degrees for the autoregressive model and $m-q$ for the moving-average model. The mixed model has then $m-p-q$ degrees of freedom, where p and q denote the order of the model (see [3.93d] and [3.93f]).

3.6.7 Spectral relationship

The detailed theoretical background of the relationship between the models and the power-spectrum cannot be presented here (see Box and Pierce, 1970, etc.). It should be sufficient to state that the B in [3.93b] and [3.94e] can be substituted by:

$$B = e^{-i2\pi f} \qquad [3.100a]$$

where $i = \sqrt{-1}$. This replacement provides one-half of the power-spectrum. Then $f_j = j/N$, and $f_1 = 1/N$ is the fundamental frequency. Thus the spectrum of a linear stationary model can be written as:

$$L(f) = 2\sigma_z^2 \varphi(e^{-i2\pi f}) \varphi(e^{i2\pi f}) \qquad [3.100b]$$

$$= 2\sigma_z^2 \, |\varphi(e^{-i2\pi f})|^2 \text{ for } 0 \le f \le 0.5 \qquad [3.100c]$$

The symbol $\|^2$ denotes the multiplication of the term inside the bars by its complex conjugate (e.g. as defined by [3.100b]) and is the absolute value squared.

For an autoregressive model we find:

$$\varphi(B) = \alpha^{-1}(B) \qquad [3.100d]$$

Therefore, the spectrum of an autoregressive model is:

$$L(f) = 2\sigma_z^2 / |1 - \alpha_1 e^{-i2\pi f} - \alpha_2 e^{-i4\pi f} - \ldots \alpha_a e^{-i2a\pi f}|^2 \text{ for } 0 \le f \le 0.5$$
$$[3.100e]$$

The first-order autoregressive process $X_t = \alpha_1 X_{t-1} + c$ has then a spectrum:

$$L(f) = 2\sigma_z^2 / |1 - \alpha_1 e^{-i2\pi f}|^2 \qquad [3.100f]$$
$$= 2\sigma_z^2 / (1 + \alpha_1^2 - 2\alpha_1 \cos 2\pi f) \qquad [3.100g]$$

for $0 \le f \le 0.5$.

The second-order autoregressive process:

$$X_t = \alpha_1 X_{t-1} + \alpha_2 X_{t-2} + c \qquad [3.101]$$

has a spectrum for $0 \le f \le 0.5$:

$$L(f) = 2\sigma_z^2 / |1 - \alpha_1 e^{-i2\pi f} - \alpha_2 e^{-i4\pi f}|^2 \qquad [3.101a]$$
$$= 2\sigma_z^2 / [1 + \alpha_1^2 + \alpha_2^2 - 2\alpha_1(1 - \alpha_2) \cos 2\pi f - 2\alpha_2 \cos 4\pi f] \qquad [3.101b]$$

In the moving-average model [3.100d] is replaced by:

$$\varphi(B) = \beta(B) \qquad [3.102]$$

The spectrum is consequently:

$$L(f) = 2\sigma_z^2 \, |1 - \beta_1 e^{-i2\pi f} - \beta_2 e^{-i4\pi f} - \ldots - \beta_b e^{-i2b\pi f}|^2 \qquad [3.102a]$$

The first-order moving-average process leads to a spectrum:

$$L(f) = 2\sigma_z^2 \, |1 - \beta_1 e^{-i2\pi f}|^2 \qquad [3.102b]$$
$$= 2\sigma_z^2 (1 + \beta_1^2 - 2\beta_1 \cos 2\pi f) \qquad [3.102c]$$

The second-order moving-average process:

$$X_t = Z_t - \beta_1 Z_{t-1} - \beta_2 Z_{t-2} \qquad [3.102d]$$

renders:

$$L(f) = 2\sigma_z^2 \, |1 - \beta_1 e^{-i2\pi f} - \beta_2 e^{-i4\pi f}|^2$$
$$= 2\sigma_z^2 [1 + \beta_1^2 + \beta_2^2 - 2\beta_1(1 - \beta_2) \cos 2\pi f - 2\beta_2 \cos 4\pi f] \qquad [3.102e]$$

for $0 \le f \le 0.5$.

The spectral relationship for the mixed model (autoregressive and moving average) follows from above:

$$L(f) = 2\sigma_z^2 |\beta(e^{-i2\pi f})|^2 / |\alpha(e^{-i2\pi f})|^2 \text{ for } 0 \le f \le 0.5 \qquad [3.103]$$

When the spectrum must be computed, the technique of the fast Fourier transform (Essenwanger, 1974a, section 4.3.6) is recommended to reduce computation expenses.

The spectrum of the mixed model can also be expressed as:

$$L(f) = 2\sigma_z^2 \frac{\left|1 - \sum_{k=1}^{n_b} \beta_k e^{-i2\pi f}\right|^2}{\left|1 - \sum_{k=1}^{n_a} \alpha_k e^{-i2\pi f}\right|^2} \qquad [3.103a]$$

Again, $0 \le f \le 0.5$, and n_a, n_b have been substituted for a and b, respectively.

An exponential model recently formulated by Bloomfield (1973) is:

$$L(f) = \sigma_z^2 \exp\left[2 \sum_{k=1}^{n_b} \beta_k \cos k 2\pi f\right] \qquad [3.104a]$$

which in logarithmic terms transforms to:

$$\ln L(f) = \ln \sigma_z^2 + 2 \sum_{k=1}^{n_b} \beta_k \cos k 2\pi f \qquad [3.104b]$$

The reader will easily recognize the summation term as a regular Fourier series.

3.6.8 *Autocorrelation functions of selected models*

We had previously discussed the relationship between the spectral behavior and the respective models. We will now ask for some expressions of autocorrelation functions for the three models. First, let us repeat that the autocorrelation function of a linear process is given by:

$$\text{Cov}_k = \sigma_z^2 \sum_{j=0}^{\infty} \varphi_j \varphi_{j+k} \qquad [3.105a]$$

The φ_j were defined as the coefficients of [3.94a]. It is evident that this resembles the definition of the covariance when φ_j is substituted for the observations (given as the deviation from the mean):

$$\text{Cov}_0 = \sigma_z^2 \sum_{j=0}^{\infty} \varphi_j^2 = \sigma_x^2 \qquad [3.105b]$$

We can further prove that the autocovariance generating function is:

$$\text{Cov}(B) = \sum_{k=-\infty}^{\infty} \text{Cov}_k B^k \qquad [3.105c]$$

The Cov_0 is then the coefficient for $B^0 = 1$.

With the definitions of the backward- and forward-shift operators B and V ([3.93c and e]) we can write:

$$\text{Cov}(B) = \sigma_z^2 \varphi(B) \varphi(V) \qquad [3.105d]$$

We define a (symmetric) autocorrelation matrix $\mathbf{P}_p$ such as:

$$\mathbf{P}_p = \begin{bmatrix} 1 & \rho_1 & \rho_2 & \rho_3 & \cdots & \rho_{p-1} \\ \rho_1 & 1 & \rho_1 & \rho_2 & \cdots & \rho_{p-2} \\ \rho_2 & \rho_1 & 1 & \rho_1 & \cdots & \rho_{p-3} \\ \cdot & \cdot & \cdot & \cdot & & \cdot \\ \cdot & \cdot & \cdot & \cdot & & \cdot \\ \cdot & \cdot & \cdot & \cdot & & \cdot \\ \rho_{p-1} & \rho_{p-2} & \rho_{p-3} & \cdots & \cdots & 1 \end{bmatrix} \qquad [3.105e]$$

We see that the main diagonal is 1 and that all other diagonals have one value of ρ_i corresponding to ρ_k in the autocorrelogram, which is the curve composed of the linear correlation coefficient as function of the lag.

A stationary process requires that the determinant of the matrix $\mathbf{P}_p$ is greater than zero. This concept is also valid for all principle minor determinants. For $p = 2$ this would require:

$$\begin{vmatrix} 1 & \rho_1 \\ \rho_1 & 1 \end{vmatrix} > 0 \text{ or } 1 - \rho_1^2 > 0$$

This condition can be readily fulfilled for any correlation except ± 1.0.

The matrix of autocorrelations leads to the Yule-Walker equations (see Box and Jenkins, 1970):

$$\boldsymbol{\alpha} = \mathbf{P}_p^{-1} \boldsymbol{\rho}_p \qquad [3.106a]$$

where $\boldsymbol{\alpha}$ is the vector of the α_i, and $\boldsymbol{\rho}$ is the vector of the autocorrelation coefficients, P_p as defined by [3.105e] (see also later [3.113a]). Based on [3.105e] we can express the autocorrelation function as:

$$\rho_k = \alpha_1 \rho_{k-1} + \alpha_2 \rho_{k-2} + \ldots + \alpha_a \rho_{k-a} \text{ for } k > 0 \qquad [3.106b]$$

Some autocorrelation functions for selected models are:

Autoregressive model. The first-order autoregressive model $X_t = \alpha_1 X_{t-1} + c$ has an autocorrelation function:

$$\rho_k = \alpha_1 \rho_{k-1} (k > 0) \qquad [3.107a]$$

The model (process) is also called a Markov model (process). The solution is:

$$\rho_k = \alpha_1^k \text{ for } k \geq 0 \qquad [3.107b]$$

This leads to $\rho_0 = 1$, as one would expect.

The second-order autoregressive model $X_t = \alpha_1 X_{t-1} + \alpha_2 X_{t-2} + c$ has a general form of the autocorrelation function:

$$\rho_k = \alpha_1 \rho_{k-1} + \alpha_2 \rho_{k-2} \text{ for } k > 0 \qquad [3.107c]$$

The starting values are $\rho_0 = 1$ and $\rho_1 = \alpha_1/(1 - \alpha_2)$. The general solution is:

$$\rho_k = C_1 R_1^k + C_2 R_2^k \qquad [3.107d]$$

where $1/R_1$ and $1/R_2$ are the roots of:

$$\alpha(B) = 1 - \alpha_1 B - \alpha_2 B^2 = 0 \qquad [3.107e]$$

Stationary solutions must lie outside the unit circle. Complex roots indicate a pseudo-periodic behavior. For more details see Box and Jenkins (1970) or Stralkowski et al. (1970). See also estimation and model identification, Section 3.6.9.

Moving-average model. The autocorrelation function of a moving-average model is:

$$\rho_k = \begin{cases} \dfrac{-\beta_k + \beta_1 \beta_{k+1} + \ldots + \beta_{b-k} \beta_b}{1 + \beta_1^2 + \ldots + \beta_b^2} & \text{for } k = 1, 2 \ldots b \\ \text{and } 0 \text{ for } k > b \end{cases} \qquad [3.108]$$

The autocorrelogram is therefore different from an autoregressive model, it indicates the order b by ρ_{k+1} being zero.

The first-order moving-average model has the autocorrelation function:

$$\rho_k = \begin{cases} -\beta_1/(1 + \beta_1^2) & \text{for } k \leq 1 \\ 0 & \text{for } k \geq 2 \end{cases} \qquad [3.108a]$$

The second-order moving-average model has the following autocorrelation form:

$$\rho_1 = -\beta_1(1 - \beta_2)/(1 + \beta_1^2 + \beta_2^2) \qquad [3.108b]$$
$$\rho_2 = -\beta_2/(1 + \beta_1^2 + \beta_2^2) \qquad [3.108c]$$
$$\rho_k = 0 \text{ for } k \geq 3$$

Mixed models. The mixed model must satisfy the difference equation:

$$\text{Cov}_k = \alpha_1 \text{Cov}_{k-1} + \ldots + \alpha_a \text{Cov}_{k-a} + \text{CCov}(k) - \beta_1 \text{CCov}(k-1) \ldots$$
$$- \beta_b \text{CCov}(k-b) \qquad [3.109a]$$

where CCov(k) is the cross-covariance function between X_t and Z_t, defined by:

$$\text{CCov}(k) = E[X_{t-k} \cdot Z_t] \qquad [3.109b]$$

(see also [3.95a]). The X_{t-k} depends only on "shocks" having occurred up to the time $t-k$, hence:

$$\text{CCov}(k) = 0 \text{ for } k > 0 \qquad [3.109c]$$

$$\text{CCov}(k) \neq 0 \text{ for } k \leq 0 \qquad [3.109d]$$

Equation [3.109a] implies:

$$\rho_k = \alpha_1 \rho_{k-1} + \alpha_2 \rho_{k-2} + \ldots + \alpha_a \rho_{k-a} \text{ for } k < b+1 \qquad [3.109e]$$

$$\text{or: } \alpha(B)\rho_k = 0 \qquad \text{for } k \geq b+1 \qquad [3.109f]$$

The autocorrelations of the mixed model depend therefore directly on the choice of the number of terms b of the moving-average model as much as they are determined by the number of autoregressive parameters.

The first-order mixed model is:

$$X_t - \alpha_1 X_{t-1} = Z_t - \beta_1 Z_{t-1} \qquad [3.110a]$$

$$\text{or: } (1 - \alpha_1 B)X_t = (1 - \beta_1 B)Z_t \qquad [3.110b]$$

The process is stationary when $-1 < \alpha_1 < 1$, and invertible for $-1 < \beta < 1$.

The autocorrelation function emerges as:

$$\text{Cov}_0 = \alpha_1 \text{Cov}_1 + \sigma_x^2 - \beta_1 \text{CCov}(-1) \qquad [3.110c]$$

$$\text{Cov}_1 = \alpha_1 \text{Cov}_0 - \beta_1 \sigma_x^2 \qquad [3.110d]$$

$$\text{Cov}_k = \alpha_1 \text{Cov}_{k-1} \text{ for } k \geq 2 \qquad [3.110e]$$

We can reformulate:

$$\text{CCov}(-1) = (\alpha_1 - \beta_1)\sigma_x^2 \qquad [3.110f]$$

Hence the autocovariance function of the model for the first two terms is:

$$\text{Cov}_0 = (1 + \beta_1^2 - 2\alpha_1\beta_1)\sigma_x^2/(1 - \alpha_1^2) \qquad [3.110g]$$

$$\text{Cov}_1 = (1 - \alpha_1\beta_1)(\alpha_1 - \beta_1)\sigma_x^2/(1 - \alpha_1^2) \qquad [3.110h]$$

The autocorrelation can be obtained by dividing by σ_x^2.

The behavior of the sample autocorrelation function of a moving-average model in the case of a non-stationary series was recently examined by Wichern (1973). He concludes that the postulation of a stationary series instead of a non-stationary series may prove of little practical consequence in individual cases, since it may be difficult to prove non-stationarity of a model for finite series. In forecasting of future behavior, however, a non-stationary series, if correct, may prove advantageous. The non-stationary model cannot be based on a mean which is no longer appropriate, e.g. a mean value calculated from a previous period under the assumption of stationarity.

3.6.9 Estimation and model identification

It is of interest that we can identify which process is reflected by the observational series or autocorrelogram. This consideration includes the determination of the number of terms a of the autoregressive and b of the moving-average model. Ideally we like to keep the number of parameters small. In turn, we must carry sufficient terms to provide a good approximation (see Section 3.6.6).

Estimation of the autocovariance can be made by the computation of the covariance:

$$\mathrm{Cov}_k = (1/N) \sum_{t=1}^{n-k} (X_t - \overline{X})(X_{t-z} - \overline{X}) \qquad k = 0, 1, 2 \ldots n_L \qquad [3.111a]$$

where n_L is the number of lags. Then an estimate for ρ_k is:

$$\rho_k = \mathrm{Cov}_k/\mathrm{Cov}_0 = \mathrm{Cov}_k/\sigma_x^2 \qquad [3.111b]$$

According to Bartlett (1946) the standard error of the correlation estimate is expressed as a variance of ρ_k, namely:

$$\epsilon^2(\rho_k) \sim (1/N) \sum_{i-\infty}^{\infty} (\rho_i^2 + \rho_{i+k}\rho_{i-k} - 4\rho_k\rho_i\rho_{i-k} + 2\rho_i^2\rho_k^2) \qquad [3.111c]$$

This reduces to:

$$\epsilon^2(\rho_k) \sim (1/N)(1 + 2 \sum_{i=1}^{q} \rho_i^2) \text{ for } k > q \qquad [3.111d]$$

where the ρ_i-terms are zero for $i > q$.

Assume $\rho_k = \alpha^k$, with k a positive number (exponential damping), then:

$$\epsilon^2(\rho_k) \sim (1/N)[\{(1 + \alpha^2)(1 - \alpha^{2k})/(1 - \alpha^2)\} - 2k\alpha^{2k}] \qquad [3.111e]$$

and: $\epsilon^2(\rho_1) \sim (1/N)(1 - \alpha^2)$ \qquad [3.111f]

Since $\alpha < 1$ the limit for large k has the simple form:

$$\lim_{k \to \infty} \epsilon^2(\rho_k) \sim (1/N)(1 + \alpha^2)/(1 - \alpha^2) \qquad [3.111g]$$

Model decision can be made by the autocorrelation function as discussed in Section 3.6.8. Box and Jenkins (1970) suggest that the number of terms be determined from the model:

$$X_t + \sum_{j=1}^{a} \alpha_j X_{t-j} + c = Z_t \qquad [3.112a]$$

where they recommend the cut-off after a terms when the partial correlation is nearly zero for $k > a$. This partial correlation R_{ki} is computed by:

ANALYSIS OF TIME SERIES

$$P_k R_{ki} = \rho_k \qquad [3.112b]$$

This is merely an expanded writing of [3.106a], the Yule-Walker equations. In expanded matrix notation we obtain for [3.112b]:

$$\begin{bmatrix} 1 & \rho_1 & \rho_2 & \cdots & \rho_{k-1} \\ \rho_1 & 1 & \rho_1 & \cdots & \rho_{k-2} \\ \vdots & & & & \vdots \\ \rho_{k-1} & \rho_{k-2} & \rho_{k-3} & \cdots & 1 \end{bmatrix} \begin{bmatrix} R_{k1} \\ R_{k2} \\ \vdots \\ R_{kk} \end{bmatrix} = \begin{bmatrix} \rho_1 \\ \rho_2 \\ \vdots \\ \rho_k \end{bmatrix} \qquad [3.113a]$$

(The P was defined by [3.105e]).
The first three coefficients are:

$$R_{11} = \rho_1 \qquad [3.113b]$$

$$R_{22} = (\rho_2 - \rho_1^2)/(1 - \rho_1^2) \qquad [3.113c]$$

$$R_{33} = \begin{vmatrix} 1 & \rho_1 & \rho_1 \\ \rho_1 & 1 & \rho_2 \\ \rho_2 & \rho_1 & \rho_3 \end{vmatrix} \Big/ \begin{vmatrix} 1 & \rho_1 & \rho_2 \\ \rho_1 & 1 & \rho_1 \\ \rho_2 & \rho_1 & 1 \end{vmatrix} \qquad [3.113d]$$

It is obvious that the determinants in the numerator and denominator have the same elements except for the last column, where the vector ρ_k is substituted in the numerator. The cut-off point for a is indicated when $R_{kk} \sim 0$. The R_{kk}-terms are the estimates in the formula:

$$\rho_j = R_{k1}\rho_{j-1} + \ldots + R_{k(k-1)}\rho_{j-k+1} + R_{kk}\rho_{j-k} \text{ for } j = 1, 2 \ldots k \qquad [3.113e]$$

where also estimates for the ρ_i can be substituted.

Durbin (1960) has given a recursive formula, which abbreviates the computation of R_{kk}, namely:

$$R_{p+1,p+1} = \frac{r_{p+1} - \sum_{j=1}^{p} R_{pj} r_{p+1-j}}{1 - \sum_{j=1}^{p} R_{pj,rj}} \qquad [3.114a]$$

where the r_i are the estimates for ρ_i. Furthermore:

$$R_{p+1,j} = R_{pj} - R_{p+1,p+1} \cdot R_{p,p-j+1} \text{ for } j = 1, 2 \ldots p \qquad [3.114b]$$

R_{11} has been defined by [3.113b].

The standard error of these partial autocorrelations can be expressed by its variance as:

$$\epsilon^2(R_{kk}) \sim 1/N \text{ for } k \geq a+1 \qquad [3.114c]$$

Levinson (1949) estimates maximum likelihood $\hat{\alpha}_j$ from a recursion formula as follows:

We expand the notation of $\hat{\alpha}_j$ to $\hat{\alpha}_j(n)$ to indicate the nth order autoregression:

$$\hat{\alpha}_n(n) = -\left[\sum_{k=0}^{n-1} \hat{\alpha}_k(n-1)\,\mathrm{Cov}(n-k)\right] \Big/ \left[\sum_{k=0}^{n-1} \hat{\alpha}_k(n-1)\,\mathrm{Cov}(k)\right] \quad [3.115]$$

where $\hat{\alpha}_0(n) = 1$.

Further: $\hat{\alpha}_j(n) = \hat{\alpha}_j(n-1) + \hat{\alpha}_n(n)\,\hat{\alpha}_{n-j}(n-1)$ for $j = 1, ..., n-1$

$$[3.115a]$$

and $\mathrm{Cov}(n-k)$ is the estimate of the covariance as usual.

The relation with the partial autocorrelation is:

$$R_{kk} = -\alpha_k(k) \quad [3.115b]$$

The partial autocorrelations from a first-order moving-average model are:

$$R_{kk} = -\beta_1^k(1-\beta_1^2)/[1-\beta_1^{2(k+1)}] \quad [3.115c]$$

Hence $R_{kk} < \beta_1^k$. This means an exponential decay for positive β_1, and for negative β_1 an oscillating correlation whose absolute value decreases exponentially.

Identification of the mixed model in the non-stationary case requires the transformation of the process to white-noise. The order of the model is then expanded to (a, b, c) where c denotes the number of differences necessary to reach stationarity (see [3.95g], etc.).

Brouwer (1971) has pointed out that Durbin's algorithm is unstable. Pagano (1972) has proposed a stable procedure for an autoregressive scheme although it is slower. The important quantities are $R_{s,s}$ and $\hat{s}_n^2(s)$. The latter is an estimate of the residual variance (see Section 3.6.6).

It is assumed that the highest order of the model has been decided, e.g. the order is p. We calculate the matrix:

$$\mathbf{P} = \mathbf{LDL}^T \quad [3.116]$$

$\mathbf{L}$ is a lower triangular matrix with unit in the diagonal and $\mathbf{D}$ a diagonal matrix:

$$\mathbf{L} = \begin{bmatrix} 1 & & & & & \\ l_{21} & 1 & & & & \\ l_{31} & l_{32} & 1 & & & \\ \cdot & \cdot & \cdot & & & \\ \cdot & \cdot & \cdot & & & \\ \cdot & \cdot & \cdot & & & \\ l_{s1} & l_{s2} & \cdot & \cdots & 1 \end{bmatrix} \quad [3.116a]$$

The elements l_{ij} will be calculated ranging from $i, j = 1 \ldots, p$.

$$\mathbf{D} = \text{diag}(d_1 \ldots d_p) \quad [3.116b]$$

whose elements are:

$$d_k = 1 - \sum_{j=1}^{k-1} b_{kj} l_{kj} \text{ with } d_1 = 1 \quad [3.116c]$$

Furthermore:

$$d_i l_{ki} = r(k-i) - \sum_{j=1}^{i-1} b_{kj} l_{ij} = b_{ki} \text{ (for } i = 1, \ldots, k-1) \quad [3.116d]$$

Notice, $r(0) = 1.0$, since the zero in the parenthesis denotes the lag and not a multiplier. The b_{kj} are defined by Pagano's algorithm above.

Let us assume that $i = 1$, $j = 1$, $k = 2$. We need $d_1, l_{21}, d_2, b_{21}, l_{11} = 1$. Then:

$$d_1 \cdot l_{21} = r(1) - b_{21} = b_{21} \quad [3.116e]$$

$$d_2 = 1 - b_{21} l_{21} \quad [3.116f]$$

If any d_k is zero the algorithm is not well defined. Under the postulation that P is positive definite, all d_k-values would be positive (see also remark after [3.105e]).

Now we determine:

$$\mathbf{P}_s = \mathbf{L}_s \mathbf{D}_s \mathbf{L}_s^T \quad (s = 1, \ldots, p) \quad [3.116g]$$

whose validity has been demonstrated by Pagano (1972). We define the vector $z^T = (z_1, \ldots, z_q)$ from:

$$\mathbf{L}z = \boldsymbol{\rho} \quad [3.117]$$

and consequently: $\mathbf{L}_s z(s) = \boldsymbol{\rho}_s \; (s = 1, \ldots, p) \quad [3.117a]$

with $z^T(s) = (z_1, \ldots, z_s)$ for $s = 1, \ldots, p$. In other words, $z(s)$ is a column vector (T stands for the transpose). Now we define:

$$v_s = z(s)/d_s \quad (\text{for } s = 1, \ldots, p) \quad [3.117b]$$

Then: $v_s = R_{s,s} \quad (s = 1, \ldots p) \quad [3.117c]$

The R_{ss} are the elements of the column vector $R_{b,s}$ of [3.113a] with the specification:

$$\mathbf{P}_s y_s = \boldsymbol{\rho}_s \quad [3.117d]$$

and: $y_s^T = (R_{s1}, \ldots R_{ss}), \quad s = 2, 3 \ldots \quad [3.117e]$

Then: $\hat{\sigma}_n^2(s) = \text{Cov}_0 \, d_{s+1} \quad \text{for } s = 1, \ldots, p-1 \quad [3.117f]$

where Cov_0 is given by [3.111a].

The behavior of the autocorrelation function for the non-stationary moving-average model has been studied by Wichen (1973). His model starts with:

$$X_t - X_{t-1} = Z_t - \beta Z_{t-1} \qquad [3.118a]$$

where $-1 < \beta \leq 1$. The corresponding sample covariance function is:

$$\rho_k = \mathrm{Cov}_k/\mathrm{Cov}_0 \qquad [3.118b]$$

His final answer is:

$$\rho_k = E(\mathrm{Cov}_k)/E(\mathrm{Cov}_0) = \frac{(N-k)[-6 + 6\lambda + (N^2 + 2k^2 - 4kN - 1)\lambda^2]}{N(N-1)[6 - 6\lambda + (N+1)\lambda^2]}$$

$$[3.118c]$$

where: $\lambda = 1 - \beta$. $\qquad [3.118d]$

As $\beta \to 1$, the non-stationary model given by [3.118a] can hardly be distinguished from the white-noise model $X_t = Z_t$. The behavior of ρ_k also depends on N.

Another aspect of estimation has been treated by Schips and Stier (1972). They have illustrated that correlations among the errors lead to differences in numerical values of the estimators, depending on the presence of correlations and the sequence of estimation. Since in atmospheric time series the correlation among errors cannot always be assumed to be zero (i.e. the errors are not independent), caution has to be exercised in all estimation procedures. This error-dependency may also influence estimations in the time-series models.

3.6.10 Inverse autocorrelation

We introduce the inverse autocorrelation (see Cleveland, 1971) from the following. Assume the autocorrelation $R(\tau)$ as usual, with $R(\tau) = \mathrm{Cov}(\tau)\sigma^2$, then:

$$\mathrm{Cov}(\tau) = \int_0^1 e^{2\pi i k f} L(f)\,df \qquad [3.119a]$$

with $i = \sqrt{-1}$ and the other symbols as introduced with [3.100b]. The f stands for the frequency (see Section 3.6.7). We may now define the reciprocal:

$$S(f) = 1/L(f) \qquad [3.119b]$$

Subsequently we write the inverse autocovariances of X_t as:

$$\mathrm{ICov}(\tau) = \int_0^1 e^{2\pi i k f} S(f)\,df \qquad [3.120a]$$

for $\tau = 0, 1 \ldots n_m$, and the inverse autocorrelation follows as:

$$IR(\tau) = ICov(\tau)/ICov(0) \qquad [3.120b]$$

An estimate of the smoothed function can be obtained by:

$$I\hat{Cov}(\tau, m) = \frac{1}{N} \sum_{j=0}^{N-1} S(\tau, m) e^{2\pi i k f_j} \qquad [3.120c]$$

where $f_j = j/N$ for all integers j. It is further assumed that $X_t = 0$ for $t = T + 1, \ldots, N$, where T is the size of the sample.

The $S(\tau, m)$ in our case is computed from:

$$\widehat{Ls}(f_j, m) = \frac{1}{2m + 1} \sum_{k=-m}^{m} L(f_j + k) \qquad [3.120d]$$

with: $\hat{S}(\tau, m) = 1/\widehat{Ls}(f_j, m)$ \qquad [3.120e]

Here m has the function of the number of line spectra involved in smoothing. It is cautioned that for f_j at and near zero round-off errors in $S(f_j, m)$ occur. We readily derive again:

$$I\hat{R}(\tau) = \hat{S}(\tau, m)/\hat{S}(0, m) \qquad [3.120f]$$

where m is an arbitrary index at the moment, but is identified in [3.120e] as the number of terms employed for smoothing.

Box and Jenkins (1970) have derived two important difference equations:

$$R(\tau) + \alpha_1 R(\tau - 1) + \ldots + \alpha_a R(\tau - a) = 0 \text{ for } t > b \qquad [3.121a]$$

and: $IR(\tau) + \beta_1 IR(\tau - 1) + \ldots + \beta_b IR(\tau - b) = 0 \text{ for } t > a$ \qquad [3.121b]

These two equations can be very helpful in the identification of the model. Cleveland (1971) recommends identification with the aid of the inverse correlation. Estimators of IR can be computed by [3.120f]. These estimates behave like autocorrelations of a moving-average model with autoregressive parameters $\beta_1 \ldots \beta_b$ and moving-average parameters $\alpha_1 \ldots \alpha_a$.

The charts by Stralkowski et al. (1970) may be used for interpretation of the model where the IR can be employed by interchanging autoregressive and moving-average.

Some of the autocorrelation functions have been given in Section 3.6.8. If we have the autocorrelation function $\varphi_k = \alpha^k$, a first-order autoregressive model is suggested. Thus $IR = \beta^k$ identifies a first-order moving-average model ($X_t + c = Z_t + \beta_1 Z_{t-1}$).

Conversely if $IR(1) = \alpha$ and $IR(k) \sim 0$ for $k > 1$, then (by [3.107d]) we recognize a moving-average model of first-order, hence we have an autoregressive model of first-order. ($X_t + \alpha_1 X_{t-1} + C = Z_t$). More details can be found in Box and Jenkins (1970).

Cleveland (1971) has applied the autoregressive model to wind-speed data. The first assumption was a first-order moving-average:

$$X_t + c = Z_t + \beta_1 Z_{t-1} \qquad [3.122a]$$

The model would have $\beta_1 = R(1)$, and $R(k) = 0$ for $k > 1$. The $IR(k) = (-\beta_1)^k$. The model proved insufficient. An expansion was recommended:

$$X_t + \alpha_1 X_{t-1} + \alpha_2 X_{t-2} + c = Z_t + \beta, Z_{t-1} \qquad [3.122b]$$

which appears to be a better representation (see remarks about the sign of α and β, [3.95b]).

The inverse autocorrelation has some advantages. Some characteristics of X_t are better recognized in the $IR(t)$ system than in the regular autocorrelation. We need no new knowledge for the interpretation. The $IR(t)$ terms are easily estimated although not readily calculated.

3.6.11 Godske's model

In earlier studies Godske (1962, 1965), Nordo (1959) and others have pointed out that meteorological time-series generally do not strictly follow the Markovian law. Godske (1966) has therefore developed a different mixed model, in which he defines a second-order autoregressive model as:

$$X_t = \alpha_1 x_{t-1} + \alpha_2 x_{t-2} + c_t \qquad [3.123a]$$

and a moving-average model of:

$$Y_{n,t} = \alpha_1 Y_{n,t-1} + \alpha_2 Y_{n,t-2} + (1/n) \sum_{i=0}^{n-1} c_{t-i} \qquad [3.123b]$$

The c_t denotes a random time-series as earlier introduced with:

$$E(c_t) = 0 \qquad [3.123c]$$

$$E(c_t^2) = \sigma_c^2 \qquad [3.123d]$$

and: $E(c_t c_{t-k}) = 0$ for $k \neq 0$ $\qquad [3.123e]$

These conditions characterize a random variable with zero means, variance σ_c^2 and intercorrelation zero (independence). A further definition is necessary, namely:

$$Y_{n,t} = (1/n) \sum_{i=0}^{n-1} x_{t-i} \qquad [3.123f]$$

the moving average of x_t. This term is also known as the overlapping mean.

The mixed series has the same autoregressive part as in [3.123a]. A definition for successive means:

$$M_{n,T} = Y_{n,nt} \qquad [3.124a]$$

is introduced. The difference:

$$\Delta x_t = x_t - x_{t-1} \ (\text{or} = \nabla X_t) \qquad [124b]$$

represents the day-to-day variation where x_t is the once-a-day observation (see also [3.93i]).

We can write for $n = 2$:

$$M_{n,T} = (\alpha_1^2 + \alpha_2)M_{n,T-1} + \alpha_1^2\alpha_2 \sum_{i=0}^{\infty} \alpha_2^i M_{n,T-2-i} + 0.5(c_t + c_{t-1})$$

$$+ 0.5\alpha_1 \sum_{i=0}^{\infty} \alpha_2^i (c_{t-i-1} + c_{t-i-2}) \quad [3.124c]$$

This result was derived by Godske (1966) who concluded that the mixed model with finite tradition length n cannot correspond to the non-Markovian autoregressive observation series x with finite tradition length.

The Δx series can be written as:

$$\Delta x_t = \alpha_1 \Delta x_{t-1} + \alpha_2 \Delta x_{t-2} + c_t - c_{t-1} \quad [3.124d]$$

The $c_t = x_t$, etc. is a random time-series as defined previously. As Godske has stated, a simple Markovian series, period 2, requires:

$$x_{2t} = \alpha_{01} x_{2t-1} + c_{zt} \quad [3.124e]$$

and: $\quad x_{2t-1} = \alpha_1, x_{2t-2} + c_{2t-1} \quad [3.124f]$

renders the simplified assumption $\sigma^2_{c(2t-1)} = 1$. His application to daily average temperatures delineates that meteorological time-series correspond to a mixed type rather than a Markovian or higher-order autoregressive model (see also Cleveland's result, Section 3.6.10).

In a later report Godske (1967) cautions particularly that the daily and yearly periodicities cannot be removed by standardization. Especially in the study of 1 or 2 hours a "periodic" model with 12 or 24 hours must be considered. For a period of three observations (e.g. one observation every eight hours) Godske (1967) has proposed a mixed model for a chain of stationary time-series:

$$X_{3t-2} = \sum_{i=3}^{3n} \alpha_{2i} X_{3t-i} + c_{3t-2} \quad [3.125a]$$

$$X_{3t-1} = \alpha_{12} X_{3t-2} + \sum_{i=3}^{3n} \alpha_{1i} X_{3t-i} + c_{3t-1} \quad [3.125b]$$

$$X_{3t} = \alpha_{01} X_{3t-1} + \alpha_{02} X_{3t-2} + \sum_{i=3}^{3n} \alpha_{01} X_{3t-i} + c_{3t} \quad [3.125c]$$

As stated earlier, the random variables c_{it} satisfy the conditions:

$E(c_t) = 0$ (mean) $\quad [3.126a]$

$E(c_{3t-1}^2) = \sigma_\epsilon^2$ for $i = 0, 1, 2$ (variance) $\quad [3.126b]$

$E(c_t c_{t-j}) = 0$ for $j \neq 0$ (correlation) $\quad [3.126c]$

$E(c_t X_{t-k}) = 0$ for $k > 0$ (correlation with observations) $\quad [3.126d]$

A simplification can be made under certain assumptions by setting:

$$X_{1,t} = X_{3t-2} \qquad [3.127a]$$

$$X_{2,t} = X_{3t-1} \qquad [3.127b]$$

$$X_{3,t} = X_{3t} \qquad [3.127c]$$

The resulting system has been called a causal chain system by Godske (1967), namely:

$$X_{1,t} = \sum_{i=1}^{n} \sum_{j=1}^{3} \alpha_{1ji} X_{j,t-i} + c_{1,t} \qquad [3.128a]$$

$$X_{2,t} = \alpha_{210} X_{1,t} + \sum_{i=1}^{n} \sum_{j=1}^{3} \alpha_{3ji} X_{j,t-i} + c_{2,t} \qquad [3.128b]$$

$$X_{3,t} = \alpha_{320} X_{2,5} + \alpha_{310} X_{1,t} + \sum_{i=1}^{n} \sum_{j=1}^{3} \alpha_{3ji} X_{j,t-i} + c_{3,t} \qquad [3.128c]$$

In this interpretation the $X_{i,t}$ may not occur contemporarily. As Godske suggests, the $X_{1,t}$ may be more a large-scale effect than $X_{2,t}$ and $X_{2,t}$ of larger scale than $X_{3,t}$. There may also be a systematic time difference in the time of the observations. The general system can be written as:

$$X_{i,t} = \sum_{j=1}^{n_1} \sum_{k=1}^{n_k} \alpha_{ijk} X_{k,t-j} + \sum_{j=0}^{n_2} \sum_{k=1}^{n_k} \beta_{ijk} c_{k,t-j} \qquad [3.128d]$$

The coefficients α and β and the c must obey conditions of stationarity. More applications, especially the aspect of prediction, may be found in the cited literature by Godske (1965, 1966, 1967). Other predictions by time-series have been discussed recently by Bloomfield (1972).

3.6.12 *Time-series and quality control*

The topic of quality control has been treated by the writer in connection with an introduction to test methods (1974a, section 6). Other tests will be presented later (1974b). This section serves primarily the purpose of describing special tests arising in time-series analysis.

Outliers. Fox (1972) has studied the problem of checking outliers. He distinguishes two types. Type I is a gross-error affecting a single observation. Type II influences a particular observation *and* the subsequent observations. Four possible configurations can be deducted, namely type I, type II, mixture and unknown.

Criteria based on likelihood-ratio methods are given by Fox but are simplified to:

$$\lambda = \Delta/\sigma_\Delta \qquad [3.129]$$

where Δ denotes the error to be tested, and σ_Δ is the estimated standard error.

It is assumed that trend or seasonal variation are either negligible or have been removed prior to the test procedure. The autoregressive model can be stated (see [3.93a]):

$$X_t = \sum_{r=1}^{n_a} \alpha_r X_{t-r} + C_t \qquad [3.130]$$

where $t = n_a + 1, ..., n$.

The observations are assumed to be:

$$Y_t = \begin{cases} X_q & \text{for } (t \neq q) \\ X_q + \Delta & \text{for } t = q \end{cases} \qquad [3.130a]$$

where q denotes the observation of the outlier.

We test H_0 for $\Delta = 0$ against H_1 for $\Delta \neq 0$. (For details in hypothesis testing see Essenwanger, 1974a, section 6.) The criterion which is asymptotically equivalent to the likelihood-ratio criterion but easier to calculate is:

$$\lambda_{q,n} = \Delta/\sigma_\Delta \qquad [3.130b]$$

with: $\text{Var}(\Delta) = \sigma_c^2 \Big/ \Big(1 + \sum_{i=1}^{n_a} \alpha_i^2\Big) = \sigma_\Delta^2 \qquad [3.130c]$

where n_a denotes the order of the model and the estimator for σ_c^2 the "residual error".

Although Fox (1972) has studied the true likelihood-ratio criterion, for practical applications the criterion [3.130b] can be treated as an equivalent replacement. If $\Delta = 0$, the following relationship between the F-test and $\lambda_{q,n}$ exists:

$$(\lambda_{q,n}^{-1} - 1)(n - 2) \sim F_{1,n-2} \qquad [3.130d]$$

When the position of the outlier is unknown we can test:

$$\lambda_m = \sup (k_{q,n}) \qquad [3.130e]$$

where $k_{q,n} = \lambda_{q,n}^{-1}$ and the maximum of the $k_{q,n}$ from $q = n_a + 1, ..., n - n_a$ is taken. Therefore:

$$(n - 2)(k_{q,n} - 1) \sim F_{1,n-2} \qquad [3.130f]$$

For the test of the type-II outlier we formulate:

$$X_t = \sum_{r=1}^{a} \alpha_r X_{t-r} + \Delta_t + C_t \qquad [3.131]$$

where the Δ_t signifies:

$$\Delta_t = \begin{cases} 0 & \text{for } t \neq q \\ \Delta & \text{for } t = q \end{cases} \qquad [3.131a]$$

The outlier affects X_q and all subsequent X_t-values $q < t \le n$ as evident from [3.131]. The hypothesis to be tested is again $H_0/\Delta = 0$ and $H_1/\Delta_1 \ne 0$.

Analogous to $\lambda_{q,n}$ Fox has derived a test criterion $\lambda^*_{q,n}$:

$$\lambda^*_{q,n} = \Delta^*/\sigma^*_c \qquad [3.131b]$$

$$\Delta^* = X_q - \sum_{r=1}^{n_a} \alpha_{r1} X_{q-r} \qquad [3.131c]$$

and:
$$(\sigma^*_c)^2 = (n - n_a)^{-1} \sum_{t=a+1}^{n} (X_t - \sum_{r=1}^{a} \alpha_{r0} X_{t-1})^2 \qquad [3.131d]$$

The α_{r0} and α_{r1} denote the estimators for α_r under H_0, H_1, respectively. The following relation with the F-test has been found by Fox:

$$\frac{(n - n_a - 1)(\lambda^*_{q,n})^2}{n - n_a - (\lambda^*_{q,n})^2} \sim F_{1, n - n_a - 1} \qquad [3.131e]$$

Fox (1972) has elaborated on the power of this test. The reader is referred to the original article.

Editing. Elimination of the gross errors has been discussed in detail by the author (1974a, section 6.10) but some modification for time-series may be presented here.

Let us assume for the purpose of introduction that the data have been brought into sequence $X_1, X_2 \ldots X_n$, where $X_1 < X_2 < \ldots < X_n$. We may compute the dispersion or an estimate (see Fisz, 1963):

$$\Delta X_k = (X_{n-k+1} - X_k)/(2\beta) \qquad [3.132a]$$

The k designates a particular, ordered observation which can be expressed in terms of the cumulative distribution, e.g. $k = 10$, $N = 200$, $F(x) = 0.05$. If the distribution form is known, the scale factor β can be determined.* Fisz has demonstrated the asymptotic efficiency of this estimate ΔX_k in the case of a Gaussian distribution. The mean can be estimated by:

$$X_m \sim (X_{N-k+1} - X_k)/2 \qquad [3.132b]$$

These estimates can be employed to set boundaries. Data points outside these limits ξ_j are considered suspicious and may imply a gross error. The boundary ξ could be set as:

$$\xi_j = \pm 3.3 \Delta X_1 + X_m \qquad [3.132c]$$

where the positive sign corresponds with the upper boundary.

In a non-symmetric probability distribution Roughan and Evans (1970) suggest a modified form based on the median:

---

* The β is not identical with the coefficient in a moving-average model!

$$\xi_{jU} = 3.3(X_{N-k+1} - X_{med})/\beta + X_{med} \qquad [3.132d]$$

and: $\xi_{jL} = 3.3(X_k - X_{med})/\beta + X_{med}$ [3.132e]

The subscript U and L denote the upper and lower boundary, respectively. The factor 3.3 corresponds to $F_{(x)} \sim 0.001$ of the Gaussian distribution. It may be replaced by a smaller or higher number if other thresholds are desired. Further methods have been listed by the author (1974a, section 6.10).

One word of caution is appropriate. In atmospheric collectives a gross-error can be created by physical cause as well as by a large mistake. Thus the outbreak of extreme cold air from the arctic into the lower latitudes could create a very low temperature reading. Other physical causes may be hurricanes, etc. These phenomena create outliers which are completely legitimate atmospheric data. They may be disproportionate in weight for the length of the time-series. It is therefore recommended to flag outliers for checking unless it is known with certainty that the time-series to be checked will be free from outliers by physical cause. Then the gross-error can be rejected. Tukey (1962) suggests winsorization of the data. In this process the outlier is replaced by the boundary.

One more hint for practical application may be useful. It was previously stated that the data should be available in ordered form. If N is large this may be quite costly even by electronic data processing. To reduce costs a partial sorting can be made either by determining the first ten observations on both ends of the frequency distribution or by grouping data into classes and find ΔX_k by interpolation from the classes. The additional error is negligible for the central classes. It is possible to list selected classes, e.g. between $\xi_1 > \bar{x} > \xi_2$ in greater detail in order to determine β and X_m. These simplifications are only a few examples. Other procedures may be left to the innovation of the reader.

Trend estimation. It is self-evident that a gross-error can drastically affect the computation of a trend in a limited data sample. Roughan and Evans (1970) have developed an algorithm for the determination of a trend curve which is independent of gross-errors. We find first in the usual manner the curve:

$$y_i = a_1 + a_2 t_i + a_3 t_i^2/2 \qquad [3.133]$$

In this case a second-order polynomial is assumed. A linear trend can be treated with the same procedure. If not already in the form where $t(N+1)/2 = 0$ the time-series can be converted to this form. Other requirements such as N being odd, and equal spacing for t aid further in simplification. The algorithm is based on the assumption that any of the $(n+1)$ sections of the total observation fitted by an nth order polynomial should have an equal number of points on either side of the curve.

We divide first the total data into $N/(n+1)$ sections. We define now:

$$S_1 = \sum_{i=n_1+1}^{n_2} g(X_i - y_i) \qquad [3.134a]$$

where $g(u)$ assumes only three values and represents a count or weight, namely:

$$g(u) = \begin{cases} 1 \text{ for } u > 0 \text{ (or } > \epsilon) \\ 0 \text{ for } u = 0 \text{ (or } |\epsilon|) \\ -1 \text{ for } u < 0 \text{ (or } < \epsilon) \end{cases} \qquad [3.134b]$$

The ϵ denotes an insignificant but finite small number, depending on the accuracy of X_i and y_i. In the case of a second-order curve as given by [3.133] the boundaries of the summations are $n_1 = N/3$ and $n_2 = 2N/3$. In this case N should be divisible by 3 because n_1 and n_2 are then integers.

Further test parameters are:

$$S_2 = -\sum_{i=1}^{n_1} g(X_i - y_i) + \sum_{i=n_2+1}^{N} g(X_i - y_i) \qquad [3.134c]$$

and: $$S_3 = \sum_{i=1}^{n_1} g(X_i - y_i) + \sum_{i=n_2+1}^{N} g(X_i - y_i) \qquad [3.134d]$$

An iterative procedure is now employed to calculate new coefficients a_i of [3.133]. We increase or decrease a_1 for positive or negative S_1, respectively. This procedure is repeated until either S_1 becomes zero or changes sign. To accelerate the iterative process Roughan and Evans (1970) recommend an increase of the addition term Δ. e.g. first we assume $a_{11} = a_1 \pm \Delta$ (sign of Δ according to S_1). Let us assume S_1 is positive. Then $a_{12} = a_{11} + 2\Delta$, etc. A geometric progression of Δ may be suitable. If the S_{1i} after the ith step changes sign, a bisecting procedure should be used, e.g. if $a_{13} = a_{12} + 4\Delta$ would lead to a change of sign we set $a_{14} = (a_{12} + a_{13})/2 = a_{12} + 2\Delta$. Since S_{13} has the opposite sign of S_{12} we could take a weighted bisecting $a_{14} = a_{12} + (S_{12}/|S_{12} - S_{13}|) 4\Delta$. Then bisecting will be repeated until $S_{1i} = 0$. This should be possible for an even number of points. An odd number of points may result in $S_{1i} = \pm 1$. Afterwards the a_2 is corrected depending on S_2 until after the jth correction $S_{2j} = 0$.

It is now necessary to check whether the modification of a_2 has affected S_{1i}. If it has, the a_1 must be readjusted again. Otherwise a_3 will be determined from S_{2k}. Roughan and Evans (1970) state that the convergence of the process has not been proven but all practical examples calculated have converged after a few cycles.

It is self-evident that the newly found curve which results from this algorithm need not be identical with the least-square fitted curve as the latter includes the points with gross-error, and the least-square fit adjustment includes the errors. If points with gross-errors have been eliminated, the described procedure by Roughan and Evans becomes superfluous. Whether elimination of outliers and computation of trends is simpler than the procedure suggested by Roughan and Evans may depend on the individual set of time-series data. The article by the cited authors includes a flow diagram for electronic computer application.

3.6.13 *Multivariate and other atmospheric models*

Several authors have recently engaged in time-series analysis techniques. Wallace and Dickinson (1972) have treated the autocorrelation matrix as the basis of eigenvectors or eigenvalues calculations. Wallace (1972) has then expanded this study to the frequency domain by including the cross-spectrum data. His application to tropical wave disturbances confirms results which have been obtained from ordinary spectral analysis except that some ambiguities have been resolved. In addition, however, he could prove the strong coupling between the mixed Rossby gravity waves of the upper troposphere and the synoptic-scale lower wave disturbance which makes the two phenomena inseparable in observational data. Thus, analysis of time-series by empirical polynomials may be an additional modern tool in analysis of time-series. Since the procedure follows the standard techniques which were described in Section 3.4 no additional equations will be presented here.

Other authors, Joseph (1973) and Davis and Rappaport (1974) have recently illustrated the merits of time-series analysis. Joseph has employed the autocorrelation function and the power-spectrum for temperature data. He concludes that a first-order Markov model seems adequate for the time-series of annual means for some Colorado stations but the maxima and minima at two of the stations rendered only white-noise.

Davis and Rappaport have analyzed an autoregressive moving-average process for Ohio drought data. Although their model for prediction emerged only as being equivalent with persistence, they nevertheless recommend the continued use of the new techniques for possible future advantages. An exponential smoothing process has been defined by them as:

$$S_t = c \sum_{k=0}^{t-1} (1-c)^k X_{t-k} + (1-c)^t X_0 \qquad [3.135a]$$

In this formula S_t is the smoothing function, the X's are the observations as usual, and the c is a smoothing constant. This model was not as successful, however, as the autoregressive moving-average model:

$$\alpha(B) X_t = \beta(B) Z_t \qquad [3.135b]$$

which is the mixed model introduced earlier by [3.95c].

Bloomfield (1973) has recommended a different exponential model for the spectrum of scalar time-series. The model was earlier presented with [3.104a and b]. The model rendered no better fit, however, than the autoregressive moving-average model as judged by the residual variance. The reader is, therefore, referred to the literature for parameter estimation within the model.

3.7 TRANSFORMATIONS

In the previous sections only a selected number of functions have been introduced. In addition to polynomials, Fourier series, and linear functions such as empirical polynomials, a variety of other interactions between variables exist which cannot all be discussed. These non-linear functions also may not yet have received a statistical treatment. Some of these functions can be transformed, however, to standard curves discussed in this book. We distinguish two groups. The first one arises in the field of curve fitting with the transformation of postulated mathematical relationships between dependent and independent variables or predictors. The second group originates in the field of frequency distributions, when we desire to convert a frequency distribution into an approximate normal distribution.

3.7.1 Transformation of special functions

With the variety of tools discussed up to this point it is logical that the reader's first reaction is to ask why we need any other special functions which have not been introduced. One simple answer can be given. Not all functional correlation is linear or in polynomial form.

It has been pointed out that the dependent variable can be expressed by a system of linear empirical functions. Although the goal in this type of manipulation is the derivation of a minimum number of predictors or factors, some systems may be better treated with a limited number of non-linear terms. This number may even be smaller than the sum of the principal factors in the factor analysis. Since statistical treatment (curve fitting, calculation of estimators, etc.) of these functions has not been given, one way is the transformation into a model which has been discussed. An example of a transformation could be as follows.

Assume we have the relationship:

$$Y = a + b \ln(X + c) \qquad [3.136]$$

The c may serve as a constant for adjusting the reference point to avoid negative numbers in the logarithm. The equation can be reduced by introducing:

$$Z = \ln(X + c) \qquad [3.136a]$$

Y becomes now: $Y = a + bZ$ $\qquad$ [3.136b]

This is a known problem in polynomial or regression analysis. Other functions may be:

$$Y = a + be^{cX+d} \qquad [3.137]$$

The Y can be converted by introducing:

$$Z = Y - a \qquad [3.137a]$$

We substitute [3.137a] into [3.137] and write:

$$Z = be^{cX+d} \qquad [3.137b]$$

This becomes: $\ln Z = \ln b + cX + d$ [3.137c]

Now replace: $\ln Z = Z_y$ [3.137d]

and: $A = \ln b + d$ [3.137e]

Then it follows: $Z_y = A + cX$ [3.137f]

which is again a known problem of curve fitting.
Other useful expressions are the power law:

$$Y = ax^b \qquad [3.138]$$

A modification is $b \neq i$ where i is an integer. The transformation:

$$\ln y = a' + b \ln x \qquad [3.138a]$$

reduces the problem to a regular polynomial fit (in double logarithmic coordinates a straight line).
The following formula is called rectangular hyperbola:

$$y = ax/(x + b) \qquad [3.139]$$

with asymptotes: $y = a$ and $x = -b$.
Another form of a rectangular hyperbola is the expression:

$$xy = c \qquad [3.140]$$

or in general form: $(x + a)(y + b) = c$ [3.140a]

The transformation: $\ln(x + a) + \ln(y + b) = c'$ [3.140b]

leads to a known solution, although the determination of a and b may be difficult (see Weibull distribution, Section 2.11).

3.7.2 *Transformation to Gaussian (normal) distribution*

Since the Gaussian distribution is by and large the best known distribution and the easiest to be calculated, not to speak of the availability of tables almost everywhere, many analysts try to relate the various types of frequency distributions by a transformation to this well established law. The transformed variable is then considered to follow the Gaussian distribution law. Use of this concept is very frequently made in applications of statistical tests, where an expression is sought for the test characteristic following the normal distribution.

One of the earliest treatments of transformations was formulated by Edgeworth in 1898; at that time he called the concept the translation method. Prior to him the translation as a method of generating frequency distributions was considered by Pearson (1895), see also Section 2.5. Johnson (1949) has later investigated the general problem of translations, and Bartlett (1947) had some comments on the use of transformations. Johnson's system is briefly summarized below.

3.7.2.1 Johnson's transformation system. The system (Johnson, 1949) is based on a transformation of a variable x to normality $y = f(x)$. We use the general relation:

$$y = \alpha + \beta f\left(\frac{x-\gamma}{\theta}\right) \qquad [3.141]$$

The f is a normalized form of the variable, with reference parameter γ and scale parameter θ, sometimes the standard deviation, but in other cases may be the range (see [3.142c]) or any other scale function.

(1) The first basic transformation is:

$$y = \alpha + \beta \ln\left(\frac{x-\gamma}{\theta}\right) \qquad [3.142]$$

$$y = \alpha + \beta \ln z \qquad [3.142a]$$

This leads to the lognormal distribution as described in Section 2.2.

(2) If we substitute $z = 1 - \gamma/x$ into [3.142] we find the expanded logarithmic form:

$$y_2 = \alpha - \beta \ln \theta + \beta \ln [z/(1-z)] \text{ for } 0 < z < 1 \qquad [3.142b]$$

A further version is:

$$y_2 = \alpha + \beta \ln [(x-\gamma)/(\gamma + \theta - x)] \text{ for } \gamma < x < (\gamma + \theta) \qquad [3.142c]$$

which is also generated by introducing $z = (x - \gamma)/\theta$ into:

$$f(z) = \ln [z/(1-z)] = 2 \operatorname{arctanh}(2z - 1) \qquad [3.142d]$$

The tanh stands for the hyperbolic tangent. While z in [3.142d] increases from 0 to 1, the $f(z)$ ranges from $-\infty$ to $+\infty$.

Fisher's transformation of the correlation coefficient (Essenwanger, 1974a, section 6.6) follows [3.142c] with $\gamma = -1$, $\theta = 2$; $\alpha \sim 0.5(n-3)^{\frac{1}{2}}$, $\ln[(1+\rho)/(1-\rho)]$ and $\beta \sim 0.5(n-3)^{\frac{1}{2}}$. The ρ represents the population coefficient while $r \equiv x$ stands for the sample coefficient.

The estimation of the parameters α, β, γ, θ is not simple as the moments estimators are not accurate enough and efficient. The best solution is probably by order statistics or percentile values. Johnson (1949) finds the following solutions:

Median: $z_{med} = 1/[1 + \exp(\alpha/\beta)]$ \qquad [3.143]

TRANSFORMATIONS

We define: $z = 2z_1 - 1$ [3.143a]

Then the intersections of the curves:

$$u_1 = z_1 - \alpha\beta$$ [3.143b]

$$u_2 = \beta^2 \ln[(1+z_1)/(1-z_1)]$$ [3.143c]

indicate whether a unimodel or bimodel distribution is transformed. The bimodel distribution causes three intersections between u_1 and u_2.

The calculation of the estimators can be divided into three groups:

(a) *Both end-points are known.* Then γ and θ are known (see [3.142c], where γ is the lower and $\gamma + \theta$ the upper boundary). Then we can readily obtain the transformation variable:

$$f_i(z) = \ln[(x_i - \gamma)/(\gamma + \theta - x_i)]$$ [3.143d]

The moments fit for the remaining parameters is identical with the maximum-likelihood fit:

$$\hat{\beta}\hat{\alpha} = -\left(\sum_{i=1}^{n} f_i\right)/N$$ [3.143e]

$$1/\hat{\beta}^2 = \sigma_f^2 \hat{N}^{-1} \sum_{i=1}^{N} (f_i - \bar{f})^2$$ [3.143f]

(unbiased estimator, divide by $N-1$, see discussion by the author, 1974a, section 1.8).

Where the data are not given in extenso, the formulae for the moments estimators would require correction by introducing the frequency density of the classes. The substitution of the percentile values (transformed percentiles of x_i) instead of the class boundaries or the central class value is better.

(b) *One end-point only is known.* Let us assume that γ is known. We estimate the median, the x_0 and the x_{100} percentile. We deduce:

$$-y_P = \hat{\alpha} + \hat{\beta}\ln[(x_0 - \gamma)/(\gamma + \hat{\theta} - x_0)]$$ [3.143g]

$$0 = \hat{\alpha} + \hat{\beta}\ln[(x_5 - \gamma)/(\gamma + \hat{\theta} - x_5)]$$ [3.143h]

$$y_P = \hat{\alpha} + \hat{\beta}\ln[(x_1 - \gamma)/(\gamma + \hat{\theta} - x_1)]$$ [3.143i]

with: $P = (1/\sqrt{2\pi}) \int_{\hat{y}_p}^{\infty} \exp(-t^2/2)\,dt$ [3.143j]

It means that the y_P is replaced by $\hat{y}_p$, the pertinent value from the Gaussian distribution. The x_0, x_5, and x_1 stand for the 0, 50, and 100% value, respectively of the cumulative distribution.

The solution for $\hat{\theta}$ follows after some manipulation:

$$\hat{\theta} = X_5(X_5 X_0 + X_5 X_1 - 2X_0 X_1)/(X_5^2 - X_0 X_1)$$ [3.143k]

where $x_j - \gamma$ has been replaced by X_j.

The remaining parameters can be found from [3.143g] through [3.143i] or from the same system as [3.143e and f].

(c) *No end-point is known.* The four parameters must be estimated. Johnson (1949) recommends that four percentile values are selected. We can write:

$$y_P = \hat{\alpha} + \hat{\beta} \ln [(x_p - \hat{\gamma})/(\hat{\gamma} + \theta - x_p)] \qquad [3.143l]$$

for the four selected percentile values. Equation [3.143] requires the replacement of y_p from:

$$P = (1/\sqrt{2\pi}) \int_{-\infty}^{y_p} \exp(-t^2/2)\, dt \qquad [3.143m]$$

(c) A third form of transformation may be based on:

$$y_3 = \alpha + \beta \operatorname{arcsinh} z \qquad [3.144]$$

$$\text{or: } y_3 = \alpha + \beta \ln [z + (z^2 + 1)^{1/2}] \qquad [3.144a]$$

where sinh stands for the hyperbolic sine and z for the standardized variable $(x - \gamma)/\theta$. The fitting is again a problem.

The median value of z is:

$$z_{\text{med}} = \sinh(\alpha/\beta) \qquad [3.144b]$$

The moments of this distribution are much more representative, as the frequency distribution shows one peak with the mode coinciding with the median or appearing either above or below. Thus the behavior of the curve is reasonably well represented by moments. The solutions according to Johnson (1949) are:

$$\bar{y} = -w^{1/2} \sinh W \qquad [3.144c]$$

$$v_2 = \sigma_y^2 = 0.5(w - 1)(w \cosh 2W + 1) \qquad [3.144d]$$

$$v_3 = 0.25 w^{1/2}(w - 1)^2 [w(w + 2) \sinh 3W + 3 \sinh W] \qquad [3.144e]$$

$$v_4 = 0.125(w - 1)^2 [w^2(w^4 + 2w^3 + 3w^2 - 3) \cosh 4W$$
$$+ 4w^2(w + 2) \cosh 2W + 3(2W + 1)] \qquad [3.144f]$$

The following abbreviations have been used in the above equations:

$$w = \exp(\beta - 2) \text{ and } W = \alpha/\beta$$

If α is positive we have $\bar{y} < y_{\text{med}} < y_{\text{mode}}$, while the direction of the inequality is reversed for negative α.

Johnson (1949) has given a graphical solution for α and β. The remaining parameters can then readily be determined.

3.7.2.2 Other systems.

The basic principle of translation remains the same, namely:

$$y = \alpha + \beta f\left(\frac{x-\gamma}{\theta}\right) \qquad [3.145]$$

We may select, however, different distribution forms for y. One of the distributions has been given by Laplace:

$$f(y) = \tfrac{1}{2}e^{-|y|} \quad -\infty < y < \infty \qquad [3.145a]$$

where $f(y)$ indicates the frequency density. If the forms of y_1, y_2 and y_3 (see [3.142], [3.142b], [3.144]) were used in conjunction with [3.145a], we could generate new systems of translations. Let these be designated y'_1, y'_2 and y'_3. Johnson (1949) has cautioned that the system y'_3 can possess moments whose numerical values become infinite.

The equation [3.145a] can be modified for a variety of distribution laws, and numerous systems could be generated in this way. Johnson and Kotz (1972b) have recently considered an application of a power transformation for the incomplete gamma distribution (Pearson, type III). They have constructed a diagram for λ, where:

$$Y = X^\lambda \qquad [3.146]$$

and $f(x)$ has a gamma-type density function. The diagram is made up for entries of $\sqrt{\beta_1}$ and β_2 (Pearson's parameters, see Section 2.5). The moments about zero are:

$$\mu_k = \Gamma(C + k\lambda)/\Gamma(C) \qquad [3.146a]$$

with k and C defined in Section 2.11, $C = c/a$, $\lambda = 1/a$.

3.7.2.3 Shenton's system.

In a detailed analysis Shenton (1965a, b) has outlined a different technique of transformation. It is well known that γ_1 and γ_2 (skewness and kurtosis) are zero for the Gaussian distribution. Pearson's parameters then become $\beta_1 = 0$ and $\beta_2 = 3$ for the Gaussian distribution.

Shenton suggests now a transformation by a polynomial series (x denotes the deviation from the mean):

$$y_p(x) = \sum_{s=0}^{n} a_s x^s \qquad [3.147]$$

where the y_p is the percentile point of the normal curve corresponding to the same cumulative distribution of x. The coefficients are then determined by minimizing the squared deviation $(y_{ep} - y_{ap})^2$, where y_{ep} is the value computed by [3.147] and y_{ap} is the equivalent value taken from the normal distribution. This is a problem of obtaining the coefficients by least-square solutions as outlined previously (see Essenwanger, 1974a, section 4.1). Although any number of percentile values can be selected, Shenton recommends the following threshold:

$P = 0.25, 0.5, 1.0, 2.5, 5.0, 10.0, 25.0, 50.0, 75.0$

$90.0, 95.0, 97.5, 99.0, 99.5, 99.7\%$

The technique requires that the theoretical values for the normal distribution be substituted in [3.147]. The calculation of the coefficients follows the established rules. The corresponding y_p-values for the given set of P-values can be found in the normal distribution tables (see Appendix or such as by Hald (1952), Beyer (1966), or Owen (1962), etc.).

Shenton has applied this regression technique to the Pearson curve system and has given tables of coefficients a_0 through a_5 for selected combinations of β_1 and β_2. In an expansion of this technique Shenton (1965b) has adapted the system to the Pearson III curve. As mentioned earlier, this distribution form is identical with the (incomplete) gamma distribution (see Section 2.11 or Essenwanger, 1974a, Section 2.5). The transformed variate becomes:

$$y_c = x^c = (X - \gamma)^c \qquad [3.148]$$

Here the γ is a reference parameter, as defined earlier. This γ can be identical with the mean. A summary of the analysis discloses that $c \sim c_s$ or c_L:

$c_s \sim 1/3$ for $\sqrt{\beta_1}$ small [3.148a]

$c_L \sim 4/\beta_1$ for $\sqrt{\beta_1}$ large [3.148b]

This result agrees with the Wilson and Hilferty (1931) transformation of the χ^2-distribution. These authors employ a cube-root transformation. The following approximation for the transformed variable can be checked against the data.

For small $\sqrt{\beta_1}$:

$\bar{y} \sim (4/\beta_1)^{1/3}$ [3.149a]

$\sigma_y \sim (1/3)(\beta_1/4)^{1/6}$ [3.149b]

$\beta_{2y} \sim 3 \ \{\text{equiv. } \beta_2(y_c)\}$ [3.149c]

For large $\sqrt{\beta_1}$:

$\bar{y} \sim 1/2$ [3.149d]

$\sigma_y \sim (\sqrt{3})/6$ [3.149e]

$\beta_{2y} \sim 9/5$ [3.149f]

This information is helpful in the judgment of whether β_1 should be considered small or large in addition to the comparison between c_s and c_L. It is

TRANSFORMATIONS

evident that a transformation with β_1 approximately zero does not necessarily render $\beta_{2y} = 3$. There exists a boundary value, where $\beta_{1y} \to 0$ and $\beta_{2y} \to 3$. The equation can be based on:

$$c = 1/3 - 4/81t - 4/243t^2 + 148/19683t^3 + (\epsilon_t) \qquad [3.150]$$

$$\text{and: } \beta_{2y} = 3 - 2/9t - 32/243t^2 + (\epsilon_t) \qquad [3.150a]$$

with $t = 4/\beta_1$. This relationship has been tabulated by Shenton (1965b).

3.7.2.4 Some original and related transformed distributions. The following Table 3.16 summarizes some of the transformations and the distribution appropriate for their application. The arctanh describes the inverse hyperbolic tangent (see definition Abramowitz and Stegun, 1964, p. 86):

$$\text{arctanh } z = \int_0^z dt/(1-t^2)$$

TABLE 3.16

Transformations

Transformation of y by	Range	Distribution property	Original type
$y = \sqrt{X}$ or $(\sqrt{X} + \sqrt{X+1})/2$	$0 \leq X \leq \infty$	$\sigma^2 = c^2\bar{x}$	incomplete gamma Poisson
ln X	$0 \leq X < \infty$	$\sigma^2 = c^2\bar{x}^2$	distribution of variance
ln $[X/(1-X)]$ = arctanh $(2X-1)$	$0 \leq X \leq 1.0$	$\sigma^2 = c^2\bar{x}(1-\bar{x})$ $\sigma^2 = c^2\bar{x}^2(1-\bar{x})^2$	beta some empirical distributions
arc sin $\sqrt{X}$ see [3.154b, c]	$0 \leq X \leq 1.0$	$\sigma^2 = c^2\bar{x}(1-\bar{x})$	beta binomial distribution
$\frac{1}{2}$ ln $[(1+X)/(1-X)]$ = arctanh X	$-1 \leq X \leq +1$	$\sigma^2 = c^2(1-\bar{x}^2)^2$	correlation coefficient

3.7.2.5 Transformation related to square root. The transformation of the Poisson, binomial and negative binomial by the expression:

$$y = \sqrt{x+c} \qquad [3.151]$$

has been suggested by various authors. Anscombe (1948) sets the constant $c = 3/8$, while Bartlett (1936, 1947) recommends $c = 0.5$ for the Poisson distribution.

Freeman and Tukey (1950) have suggested:

$$y = (\sqrt{x} + \sqrt{x+1})/2 \qquad [3.152a]$$

for the transformation of Poisson-type data. This would lead to a variance of unity. They have further pointed out that, for significance tests or the setting of confidence limits, the probability of exceedance for the normal deviate t is the same as the number of successes x from n in the binomial distribution when a transformation of the form:

$$T = 2[\sqrt{(k+1)q} - \sqrt{(n-k)p}] \qquad [3.152b]$$

is made. Here n and p denote the parameters of the binomial distribution and k is the threshold of x to be exceeded in the binomial distribution. Other transformations are:

$$T_1 = T + (T + 2p - 1)/(12\sqrt{E_1}) \qquad [3.152c]$$

$$T_2 = T + (T^2 - 4)/(12E_2) \qquad [3.152d]$$

$$T_3 = T_2 + (T_2 + 2p - 1)/(12\sqrt{E_1}) \qquad [3.152e]$$

The following abbreviations have been made:

$$E_1 = np \text{ or } nq, \text{ whichever is less} \qquad [3.153a]$$

and: $$E_2 = (np + 1)^{-1/2} - (nq + 1)^{-1/2} \qquad [3.153b]$$

Freeman and Tukey (1950) recommend for variance stabilization the angular transformation:

$$y = [\arcsin\sqrt{x/(n+1)} + \arcsin\sqrt{(x+1)/(n+1)}]/2 \qquad [3.154a]$$

This transformation has a variance within ± 6% of:

$$\sigma_y^2 = 1/(n + 0.5) \text{ (radians)} \qquad [3.154b]$$

or: $$\sigma_y^2 = 821/(n + 0.5) \text{ (degrees)} \qquad [3.154c]$$

CHAPTER 4

CALCULATION OF EIGENVALUES AND EIGENVECTORS

4.1 *MATRICES AND OPERATIONS*

Definitions

The following discussions cannot replace any regular text on matrix algebra but should give the reader some knowledge of the treatment of matrices. The special goal is the preparation of the background for the utilization of matrix algebra in this textbook.

It is common knowledge that a matrix is a rectangular array of numbers with columns and rows. In the matrix*:

$$\begin{bmatrix} x_{11} & x_{12} & \ldots & x_{1n} \\ x_{21} & x_{22} & \ldots & x_{2n} \\ \cdot & \cdot & \cdot & \\ \cdot & \cdot & \cdot & \\ \cdot & \cdot & \cdot & \\ x_{m1} & x_{m2} & & x_{mn} \end{bmatrix}$$ [4.1]

the x_{ij} are called elements and according to convention the first subscript signifies the row, the second the column. A matrix of m rows and n columns is said to be of order $m \times n$ or m by n. Subsequently, capital letters are used for matrices, either **X** or $\mathbf{M}_X$. The order can be indicated $\mathbf{X}_{nm}$ or $\mathbf{M}_{Xnm}$ but will be omitted when not ambiguous.

It is trivial that a matrix of order $n \times n$ is called a square matrix, one of order $m \times n$ or $n \times m$ with $m \neq n$ a rectangular matrix. The elements x_{11}, $x_{22} \ldots x_{nn}$ are called diagonal elements and their sum is called the trace:

$$\text{tr}(\mathbf{X}) = \sum_{1}^{n} x_{ii}$$ [4.2]

If two matrices $A = [a_{ij}]$ and $B = [b_{ij}]$ have the identical elements, then and only then is:

$a_{ij} \equiv b_{ij}$ (for $i = 1, \ldots n, j = 1, \ldots n$) [4.3]

and: **A** = **B** [4.3a]

---

* In many notations the brackets are replaced by parentheses.

The equal statement thus expresses a duplication of one matrix. When all elements $x_{ij} = 0$ for all i- or j-values, the matrix is called a zero matrix which sometimes is written $X = 0$, provided no confusion of the order exists.

The matrix of order $1 \times m$ or $m \times 1$ is called a vector and shall be symbolized by small letters such as x.

Sums

If $X = [x_{ij}]$ and $Y = [y_{ij}]$, then:

$$Z = X \pm Y \qquad [4.4]$$

is defined as the operation of the addition of elements, i.e.:

$$x_{ij} \pm y_{ij} = z_{ij} \qquad [4.4a]$$

Hence: $Z = [z_{ij}] = [x_{ij} \pm y_{ij}] \qquad [4.4b]$

Example:

$$X = \begin{bmatrix} 3 & 4 & 7 \\ 2 & 1 & 3 \end{bmatrix} \text{ and } Y = \begin{bmatrix} 2 & 0 & 3 \\ 4 & 1 & 2 \end{bmatrix}; \quad Z = X + Y = \begin{bmatrix} 5 & 4 & 10 \\ 6 & 2 & 5 \end{bmatrix};$$

$$Z = X - Y = \begin{bmatrix} 1 & 4 & 4 \\ -2 & 0 & 1 \end{bmatrix}$$

Consequently the two matrices must be of the same order. They are said to be conformable for addition (subtraction).

A matrix multiplied by a scalar factor α involves the following operation:

$$Z = \alpha X = [\alpha x_{ij}] = \alpha[x_{ij}] \qquad [4.5]$$

All elements of X are multiplied by α. Note:

$$\alpha X = X \alpha \qquad [4.5a]$$

The operation:

$$Y = -X \qquad [4.5b]$$

means we change the sign of *all* elements in X.

For conformable matrices we find:

Commutative law: $X + Y = Y + X$ [4.5c]

Associative law: $X + (Y + Z) = (X + Y) + Z$ [4.5d]

Further: $\alpha(X + Y) = \alpha X + \alpha Y$ [4.5e]

Multiplication

We define the product:

$$Z = XY \qquad [4.6]$$

where X is of order $n_i \times p$ and Y of $p \times n_j$. The matrix Z has the following elements:

$$z_{ij} = \sum_{k=1}^{p} x_{ik} y_{kj} \qquad [4.6a]$$

We recognize that a row element of X is multiplied with a column element of Y. The expression conformable for multiplication refers now to equal elements in rows of X and columns of Y. The elements of Z are obtained by summation of all pertinent terms according to the multiplication rule of [4.6a]
Example:

$$X = \begin{bmatrix} 2 & 3 & 4 \\ 0 & 2 & 1 \end{bmatrix}, \quad Y = \begin{bmatrix} 1 \\ 2 \\ 3 \end{bmatrix}. \quad X \text{ order } 2 \times 3, Y \text{ order } 3 \times 1,$$

$$n_i = 2, n_j = 1, p = 3$$

The X and Y are conformable for multiplication. Z has the elements z_{11} and z_{21}:

$$z_{11} = 2 \cdot 1 + 3 \cdot 2 + 4 \cdot 3 = 20; \quad z_{21} = 0 \cdot 1 + 2 \cdot 2 + 1 \cdot 3 = 7$$

Thus: $Z = \begin{bmatrix} 20 \\ 7 \end{bmatrix}$

Under the assumption of conformality, we find:

$$X(Y + Z) = XY + XZ \text{ (distributive law)} \qquad [4.6b]$$

$$(X + Y)Z = XZ + YZ \text{ (distributive law)} \qquad [4.6c]$$

$$X(YZ) = (XY)Z \text{ (associative law)} \qquad [4.6d]$$

Note, in general:

$XY \neq YX$

$XY = 0$ does not necessarily imply $X = 0$ and/or $Y = 0$

$XY = XZ$ does not necessarily imply $Y = Z$

The product of a matrix can also be found by partitioning.

$$\mathbf{A} = \begin{bmatrix} \mathbf{A}_{11} & \mathbf{A}_{12} \\ \mathbf{A}_{21} & \mathbf{A}_{22} \end{bmatrix} = \left[\begin{array}{cc|c} a_{11} & a_{12} & a_{13} \\ a_{21} & a_{22} & a_{23} \\ \hline a_{31} & a_{32} & a_{33} \end{array}\right] = \left[\begin{array}{cc|c} 2 & 1 & 0 \\ 1 & 3 & -1 \\ \hline 0 & 2 & 1 \end{array}\right]$$

$$\mathbf{B} = \begin{bmatrix} \mathbf{B}_{11} & \mathbf{B}_{12} \\ \mathbf{B}_{21} & \mathbf{B}_{22} \end{bmatrix} = \left[\begin{array}{ccc|c} b_{11} & b_{12} & b_{13} & b_{14} \\ b_{21} & b_{22} & b_{23} & b_{24} \\ \hline b_{31} & b_{32} & b_{33} & b_{34} \end{array}\right] = \left[\begin{array}{ccc|c} 3 & 2 & 1 & 0 \\ 0 & 2 & -1 & 2 \\ \hline 3 & -1 & 2 & 1 \end{array}\right]$$

$\mathbf{C} = \mathbf{AB}$, with elements:

$$\begin{bmatrix} \mathbf{C}_{11} & \mathbf{C}_{12} \\ \mathbf{C}_{21} & \mathbf{C}_{22} \end{bmatrix} = \left[\begin{array}{c|c} \mathbf{A}_{11}\mathbf{B}_{11} + \mathbf{A}_{12}\mathbf{B}_{21} & \mathbf{A}_{11}\mathbf{B}_{12} + \mathbf{A}_{12}\mathbf{B}_{22} \\ \hline \mathbf{A}_{21}\mathbf{B}_{11} + \mathbf{A}_{22}\mathbf{B}_{21} & \mathbf{A}_{21}\mathbf{B}_{12} + \mathbf{A}_{22}\mathbf{B}_{22} \end{array}\right]$$

$$\mathbf{C} = \left[\begin{array}{c|c} \begin{bmatrix} 2 & 1 \\ 1 & 3 \end{bmatrix} \begin{bmatrix} 3 & 2 & 1 \\ 0 & 2 & -1 \end{bmatrix} + \begin{bmatrix} 0 \\ -1 \end{bmatrix} [3 \;\; -1 \;\; 2] & \begin{bmatrix} 2 & 1 \\ 1 & 3 \end{bmatrix} \begin{bmatrix} 0 \\ 2 \end{bmatrix} + \begin{bmatrix} 0 \\ -1 \end{bmatrix} [1] \\ \hline [0 \;\; 2] \begin{bmatrix} 3 & 2 & 1 \\ 0 & 2 & -1 \end{bmatrix} + [1][3 \;\; -1 \;\; 2] & [0 \;\; 2] \begin{bmatrix} 0 \\ 2 \end{bmatrix} + [1][1] \end{array}\right]$$

$$\mathbf{C} = \left[\begin{array}{c|c} \begin{bmatrix} 6 & 6 & 1 \\ 3 & 8 & -2 \end{bmatrix} + \begin{bmatrix} 0 & 0 & 0 \\ -3 & 1 & -2 \end{bmatrix} & \begin{bmatrix} 2 \\ 6 \end{bmatrix} + \begin{bmatrix} 0 \\ -1 \end{bmatrix} \\ \hline [0 \;\; 4 \;\; -2] + [3 \;\; -1 \;\; 2] & [4] + [1] \end{array}\right] = \left[\begin{array}{ccc|c} 6 & 6 & 1 & 2 \\ 0 & 9 & -4 & 5 \\ \hline 3 & 3 & 0 & 5 \end{array}\right]$$

For a check, the element:

$$c_{11} = 2 \cdot 3 + 1 \cdot 0 + 0 \cdot 3 = 6$$

$$c_{22} = 1 \cdot 2 + 3 \cdot 2 - 1 \cdot (-1) = 9, \text{ etc.}$$

The $n \times 1$ matrix is called a vector (e.g. denoted by small letters). Then x is a column vector and x^T a row vector.

The scalar product between two vectors x and y is defined as:

$$z_s = (x^T \cdot y) = (x \cdot y^T) \qquad [4.6e]$$

$$\text{with: } z_s = \Sigma\, z_{ij} = \Sigma\, x_{ij} y_{ij} \qquad [4.6f]$$

TYPES OF MATRICES

4.2 TYPES OF MATRICES (DEFINITION)

Triangular, diagonal matrices

We have previously distinguished between a rectangular and a square matrix. A square matrix X whose elements x_{ij} are zero for $i > j$ or $i < j$ is called an upper or lower triangular matrix, respectively:

Upper triangular Lower triangular Diagonal

$$\begin{bmatrix} x_{11} & x_{12} & x_{13} \\ 0 & x_{22} & x_{23} \\ 0 & 0 & x_{33} \end{bmatrix} \quad \begin{bmatrix} x_{11} & 0 & 0 \\ x_{12} & x_{22} & 0 \\ x_{13} & x_{23} & x_{33} \end{bmatrix} \quad \begin{bmatrix} x_{11} & 0 & 0 \\ 0 & x_{22} & 0 \\ 0 & 0 & x_{33} \end{bmatrix}$$

When x_{ij} is zero for both $i > j$ and $i < j$, and $x_{ij} \neq 0$ for $i = j$, it is called diagonal matrix, we shall generally write it D_X. The diagonal matrix D_X is called a scalar when $x_{11} = x_{22} = \ldots x_{nn} = \alpha$. If $\alpha = 1$ then the matrix is called the Identity matrix I_n or I, where n denotes the order:

$$I = \begin{bmatrix} 1 & 0 & 0 \\ 0 & 1 & 0 \\ 0 & 0 & 1 \end{bmatrix} \qquad [4.7]$$

$$pI_n = \sum_1^p I_n \qquad [4.7a]$$

e.g.: $pI_3 = \begin{bmatrix} p & 0 & 0 \\ 0 & p & 0 \\ 0 & 0 & p \end{bmatrix}$

but: $I^p = I$ [4.7b]

If X is of the order 2×3, then:

$$I_2 X = X I_3 = I_2 X I_3 = X \qquad [4.7c]$$

If $XY = YX$, then X and Y are called commutative.

$XY = -YX$, the matrices anti-commute.

$X^{p+1} = X$ is called periodic (p positive integer).

$X^2 = X$, then X is called idempotent.

$X^p = 0$ is called nilpotent (p positive integer).

Inverse matrix

If $XY = YX = I$, then Y is called the inverse of X. This is written:

$$X = Y^{-1} \text{ or } M_X = M_Y^{-1} \qquad [4.8]$$

Not every square matrix has an inverse. Further, assuming that X^{-1} and Y^{-1} exist:

$$(XY)^{-1} = Y^{-1}X^{-1}. \qquad [4.8a]$$

(Note the reversal of X and Y on the right side.)

$$X^2 = I, \text{ the } X \text{ is called involutory} \qquad [4.8b]$$

The computation of an inverse matrix is discussed in Section 4.6.

Transpose matrix

When the rows and columns of an $m \times n$ matrix are interchanged we obtain the transpose of a matrix:

$$Y = X^T, \text{ sometimes written with prime } Y = X' \qquad [4.9]$$

Example:

$$X = \begin{bmatrix} 1 & 3 & 5 \\ 2 & 4 & 6 \end{bmatrix}; \quad X^T = \begin{bmatrix} 1 & 2 \\ 3 & 4 \\ 5 & 6 \end{bmatrix}$$

The element x_{ij} becomes the element x_{ji} of the transpose.

$$(X^T)^T = X \qquad [4.9a]$$
$$(\alpha X)^T = \alpha X^T \qquad [4.9b]$$
$$(X + Y)^T = X^T + Y^T \qquad [4.9c]$$
$$(XY)^T = Y^T \cdot X^T \qquad [4.9d]$$

Symmetric matrix

A square matrix is called symmetric when:

$$X^T = X \qquad [4.10]$$

Then: $x_{ij} = x_{ji}$ \qquad [4.10a]

If X is a square matrix, then the matrix Y is symmetric for:

$$Y = X + X^T \qquad [4.10b]$$

TYPES OF MATRICES

$X^T = -X$ (skewed symmetric)   [4.10c]

and: $x_{ij} = -x_{ji}$ [4.10d]

$Y = X - X^T$ is skewed symmetric   [4.10e]

Conjugate matrix

Complex numbers $a + bi$ and $a - bi$ are called conjugate ($i = \sqrt{-1}$). If $x = a + bi$ then the conjugate is denoted by a bar. $\bar{x} = a - bi = \overline{a + bi}$. If $z_2 = \bar{z}_1$ then $\bar{z}_2 = \bar{\bar{z}}_1 = z_1$. Further if z_1 and z_2 are complex numbers then $\overline{z_1 + z_2} = \bar{z}_1 + \bar{z}_2$. Further $\overline{z_1 \cdot z_2} = \bar{z}_1 \cdot \bar{z}_2$.

When the matrix X has complex numbers as elements then $\bar{X}$ is the conjugate matrix when the elements are replaced by the conjugate element. Example:

$$X = \begin{bmatrix} 2+i & i \\ 4 & 2-i \end{bmatrix} \quad \text{and} \quad \bar{X} = \begin{bmatrix} 2-i & -i \\ 4 & 2+i \end{bmatrix}$$

$\bar{\bar{X}} = X$ [4.11a]

$\overline{(\alpha X)} = \bar{\alpha}\bar{X}$ [4.11b]

$\overline{(X + Y)} = \bar{X} + \bar{Y}$ [4.11c]

$\overline{(XY)} = \bar{X} \cdot \bar{Y}$ [4.11d]

$(\bar{X})^T = \overline{(X^T)}$ [4.11e]

Hermetian matrix

$X = \overline{X^T}$ is called Hermetian.   [4.12]

Then: $x_{ij} = \bar{x}_{ji}$ [4.12a]

$\overline{X^T} = -X$ is skew Hermetian.   [4.12b]

The diagonal elements of a skew-Hermetian matrix are either zero or pure imaginaries. Then:

$x_{ij} = -\bar{x}_{ji}$

4.3 DETERMINANTS

Product rule

The determinants are based on square matrices. We consider the product:

$$P = x_{1j_1} \cdot x_{2j_2} \cdot x_{3j_3} \ldots x_{nj_n} \qquad [4.13]$$

In this product one and only one element comes from any row and one and only one element from any column. The first subscripts i are in sequence, the second subscripts can take any sequence of permutations. Permutations are the possible exchanges in the sequence of numbers. E.g. take 123. Then 231, 312, 132, 321, 213 are all possible permutations with three numbers. The possible permutations of any number n of (different) numerals are $n!$.

The determinant of a matrix $\mathbf{X}$ (denoted $\|\mathbf{X}\|$)* is now:

$$\|\mathbf{X}\| = \sum_N (-1)^k P \qquad [4.13a]$$

where $N = n!$. This means we sum all possible products of permutations. The k is the number of inversions and determines the sign of the product.

When a larger integer precedes a smaller integer in a permutation it is called an inversion. The number of inversions can be even or odd, e.g., 123 would be even, 132 is odd. Further 312 is even since 3 precedes 1 and 2, 321 is odd since 3-2, 3-1, 2-1 are inversions, etc. In 4312 the inversions are 4-3, 4-1, 4-2, 3-1, 3-2, hence an odd number of inversions. When the number of inversions of the permutation is even, the sign is $+1$, otherwise -1.

Example:

$$\mathbf{X} = \begin{bmatrix} x_{11} & x_{12} \\ x_{21} & x_{22} \end{bmatrix}$$

$$\|\mathbf{X}\| = (-1)^k x_{11}x_{22} + (-1)^k x_{12}x_{21} = x_{11}x_{22} - x_{12}x_{21} \qquad [4.13b]$$

The first product has no inversion, j_n: 1-2, the second product shows one, j_n: 2-1. (Note, we judge only the j_n.)

Partition rule

For larger determinants the permutations of P may not be very easy to compute. We apply, therefore, a rule which provides the determinant by breaking down the matrix into matrices of smaller order until finally only a 2 × 2 matrix is left whose solution has been given by [4.13b].

This breakdown is quite easy. We take every element of one row or column

---

* The double bar is used to distinguish from the absolute value of any variate.

DETERMINANTS

and multiply by its "minor determinant". The minor determinant is the determinant of the minor matrix which originates when the column and row in which the element occurs has been cancelled. The minor matrix is (at least) one order lower than the original matrix. When the minor determinant has a sign attached it is called a "co-factor" in some sources. The sign follows again $(-1)^k$, where $k = i + j$, i.e., we add the indices of the element. When k is odd, the sign is negative.

Example:

$$X = \begin{bmatrix} x_{11} & x_{12} & x_{13} \\ x_{21} & x_{22} & x_{23} \\ x_{31} & x_{32} & x_{33} \end{bmatrix}$$

$$\|X\| = x_{11} \begin{vmatrix} x_{22} & x_{23} \\ x_{32} & x_{33} \end{vmatrix} - x_{21} \begin{vmatrix} x_{12} & x_{13} \\ x_{32} & x_{33} \end{vmatrix} + x_{31} \begin{vmatrix} x_{12} & x_{13} \\ x_{22} & x_{23} \end{vmatrix} \quad [4.14a]$$

or:

$$\|X\| = x_{11} \begin{vmatrix} x_{22} & x_{23} \\ x_{32} & x_{33} \end{vmatrix} - x_{12} \begin{vmatrix} x_{21} & x_{23} \\ x_{31} & x_{33} \end{vmatrix} + x_{13} \begin{vmatrix} x_{21} & x_{22} \\ x_{31} & x_{32} \end{vmatrix} \quad [4.14b]$$

The reader can prove very easily that [4.14a] and [4.14b] lead to a result which is identical with [4.13a].

Numerical example:

$$X = \begin{bmatrix} 1 & 4 & -2 \\ 3 & 2 & 0 \\ 0 & 3 & 2 \end{bmatrix}$$

$$\|X\| = 1 \begin{vmatrix} 2 & 0 \\ 3 & 2 \end{vmatrix} - 3 \begin{vmatrix} 4 & -2 \\ 3 & 2 \end{vmatrix} + 0 \begin{vmatrix} 4 & -2 \\ 2 & 0 \end{vmatrix}$$

$$= 1(4 - 3 \cdot 0) - 3(8 - (-6)) + 0(0 - (-4)) = 4 - 3 \cdot 14 = -38$$

Some properties of determinants

(1) If every element in one row (or column) is zero,

$$\|X\| = 0. \quad [4.14c]$$

(2) $\|X^T\| = \|X\|$. $\quad [4.14d]$

(3) If one row (or column) of X is multiplied by a factor α to form a matrix Y and the other elements are the same, then:

$$\|Y\| = \alpha\|X\| \quad [4.14e]$$

Example:

$$\|\mathbf{X}\| = \begin{bmatrix} x_{11} & x_{12} & x_{13} \\ x_{21} & x_{22} & x_{23} \\ x_{31} & x_{32} & x_{33} \end{bmatrix}; \quad \|\mathbf{Y}\| = \begin{bmatrix} x_{11} & \alpha x_{12} & x_{13} \\ x_{21} & \alpha x_{22} & x_{23} \\ x_{31} & \alpha x_{32} & x_{33} \end{bmatrix} = \alpha \begin{bmatrix} x_{11} & x_{12} & x_{13} \\ x_{21} & x_{22} & x_{23} \\ x_{31} & x_{32} & x_{33} \end{bmatrix}$$

(4) If **Y** is obtained from **X** by exchange of adjacent rows (columns) then:

$$\|\mathbf{Y}\| = -\|\mathbf{X}\| \qquad [4.14\text{f}]$$

This relationship is valid for the exchange of any two rows (columns).

(5) If two rows (columns) of **X** are identical $\|\mathbf{X}\| = 0$. [4.14g]

Example:

$$\mathbf{X} = \begin{bmatrix} 1 & 2 & 3 \\ 1 & 2 & 3 \\ 4 & 5 & 6 \end{bmatrix}$$

$$\|\mathbf{X}\| = 1 \begin{vmatrix} 2 & 3 \\ 5 & 6 \end{vmatrix} - 1 \begin{vmatrix} 2 & 3 \\ 5 & 6 \end{vmatrix} + 4 \begin{vmatrix} 2 & 3 \\ 2 & 3 \end{vmatrix} = 0$$

The first two terms are identical but positive and negative, in the last term the determinant of the matrix is zero ($2 \cdot 3 - 2 \cdot 3$).

(6) The determinants of $\|\mathbf{X}\|$ and $\|\mathbf{Y}\|$ are equal when the second matrix **Y** contains the same elements plus an added row (or column) multiplied by a scalar.

Example:

$$\begin{vmatrix} x_{11} & x_{12} \\ x_{21} & x_{22} \end{vmatrix} = \begin{vmatrix} x_{11} + \alpha x_{12} & x_{12} \\ x_{21} + \alpha x_{22} & x_{22} \end{vmatrix} = \begin{vmatrix} x_{11} & x_{12} \\ x_{21} + \alpha x_{11} & x_{22} + \alpha x_{12} \end{vmatrix} \qquad [4.14\text{h}]$$

Laplace expansion

Instead of the partitioning by one element and the associated minor determinant the matrix can also be broken into determinants by cancelling the rows and columns of more than one element. We first develop a notation. Assume we write the index of the element as $x_{i_k j_k}$ and let i go from i_1 to i_n (columns), j_k from j_1 to j_n (rows). Then (for $n = 2$ as example) the following notation is adopted:

DETERMINANTS

$$\mathbf{X}\begin{vmatrix}\text{column index}\\ \text{row index}\end{vmatrix} = \mathbf{X}\begin{vmatrix}j_1 & j_2\\ i_1 & i_2\end{vmatrix}; \quad \text{the determinant } \|\mathbf{X}\| = \begin{vmatrix}x_{i_1 j_1} & x_{i_1 j_2}\\ x_{i_2 j_1} & x_{i_2 j_2}\end{vmatrix} \quad [4.15a]$$

$$\text{or (e.g.): } \mathbf{X}\begin{vmatrix}j_2 & j_4\\ i_1 & i_3\end{vmatrix} \text{ is } \|\mathbf{X}\| = \begin{vmatrix}x_{i_1 j_2} & x_{i_1 j_4}\\ x_{i_3 j_2} & x_{i_3 j_4}\end{vmatrix} \quad [4.15b]$$

Determinants are called complementary minors when they supplement each other. Assume $i_k = i_1 \ldots i_5$ and $j_k = j_1 \ldots j_5$ then:

$$\mathbf{X}\begin{vmatrix}j_2 & j_4\\ i_1 & i_3\end{vmatrix} \text{ and } \mathbf{X}\begin{vmatrix}j_1 & j_3 & j_5\\ j_2 & j_4 & j_5\end{vmatrix} \quad [4.15c]$$

are complementary minors.
Laplace has shown that:

$$\|\mathbf{X}\| = \sum_p (-1)^k \left| \mathbf{X}\begin{matrix}j_1 j_2 \ldots j_m\\ i_2 i_2 \ldots i_m\end{matrix} \right| \cdot \left| \mathbf{X}\begin{matrix}j_{m+1} j_{m+2} \ldots j_n\\ i_{m+1} i_{m+2} \ldots i_n\end{matrix} \right| \quad [4.16]$$

$k = i_1 + i + \ldots i_m + j_1 + j_2 + \ldots j_m$ (again, odd minus and even plus); p stands for the number of permutations possible with the given number of elements (e.g. $p = n(n-1) \ldots (n-m+1)/m!$ where the index $i = 1 \ldots m \ldots n$).

Example 4.1:

$$\mathbf{X} = \begin{bmatrix} 2 & 3 & 3 & -2\\ -2 & 4 & 1 & 3\\ 0 & 2 & 3 & 1\\ 4 & 0 & 1 & 2 \end{bmatrix}$$

We take the first two rows for Laplace's rule. Then $m = 2$, $n = 4$. We compute $p = 4 \cdot 3/2! = 6$, we must have six terms. The sign is determined by the summation of indices:

$k_1 = 1+2+1+2, k_2 = 1+2+1+3, k_3 = 1+2+2+3,$
$k_4 = 1+2+2+3, k_5 = 1+2+2+4, k_6 = 1+2+3+4$

$$\|\mathbf{X}\| = +\begin{vmatrix}2 & 3\\ -2 & 4\end{vmatrix}\begin{vmatrix}3 & 1\\ 1 & 2\end{vmatrix} - \begin{vmatrix}2 & 3\\ 2 & 1\end{vmatrix}\begin{vmatrix}2 & 1\\ 0 & 2\end{vmatrix} + \begin{vmatrix}2 & -2\\ -2 & 3\end{vmatrix}\begin{vmatrix}2 & 3\\ 0 & 1\end{vmatrix}$$

$$+\begin{vmatrix}3&3\\4&1\end{vmatrix}\begin{vmatrix}0&1\\4&2\end{vmatrix}-\begin{vmatrix}3&-2\\4&3\end{vmatrix}\begin{vmatrix}0&3\\4&1\end{vmatrix}+\begin{vmatrix}3&-2\\1&3\end{vmatrix}\begin{vmatrix}0&2\\4&0\end{vmatrix}$$

$$= (8+6)(6-1) - (2-6)(4-0) + (6-4)(2-0) + (3-12)(0-4)$$
$$- (9+8)(0-12) + (9+2)(0-8)$$
$$= 14 \cdot 5 - (-4)4 + 2 \cdot 2 + (-9)(-4) - (17)(-12) + 11(-8)$$
$$= 70 + 16 + 4 + 36 + 204 - 88 = 242$$

Other operations with determinants

$$\|\mathbf{XY}\| = \|\mathbf{X}\| \cdot \|\mathbf{Y}\| \tag{4.17}$$

4.4 EQUIVALENCE OF MATRICES

Rank of matrices

For better understanding some matrix operations a brief summary of the equivalence of matrices is appropriate. We first define the rank of a matrix. A matrix **X** is said to be of rank r if at least one of the minor determinants (dimension r-square) differs from zero, while every $(r+1)$-square minor determinant is zero. A zero matrix has the rank 0.

Example 4.2:

$$\mathbf{X} = \begin{bmatrix} 1 & 2 & 3 & 4 \\ 1 & 2 & 3 & 4 \\ 2 & 6 & 0 & 7 \\ 1 & 2 & 1 & 3 \end{bmatrix} \text{ has rank } r = 3, \text{ since } \|\mathbf{X}_{11}\| = \begin{vmatrix} 2 & 3 & 4 \\ 6 & 0 & 7 \\ 2 & 1 & 3 \end{vmatrix} = -2 \neq 0$$

$\|\mathbf{X}\| = 0$

An n-square matrix is called non-singular for $r = n$ (or $\|\mathbf{X}_n\| \neq 0$). Otherwise the matrix is called singular. The product of two or more non-singular matrices is non-singular. The product is singular if one of the matrices is singular. Every non-singular matrix can be expressed as a product of elementary matrices (see following section).

Transformations

An elementary transformation of a matrix is the interchange of rows or columns, the multiplication of every row (column) element by a (non-zero) scalar factor, or the addition of any other row (column) to one column, even when the added column (row) is multiplied by a scalar factor. These do not change the order nor the rank of the matrix. When elementary transformations are applied to the identity matrix, the result is an elementary matrix.

The inverse of an elementary transformation is an operation which restores the matrix to its original form.

Equivalence and canonical matrix

We define two matrices to be equivalent ($\mathbf{X} \sim \mathbf{Y}$) when the first matrix can be converted into the second matrix by one or more elementary transformations. We distinguish row and column equivalence. Any non-zero matrix is row-equivalent to a canonical matrix. This canonical matrix is defined as a matrix where:

(1) one or more elements of each of the first r rows are non-zero (rank r):

(2) all other rows have zero elements:
(3) the first non-zero element is 1 in the ith row ($i = 1, 2, \ldots r$);
(4) if the column in which this unit element stands is denoted by j_i then $j_1 < j_2 < \ldots j_r$; and
(5) the only non-zero element in the columns with the unit elements is this element.

Example:

$$X_C = \begin{bmatrix} 1 & 2 & 0 & 0 & 3 \\ 0 & 0 & 1 & 0 & -3 \\ 0 & 0 & 0 & 1 & 0 \\ 0 & 0 & 0 & 0 & 0 \end{bmatrix} \text{ is a canonical matrix form}$$

A matrix X can be reduced into its canonical form by the following steps:
(1) Look for the first non-zero element in the column j_1.
(2) Divide the column by this first non-zero element, e.g. $a_{2j_1} \neq 0$. Then $a'_{1j_1} = 0, a'_{2j_1} = 1, a'_{3j_1} = a_{3j_1}/a_{2j_1}$, etc.
(3) Use row transformation to obtain zero elsewhere in the first column.

Example 4.3:

$$X = \begin{bmatrix} 3 & 12 & 6 & 9 \\ 2 & 8 & 3 & -2 \\ -1 & -4 & 3 & 1 \end{bmatrix} \sim \begin{bmatrix} 1 & 4 & 2 & 3 \\ 2 & 8 & 3 & -2 \\ -1 & -4 & 3 & 1 \end{bmatrix} \sim \begin{bmatrix} 1 & 4 & 2 & 3 \\ 0 & 0 & -1 & -8 \\ 0 & 0 & 5 & 4 \end{bmatrix}$$

$$\sim \begin{bmatrix} 1 & 4 & 2 & 3 \\ 0 & 0 & +1 & +8 \\ 0 & 0 & 5 & 4 \end{bmatrix} \sim \begin{bmatrix} 1 & 4 & 0 & -13 \\ 0 & 0 & 1 & 8 \\ 0 & 0 & 0 & -36 \end{bmatrix} \sim \begin{bmatrix} 1 & 4 & 0 & -13 \\ 0 & 0 & 1 & 8 \\ 0 & 0 & 0 & 1 \end{bmatrix}$$

The last matrix is the canonical form.

$$X \sim \begin{bmatrix} 1 & 4 & 0 & 0 \\ 0 & 0 & 1 & 0 \\ 0 & 0 & 0 & 1 \end{bmatrix}$$

Normal form

Any matrix of rank $r > 0$ can be converted by elementary transformation into the form I_r, $\begin{bmatrix} I_r & 0 \\ 0 & 0 \end{bmatrix}$; $[I_r \; 0]$ or $\begin{bmatrix} I_r \\ 0 \end{bmatrix}$ which is called the normal form. A zero matrix is its own normal form.

Orthogonal matrices

A square matrix is called orthogonal if:

$$XX^T = X^TX = I \qquad [4.18a]$$

Consequently: $X^T = X^{-1}$ [4.18b]

Any given arbitrary matrix can be transformed into an orthogonal matrix by standard procedures. Assume we have a square matrix A. We call the column vectors a_1, a_2, a_3, etc. with each vector having the elements a_{ij}. In the process of orthogonalization we must convert the orthogonalized vector to unity length. Assume the vector has the elements b_{ij}. First we calculate:

$$l^2(b_i) = \sum_1^j b_{ij}^2 \qquad [4.19]$$

$$e_{ij} = b_{ij}/l(b_i) \qquad [4.19a]$$

This vector has now the length 1 or is said to be normalized. We continue now with the process of orthogonalizing the matrix A. First we assume a_{1j} is the first vector and convert to:

$$e_{1j} = a_{1j}/l(a_1) \qquad [4.20a]$$

The next vector is a_{2j}. We remember that:

$$\sum_1^j a_{ij} \cdot a_{kj} = \delta_{ik} \qquad [4.20b]$$

where δ_{ik} is the Kronecker delta, i.e. $\delta_{ik} \begin{cases} = 1 \text{ for } k = i \\ = 0 \text{ for } k \neq i \end{cases}$ [4.20c]

We form now a vector:

$$v_{2j} = a_{2j} - ce_{1j} \qquad [4.20d]$$

In order to fulfill the orthogonality condition, we have:

$$c = \sum_1^j (a_{2j}e_{1j}) \qquad [4.20e]$$

where the term in parenthesis is the scalar product of a vector, i.e. every single element is multiplied by the respective term of the second vector, as row by column in matrix computation. When c has been determined by means of [4.20e], the operation of [4.20d] can be performed. A final step is:

$$e_{2j} = v_{2j}/l(v_2) \qquad [4.20f]$$

General: $v_{sj} = a_{sj} - \sum_{k=1}^{s=1} c_k e_{kj}$ [4.20g]

$$c_k = \sum_1^j (a_{sj}e_{kj}) \qquad [4.20h]$$

CALCULATION OF EIGENVALUES AND EIGENVECTORS

and: $e_{sj} = v_{sj}/l(v_s)$ [4.20i]

Example 4.4:

Assume that $A = \begin{bmatrix} 1 & 1 & 1 \\ 1 & 0 & 2 \\ 0 & 2 & 1 \end{bmatrix}$ which was arbitrarily chosen. Then:

$l^2(a_1) = 1 + 1 + 0 = 2;\ \ell(a_1) = \sqrt{2}$

$e_{1j}^T = (1/\sqrt{2};\ 1/\sqrt{2};\ 0)$

$a_{2j}^T = 1, 0, 2$

$c_1 = a_{2j} \cdot e_{1j}$

$1 \cdot 1/\sqrt{2} = 1/\sqrt{2}$
$0 \cdot 1/\sqrt{2} = 0$
$\underline{2 \cdot 0\ \ \ \ = 0}$
$c_1 = 1/\sqrt{2}$

$v_{2j} = a_{2j} - c_1 e_{1j}$

$1 - 1/2 =\ \ \ \ 1/2$
$0 - 1/2 = -1/2$
$2 - 0\ \ \ =\ \ \ \ 2$

$l^2(v_2) = 1/4 + 1/4 + 4 = 9/2;\ \ l(v_2) = 3/\sqrt{2}$

$e_{2j}^T = \dfrac{1/\sqrt{2}}{2 \cdot 3},\ -\dfrac{1/\sqrt{2}}{2 \cdot 3},\ \dfrac{2 \cdot \sqrt{2}}{3}$, hence $[1/(3\sqrt{2}),\ -1/(3\sqrt{2}),\ 4/(3\sqrt{2})]$

The third orthogonalized vector is:

$a_{3j}^T = 1, 2, 1$

$c_1 = a_{3j} \cdot e_{1j}$ $\ \ \ \ \ \ \ \ \ \ \ \ c_2 = a_{3j} e_{2j}$

$1 \cdot 1/\sqrt{2} = 1/\sqrt{2}$ $\ \ \ \ \ 1 \cdot [1/(3\sqrt{2})] = 1/(3\sqrt{2})$
$2 \cdot 1/\sqrt{2} = 2/\sqrt{2}$ $\ \ \ \ \ 2[-1/(3\sqrt{2})] = -2/(3\sqrt{2})$
$\underline{1 \cdot 0\ \ \ \ \ = 0}$ $\ \ \ \ \ \ \ \underline{1[4/(3\sqrt{2})] = 4/(3\sqrt{2})}$
$\ \ \ \ \ c_1 = 3/\sqrt{2}$ $\ \ \ \ \ \ \ \ \ \ c_2 = 3/(3\sqrt{2}) = 1/\sqrt{2}$

$v_3 = a_{3j} - c_1 e_{1j} - c_2 e_{2j}$

$1 - 3/2 -\ \ \ \ \ \ 1/6\ \ = -2/3$
$2 - 3/2 - (-1/6) =\ \ \ \ 2/3$
$1 - 0\ \ -\ \ \ \ \ \ 2/3\ =\ \ \ \ 1/3$

$l^2(v_2) = 4/9 + 4/9 + 1/9 = 1;\ l(v_2) = 1$

EQUIVALENCE OF MATRICES

$e_{3j}^T = (-2/3, 2/3, 1/3)$

The final result is an orthogonal matrix:

$$E = \begin{bmatrix} \frac{1}{\sqrt{2}} & \frac{1}{3\sqrt{2}} & -\frac{2}{3} \\ \frac{1}{\sqrt{2}} & \frac{1}{3\sqrt{2}} & \frac{2}{3} \\ 0 & \frac{4}{3\sqrt{2}} & \frac{1}{3} \end{bmatrix}$$

It can be readily checked that $E \cdot E^T = I$ and $E^{-1} \equiv E^T$.

It is immaterial for the above outlined method whether the components are orthogonalized a priori. The system adjusts by itself.

Example 4.5:

Assume that $u_1^T = (1, 1, 0)$ and $u_2^T = (-3, -2, 1)$ are eigenvectors. Then a linear combination is also an eigenvector, e.g., $u_3^T = (-2, -1, 1)$.
Now:

$e_{1j}^T = (1/\sqrt{2}, 1/\sqrt{2}, 0)$ as previously

$c_1 = -3/\sqrt{2} - 2/\sqrt{2} + 0 = -5/\sqrt{2}$

$v_{2j} = u_{2j} - c_1 e_{1j}$

$-3 - (-5/2) = -1/2$
$-2 - (-5/2) = +1/2$
$1 - (0) = 1$

$l^2(v_2) = 1/4 + 1/4 + 1 = 3/2;\quad l(v_2) = \sqrt{3/2}$

$e_{2j}^T = (-1/\sqrt{6}, +1/\sqrt{6}, +2/\sqrt{6})$

$u_3^T = -2, -1, 1$

$c_1 = (e_1 u_3) = -2/\sqrt{2} - 1/\sqrt{2} + 0 = -3/\sqrt{2}$
$c_2 = (e_2 u_3) = +2/\sqrt{6} - 1/\sqrt{6} + 2/\sqrt{6} = +3/\sqrt{6}$

$v_{3j} = u_{3j} - c_1 e_{1j} - c_2 e_{2j}$

$-2 - (-3/2) - (-3/6) = 0$
$-1 - (-3/2) - (+3/6) = 0$
$1 - (0) - (+6/6) = 0$

This means the u_3 is not an independent vector to be added. Let us assume we had $u_3^T = (1, -1, 1)$ which is an orthogonal component to e_1^T and e_2^T, then:

$$c_1 = 1/\sqrt{2} - 1/\sqrt{2} + 0 = 0$$
$$c_2 = -1/\sqrt{6} - 1/\sqrt{6} + 2/\sqrt{6} = 0$$

which means $v_{3j} = u_{3j}$ and $l^2(v_3) = 3$, $l(v_3) = \sqrt{3}$.

$$e_{3j}^T = (1/\sqrt{3}, -1/\sqrt{3}, 1/\sqrt{3})$$

i.e. an orthogonal component needs only to be normalized.

4.5 ADJOINTS

The co-factor of a square matrix was defined as a signed minor determinant, let us write it $(-1)^{i+j} \|M\|$ or with the Greek symbol ξ_{ij} for the element x_{ij}. We may define a matrix consisting entirely of co-factors and call it the adjoint.*

$$\text{adj } X = \begin{bmatrix} \xi_{11} & \xi_{21} & \cdots & \xi_{n1} \\ \xi_{12} & \xi_{22} & \cdots & \xi_{n2} \\ \xi_{1n} & \xi_{2n} & \cdots & \xi_{nn} \end{bmatrix} \qquad [4.21]$$

It should be noted that the index of the co-factors from the matrix i (rows) and j (columns) has columns and rows exchanged, i.e. the co-factors of the ith row elements stand in the ith column of the adjoint, the co-factors of the jth column elements stand in the jth row.

Example:

$$X = \begin{bmatrix} 1 & 0 & 2 \\ 3 & -1 & 1 \\ 2 & -2 & 0 \end{bmatrix} \quad \text{adj } X = \begin{bmatrix} +\begin{vmatrix} -1 & 1 \\ -2 & 0 \end{vmatrix} & -\begin{vmatrix} 0 & 2 \\ -2 & 0 \end{vmatrix} & +\begin{vmatrix} 0 & 2 \\ -1 & 1 \end{vmatrix} \\ -\begin{vmatrix} 3 & 1 \\ 2 & 0 \end{vmatrix} & +\begin{vmatrix} 1 & 2 \\ 2 & 0 \end{vmatrix} & -\begin{vmatrix} 1 & 2 \\ 3 & 1 \end{vmatrix} \\ +\begin{vmatrix} 3 & -1 \\ 2 & -2 \end{vmatrix} & -\begin{vmatrix} 1 & 0 \\ 2 & -2 \end{vmatrix} & +\begin{vmatrix} 1 & 0 \\ 3 & -1 \end{vmatrix} \end{bmatrix}$$

$$\text{adj } X = \begin{bmatrix} +2 & -4 & +2 \\ +2 & -4 & +5 \\ -4 & +2 & -1 \end{bmatrix}$$

Since the top row of the adjoint comprises the co-factors of the first column, the determinant of X is the sum of the elements of the first row, multiplied by the elements of the first column. This is valid for any corresponding row and column (e.g. $1 \cdot 2 - 3 \cdot 4 + 2 \cdot 2 = -6$).

Further:

$$X(\text{adj } X) = \|X\| \cdot I_n = (\text{adj } X)X \qquad [4.21a]$$

$$\|X\| \cdot \|\text{adj } X\| = \|\text{adj } X\| \cdot \|X\| = \|X\|^n \qquad [4.21b]$$

$$\|\text{adj } X\| = \|X\|^{n-1} \quad \text{(non-singular } n\text{-square matrix } X) \qquad [4.21c]$$

---

* Although in algebra the expression adjoint (or adjugate) is commonly used for the matrix of co-factors, in differential equations or quantum mechanics the adjoint means the Hermetian conjugate.

For a singular matrix

$$X(\text{adj } X) = (\text{adj } X)X = 0 \quad [4.21d]$$

If the rank of a matrix $r < n - 1$, the adj $X = 0$. If X is of rank $n - 1$, then adj X is of rank 1.

Product rule: $\text{adj}(XY) = (\text{adj } X) \cdot (\text{adj } Y)$ [4.21e]

Further: $X = \begin{bmatrix} a & b \\ c & d \end{bmatrix}, \text{adj}(X) = \begin{bmatrix} d & -b \\ -c & a \end{bmatrix}$ [4.21f]

4.6 THE INVERSE OF A MATRIX

We will now discuss a few facts about the inverse which was defined by [4.8]. An n-square matrix has an inverse if and only if it is a non-singular matrix. This inverse is unique. Recall that $XX^{-1} = I$. If X is non-singular, then:

$$XY = XZ \quad [4.22]$$

implies that $Y = Z$.

The inverse of a diagonal matrix with elements $\alpha_1, \alpha_2 \ldots \alpha_n$ is the diagonal matrix with elements $1/\alpha_1, 1/\alpha_2 \ldots 1/\alpha_n$. The inverse is not always trivial to compute.

Inverse from elementary matrices

We denote an elementary transformation of row inter-changes from the identity matrix by R_i and of columns by C_i. Then R_i and C_i are elementary matrices. When:

$$P \times Q = I \quad [4.22a]$$

where $P = R_k \ldots R_2 \cdot R_1 ; Q = C_1 \cdot C_2 \ldots C_k$, then:

$$X^{-1} = Q \cdot P = C_1 \cdot C_2 \ldots C_k \cdot R_k \ldots R_2 \cdot R_1 \quad [4.22b]$$

The problem is to find these elementary matrices.

Inverse from the adjoint

$$X^{-1} = \frac{\text{adj } X}{\|X\|} \quad [4.23]$$

i.e. every element of the adjoint is divided by $\|X\|$.

Example 4.6:

Let us take the matrix of the adjoint, [4.21]:

THE INVERSE OF A MATRIX

$$X = \begin{bmatrix} 1 & 0 & 2 \\ 3 & -1 & 1 \\ 2 & -2 & 0 \end{bmatrix}; \text{adj } X = \begin{bmatrix} 2 & -4 & +2 \\ 2 & -4 & +5 \\ -4 & 2 & -1 \end{bmatrix}; \|X\| = -6;$$

$$X^{-1} = \begin{bmatrix} -1/3 & +2/3 & -1/3 \\ -1/3 & +2/3 & -5/6 \\ +2/3 & -1/3 & 1/6 \end{bmatrix}$$

This is probably the easiest way how to compute an inverse.

Inverse by partitioning

We assume a matrix X with partitions $X_{11}(k \times k)$, $X_{12}(k \times l)$, $X_{21}(l \times k)$, $X_{22}(l \times l)$ and its inverse with partition A_{ij} and the $k + l = n$.

$$X = \left[\begin{array}{c|c} X_{11} & X_{12} \\ k \times k & k \times l \\ \hline X_{21} & X_{22} \\ l \times k & l \times l \end{array}\right] \text{ and } X^{-1} = A = \left[\begin{array}{c|c} A_{11} & A_{12} \\ k \times k & k \times l \\ \hline A_{21} & A_{22} \\ l \times k & l \times l \end{array}\right]$$

(1) $X_{11}A_{11} + X_{12}A_{21} = I_k$ [4.24a]

(2) $X_{11}A_{12} + X_{12}A_{22} = 0$ [4.24b]

(3) $A_{21}X_{11} + A_{22}X_{21} = 0$ [4.24c]

(4) $A_{21}X_{12} + A_{22}X_{22} = I$ (proof from $XA = AX = I_n$)   [4.24d]

Provided X_{11} is non-singular, we have:

$A_{11} = X_{11}^{-1} + (X_{11}^{-1}X_{12})\alpha^{-1}(X_{21}X_{11}^{-1})$ [4.24e]

$A_{12} = -(X_{11}^{-1}X_{12})\alpha^{-1}$ [4.24f]

$A_{21} = -\alpha^{-1}(X_{21}X_{11}^{-1})$ [4.24g]

$A_{22} = 1/\alpha = \alpha^{-1}$ [4.24h]

$\alpha = X_{22} - X_{21}(X_{11}^{-1}X_{12})$ [4.24i]

Example 4.7:

$$X = \begin{bmatrix} 1 & 0 & 2 \\ 3 & -1 & 1 \\ \hline 2 & -2 & 0 \end{bmatrix}, X_{11} = \begin{bmatrix} 1 & 0 \\ 3 & -1 \end{bmatrix}; X_{12} = \begin{bmatrix} 2 \\ 1 \end{bmatrix}; X_{21} = [2\ -2];$$

$$X_{22} = [0]$$

$$\text{adj } X_{11} = \begin{vmatrix} -1 & 0 \\ -3 & 1 \end{vmatrix}, \|X_{11}\| = -1; X_{11}^{-1} = \begin{vmatrix} 1 & 0 \\ 3 & -1 \end{vmatrix} \quad X_{11}X_{11}^{-1} = \begin{vmatrix} 1 & 0 \\ 0 & 1 \end{vmatrix}$$

$$X_{11}^{-1}X_{12} = \begin{bmatrix} 1 & 0 \\ 3 & -1 \end{bmatrix} \begin{bmatrix} 2 \\ 1 \end{bmatrix} = \begin{bmatrix} 2 \\ 5 \end{bmatrix} \quad X_{21}X_{11}^{-1} = [2\ -2] \begin{bmatrix} 1 & 0 \\ 3 & -1 \end{bmatrix} = [-4 + 2]$$

$$\alpha = [0] - [2 - 2] \begin{bmatrix} 2 \\ 5 \end{bmatrix} = 6$$

$A_{22} = 1/\alpha = 1/6$

$A_{21} = -1/6[-4 + 2] = [+2/3\ -1/3]$

$$A_{12} = -\begin{bmatrix} 2 \\ 5 \end{bmatrix} 1/6 = \begin{bmatrix} -1/3 \\ -5/6 \end{bmatrix}$$

$$A_{11} = \begin{bmatrix} 1 & 0 \\ 3 & -1 \end{bmatrix} + \begin{bmatrix} 2 \\ 5 \end{bmatrix} 1/6\ [-4 + 2] = \begin{bmatrix} 1 & 0 \\ 3 & -1 \end{bmatrix} + \begin{bmatrix} -1\tfrac{1}{3} & 2/3 \\ -3\tfrac{1}{3} & 1\tfrac{2}{3} \end{bmatrix} =$$

$$= \begin{bmatrix} -1/3 & 2/3 \\ -1/3 & +2/3 \end{bmatrix}$$

This renders the same inverse as previously calculated from the adjoint.

This method of partitioning is most efficiently used when the original matrix X is symmetric. Then $A_{12} = A_{21}^T$.

One final equation may be helpful

$$X^{-1} = (X^T X)^{-1} X^T \qquad [4.25]$$

Since the matrix $Y = X^T X$ is a symmetric matrix, the partitioning method can be applied to find the inverse of Y. The inverse of X is then a plain multiplication of the inverse of Y times the transpose of X.

THE INVERSE OF A MATRIX

Example 4.8:

$$X = \begin{bmatrix} 1 & 0 & 2 \\ 3 & -1 & 1 \\ 2 & -2 & 0 \end{bmatrix}; X^T = \begin{bmatrix} 1 & 3 & 2 \\ 0 & -1 & -2 \\ 2 & 1 & 0 \end{bmatrix};$$

$$Y = X^T X = \left[\begin{array}{cc|c} -14 & -7 & 5 \\ -7 & 5 & -1 \\ \hline 5 & -1 & 5 \end{array}\right]$$

$$Y_{11} = \begin{bmatrix} 14 & -7 \\ -7 & 5 \end{bmatrix}; Y_{12} = \begin{bmatrix} 5 \\ -1 \end{bmatrix}; Y_{21} = [5 \; -1]; Y_{22} = 5$$

$$\text{Adj } Y_{11} = \begin{bmatrix} 5 & 7 \\ 7 & 14 \end{bmatrix}; \|Y_{11}\| = 21; Y_{11}^{-1} = \begin{bmatrix} 5/21 & 1/3 \\ 1/3 & 2/3 \end{bmatrix}; Y_{11} Y_{11}^{-1} = \begin{bmatrix} 1 & 0 \\ 0 & 1 \end{bmatrix}$$

$$Y_{11}^{-1} Y_{12} = \begin{bmatrix} 5/21 & 1/3 \\ 1/3 & 2/3 \end{bmatrix} \begin{bmatrix} 5 \\ -1 \end{bmatrix} = \begin{bmatrix} 6/7 \\ 1 \end{bmatrix}; Y_{21} Y_{11}^{-1} = [5 \; -1] \begin{bmatrix} 5/21 & 1/3 \\ 1/3 & 2/3 \end{bmatrix}$$

$$= [6/7 \quad 1]$$

$$\alpha = 5 - [5 - 1] \begin{bmatrix} 6/7 \\ 1 \end{bmatrix} = 12/7; 1/\alpha = 7/12$$

$$A_{22} = 7/12$$

$$A_{21} = -[6/7 \quad 1](7/12) = [-1/2 \quad -7/12]$$

$$A_{12} = \begin{bmatrix} -1/2 \\ -7/12 \end{bmatrix}$$

$$A_{11} = \begin{bmatrix} 5/21 & 1/3 \\ 1/3 & 2/3 \end{bmatrix} + \begin{bmatrix} 6/7 \\ 1 \end{bmatrix}(7/12)[6/7, \; 1] = \begin{bmatrix} 5/21 & 1/3 \\ 1/3 & 2/3 \end{bmatrix} +$$

$$+ \begin{bmatrix} 3/7 & 1/2 \\ 1/2 & 7/12 \end{bmatrix} = \begin{bmatrix} 2/3 & 5/6 \\ 5/6 & 5/4 \end{bmatrix}$$

$$\mathbf{Y}^{-1} = \begin{bmatrix} 2/3 & 5/6 & -1/2 \\ 5/6 & 5/4 & -7/12 \\ -1/2 & -7/12 & 7/12 \end{bmatrix};$$

$$\mathbf{Y}^{-1}\mathbf{X}^T = \begin{bmatrix} 2/3 & 5/6 & -1/2 \\ 5/6 & 5/4 & -7/12 \\ -1/2 & -7/12 & 7/12 \end{bmatrix} \begin{bmatrix} 1 & 3 & 2 \\ 0 & -1 & -2 \\ 2 & 1 & 0 \end{bmatrix} = \begin{bmatrix} -1/3 + 2/3 & -1/3 \\ -1/3 & 2/3 & 5/6 \\ 2/3 & -1/3 & 1/6 \end{bmatrix}$$

4.7 SIMILAR MATRICES

Two matrices are called "similar" when a non-singular square matrix **R** exists so that:

$$\mathbf{X} = \mathbf{R}^{-1}\mathbf{Y}\mathbf{R} \qquad [4.26]$$

It can be shown that the characteristic equations (see Section 4.8) of similar matrices are the same.

If v is an invariant vector of **X**, and **X** is defined by [4.26], then:

$$w = \mathbf{R}v \qquad [4.26a]$$

is an invariant vector of **Y**, corresponding to the same characteristic root (see Section 4.8).

4.8 CHARACTERISTIC EQUATIONS, EIGENVALUES AND EIGENVECTORS

On various pages of this textbook the expressions "eigenvalue" and "eigenvector" have appeared. Their computation is a mathematical problem and a brief summary of some methods will be given in the following.

It is assumed that a square matrix of order n is given. The problem to find an eigenvector can be mathematically formulated as:

$$\mathbf{A}x = \lambda x \qquad [4.27]$$

where x is a vector of order $n \times 1$. This vector is an invariant vector under the transformation. We can rewrite [4.27] as:

$$\mathbf{A}x - \lambda x = 0 \qquad [4.27a]$$

or: $(\mathbf{A} - \lambda \mathbf{I}) = 0$ \qquad [4.27b]

In this form the equation represents a set of (homogeneous) linear equations.

CHARACTERISTIC EQUATIONS, EIGENVALUES AND EIGENVECTORS 349

It is obvious that a trivial solution is $x_1 = x_2 = \ldots x_n = 0$. Besides this trivial solution a non-zero vector x exists only when $(A - \lambda I) = 0$ which means the rank of the matrix is less than n and the determinant $\|A - \lambda I\| = 0$.
In extended form we can write:

$$\begin{vmatrix} a_{11}-\lambda & a_{12} & \ldots a_{1n} \\ a_{21} & a_{22}-\lambda & \ldots a_{2n} \\ \vdots & \vdots & \vdots \\ a_{n1} & a_{n2} & \ldots a_{nn}-\lambda \end{vmatrix} = 0 \qquad [4.28]$$

The matrix is merely the summation of the two matrices A and λI, and is associated with a characteristic polynomial $\phi(\lambda)$ of the degree n in λ. The equation $\phi(\lambda) = 0$ is termed the characteristic equation of A and the roots λ_1, $\lambda_2 \ldots \lambda_n$ are called the characteristic roots of A. Characteristic roots are also named eigenvalues or latent roots, and their associated vectors are called the eigenvectors or latent vectors.

The characteristic polynomial of [4.28] can be written as:

$$\phi(\lambda) = (-1)^\nu(\lambda^\nu - C_1 \lambda^{\nu-1} - C_2 \lambda^{\nu-2} - \ldots - C_{\nu-1}\lambda + C_n) = 0 \qquad [4.28a]$$

The identity exists:

$$\phi(\lambda) = \|A - \lambda I\| \qquad [4.28b]$$

The constants C_i are rather complicated expressions of the elements a_{ij} of the matrix A (see later Section 4.9).

Equation [4.28a] is called the characteristic equation. It is possible to transform $\phi(\lambda)$ into the form:

$$(\lambda - c_1)(\lambda - c_2) \ldots (\lambda - c_{\nu-1})(\lambda - c_\nu) = 0 \qquad [4.28c]$$

It can be readily seen that the sought eigenvalues λ_i in this form are:

$$\lambda_i = c_i \qquad [4.28d]$$

which are also called the characteristic roots. These can be real or complex. The eigenvector for a particular λ_i is then determined from the matrix equation:

$$\begin{bmatrix} a_{11}-\lambda_i & a_{12} & \ldots a_{1n} \\ a_{21} & a_{22}-\lambda_i & \ldots a_{2n} \\ \vdots & \vdots & \vdots \\ a_{n1} & a_{n2} & \ldots a_{nn}-\lambda_i \end{bmatrix} \begin{bmatrix} x_1 \\ x_2 \\ \vdots \\ x_n \end{bmatrix} = 0 \qquad [4.29]$$

Although n eigenvalues can be found, the n corresponding eigenvectors do exist when all n eigenvalues are distinct. They may not exist when some eigenvalues are equal. This can be expressed in the form of Jordan submatrices. We may designate $C_m(\lambda_1)$ as a matrix with one eigenvalue of multiplicity m, but only one eigenvector. The relationship between multiplicity of eigenvalues and associated eigenvectors is sometimes expressed in the Jordan canonical form, which can be written symbolically as:

$$J = \begin{bmatrix} C_2(\lambda_1) & & & \\ & C_2(\lambda_1) & & \\ & & C_1(\lambda_1) & \\ & & & C_1(\lambda_2) \end{bmatrix} \quad [4.30]$$

The interpretation in this case would be that five eigenvalues have the numerical value of λ_1 (i.e. for λ_1 the $m = 2, 2, 1$ for C), one has λ_2 but only four distinct eigenvectors exist. The form:

$$J = \begin{bmatrix} C_3(\lambda_1) & & \\ & C_2(\lambda_1) & \\ & & C_1(\lambda_2) \end{bmatrix} \quad [4.30a]$$

although defining the same two eigenvalues λ_1 and λ_2 is not "similar" to [4.29] as only three eigenvectors appear. Only similar matrices have the same number of eigenvectors (see Section 4.7 on similarity transformation).

When A is real and symmetric (i.e. $A = A^T$) the roots are all real. A characteristic equation can also be established for a non-symmetric square matrix.

The eigenvectors of a symmetric matrix are orthogonal. In the Jordan normal form the multiple matrices can be written symbolically as:

$$C_2 = \begin{bmatrix} \lambda & 1 \\ 0 & \lambda \end{bmatrix} \quad [4.30b]$$

$$C_3 = \begin{bmatrix} \lambda & 1 & 1 \\ 0 & \lambda & 1 \\ 0 & 0 & \lambda \end{bmatrix} \quad [4.30c]$$

etc.

This scheme is sometimes helpful in the calculation of the remaining linearly independent eigenvectors (see Example 4.9). The sequence of eigenvalues in the triangular or diagonal matrix is immaterial as long as eigenvalues and eigenvectors are coordinated.

CHARACTERISTIC EQUATIONS, EIGENVALUES AND EIGENVECTORS

The Jordan canonical form has a significant application. When matrices of multiplicity higher than 1 exist, in most cases it is not possible to transform the matrix **A** into a diagonal form by a similarity transformation but it can be brought into triangular form as given by the Jordan form. More details can be found in the subsequent Section 4.9.

Example 4.9:

Given $\mathbf{A} = \begin{bmatrix} 1 & 2 \\ -1 & 4 \end{bmatrix}$, characteristic equation $\begin{vmatrix} 1-\lambda & 2 \\ -1 & 4-\lambda \end{vmatrix} = 0$, or:

$\lambda^2 - 5\lambda + 4 - (-2) = 0$

$\lambda^2 - 5\lambda + 6 = 0$

$(\lambda - 3)(\lambda - 2) = 0$

$\lambda_1 = 3; \lambda_2 = 2$

The characteristic equation is fulfilled for any of the individual λ values, either λ_1 or λ_2.

Eigenvector $\lambda_1 = 3$: $\begin{pmatrix} -2 & 2 \\ -1 & 1 \end{pmatrix} \begin{pmatrix} x_1 \\ x_2 \end{pmatrix} = 0$

$-2x_1 + 2x_2 = 0$

$-x_1 + 1x_2 = 0$

$x_1 = x_2 = 1$

Eigenvector $\lambda_2 = 2$: $\begin{pmatrix} -1 & 2 \\ -1 & 2 \end{pmatrix} \begin{pmatrix} x_1 \\ x_2 \end{pmatrix} = 0$

$-x_1 + 2x_2 = 0$

$x_1 = 2x_2$

$x_1 = 2, x_2 = 1$

The eigenvectors are $\begin{pmatrix} 1 \\ 1 \end{pmatrix}$ and $\begin{pmatrix} 2 \\ 1 \end{pmatrix}$.

It is obvious that solutions for the two eigenvectors are also:

$k_1 \begin{pmatrix} 1 \\ 1 \end{pmatrix}$ and $k_2 \begin{pmatrix} 2 \\ 1 \end{pmatrix}$

where the k_1 and k_2 are arbitrary constants. The eigenvector is only determined up to an arbitrary factor.

4.9 EIGENVALUES AND DIAGONAL MATRIX

The similarity transformation was previously introduced. It was stated that the characteristic equations of similar matrices are the same. Let us assume that **X** is similar to **A**, i.e. by [4.26] we can write:

$$\mathbf{X} = \mathbf{E}^{-1}\mathbf{A}\mathbf{E} \qquad [4.31]$$

We know further that $\lambda \mathbf{I}$ has the same characteristic equations as **A**. Hence:

$$\lambda \mathbf{I} = \mathbf{E}^{-1}\mathbf{A}\mathbf{E} \qquad [4.31a]$$

Since the matrix on the left side contains all the eigenvalues $\lambda_1, \ldots, \lambda_n$ in the diagonal and all other elements are zero, we can write it as a diagonal matrix:

$$\mathbf{D}_\lambda = \mathbf{E}^{-1}\mathbf{A}\mathbf{E} \qquad [4.31b]$$

This statement can be so interpreted that a matrix **E** exists which transforms **A** into a diagonal matrix whose (diagonal) elements are the eigenvalues. This matrix **E** is the matrix of eigenvectors. When these are known, the eigenvalues can be obtained. Unfortunately in most cases the eigenvectors are not known and must be calculated from the eigenvalues.

When **A** does not have n independent linearly invariant vectors, the **A** cannot be transformed by similarity into a diagonal matrix. It can be proven, however, that every square matrix is similar to a triangular matrix **T**, whose diagonal elements are the characteristic roots. Then:

$$\mathbf{T}_\lambda = \mathbf{Z}^{-1}\mathbf{A}\mathbf{Z} \qquad [4.31c]$$

When **A** is any real n-square symmetric matrix with real characteristic roots then an orthogonal matrix **Z** exists so that:

$$\mathbf{T}_\lambda = \mathbf{Z}^{-1}\mathbf{A}\mathbf{Z} = \mathbf{Z}^{\mathrm{T}}\mathbf{A}\mathbf{Z} \qquad [4.31d]$$

The principle can be expanded to a square matrix with complex elements or complex characteristic roots, and:

$$\mathbf{T}_\lambda = \mathbf{Z}^{-1}\mathbf{A}\mathbf{Z} = \overline{\mathbf{Z}^{\mathrm{T}}}\mathbf{A}\mathbf{Z} \qquad [4.31e]$$

Although it is known that $\|\mathbf{Z}\| = \pm 1$, only in exceptional cases or for square matrices less than $n = 4$ can the **Z** be determined by intuition.

Example 4.10: Distinct eigenvalues

Given $\mathbf{A} = \dfrac{1}{4}\begin{pmatrix} 20 & -7 & 5 \\ 28 & -9 & 11 \\ 20 & -11 & 17 \end{pmatrix}$

The introduction of λ leads to the characteristic equation:

EIGENVALUES AND DIAGONAL MATRIX

$64(-\lambda^3 + 7\lambda^2 - 14\lambda + 8) = 0$

or: $(\lambda - 4)(\lambda - 2)(\lambda - 1) = 0$

Eigenvalues: $\lambda_1 = 4, \lambda_2 = 2, \lambda_3 = 1$.

Eigenvectors matrix:

$$R = \begin{pmatrix} 1 & 1 & 1 \\ 2 & 1 & 3 \\ 2 & -1 & 1 \end{pmatrix}; \quad R^{-1} = \begin{pmatrix} +1 & -1/2 & +1/2 \\ +1 & -1/4 & -1/4 \\ -1 & +3/4 & -1/4 \end{pmatrix}$$

$$R^{-1}A = \begin{pmatrix} +1 & -1/2 & +1/2 \\ +1 & -1/4 & -1/4 \\ -1 & +3/4 & -1/4 \end{pmatrix} \frac{1}{4}\begin{pmatrix} 20 & -7 & 5 \\ 28 & -9 & 11 \\ 20 & -11 & 17 \end{pmatrix} = \frac{1}{4}\begin{pmatrix} +16 & -8 & +8 \\ +8 & -2 & -2 \\ -4 & +3 & -1 \end{pmatrix}$$

$$R^{-1}AR = \frac{1}{4}\begin{pmatrix} +16 & -8 & +8 \\ +8 & -2 & -2 \\ -4 & +3 & -1 \end{pmatrix}\begin{pmatrix} 1 & 1 & 1 \\ 2 & 1 & 3 \\ 2 & -1 & 1 \end{pmatrix} = \frac{1}{4}\begin{pmatrix} +16 & 0 & 0 \\ 0 & +8 & 0 \\ 0 & 0 & 0 \end{pmatrix} =$$

$$= \begin{pmatrix} 4 & 0 & 0 \\ 0 & 2 & 0 \\ 0 & 0 & 1 \end{pmatrix}$$

The eigenvector matrix is not orthogonal (non-symmetric matrix A), but can be made orthogonal.

$u_1^T = 1, 2, 2; l^2(u_1) = 9, l(u_1) = 3$

$e_1^T = 1/3, 2/3, 2/3$

$u_2^T = 1, 1, -1$

$c = (u_2 e_1^T) = 1/3 + 2/3 - 2/3 = 1/3$

$v_2^T = u_2^T - c e_1^T = 1 - 1/9; 1 - 2/9; -1 - 2/9$

$v_2^T = 8/9, 7/9, -11/9; l^2(v_2) = 234/81 = 26/9; l = \sqrt{26}/3$

$e_2^T = 8/3\sqrt{26}, 7/3\sqrt{26}, -11/3\sqrt{26}$

$u_3^T = 1, 3, 1$

$c_1 = (u_3 e_1^T) = 1/3 + 6/3 + 2/3 = 3$

$c_2 = (u_3 e_2^T) = 8/3\sqrt{26} + 21/3\sqrt{26} - 11/3\sqrt{26} = 6/\sqrt{26}$

$v_3^T = u_3^T - c_1 e_1^T - c_2 e_2^T$

$$= 1 - 1 - 16/26; 3 - 2 - 14/26; 1 - 2 + 22/26$$
$$v_3^T = -8/13, 6/13, -2/13; \ l^2(v_3) = 104/169 = 8/13; l = 4/\sqrt{26}$$
$$e_3^T = -4/\sqrt{26}, 3/\sqrt{26}, -1/\sqrt{26}$$

The orthogonalized eigenvector matrix is:

$$E = \begin{pmatrix} \dfrac{1}{3} & \dfrac{8}{3\sqrt{26}} & \dfrac{-4}{\sqrt{26}} \\ \dfrac{2}{3} & \dfrac{7}{3\sqrt{26}} & \dfrac{3}{\sqrt{26}} \\ \dfrac{2}{3} & \dfrac{-11}{3\sqrt{26}} & \dfrac{-1}{\sqrt{26}} \end{pmatrix}$$

It is readily seen that $(e_1^T e_2)$, $(e_1^T e_3)$, $(e_2^T e_3)$ is zero and $(e_1^T e_1)$ etc. $= 1$. Further $EE^T = I$ and $E^T = E^{-1}$.

It should be noted that:

$$R^{-1} A R = D_\lambda$$

but $E^{-1} A E \neq D_\lambda$, since A is a non-symmetric square matrix. However, $E^{-1} A E = T_\lambda$ whose diagonal elements contain the eigenvalues. For a symmetric square matrix, as previously stated, the eigenvectors are orthogonal. Consequently, both $R^{-1} A R$ and $E^{-1} A E$ lead to D_λ.

Example 4.11: Double root, distinct eigenvectors

Given $A = \begin{pmatrix} 23/18 & 1/9 & 13/18 \\ 0 & 1 & 0 \\ 5/18 & 1/9 & 31/18 \end{pmatrix}$

It can be shown that the characteristic polynomial is:

$$(\lambda - 2)(\lambda - 1)^2 = 0$$

Hence: $\lambda_1 = \lambda_2 = 1, \lambda_3 = 2$.

We have for $\lambda_3 = 2$ the matrix:

$$\begin{pmatrix} -13/18 & 1/9 & 13/18 \\ 0 & -1 & 0 \\ 5/18 & 1/9 & -5/18 \end{pmatrix} \begin{pmatrix} x_{31} \\ x_{32} \\ x_{33} \end{pmatrix} = 0$$

with eigenvector $x_3 = (1, 0, 1)$.
For $\lambda_1 = \lambda_2 = 1$ we find:

EIGENVALUES AND DIAGONAL MATRIX

$$\begin{pmatrix} 5/18 & 1/9 & 13/18 \\ 0 & 0 & 0 \\ 5/18 & 1/9 & 13/18 \end{pmatrix} \begin{pmatrix} x_{11} \\ x_{12} \\ x_{13} \end{pmatrix} = 0$$

Hence $5x_{11} + 2x_{12} + 13x_{13} = 0$.

As will be explained later, two independent eigenvectors are:

$x_1^T = 2, -5, 0, \quad x_2^T = 13, 0, -5$.

Consequently:

$$\mathbf{R} = \begin{pmatrix} 1 & 2/5 & 13/5 \\ 0 & -1 & 0 \\ 1 & 0 & -1 \end{pmatrix} \quad \text{and} \quad \|\mathbf{R}\| = 1 + 13/5 = 18/5 \neq 0$$

The diagonalization process leads to the following:

$$\frac{1}{18}\begin{pmatrix} 5 & 2 & 13 \\ 0 & -18 & 0 \\ 5 & 2 & -5 \end{pmatrix} \frac{1}{18}\begin{pmatrix} 23 & 2 & 13 \\ 0 & 18 & 0 \\ 5 & 2 & 31 \end{pmatrix} \frac{1}{5}\begin{pmatrix} 5 & 2 & 13 \\ 0 & -5 & 0 \\ 5 & 0 & -5 \end{pmatrix} = \begin{pmatrix} 2 & 0 & 0 \\ 0 & 1 & 0 \\ 0 & 0 & 1 \end{pmatrix}$$

The orthogonal eigenvector matrix is:

$$\mathbf{M}_e = \begin{pmatrix} 1/\sqrt{2} & 1/3\sqrt{3} & 5/3\sqrt{6} \\ 0 & -5/3\sqrt{3} & +2/3\sqrt{6} \\ 1/\sqrt{2} & -1/3\sqrt{3} & -5/3\sqrt{6} \end{pmatrix}$$

In this case two linearly independent eigenvectors for $\lambda_1 = \lambda_2 = 1$ have been found.

Example 4.12: Double root and only one eigenvector

Given $\mathbf{A} = \begin{pmatrix} 1 & 0 & 0 \\ -1 & 2 & 3 \\ 1 & 1/3 & 2 \end{pmatrix}$

The characteristic polynomial is $(\lambda - 1)^2(\lambda - 3)$.

For $\lambda = 3$ we find the equations:

$$\begin{pmatrix} -2 & 0 & 0 \\ -1 & -1 & 3 \\ 1 & 1/3 & -1 \end{pmatrix} \begin{pmatrix} x_1 \\ x_2 \\ x_3 \end{pmatrix} = 0$$

with vector $x^T = (0, 3, 1)$. The $\lambda = 1$ provides:

$$\begin{pmatrix} 0 & 0 & 0 \\ -1 & 1 & 3 \\ 1 & 1/3 & 1 \end{pmatrix} \begin{pmatrix} x_1 \\ x_2 \\ x_3 \end{pmatrix} = 0$$

with a solution $x^T = (0, -3, 1)$. ($x_2 = 0$ and $x_3 = 0$ lead to contradictory solutions of the system, as only one row can be satisfied).

We know now that no similarity transformation exists to convert **A** into a diagonal matrix, since only two linearly independent eigenvectors exist. In the Jordan normal form we would find the triangular matrix:

$$\mathbf{T} = \mathbf{J} = \begin{pmatrix} C_2(\lambda_1) & \\ & C(\lambda_3) \end{pmatrix}$$

Consequently: $\mathbf{T} = \begin{pmatrix} 1 & 1 & 0 \\ 0 & 1 & 0 \\ \hline 0 & 0 & 3 \end{pmatrix}$

The eigenvalues appear in the diagonal (1, 1, 3). Now we can write:

$$\mathbf{A} \cdot \mathbf{Z} = \mathbf{Z} \cdot \mathbf{T}$$

where **A**, **T**, and two columns of **Z** are known. We find the missing independent vector by:

$$\begin{pmatrix} 1 & 0 & 0 \\ -1 & 2 & 3 \\ 1 & 1/3 & 2 \end{pmatrix} \begin{pmatrix} 0 & z_1 & 0 \\ -3 & z_2 & 3 \\ 1 & z_3 & 1 \end{pmatrix} = \begin{pmatrix} 0 & z_1 & 0 \\ -3 & z_2 & 3 \\ 1 & z_3 & 1 \end{pmatrix} \begin{pmatrix} 1 & 1 & 0 \\ 0 & 1 & 0 \\ 0 & 0 & 3 \end{pmatrix}$$

(Note the proper association of the known and unknowns with λ in **T**.)

By matrix multiplication we derive:

$$\begin{pmatrix} 0 & z_1 & 0 \\ -3, & -z_1 + 2z_2 + 3z_3, & 9 \\ 1, & z_1 + z_2/3 + 2z_3, & 3 \end{pmatrix} = \begin{pmatrix} 0 & z_1 & 0 \\ -3, & -3 + z_2, & 9 \\ 1, & 1 + z_2, & 3 \end{pmatrix}$$

This renders the equations:

$$z_1 = z_1$$
$$-z_1 + 2z_2 + 3z_3 = -3 + z_2 \text{ or } -z_1 + z_2 + 3z_3 = -3$$
$$z_1 + z_2/3 + 2z_3 = 1 + z_3 \text{ or } z_1 + z_2/3 + z_3 = 1$$

It can be seen that the vector $z^T = (3/2, -3/2, 0)$ is a solution.

Hence: $\mathbf{Z} = \begin{pmatrix} 0 & 3/2 & 0 \\ -3 & -3/2 & 3 \\ 1 & 0 & 1 \end{pmatrix}$

Since $\|\mathbf{Z}\| = 9$, the inverse $\mathbf{Z}^{-1}$ exists. Therefore $\mathbf{Z}^{-1}\mathbf{A}\mathbf{Z} = \mathbf{T}$ as can readily be checked.

$$\mathbf{Z}^{-1} = \begin{pmatrix} -1/6 & -1/6 & 1/2 \\ 2/3 & 0 & 0 \\ 1/6 & 1/6 & 1/2 \end{pmatrix}$$

It is self-evident that the computation of $\mathbf{Z}$ is not a trivial matter for $n > 3$. More details can be found in the literature (e.g. Zurmuehl, 1958, etc.).

4.10 LARGEST AND SMALLEST EIGENVALUES BY ITERATION

The method of the characteristic equation may be very cumbersome for $n > 3$. We have also seen that the conversion to a diagonal matrix is not readily performed. A procedure which leads to the largest eigenvalue and which can readily be performed by electronic computers is as follows:

The eigenvalue problem has been stated by [4.27] as $\mathbf{A}x = \lambda x$. Now let us assume that we start with:

$$\mathbf{A}x = y = cz \qquad [4.32]$$

Since we know that $x \equiv y$ is the solution for the eigenvalue we could make a guess at x and compute y. This procedure succeeds when $\mathbf{A}$ is real and non-singular (i.e. $\|\mathbf{A}\| \neq 0$), and λ is real and the dominant eigenvalue. ($\|\mathbf{A}\| = 0$ will be treated later.)

The first x can be selected rather arbitrarily and conventionally the unit vector $x = (1, 1, 1 \ldots)$ will be taken. Since x is a column vector with elements $x_1, x_2 \ldots x_n$ the summation for every element i of the first y_1 is:

$$y_{1i} = \sum_{k=1}^{n} a_{ik} x_k \qquad [4.32a]$$

We select then the maximum y_{1i} denoted $y_{\max,i}$, and build:

$$c_1 = y_{\max,i}/x_i \qquad [4.32b]$$

The c_1 is the first approximation for the largest eigenvalue λ_1. We divide every element of y by c_1 and obtain a new vector x with elements:

$$x_i = y_{1i}/c_1 \qquad [4.32c]$$

This new vector is substituted for x and the procedure [4.32a] repeated. We generalize:

$$x_i = y_{hi}/c_h \qquad [4.32d]$$

The c_h-values approximate λ_1 with $c_h \to \lambda_1$ for $h \to \infty$. For practical purposes h cannot go to infinity. We check therefore:

$$c_{h+1}/c_h = r \to 1 \text{ as } h \to \infty \qquad [4.32e]$$

The approximation of $r \to 1$ can be discontinued at any desired level of accuracy. The process will by and large converge very rapidly.

Should we desire the smallest eigenvalue, we can start with:

$$\mathbf{A}^{-1}x = y \qquad [4.33]$$

and proceed exactly as described previously. The only difference is now that $(1/c_h) \to \lambda$.

When $\|\mathbf{A}\| = 0$ then $\lambda = 0$ is a characteristic root of $\mathbf{A}$. The characteristic

LARGEST AND SMALLEST EIGENVALUES BY ITERATION

polynomial $\phi(\lambda)$ does not have a general term in this case, e.g. for a third degree polynomial: $\lambda^3 + b\lambda^2 + c\lambda = 0$. Immediately we see that $\phi(\lambda) = \lambda(\lambda^2 + b\lambda + c) = 0$ with $\lambda = 0$ being a solution. The characteristic vector for $\lambda = 0$ is obtained in the usual manner, i.e. from A itself since $\lambda = 0$ in [4.27]. Usually a particular eigenvector is then selected by an additional condition, e.g. that a special value x_i be 1 or a similar requirement. The calculation of the eigenvector is presented later in Section 4.13.

Sometimes the eigenvector may be known but not the eigenvalue. Then the x can be replaced by the eigenvector, which leads to a quick answer for the λ.

Example 4.13: Iteration of eigenvector

$$A = \begin{pmatrix} 2 & 1 \\ 3 & 4 \end{pmatrix}; \quad \|A\| = 5 \neq 0$$

We assume $x^T = [1, 1]$ and calculate:

$$\begin{pmatrix} 2 & 1 \\ 3 & 4 \end{pmatrix} \begin{pmatrix} 1 \\ 1 \end{pmatrix} = \begin{pmatrix} 3 \\ 7 \end{pmatrix}$$

Solution: $\lambda_{11} = 7$, $x_1^T = (3/7 \quad 1)$.

Now:
$$\begin{pmatrix} 2 & 1 \\ 3 & 4 \end{pmatrix} \begin{pmatrix} 3/7 \\ 1 \end{pmatrix} = \begin{pmatrix} 13/7 \\ 37/7 \end{pmatrix}$$

Second solution: $\lambda_{12} = 37/7 = 5\frac{2}{7}$, $x_2^T = (13/37, \quad 1)$.

Again:
$$\begin{pmatrix} 2 & 1 \\ 3 & 4 \end{pmatrix} \begin{pmatrix} 13/37 \\ 1 \end{pmatrix} = \begin{pmatrix} 63/37 \\ 187/37 \end{pmatrix}$$

Solution: $\lambda_{13} = 5\frac{2}{37}$.

One can already see that λ_1 converges to 5. It can be readily checked that $\lambda_1 = 5$ is a solution.

The smallest λ we obtain from the inverse A^{-1}:

$$A^{-1} = \begin{pmatrix} 4 & -3 \\ -1 & 2 \end{pmatrix}, \quad \begin{pmatrix} 4 & -3 \\ -1 & 2 \end{pmatrix} \begin{pmatrix} 1 \\ 1 \end{pmatrix} = \begin{pmatrix} 1 \\ 1 \end{pmatrix}, \quad \lambda_2 = 1.$$

The two solutions are $\lambda_1 = 5$ and $\lambda_2 = 1$. The characteristic equation is $(\lambda - 5)(\lambda - 1) = 0$.

The eigenvectors:

λ_1: $\begin{pmatrix} -3 & 1 \\ 3 & 1 \end{pmatrix} (x) = 0 \quad x_1^T = (1/3, \quad 1)$

check $\begin{pmatrix} 2 & 1 \\ 3 & 4 \end{pmatrix} \begin{pmatrix} +1/3 \\ 1 \end{pmatrix} = 5 \begin{pmatrix} +1/3 \\ 1 \end{pmatrix}$.

λ_2: $\begin{pmatrix} 1 & 1 \\ 3 & 3 \end{pmatrix} (x) = 0$, $x_2^T = (1, -1)$.

It can be seen that the eigenvectors are approximated in the above scheme for $\lambda_2 = 1$. The second time we work with the inverse of $\mathbf{A}$, hence the eigenvectors are different.

4.11 COMPUTATION OF THE CHARACTERISTIC POLYNOMIAL

The general concepts of eigenvalues and eigenvectors have been discussed in the previous section and a method was introduced to compute the largest or smallest eigenvalue and its related eigenvector. In other problems we need all eigenvalues, however. Some of the methods presented in the preceding sections are not very suitable for electronic computer adaptation. Some techniques for this purpose will now be given.

The general solution of eigenvalues and eigenvectors necessitates three steps. We must first find the characteristic polynomial. This means the determination of the constants α_i of:

$$\phi(\lambda) = (\alpha_0)\lambda^\nu - \alpha_1 \lambda^{\nu-1} - \alpha_2 \lambda^{\nu-2} - \ldots - \alpha_{n-1}\lambda - \alpha_n = 0 \quad [4.34]$$

The second step is the finding of the roots of $\phi(\lambda)$. This may not always be easy especially when ν is > 3 and the roots are complex. The last step is the determination of the eigenvectors from the known roots. Solutions for these three steps will now be given.

Determination of the characteristic polynomial

We assume a matrix $\mathbf{A}$ to be given with dimension $\nu \times \nu$, and we need the characteristic polynomial. It can be proven that the following procedure leads to the constants α_i of the characteristic polynomial (with $\mathbf{A}_1 = \mathbf{A}$):

$\alpha_1 = \text{tr}(\mathbf{A}_1)$ [4.35a]
$\mathbf{A}_2 = \mathbf{A}(\mathbf{A}_1 - \alpha_1 \mathbf{I})$ [4.35b]
$\alpha_2 = (1/2)\,\text{tr}(\mathbf{A}_2)$ [4.35c]
$\mathbf{A}_3 = \mathbf{A}(\mathbf{A}_2 - \alpha_2 \mathbf{I})$ [4.35d]
$\alpha_3 = (1/3)\,\text{tr}(\mathbf{A}_3)$ [4.35e]

.
.
.

COMPUTATION OF THE CHARACTERISTIC POLYNOMIAL

general: $\mathbf{A}_n = \mathbf{A}(\mathbf{A}_{n-1} - \alpha_{n-1}\mathbf{I})$ [4.35f]

$\alpha_n = (1/n) \operatorname{tr}(\mathbf{A}_n) = \|\mathbf{A}\|$ [4.35g]

where $n = 1, 2, \ldots \nu$ and $\alpha_0 = 1$ is the coefficient of λ^ν.

It should be noted that the last matrix $\mathbf{A}_\nu$ is a diagonal matrix:

$$\mathbf{A}_\nu = \mathbf{D}\alpha_\nu = \begin{bmatrix} \alpha_\nu & 0 & \cdots & 0 \\ 0 & \alpha_\nu & \cdots & 0 \\ \cdot & \cdot & \cdots & \cdot \\ \cdot & \cdot & \cdots & \cdot \\ \cdot & \cdot & \cdots & \cdot \\ 0 & 0 & \cdots & \alpha_\nu \end{bmatrix}$$ [4.35h]

The preceding technique involves only multiplication, subtraction and the trace of matrices. These are operations which can be easily performed on electronic computers.

It should be mentioned that one by-product of this technique is the inverse. It can be readily verified that:

$\mathbf{A}^{-1} = (1/\alpha_\nu)(\mathbf{A}_{\nu-1} - \alpha_{\nu-1}\mathbf{I})$ [4.36]

This means the inverse of a 4×4 matrix is:

$\mathbf{A}^{-1} = (1/\alpha_4)(\mathbf{A}_3 - \alpha_3\mathbf{I})$ [4.36a]

Example 4.14: Determination of constants for the characteristic polynomial

Let us assume we have the matrix $\mathbf{A}$ from the Example 4.10:

$$\mathbf{A} = \frac{1}{4}\begin{pmatrix} 20 & -7 & 5 \\ 28 & -9 & 11 \\ 20 & -11 & 17 \end{pmatrix}$$

$\alpha_1 = (20 - 9 + 17)/4 = 28/4 = 7$

$$\mathbf{A}_2 = \frac{1}{4}\begin{pmatrix} 20 & -7 & 5 \\ 28 & -9 & 11 \\ 20 & -11 & 17 \end{pmatrix} \frac{1}{4}\begin{pmatrix} -8 & -7 & 5 \\ 28 & -37 & 11 \\ 20 & -11 & -11 \end{pmatrix} =$$

$$= \frac{1}{16}\begin{pmatrix} -256 + 64 & -32 \\ -256 + 16 & -80 \\ -128 + 80 & -208 \end{pmatrix} = \begin{pmatrix} -16 & 4 & -2 \\ -16 & 1 & -5 \\ -8 & 5 & -13 \end{pmatrix}$$

$\alpha_2 = (-16 + 1 - 13)/2 = -14$

It should be noted that $(\mathbf{A}_1 - \alpha_1 \mathbf{I})$ was computed for $\alpha_1 = 28$, since 1/4 stands before the matrix and all values in the matrix have been multiplied by 4.

$$\mathbf{A}_3 = \frac{1}{4}\begin{pmatrix} 20 - 7 & 5 \\ 28 - 9 & 11 \\ 20 - 11 & 17 \end{pmatrix}\begin{pmatrix} -2 & 4 & -2 \\ -16 & 15 & -5 \\ -8 & 5 & +1 \end{pmatrix} = \frac{1}{4}\begin{pmatrix} 32 & 0 & 0 \\ 0 & 32 & 0 \\ 0 & 0 & 32 \end{pmatrix}$$

$$= \begin{pmatrix} 8 & 0 & 0 \\ 0 & 8 & 0 \\ 0 & 0 & 8 \end{pmatrix}$$

$\alpha_3 = (8 + 8 + 8)/3 = 8$

The characteristic polynomial is:

$$\lambda^3 - 7\lambda^2 + 14\lambda - 8 = 0$$

which can be readily verified from Example 4.10. Further it can be checked that $\|\mathbf{A}\| = \alpha_n = 8$, and:

$$\mathbf{A}^{-1} = \frac{1}{8}\begin{pmatrix} -2 & 4 & -2 \\ -16 & 15 & -5 \\ -8 & 5 & 1 \end{pmatrix}$$

which is $\mathbf{A}_2 - \alpha_2 \mathbf{I}$ as required by [4.36a]. In addition, $\mathbf{A}_3 - \alpha_3 \mathbf{I}$ would give a matrix of zeroes.

It is evident that the procedure for $\nu = 3$ will not save much time for calculations by hand compared with other methods, as it is usually not difficult to compute the characteristic equation of a 3 × 3 matrix. Beyond $\nu = 3$, however, the technique has ostensible advantages.

4.12 THE DETERMINATION OF THE ROOTS

Explicit expressions

This becomes now a problem of higher algebra. It is well known that explicit solutions exist for ν up to 4. For $\nu > 4$ no general formulae have been derived. Some general theorems on polynomial equations will be repeated here.

Every polynomial equation with arbitrarily given real or complex coefficients has at least one root (real or imaginary).

If λ_i is a root of $\phi(\lambda)$, then $\phi(\lambda)$ is divisible by $(\lambda - \lambda_i)$, i.e.:

$$\phi(\lambda) = (\lambda - \lambda_i)\phi_1(\lambda) \qquad [4.37]$$

This polynomial is of degree $\nu - 1$.

If the coefficients are real, then every polynomial with ν being odd has at least one real root.

The number of positive (real) roots with real coefficients is smaller or equal to the number of sign changes. The characteristic equation in Example 4.14 has three sign changes (zero terms are ignored). Hence there are only up to three positive roots (see Example 4.10).

Solutions of quadratic equations are trivial. For:

$$ax^2 + bx + c = 0 \qquad [4.37a]$$

we find the two roots:

$$x = (-b \pm \sqrt{b^2 - 4ac})/(2a) \qquad [4.37b]$$

Notice the condition $b^2 \geqslant 4ac$ for real roots. In electronic computer handling the absolute term $|b^2 - 4ac|$ should be taken for $b^2 < 4ac$ (i.e. imaginary part of complex root).

In order to reduce polynomials to the form of [4.37] we may adopt the regular procedure of division, e.g. we know one root, e.g. $\lambda_1 = 4$. Then, for example:

$$\lambda^3 - 9\lambda^2 + 26\lambda - 24 = 0 \quad \text{divided by } (\lambda - 4)$$

Step 1: $\qquad\qquad\qquad\qquad\qquad\qquad\qquad \lambda^3/\lambda = \lambda^2$

Step 2: $\quad \lambda^2(\lambda - 4) = \underline{\lambda^3 - 4\lambda^2}$

Remainder $\qquad\qquad -5\lambda^2 + 26\lambda \qquad$ Step 3: $-5\lambda^2/\lambda = -5\lambda$

Step 4: $-5\lambda(\lambda - 4) = \underline{-5\lambda^2 + 20\lambda}$

Remainder $\qquad\qquad\qquad 6\lambda - 24 \quad$ Step 5: $\quad 6\lambda/\lambda = 6$

Step 6: $\quad 6(\lambda - 4) = \qquad 6\lambda - 24 \quad$ Answer: $\phi_1(\lambda) = \lambda^2 - 5\lambda + 6$

These steps can be programmed exactly for electronic computer processing. When more factors are known, either a step-by-step reduction or a full-term division can be performed.

Provision should be made that the term of the highest degree in λ is not zero before division. We must also take care to add any remainder when the factor is not exactly a solution.

Degrees higher than six

When one eigenvalue is known (and we have seen that we can determine the largest and smallest one by iteration under some conditions) then a division can be performed to reduce the degree of the polynomial. When the highest degree for $\phi(\lambda)$ is still above six after division again no explicit formula can be given as $\nu > 4$. Then again the largest and smallest eigenvalue could be determined, etc. until finally the highest degree is $\nu \leqslant 4$. This iterative sequence may sometimes be quite tedious. Consequently, we need a different and simpler routine to find the largest root of an equation. Such a method is due to Graeffe. (See Whittaker and Robinson, 1944.) The given polynomial $\phi(\lambda)$ is replaced by $\psi(\lambda)$, whose roots are the squares of the original polynomial $\phi(\lambda)$. We assume:

$$\phi(\lambda) = a_1\lambda^\nu + a_2\lambda^{\nu-1} + \ldots + a_n\lambda + a_{n+1} = 0 \qquad [4.38]$$

The new polynomial is then:

$$\psi(x) = b_1 x^n + b_2 x^{n-1} + \ldots + b_n x + b_{n+1} = 0 \qquad [4.39]$$

where the roots are now $x_1 = \lambda_1^2$, $x_2 = \lambda_2^2$, etc.

The coefficients b_i can be computed as follows:

$$b_1 = a_1^2 \qquad [4.40a]$$

$$b_2 = -a_2^2 + 2a_1 a_3 \qquad [4.40b]$$

$$b_3 = a_3^2 - 2a_2 a_4 + 2a_1 a_5 \qquad [4.40c]$$

General: $b_i = (-1)^{i-1}(a_i^2 - 2a_{i-1}a_{i+1} + 2a_{i-2}a_{i+2} + \ldots)$ [4.40d]

and: $b_n = (-1)^{n-1}(a_n^2 - 2a_{n-1}a_{n+1})$ [4.40e]

$$b_{n+1} = (-1)^n a_{n+1}^2 \qquad [4.40f]$$

What has been accomplished? Assume the root-squaring procedure has been applied k times. Then the new polynomial can be written:

$$\psi_k(\lambda) = c_1(x^k)^n + c_2(x^k)^{n-1} + \ldots + c_n(x^k) + c_{n-1} = 0 \qquad [4.41]$$

We know from the general theory:

$$c_2/c_1 = -(x_1^k + x_2^k + \ldots + x_n^k)$$

$$= -x_1^k \left(1 + \frac{x_2^k}{x_1^k} + \ldots + \frac{x_n^k}{x_1^k}\right) \qquad [4.41a]$$

The part in parenthesis from x_2 on is negligible (i.e. $x_n^k/x_1^k \ll 1$ for $n > 1$ by definition of x_1) and we have an approximate solution for x_1^k. Further:

$$c_3/c_1 = x_1^k x_2^k \left(1 + \frac{x_3^k}{x_2^k} + \ldots\right) \qquad [4.41b]$$

$$c_4/c_1 = -x_1^k x_2^k x_3^k (1 + \ldots) \qquad [4.41c]$$

$$c_{n+1}/c_1 = (-1)^n x_1^k x_2^k \ldots x_n^k \qquad [4.41d]$$

It does not really matter which factor x_i contributes the largest part since this largest part can be factored out (i.e. [4.41a] can be written for any x_i in this form). When sufficient squaring has been performed it is obvious that $c_2/c_1 \sim x_1^k$, $c_3/c_1 \sim x_1^k x_2^k$, etc. When x_i^k is known, the λ_i can readily be obtained. The technique is suitable for $\nu > 2$. Therefore no explicit expressions are given in this text for third- and fourth-degree polynomials.

Multiple and complex roots

The method introduced in the preceding section will not be valid for multiple and complex roots. For multiple (real) roots we find:

$$-c_2/c_1 \sim m \cdot x_1^k \qquad [4.42]$$

since $x_2^k/x_1^k \sim 1$, etc. in [4.41a]. If $m = 2$, then $c_3/c_1 \sim x_1^{k+2}$. One can thus recognize the contribution of multiple roots from the sequence of coefficients.

If complex roots for polynomial equations with real coefficients exist then they are conjugate. Assume:

$$x_2 = p \cdot e^{i\theta} \qquad [4.43a]$$

and: $x_3 = p \cdot e^{-i\theta} \qquad [4.43b]$

then: $x_2 \cdot x_3 = p^2 \qquad [4.43c]$

while the influence in [4.41a] would be unpredictable. By and large, the contribution of complex roots is of the form p^{jk}. Hence:

$$p_1^{jk} \sim (-1)^j c_{j+1}/c_1 \qquad [4.44]$$

This term p_1^{jk} would be the absolute value of the first j roots. When root-squaring takes place, then:

$$p_1^{2jk} \sim (-1)^j d_{j+1}/d_1 \qquad [4.44a]$$

and: $d_{j+1} \sim c_{j+1}^2 \qquad [4.44b]$

It can be interpreted that the coefficient containing a pivotal element p will behave differently from the other coefficients. We can, therefore, recognize by the behavior of the coefficients whether a complex root is involved. If c_r

and c_{r+s} are two adjacent pivotal coefficients then there are exactly s roots having an absolute value p given by:

$$p^{sk} = (-1)^s c_{r+s}/c_r \qquad [4.44c]$$

When the x_1 is obtained from the x_1^k it is not exactly known whether the root is real or not when multiple roots have been discovered. The answer can be found by trial. One substitutes a real value into $\phi(\lambda)$. This replacement reduces the $\phi(\lambda)$ by one degree should the real root be correct. Then one can try again, etc. Both positive and negative λ must be tried since $x = \lambda^2$.

After it has been determined that the root is complex, a quadratic factor:

$$\lambda^2 + b\lambda + c = 0 \qquad [4.45]$$

is appropriate since $\lambda_1 = \alpha + \beta i$ and $\lambda_2 = \alpha - \beta i$ are both roots of a polynomial with real coefficients. (See [4.37b].) The quadratic equation $[\lambda - (\alpha + \beta i)] \cdot [\lambda - (\alpha - \beta i)]$ results in:

$$\lambda^2 - 2\alpha\lambda + (\alpha^2 + \beta^2) = 0 \qquad [4.45a]$$

Another version is:

$$(\lambda - pe^{i\theta})(\lambda - pe^{-i\theta}) = \lambda^2 - p\lambda(e^{i\theta} + e^{-i\theta}) + p^2$$

This shows that $c = p^2 = \alpha^2 + \beta^2$, i.e. c is the pivotal element ([4.43c]). The b is unknown, however, and must be found.

The process is elaborate and technical details will be omitted here. The theoretical background has been discussed in the literature (e.g. Pennington, 1970; Graybill, 1961, 1969, etc.). Only the answer for the solution will be given here.

We assume the existence of the characteristic polynomial ([4.39]) with real coefficients. The absolute value p of the complex root is also known ([4.43c]). Then the b for [4.45] can be determined from a set of polynomial equations, which have been derived by dividing the $\phi(\lambda)$ by [4.45]. The result provides:

$$R_{1,i+1}(b) = R_{2,i}(b) - bR_{1,i}(b) \qquad [4.46a]$$

$$\text{and: } R_{2,i+1}(b) = a_{i+3} - cR_{1,i}(b) \qquad [4.46b]$$

where $c = p^2$.

The functions of $R_{h,k}$ as polynomials of b are defined by:

$$R_{1,i}(b) = \sum_{j=0}^{i} \gamma_{k+1,i} b^j \qquad [4.46c]$$

$$\text{and: } R_{2,i}(b) = \sum_{j=0}^{i-1} \delta_{k,i} b^j \qquad [4.46d]$$

where $k = i - j$ and the $i = 1, 2 \ldots \nu - 1$ for the degree ν of $\phi(\lambda)$. It is obvious from [4.46c and d] that R_1 will be a polynomial of degree i and the R_2 is one

THE DETERMINATION OF THE ROOTS

with degree $i-1$. The highest common factor between R_1 and R_2 is then the the sought value of b. The procedure to determine the highest common factor will be explained in the section 'Highest common factor of two polynomials'. A check can be made in dividing $\phi(\lambda)$ by [4.45] which should leave no remainder (except for rounding accuracy).

The sequence of coefficients for R_1 and R_2 can be computed as follows:

$$\gamma_{1,1} = -a_1 \qquad [4.47a]$$

$$\gamma_{2,1} = a_2 \qquad [4.47b]$$

$$\delta_{1,1} = a_3 - a_1 c \qquad [4.47c]$$

$$\gamma_{1,i+1} = -\gamma_{1,i} \qquad [4.47d]$$

$$\gamma_{2,i+1} = -\gamma_{2,i} \qquad [4.47e]$$

These are basic coefficients. Others are provided by:

$$\gamma_{k+2,i+1} = \delta_{k,i} - \gamma_{k+2,i} \qquad [4.47f]$$

$$\delta_{i+1,i+1} = a_{i+3} - c\gamma_{i+1,i} \quad (\text{for } j=0) \qquad [4.47g]$$

$$\delta_{k+1,i+1} = -c\gamma_{k+1,i} \quad (\text{for } j>0) \qquad [4.47h]$$

The coefficients follow from the substitution of [4.46c and d] into [4.46a and b]. It should be noted that several of the undefined coefficients called for by [4.47f] through [4.47h] are zero since the last non-zero terms in [4.46c and d] are $\delta_{i,i}$ and $\gamma_{i+1,i}$. This leads to the coefficient matrix as:

$$
\begin{array}{c|llll}
i & & & & \\
0 & \gamma_{11} = -a_1 & 0 & & \\
1 & \gamma_{11} & \gamma_{2,1} & 0 & \\
2 & \gamma_{12} & \gamma_{2,2} & \gamma_{3,2} & 0 \\
3 & \gamma_{13} & \gamma_{2,3} & \gamma_{3,3} & \gamma_{4,3} & 0
\end{array}
\qquad [4.48a]
$$

etc.
and:

$$
\begin{array}{c|llll}
i & & & & \\
1 & \delta_{11} = a_3 - a_1 c & 0 & & \\
2 & \delta_{1,2} & \delta_{2,2} & 0 & \\
3 & \delta_{1,3} & \delta_{2,3} & \delta_{3,3} & 0
\end{array}
\qquad [4.48b]
$$

A fourth-degree polynomial ($\nu = 4$) for λ would have the following coefficients where $i = \nu - 1 = 3$ and $k = 3 - j$:

$$R_{1,3} = \gamma_{4,3} + \gamma_{3,3}b + \gamma_{2,3}b^2 + \gamma_{1,3}b^3 \qquad [4.49a]$$
$$R_{2,3} = \delta_{3,3} + \delta_{2,3}b + \delta_{1,3}b^2 \qquad [4.49b]$$

The coefficients to be computed are as follows:

(1) $\gamma_{13} = -\gamma_{12}$ with $\gamma_{12} = -\gamma_{11} = a_1$.

(2) $\gamma_{23} = -\gamma_{22}$ with $\gamma_{22} = -\gamma_{21} = -a_2$.

(3) $\gamma_{33} = \delta_{12} - \gamma_{32}$ with $\gamma_{32} = \delta_{11} - (\gamma_{3,1})$.

(4) $\gamma_{43} = \delta_{22} - (\gamma_{42})$

(1) $\delta_{13} = -c\gamma_{12}$.

(2) $\delta_{23} = -c\gamma_{22}$.

(3) $\gamma_{33} = a_5 - c\gamma_{32}$.

Further: $\delta_{22} = a_4 - c\gamma_{21}$, $\delta_{12} = -c\gamma_{21}$, $\delta_{11} = a_3 - a_1c$.

The γ-values in parenthesis are zero. The sequence of coefficients to be computed can be arranged in a cycle whose next term requires only determined values:

$$\gamma_{1,i+1} = -\gamma_{1,i}$$

$$\gamma_{2,i+1} = -\gamma_{2,i}$$

$$\delta_{i+1,i+1} = a_{i+3} - \gamma_{i+1,i}$$

$$\delta_{i+1-j,i+1} = -c\gamma_{i+1-j,i}$$

$$\gamma_{i+2-j} = \delta_{i-j,i} - \gamma_{i+2-j,i}$$

This should be sufficient explanation for the determination of the roots. More details can be found in the literature such as Pennington (1970), Graybill (1961, 1969) etc.

Highest common factor of two polynomials

When the highest common factor of two polynomials is sought a sequential procedure can be obtained for the highest factor. Assume we have $\phi(\lambda)$ and divide by $g(\lambda)$. Then:

$$\phi(\lambda) = g(\lambda) \cdot q_1(\lambda) + R_1(\lambda) \qquad [4.50]$$

where $q_1(\lambda)$ is the quotient for division by $g(\lambda)$, and $R_1(\lambda)$ the remainder. Now divide $g(\lambda)$ by $R_1(\lambda)$ such as:

$$g(\lambda) = R_1(\lambda)q_2(\lambda) + R_2(\lambda) \qquad [4.50a]$$

THE DETERMINATION OF THE ROOTS

The next step is:

$$R_1(\lambda) = R_2(\lambda)q_3(\lambda) + R_3(\lambda) \qquad [4.50b]$$

The continuation will finally lead to a remainder zero or close to zero. The last two steps can be written as:

$$R_{m-2} = R_{m-1}q_m + R_m \qquad [4.50c]$$

$$R_{m-1} = R_m q_{m+1} + 0 \qquad [4.50d]$$

Then R_m is the highest common factor of $\phi(\lambda)$ and $g(\lambda)$. The highest common factor is not unique as $\alpha \cdot R_m$ is also the highest common factor. The method is illustrated in Example 4.15.

Example 4.15: Complex-root determination and highest common factor

We postulate $\lambda^6 - 11\lambda^5 + 50\lambda^4 - 120\lambda^3 + 159\lambda^2 - 109\lambda + 30 = 0$. It has a complex root. The complex factor is $\lambda^2 + b\lambda + c$. We have determined $c = 5$ from the square-root method. We now apply [4.46] with coefficients [4.47] and obtain two equations for b:

$$\phi(b) = b^5 + 11b^4 + 30b^3 - 45b^2 - 266b - 216 = 0$$

$$g(b) = b^4 + 11b^3 + 35b^2 + 10b - 72 = 0$$

(The detailed computation of the coefficients follows an expression of the fourth-degree polynomial example and is left to the reader.)

We must now determine the highest factor (see [4.50]). First we divide:

$$\phi(b)/g(b) = q_1(b) + R_1(b)$$

$q_1(b) = b$, which is of little interest

$$R_1(b) = -5b^3 - 55b^2 - 194b - 216$$

as can readily be obtained. We divide now:

$g(b)/R_1(b)$ with $q_2(b) = -b/5$

$$R_2(b) = -19b^2/5 - 166b/5 - 72$$

For convenience we can multiply $R_2(b)$ by -5 and divide:

$-R_1(b)/[-5R_2(b)]$ with $q_3(b) = 5b/q_1 + 215/361$

$$R_3 = 144b/361 + 576/361$$

Again we divide:

$$[-5R_2(b)]/[361R_3 q_4(b)] = 19b/144 + 90/144$$

$$R_4 = 0$$

The highest common factor is now the last non-zero remainder. This was:

$361R_3 = 144b + 576 = 0$

Consequently: $b = -4$

It can be verified that $\lambda^2 - 4\lambda + 5$ has a complex root; $\lambda_1 = 2 + i$, $\lambda_2 = 2 - i$ and is a factor of the characteristic equation $\phi(\lambda) = (\lambda^2 - 4\lambda + 5)(x - 1)^2 \cdot (x - 2)(x - 3)$.

The Newton-Raphson method of iteration

Although other iteration procedures have been described in the literature (e.g. Whittaker and Robinson, 1944) this method is very attractive for its simplicity and has found widespread utilization. The real root of a polynomial equation can be obtained by this iteration procedure when $f(x)$, the function for which the root λ is sought, has a simple derivation.

We denote the root with λ and the correction by c. Then the root can be written as:

$$\lambda = \lambda_0 + c_1 \qquad [4.51]$$

From Taylor's formula we derive after some transformation:

$$f(\lambda_0) + c_1 f'(\lambda_0) + (c_1^2/2) f''(\lambda_0 + \theta c_1) = 0 \qquad [4.52]$$

where the primes denote the derivatives.
For sufficiently small c we can approximate:

$$c_1 = \frac{-f(\lambda_0)}{f'(\lambda_0)} \qquad [4.51a]$$

Then: $\lambda_1 = \lambda_0 + c_1 \qquad [4.51b]$

is the better estimate for λ.

This procedure can be repeated which leads to:

$$\lambda_2 = \lambda_1 + c_2 \qquad [4.51c]$$

where $c_2 = -f(\lambda_1)/f'(\lambda_1)$

We can stop when $c_i < \epsilon$, where ϵ is a predetermined error threshold of accuracy. The method converges when:

$$[f(x)f''(x)]/[f'(x)]^2 < 1 \qquad [4.52a]$$

4.13 DETERMINATION OF THE EIGENVECTORS

The procedures for the determination of eigenvalues have been outlined in the previous sections. When an eigenvalue λ has been obtained the task is the finding of the associated eigenvector. This problem goes back to the situation of solving a system of linear equations. We treat this topic now. The problem is also identical with diagonalizing a matrix.

Let us assume we have a system of equations:

$$
\begin{vmatrix}
a_{11}x_1 + a_{12}x_2 + a_{13}x_3 + \ldots a_{1n}x_n = b_1 \\
a_{21}x_1 + a_{22}x_2 + \ldots \quad\quad a_{2n}x_n = b_2 \\
\cdot \quad\quad \cdot \quad\quad \cdot \quad\quad\quad\quad \cdot \quad\quad \cdot \\
\cdot \quad\quad \cdot \quad\quad \cdot \quad\quad\quad\quad \cdot \quad\quad \cdot \\
\cdot \quad\quad \cdot \quad\quad \cdot \quad\quad\quad\quad \cdot \quad\quad \cdot \\
a_{n1}x_1 + a_{n2}x_2 + \ldots \quad\quad a_{nn}x = b_n
\end{vmatrix}
\quad [4.53]
$$

This system can be written in matrix form:

$$\mathbf{A}x = b \quad\quad [4.53a]$$

where $\mathbf{A}$ is the matrix containing all the coefficients and x is a vector with components (elements) x_1 through x_n. The vector b contains the constants on the right side. If all b_i are zero, the system is said to be homogeneous.

The rank of the matrix is r, and when $r = n$, a unique solution vector x exists. If $r < n$, then at least one solution vector can be found. In addition, $n - r$ linearly independent vectors can be found which are solutions of the homogeneous set of equations:

$$\mathbf{A}x = 0 \quad\quad [4.53b]$$

Let these vectors be called $y_1 \ldots y_{n-r}$.

It is known that the vector x plus any linear combination of these y_i is also a solution. If $b = 0$ then $x = 0$ is the solution for the homogeneous case and only the y_i vectors remain. This is the situation for the eigenvectors. We need now a scheme to solve this system of linear equations. For every λ the vectors y will be determined. One of the best methods to solve a linear equation with the aid of electronic computers is the elimination method. This procedure requires the multiplication, division, addition and subtraction of columns, all operations which do not change the rank or order of the matrix (see Transformations, Section 4.4b), and are readily performed by the computer.

We select first a non-zero element of the first column. (There should be one at least, else $\|\mathbf{A} - \lambda \mathbf{I}\| = 0$.) Let us assume that a_{11} is non-zero. We divide the first row by a_{11}, which leaves the first row with:

$$1, \frac{a_{12}}{a_{11}}, \frac{a_{13}}{a_{11}}, \ldots, \frac{a_{1n}}{a_{11}}, \frac{b_1}{a_{11}}$$

We multiply now the first row by a_{21} and subtract this row from the second row (or multiply by $-a_{21}$ and add the rows). We continue by multiplying the first row by a_{31} and subtracting the row from the third row, etc.

The respective element of the first column from each row appears in the subtraction term. After the operation has been carried out the first column contains only zero values except for the first row where the value is one. The matrix takes now the following form:

$$\begin{bmatrix} 1 & a'_{12}, & a'_{13} \ldots a'_{1n} & b'_1 \\ 0 & a'_{22}, & a'_{23} \ldots a'_{2n} & b' \\ \cdot & \cdot & \cdots \cdot & \cdot \\ \cdot & \cdot & \cdots \cdot & \cdot \\ 0 & a'_{n2}, & a'_{n3} \ldots a'_{nn} & b'_n \end{bmatrix}$$ [4.54]

The prime denotes the new elements, i.e. $a'_{22} = a_{22} - (a_{12}/a_{11}) \cdot a_{21}$, etc.

We repeat the procedure with the second row. We divide by the first non-zero value of the second row other than that of the first row where the number 1 is in the first column. This means we must have an element where every other element up to the one selected in the row is zero. Let us assume that a_{42} is non-zero. Then we divide row 4 by a_{42} and multiply this row by the respective row elements of column 2 as previously outlined. We subtract the row. This provides:

$$\begin{bmatrix} 1 & 0 & a''_{13} \ldots a''_{1n} & b'' \\ 0 & 0 & a''_{23} \ldots a''_{2n} & b''_2 \\ 0 & 0 & a''_{33} \ldots a''_{3n} & b''_3 \\ 0 & 1 & a_{43} \ldots a''_{4n} & b''_4 \\ \cdot & \cdot & \cdot & \cdot \\ \cdot & \cdot & \cdot & \cdot \\ 0 & 0 & a''_{n3} \ldots a''_{nn} & b''_n \end{bmatrix}$$ [4.54a]

The double prime denotes the new elements again, e.g. $a''_{33} = a'_{33} - (a'_{43}/a'_{42}) \cdot a'_{32}$, etc.

Since a'_{22} and a'_{32} were assumed to be zero, the unit element stands now in the fourth row. It is now recommended by some authors to exchange rows so that this row with the unit element is placed into the second row. However, this is not a necessity when the rule is adopted that the next element of the selected column must be non-zero *and* all other elements of the previous columns must be zero.

DETERMINATION OF THE EIGENVECTORS

The final process would then render:

$$\begin{bmatrix} 1 & 0 & 0 & \ldots & c_1 \\ 0 & 0 & 1 & \ldots & c_2 \\ 0 & 0 & 0 & \ldots & c_3 \\ 0 & 1 & 0 & \ldots & c_4 \\ \cdot & \cdot & \cdot & & \cdot \\ \cdot & \cdot & \cdot & & \cdot \\ \cdot & \cdot & \cdot & & \cdot \\ 0 & 0 & 0 & \ldots & c_n \end{bmatrix} \begin{bmatrix} x_1 \\ x_2 \\ x_3 \\ x_4 \\ \cdot \\ \cdot \\ \cdot \\ x_n \end{bmatrix}$$ [4.54b]

where the c vector symbolizes the combined operations made on the vector b. In the sorted version we have:

$$\begin{bmatrix} 1 & 0 & 0 & \ldots & c_1 \\ 0 & 1 & 0 & \ldots & c_2 \\ 0 & 0 & 1 & \ldots & c_3 \\ \cdot & \cdot & \cdot & & \cdot \\ \cdot & \cdot & \cdot & & \cdot \\ \cdot & \cdot & \cdot & \ldots & \cdot \end{bmatrix} \begin{bmatrix} x_{k1} \\ x_{k2} \\ x_{k3} \\ \cdot \\ \cdot \\ \cdot \end{bmatrix}$$ [4.54c]

where the sequence k_1 is not by ascending integers. It is obvious that in:

$\mathbf{A}x = c$ [4.55]

the $\mathbf{A}$ is a diagonal matrix with all unit elements, i.e. we find now:

$\mathbf{I}x = c$ [4.55a]

with: $x_i = c_i$ [4.55b]

in which exactly one x_i-value corresponds to one element of c_i, either in the form of [4.54c] or in the original form [4.54b], which cannot be written in this simple solution as the unit values do not all appear on the diagonal. This elimination method was developed for the solution of linear equations and is also known as the Gauss-Jordan method.

Another method of substitution is the Gauss-Seidel process which will not be covered here. The reader may refer to the literature (see Pennington, 1970, etc.).

It is self-evident that in the case of the homogeneous equation the vector $b = 0$, and we would be left with a system of a matrix $\mathbf{I}$ and the equation $\mathbf{I}x = 0$, which has a vector x with all elements zero as already discussed. This solution is trivial and known a priori. Solutions in which the system is homogeneous as in the eigenvector problem require some modification.

We go back to the prior discussion that a unique solution exists when the rank $r = n$. This is equivalent to saying that the set of equations is consistent. Actually consistency is not alone defined by the matrix **A**. We can form an "augmented" matrix by including b with the coefficients ([4.54a]). Then the set is consistent when both **A** *and* the "augmented" matrix have the same rank. In that case a solution exists. The consistency consequently is independent of whether the set comprises more equations than unknowns or vice versa. In the case of a homogeneous system both matrices have the same rank and the system is consistent.

Consistent systems may have infinitely many solutions. If $r = n$, a unique solution can be found; if $r < n$, then $n - r$ linearly independent vectors y_i can be derived. The vector x is called a particular solution. The homogeneous set has no particular solution ($x = 0$). The vectors y_i are solutions of the homogeneous set $Ax = 0$.

Linear independence can be defined from the equation:

$$\alpha_1 y_1 + \alpha_2 y_2 + \ldots \alpha_{n-r} y_{n-r} = 0 \qquad [4.56]$$

When no set of coefficients other than $\alpha_1 = \alpha_2 = \ldots \alpha_{n-r} = 0$ fulfills this equation the vectors y_i are called linearly independent.

The solution for eigenvectors goes now back to the complete system of linear equations. Since the homogeneous system is consistent, at least one row will appear with zeros after the elimination process ($n - r = 1$, since $n = r$ has no linearly independent vector). This means that the matrix **A** cannot be completely reduced to the identity matrix. The x_i-values of the zero rows can be arbitrarily chosen. The solution is therefore not unique. In the solution of linear equations this corresponds to the general solution.

Example 4.16: Determination of eigenvectors

We discuss now in detail the determination of the eigenvectors of Example 4.10. The matrix:

$$A = \frac{1}{4} \begin{pmatrix} 20 & -7 & 5 \\ 28 & -9 & 11 \\ 20 & -11 & 17 \end{pmatrix}$$

has the eigenvalues $\lambda_1 = 4, \lambda_2 = 2, \lambda_3 = 1$.

The first eigenvalue leads to the matrix $(A - \lambda_1 I)$, i.e.:

$$\frac{1}{4} \begin{pmatrix} 4 & -7 & 5 \\ 28 & -25 & 11 \\ 20 & -11 & 1 \end{pmatrix} \begin{pmatrix} x_1 \\ x_2 \\ x_3 \end{pmatrix} = 0$$

According to the rules we divide the first row by 4 to obtain $1, -7/4, 5/4$.

DETERMINATION OF THE EIGENVECTORS

We multiply by -28 and add the terms to the second row, multiply by -20 and add to the third row. The matrix reads after these steps (by omitting the divisor $1/4$ in front):

$$\begin{pmatrix} 1 & -7/4 & 5/4 \\ 0 & 24 & -24 \\ 0 & 24 & -24 \end{pmatrix}$$

It is immediately seen that the last line becomes zero by subtraction of the second row. Division by 24, multiplication by $7/4$ of the second row and addition to the first line leaves finally the eigenvector matrix. It has the rank 2, with $n - r = 3 - 2 = 1$ eigenvector:

$$\begin{pmatrix} 1 & 0 & -1/2 \\ 0 & 1 & -1 \\ 0 & 0 & 0 \end{pmatrix}$$

Common practice is now to supplement the rows with all zeros by -1 in the diagonal:

$$\begin{pmatrix} 1 & 0 & -1/2 \\ 0 & 1 & -1 \\ 0 & 0 & -1 \end{pmatrix}$$

The last column (whose diagonal element was filled in) is the eigenvector. Since the eigenvector is not unique, $-2x_{\lambda 1}$ is also an eigenvector. In this form it was used previously.

The $\lambda_2 = 2$ provides $(\mathbf{A} - \lambda_2 \mathbf{I})$, i.e.:

$$\frac{1}{4}\begin{pmatrix} 12 & -7 & 5 \\ 28 & -17 & 11 \\ 20 & -11 & 9 \end{pmatrix}$$

The first operation making all elements of column 1 equal to zero (except a_{11}) gives:

$$\begin{pmatrix} 1 & -7/12 & 5/12 \\ 0 & -2/3 & -2/3 \\ 0 & 2/3 & 2/3 \end{pmatrix} \text{ which converts to } \begin{pmatrix} 1 & 0 & 1 \\ 0 & +1 & +1 \\ 0 & 0 & 0 \end{pmatrix}$$

after the second step. Finally:

$$\begin{pmatrix} 1 & 0 & 1 \\ 0 & 1 & 1 \\ 0 & 0 & -1 \end{pmatrix}$$

with the last column as the eigenvector. The last $\lambda_3 = 1$ renders:

$$\frac{1}{4}\begin{pmatrix} 16 & -7 & 5 \\ 28 & -13 & 11 \\ 20 & -11 & 13 \end{pmatrix}$$

The operation leads in sequence to:

$$\begin{pmatrix} 1 & -7/16 & 5/16 \\ 0 & -3/4 & 9/4 \\ 0 & -9/4 & 27/4 \end{pmatrix} \to \begin{pmatrix} 1 & 0 & -1 \\ 0 & -1 & -3 \\ 0 & 0 & 0 \end{pmatrix} \to \begin{pmatrix} 1 & 0 & -1 \\ 0 & -1 & -3 \\ 0 & 0 & -1 \end{pmatrix}$$

with the eigenvector in the last column.

Example 4.17: Eigenvalue computation for Example 4.12, double root, one eigenvalue.

Given is:

$$A = \begin{pmatrix} 1 & 0 & 0 \\ -1 & 2 & 3 \\ 1 & 1/3 & 2 \end{pmatrix}$$

with $\lambda_1 = 3$, $\lambda_2 = \lambda_3 = 1$.

First we have for $\lambda_1 = 3$:

$$\begin{pmatrix} -2 & 0 & 0 \\ -1 & -1 & 3 \\ 1 & 1/3 & -1 \end{pmatrix} \to \begin{pmatrix} 1 & 0 & 0 \\ 0 & -1 & 3 \\ 0 & 1/3 & -1 \end{pmatrix} \to \begin{pmatrix} 1 & 0 & 0 \\ 0 & 1 & -3 \\ 0 & 0 & 0 \end{pmatrix} \to \begin{pmatrix} 1 & 0 & 0 \\ 0 & 1 & -3 \\ 0 & 0 & -1 \end{pmatrix}$$

Eigenvector $x^T = 0, 3, 1$ as stated in Example 4.12 (rank of matrix $r = 2$).

Now we treat $\lambda_1 = 1$:

$$\begin{pmatrix} 0 & 0 & 0 \\ -1 & 1 & 3 \\ 1 & 1/3 & 1 \end{pmatrix} \to \begin{pmatrix} 0 & 0 & 0 \\ 1 & -1 & -3 \\ 0 & 4/3 & 4 \end{pmatrix} \to \begin{pmatrix} 0 & 0 & 0 \\ 1 & 0 & 0 \\ 0 & 1 & 3 \end{pmatrix} \to \begin{pmatrix} 1 & 0 & 0 \\ 0 & 1 & 3 \\ 0 & 0 & 0 \end{pmatrix} \to$$

$$\to \begin{pmatrix} 1 & 0 & 0 \\ 0 & 1 & 3 \\ 0 & 0 & -1 \end{pmatrix}$$

DETERMINATION OF THE EIGENVECTORS

Since the rank of the matrix is $r = 2$, as:

$$\begin{Vmatrix} 1 & 0 \\ 0 & 1 \end{Vmatrix} \neq 0 \text{ (see the fourth matrix above)}$$

only one linearly independent eigenvector exists, appearing in the last column. Another vector for the transformation to triangular form must be obtained from the Jordan normal form (see Example 4.12).

Example 4.18: Double root with distinct eigenvalues (Example 4.11)

Given:

$$A = \frac{1}{18} \begin{pmatrix} 23 & 2 & 13 \\ 0 & 18 & 0 \\ 5 & 2 & 31 \end{pmatrix}$$

and: $\lambda_3 = 2, \lambda_2 = \lambda_1 = 1$

$$\lambda_3 = 2: \begin{pmatrix} -13 & 2 & 13 \\ 0 & -18 & 0 \\ 5 & 2 & -5 \end{pmatrix} \to \begin{pmatrix} 1 & -2/13 & -1 \\ 0 & -18 & 0 \\ 0 & 36/13 & 0 \end{pmatrix} \to \begin{pmatrix} 1 & 0 & -1 \\ 0 & 1 & 0 \\ 0 & 0 & 0 \end{pmatrix} \to$$

$$\to \begin{pmatrix} 1 & 0 & -1 \\ 0 & 1 & 0 \\ 0 & 0 & -1 \end{pmatrix}$$

$$\lambda_2 = \lambda_1 = 1: \begin{pmatrix} 5 & 2 & 13 \\ 0 & 0 & 0 \\ 5 & 2 & 13 \end{pmatrix} \to \begin{pmatrix} 1 & 2/5 & 13/5 \\ 0 & 0 & 0 \\ 0 & 0 & 0 \end{pmatrix}$$

This matrix has the rank $r = 1$, consequently $n - r = 3 - 1 = 2$, and two eigenvectors can be found:

$$\begin{pmatrix} 1 & 2/5 & 13/5 \\ 0 & -1 & 0 \\ 0 & 0 & -1 \end{pmatrix}$$

The eigenvectors $x_{\lambda_2}^T = (2, -5, 0)$ and $x_{\lambda_1}^T = (13, 0, -5)$ fulfill $x_1 + 2/5 x_2 + 13/5 x_3 = 0$.

$$R = \begin{pmatrix} 1 & 2/5 & 13/5 \\ 0 & -1 & 0 \\ 1 & 0 & -1 \end{pmatrix}$$

Example 4.19: Eigenvalue of a matrix with $\|A\| = 0$

$$A = \begin{pmatrix} 1 & 1 & -0.5 \\ 0 & 1 & 0 \\ -2 & 2 & 1 \end{pmatrix} \text{ with } \|A\| = 0.$$

The characteristic polynomial is:

$$\lambda^3 - 3\lambda^2 + 2\lambda = 0$$

consequently $\lambda(\lambda - 1)(\lambda - 2) = 0$. The roots are $\lambda_1 = 2$, $\lambda_2 = 1$, $\lambda_3 = 0$ in decreasing order.

Eigenvectors:

$$\lambda_1 = 2: \begin{pmatrix} -1 & 1 & -0.5 \\ 0 & -1 & 0 \\ -2 & 2 & -1 \end{pmatrix} \to \begin{pmatrix} +1 & -1 & +0.5 \\ 0 & -1 & 0 \\ 0 & 0 & 0 \end{pmatrix} \to \begin{pmatrix} 1 & 0 & 0.5 \\ 0 & -1 & 0 \\ 0 & 0 & -1 \end{pmatrix}$$

Eigenvector $(0.5, 0, -1)$.

$$\lambda_2 = 1: \begin{pmatrix} 0 & 1 & -0.5 \\ 0 & 0 & 0 \\ -2 & 2 & 0 \end{pmatrix} \to \begin{pmatrix} 1 & -1 & 0 \\ 0 & 1 & -0.5 \\ 0 & 0 & 0 \end{pmatrix} \to \begin{pmatrix} 1 & 0 & -0.5 \\ 0 & 1 & -0.5 \\ 0 & 0 & 0 \end{pmatrix} \to$$

$$\to \begin{pmatrix} 1 & 0 & -0.5 \\ 0 & 1 & -0.5 \\ 0 & 0 & -1 \end{pmatrix}$$

Eigenvector $(0.5, 0.5, 1)$.

$$\lambda_3 = 0: \begin{pmatrix} 1 & 1 & -0.5 \\ 0 & 1 & 0 \\ -2 & 2 & 1 \end{pmatrix} \to \begin{pmatrix} 1 & 1 & -0.5 \\ 0 & 1 & 0 \\ 0 & 4 & 0 \end{pmatrix} \to \begin{pmatrix} 1 & 0 & -0.5 \\ 0 & 1 & 0 \\ 0 & 0 & 0 \end{pmatrix} \to$$

$$\to \begin{pmatrix} 1 & 0 & -0.5 \\ 0 & 1 & 0 \\ 0 & 0 & -1 \end{pmatrix}$$

Eigenvector $(0.5, 0, 1)$.

4.14 LINEAR EQUATIONS

The section on matrices would not be complete without some mention of solutions for sets of linear equations. One technique, the elimination method, has already been discussed (see Section 4.13). Several other procedures exist.

The square-root method

One classical technique is the splitting of the matrix **A** into two triangular matrices **T**, such that:

$$\mathbf{A} = \mathbf{T}^T \cdot \mathbf{T} \qquad [4.57]$$

This is sometimes called the "square root" of the matrix, and the procedure is also applied to solve a set of linear equations (see Dwyer, 1941, 1945). It is self-evident that again $\|\mathbf{A}\| \neq 0$. It is further required that **A** is a symmetric matrix.

We go back to the set of linear equations:

$$\mathbf{A}x = b \qquad [4.57a]$$

By substitution we get:

$$\mathbf{T}^T \mathbf{T} x = b \qquad [4.57b]$$

This can be divided into two product terms:

$$\mathbf{T}x = z \qquad [4.57c]$$

and: $\mathbf{T}^T z = b \qquad [4.57d]$

Both matrices **T** and **T**^{T} are triangular matrices.

We find an explicit expression for the elements of the matrix **T** by setting:

$$t_{11}^2 = a_{11} \qquad [4.58a]$$

$$t_{12} = a_{12}/t_{11} \qquad [4.58b]$$

$$t_{22} = (a_{22} - t_{12}^2) \qquad [4.58c]$$

general: $t_{jj}^2 = (a_{jj} - t_{1j}^2 - t_{2j}^2 - \ldots - t_{j-1, j}^2) \qquad [4.58d]$

$$t_{jk}^2 = (a_{jk} - t_{1j}t_{1k} - t_{2j}t_{2k} - \ldots - t_{j-1, j} t_{j-1, k})/t_{jj} \qquad [4.58e]$$

The next steps are simple arithmetic:

$$t_{11}z_1 = b_1 \qquad [4.59a]$$

$$t_{12}z_1 + t_{22}z_2 = b_2 \qquad [4.59b]$$

general: $t_{ij}z_1 + t_{2j}z_2 + \ldots + t_{jj}z_j = b_j \qquad [4.59c]$

In this set of equations only one unknown per line appears, which can be determined. The set of [4.59a] can be transformed to explicit expressions for z_j, namely:

$$z_1 = b_1/t_{11} \qquad [4.60a]$$

$$z_2 = (b_2 - t_{12}t_1)/t_{22} \qquad [4.60b]$$

general: $z_j = (b_j - t_{1j}z_1 - t_{2j}z_2 - \ldots - t_{j-1}z_{j-1})/t_{ij}$ [4.60c]

Finally we calculate x from:

$$\mathbf{T}x = z \qquad [4.61]$$

$$x_n = z_n/t_{nn} \qquad [4.61a]$$

$$x_{n-1} = (z_{n-1} - t_{n-1,n}x_n)/t_{n-1,n-1} \qquad [4.61b]$$

general: $x_j = (z_j - t_{j,j-1}x_{j+1} - t_{j,j-2}x_{j+2} - \ldots - t_{jn}x_n)/t_{jj}$ [4.61c]

It should be noted that $t_{j,j+i} = 0$ for $j + i > n$, see first [4.61a].

It was previously assumed that $\mathbf{A}$ must be a symmetric matrix. When the original matrix is non-symmetric, let us say $\mathbf{B}$, we would start with a system:

$$\mathbf{B}x = b \qquad [4.62]$$

We set now: $\mathbf{S} = \mathbf{B}^T \cdot \mathbf{B}$ [4.62a]

which produces a symmetric matrix. Then we find:

$$\mathbf{Y} = \mathbf{B}^T b \qquad [4.62b]$$

and set: $\mathbf{S}x = y$ [4.62c]

This system is now equivalent to [4.57a] for which the solution has been given.

Cramer's rule

The system of linear equations has been introduced as:

$$\mathbf{A}x = b \qquad [4.63a]$$

To solve these equations we reformulate the problem:

$$x = \mathbf{A}^{-1}b \qquad [4.63b]$$

Now: $x = (\text{adj } \mathbf{A} \cdot b)/\|\mathbf{A}\|$ [4.63c]

In terms of the elements of $\mathbf{A}$ this can be rephrased as:

$$x_j = \sum_i v_{ji} b_i \qquad [4.63d]$$

$$x_j = (\sum_i w_{ij} b_i)/\|\mathbf{A}\| \qquad [4.63e]$$

where v_{ji} denotes the elements of the inverse $\mathbf{A}^{-1}$ and w_{ij} the elements of the adjoint $\mathbf{A}$.

This solution was first stated by G. Cramer and has entered the literature as Cramer's rule.

The numerator of [4.63e] is the determinant of the matrix A when the jth column is replaced by the elements of b. In other words:

$$x_j = \|A_j\|/\|A\| \qquad [4.63f]$$

where A_j is the determinant after the jth column is replaced by b.

4.15 CONCLUSIONS

It is obvious that in the preceding sections no detailed course on matrix algebra has been given nor was there any attempt to describe all the various sophisticated details of solving linear equations, roots, complex roots or eigenvalues. The author intended, however, to give the reader, who is unfamiliar with matrix algebra and numerical methods, a starting point. The reader should be able to apply some of the advanced methods of statistical analysis as discussed in this textbook where matrix algebra is a prerequisite. In many cases the given examples here suffice to illustrate the solutions. Should more information be needed the reader is referred to the literature (e.g. Pennington, 1970; Ralston and Wilt, 1967; Graybill, 1969; Wilt, 1962; Boas, 1966; Wilkinson, 1965; Stigant, 1959, etc.).

APPENDIX

INTEGRAL AND ORDINATE OF THE GAUSSIAN DISTRIBUTION

$$F(a) = (\sqrt{2\pi})^{-1} \int_{-\infty}^{a} \exp(-0.5a^2) \, da$$

with $a = (x - \bar{x})/\sigma$ or $x - \bar{x} = a \cdot \sigma$.

Consequently for $x < \bar{x}$ the a would be negative and:

$$F(-a) = (\sqrt{2\pi})^{-1} \int_{-\infty}^{-a} \exp(-0.5a^2) \, da = 1.0 - F(a)$$

Furthermore: $f(a) = (\sqrt{2\pi})^{-1} \exp(-0.5a^2) = f(-a)$

Example: $\sigma = 2, x = 4.2, \bar{x} = 3, a = 0.6$.

$F(a) = 0.7257, f(a) = 0.3332$.

$\sigma = 2, x = 1.8, \bar{x} = 3, a = -0.6$.

$F(-a) = 1.0 - 0.7257 = 0.2743, f(-a) = 0.3332$.

a	$F(a)$	$f(a)$	a	$F(a)$	$f(a)$	a	$F(a)$	$f(a)$
0.00	0.5000	0.3989	0.46	0.6772	0.3589	0.92	0.8212	0.2613
0.01	0.5040	0.3989	0.47	0.6808	0.3572	0.93	0.8238	0.2589
0.02	0.5080	0.3989	0.48	0.6844	0.3555	0.94	0.8264	0.2565
0.03	0.5120	0.3988	0.49	0.6879	0.3538	0.95	0.8289	0.2541
0.04	0.5160	0.3986	0.50	0.6915	0.3520	0.96	0.8315	0.2516
0.05	0.5199	0.3984	0.51	0.6950	0.3503	0.97	0.8340	0.2492
0.06	0.5239	0.3982	0.52	0.6985	0.3485	0.98	0.8365	0.2468
0.07	0.5279	0.3980	0.53	0.7019	0.3467	0.99	0.8389	0.2444
0.08	0.5319	0.3977	0.54	0.7054	0.3448	1.00	0.8413	0.2420
0.09	0.5359	0.3973	0.55	0.7088	0.3429	1.01	0.8438	0.2396
0.10	0.5398	0.3970	0.56	0.7123	0.3410	1.02	0.8461	0.2371
0.11	0.5438	0.3965	0.57	0.7157	0.3391	1.03	0.8485	0.2347
0.12	0.5478	0.3961	0.58	0.7190	0.3372	1.04	0.8508	0.2323
0.13	0.5517	0.3956	0.59	0.7224	0.3352	1.05	0.8531	0.2299
0.14	0.5557	0.3951	0.60	0.7257	0.3332	1.06	0.8554	0.2275
0.15	0.5596	0.3945	0.61	0.7291	0.3312	1.07	0.8577	0.2251
0.16	0.5636	0.3939	0.62	0.7324	0.3292	1.08	0.8599	0.2227
0.17	0.5675	0.3932	0.63	0.7357	0.3271	1.09	0.8621	0.2203
0.18	0.5714	0.3925	0.64	0.7389	0.3251	1.10	0.8643	0.2179
0.19	0.5753	0.3918	0.65	0.7422	0.3230	1.11	0.8665	0.2155
0.20	0.5793	0.3910	0.66	0.7454	0.3209	1.12	0.8686	0.2131
0.21	0.5832	0.3902	0.67	0.7486	0.3187	1.13	0.8708	0.2107
0.22	0.5871	0.3894	0.68	0.7517	0.3166	1.14	0.8729	0.2083
0.23	0.5910	0.3885	0.69	0.7549	0.3144	1.15	0.8749	0.2059
0.24	0.5948	0.3876	0.70	0.7580	0.3123	1.16	0.8770	0.2036
0.25	0.5987	0.3867	0.71	0.7611	0.3101	1.17	0.8790	0.2012
0.26	0.6026	0.3857	0.72	0.7642	0.3079	1.18	0.8810	0.1989
0.27	0.6064	0.3847	0.73	0.7673	0.3056	1.19	0.8830	0.1965
0.28	0.6103	0.3836	0.74	0.7704	0.3034	1.20	0.8849	0.1942
0.29	0.6141	0.3825	0.75	0.7734	0.3011	1.21	0.8869	0.1919
0.30	0.6179	0.3814	0.76	0.7764	0.2989	1.22	0.8888	0.1895
0.31	0.6217	0.3802	0.77	0.7794	0.2966	1.23	0.8907	0.1872
0.32	0.6255	0.3790	0.78	0.7823	0.2943	1.24	0.8925	0.1849
0.33	0.6293	0.3778	0.79	0.7852	0.2920	1.25	0.8944	0.1826
0.34	0.6331	0.3765	0.80	0.7881	0.2897	1.26	0.8962	0.1804
0.35	0.6368	0.3752	0.81	0.7910	0.2874	1.27	0.8980	0.1781
0.36	0.6406	0.3739	0.82	0.7939	0.2850	1.28	0.8997	0.1758
0.37	0.6443	0.3726	0.83	0.7967	0.2827	1.29	0.9015	0.1736
0.38	0.6480	0.3712	0.84	0.7995	0.2803	1.30	0.9032	0.1714
0.39	0.6517	0.3697	0.85	0.8023	0.2780	1.31	0.9049	0.1691
0.40	0.6554	0.3683	0.86	0.8051	0.2756	1.32	0.9066	0.1669
0.41	0.6591	0.3668	0.87	0.8079	0.2732	1.33	0.9082	0.1647
0.42	0.6628	0.3653	0.88	0.8106	0.2709	1.34	0.9099	0.1626
0.43	0.6664	0.3637	0.89	0.8133	0.2685	1.35	0.9115	0.1604
0.44	0.6700	0.3621	0.90	0.8159	0.2661	1.36	0.9131	0.1582
0.45	0.6736	0.3605	0.91	0.8186	0.2637	1.37	0.9147	0.1561

APPENDIX

a	$F(a)$	$f(a)$	a	$F(a)$	$f(a)$	a	$F(a)$	$f(a)$
1.38	0.9162	0.1539	1.84	0.9671	0.0734	2.30	0.9893	0.0283
1.39	0.9177	0.1518	1.85	0.9678	0.0721	2.31	0.9896	0.0277
1.40	0.9192	0.1497	1.86	0.9686	0.0707	2.32	0.9898	0.0270
1.41	0.9207	0.1476	1.87	0.9693	0.0694	2.33	0.9901	0.0264
1.42	0.9222	0.1456	1.88	0.9699	0.0681	2.34	0.9904	0.0258
1.43	0.9236	0.1435	1.89	0.9706	0.0669	2.35	0.9906	0.0252
1.44	0.9251	0.1415	1.90	0.9713	0.0656	2.36	0.9909	0.0246
1.45	0.9265	0.1394	1.91	0.9719	0.0644	2.37	0.9911	0.0241
1.46	0.9279	0.1374	1.92	0.9726	0.0632	2.38	0.9913	0.0235
1.47	0.9292	0.1354	1.93	0.9732	0.0620	2.39	0.9916	0.0229
1.48	0.9306	0.1334	1.94	0.9738	0.0608	2.40	0.9918	0.0224
1.49	0.9319	0.1315	1.95	0.9744	0.0596	2.41	0.9920	0.0219
1.50	0.9332	0.1295	1.96	0.9750	0.0584	2.42	0.9922	0.0213
1.51	0.9345	0.1276	1.97	0.9756	0.0573	2.43	0.9925	0.0208
1.52	0.9357	0.1257	1.98	0.9761	0.0562	2.44	0.9927	0.0203
1.53	0.9370	0.1238	1.99	0.9767	0.0551	2.45	0.9929	0.0198
1.54	0.9382	0.1219	2.00	0.9773	0.0540	2.46	0.9931	0.0194
1.55	0.9394	0.1200	2.01	0.9778	0.0529	2.47	0.9932	0.0189
1.56	0.9406	0.1182	2.02	0.9783	0.0519	2.48	0.9934	0.0184
1.57	0.9418	0.1163	2.03	0.9788	0.0508	2.49	0.9936	0.0180
1.58	0.9429	0.1145	2.04	0.9793	0.0498	2.50	0.9938	0.0175
1.59	0.9441	0.1127	2.05	0.9798	0.0488	2.51	0.9940	0.0171
1.60	0.9452	0.1109	2.06	0.9803	0.0478	2.52	0.9941	0.0167
1.61	0.9463	0.1092	2.07	0.9808	0.0468	2.53	0.9943	0.0163
1.62	0.9474	0.1074	2.08	0.9812	0.0459	2.54	0.9945	0.0158
1.63	0.9484	0.1057	2.09	0.9817	0.0449	2.55	0.9946	0.0155
1.64	0.9495	0.1040	2.10	0.9821	0.0440	2.56	0.9948	0.0151
1.65	0.9505	0.1023	2.11	0.9826	0.0431	2.57	0.9949	0.0147
1.66	0.9515	0.1006	2.12	0.9830	0.0422	2.58	0.9951	0.0143
1.67	0.9525	0.0989	2.13	0.9834	0.0413	2.59	0.9952	0.0139
1.68	0.9535	0.0973	2.14	0.9838	0.0404	2.60	0.9953	0.0136
1.69	0.9545	0.0957	2.15	0.9842	0.0396	2.61	0.9955	0.0132
1.70	0.9554	0.0940	2.16	0.9846	0.0387	2.62	0.9956	0.0129
1.71	0.9564	0.0925	2.17	0.9850	0.0379	2.63	0.9957	0.0126
1.72	0.9573	0.0909	2.18	0.9854	0.0371	2.64	0.9959	0.0122
1.73	0.9582	0.0893	2.19	0.9857	0.0363	2.65	0.9960	0.0119
1.74	0.9591	0.0878	2.20	0.9861	0.0355	2.66	0.9961	0.0116
1.75	0.9599	0.0863	2.21	0.9864	0.0347	2.67	0.9962	0.0113
1.76	0.9608	0.0848	2.22	0.9868	0.0339	2.68	0.9963	0.0110
1.77	0.9616	0.0833	2.23	0.9871	0.0332	2.69	0.9964	0.0107
1.78	0.9625	0.0818	2.24	0.9875	0.0325	2.70	0.9965	0.0104
1.79	0.9633	0.0804	2.25	0.9878	0.0317	2.71	0.9966	0.0101
1.80	0.9641	0.0790	2.26	0.9881	0.0310	2.72	0.9967	0.0099
1.81	0.9649	0.0775	2.27	0.9884	0.0303	2.73	0.9968	0.0096
1.82	0.9656	0.0761	2.28	0.9887	0.0297	2.74	0.9969	0.0093
1.83	0.9664	0.0748	2.29	0.9890	0.0290	2.75	0.9970	0.0091

a	$F(a)$	$f(a)$	a	$F(a)$	$f(a)$	a	$F(a)$	$f(a)$
2.76	0.9971	0.0088	3.22	0.9994	0.0022	3.68	0.9999	0.0005
2.77	0.9972	0.0086	3.23	0.9994	0.0022	3.69	0.9999	0.0004
2.78	0.9973	0.0084	3.24	0.9994	0.0021	3.70	0.9999	0.0004
2.79	0.9974	0.0081	3.25	0.9994	0.0020	3.71	0.9999	0.0004
2.80	0.9974	0.0079	3.26	0.9994	0.0020	3.72	0.9999	0.0004
2.81	0.9975	0.0077	3.27	0.9995	0.0019	3.73	0.9999	0.0004
2.82	0.9976	0.0075	3.28	0.9995	0.0018	3.74	0.9999	0.0004
2.83	0.9977	0.0073	3.29	0.9995	0.0018	3.75	0.9999	0.0004
2.84	0.9977	0.0071	3.30	0.9995	0.0017	3.76	0.9999	0.0003
2.85	0.9978	0.0069	3.31	0.9995	0.0017	3.77	0.9999	0.0003
2.86	0.9979	0.0067	3.32	0.9996	0.0016	3.78	0.9999	0.0003
2.87	0.9979	0.0065	3.33	0.9996	0.0016	3.79	0.9999	0.0003
2.88	0.9980	0.0063	3.34	0.9996	0.0015	3.80	0.9999	0.0003
2.89	0.9981	0.0061	3.35	0.9996	0.0015	3.81	0.9999	0.0003
2.90	0.9981	0.0060	3.36	0.9996	0.0014	3.82	0.9999	0.0003
2.91	0.9982	0.0058	3.37	0.9996	0.0014	3.83	0.9999	0.0003
2.92	0.9983	0.0056	3.38	0.9996	0.0013	3.84	0.9999	0.0003
2.93	0.9983	0.0055	3.39	0.9997	0.0013	3.85	0.9999	0.0002
2.94	0.9984	0.0053	3.40	0.9997	0.0012	3.86	0.9999	0.0002
2.95	0.9984	0.0051	3.41	0.9997	0.0012	3.87	0.9999	0.0002
2.96	0.9985	0.0050	3.42	0.9997	0.0012	3.88	0.9999	0.0002
2.97	0.9985	0.0048	3.43	0.9997	0.0011	3.89	1.0000	0.0002
2.98	0.9986	0.0046	3.44	0.9997	0.0011	3.90	1.0000	0.0002
2.99	0.9986	0.0046	3.45	0.9997	0.0010	3.91	1.0000	0.0002
3.00	0.9987	0.0044	3.46	0.9997	0.0010	3.92	1.0000	0.0002
3.01	0.9987	0.0043	3.47	0.9997	0.0010	3.93	1.0000	0.0002
3.02	0.9987	0.0042	3.48	0.9997	0.0009	3.94	1.0000	0.0002
3.03	0.9988	0.0040	3.49	0.9998	0.0009	3.95	1.0000	0.0002
3.04	0.9988	0.0039	3.50	0.9998	0.0009	3.96	1.0000	0.0002
3.05	0.9989	0.0038	3.51	0.9998	0.0008	3.97	1.0000	0.0002
3.06	0.9989	0.0037	3.52	0.9998	0.0008	3.98	1.0000	0.0001
3.07	0.9989	0.0036	3.53	0.9998	0.0008	3.99	1.0000	0.0001
3.08	0.9990	0.0035	3.54	0.9998	0.0008	4.00	1.0000	0.0001
3.09	0.9990	0.0034	3.55	0.9998	0.0007			
3.10	0.9990	0.0033	3.56	0.9998	0.0007		$F(a)$	a
3.11	0.9991	0.0032	3.57	0.9998	0.0007		0.50	0.0000
3.12	0.9991	0.0031	3.58	0.9998	0.0007		0.60	0.2533
3.13	0.9991	0.0030	3.59	0.9998	0.0006		0.70	0.5244
3.14	0.9992	0.0029	3.60	0.9998	0.0006		0.80	0.8416
3.15	0.9992	0.0028	3.61	0.9998	0.0006		0.90	1.2816
3.16	0.9992	0.0027	3.62	0.9999	0.0006		0.95	1.6449
3.17	0.9992	0.0026	3.63	0.9999	0.0005		0.975	1.9600
3.18	0.9993	0.0025	3.64	0.9999	0.0005		0.99	2.3263
3.19	0.9993	0.0025	3.65	0.9999	0.0005		0.995	2.5758
3.20	0.9993	0.0024	3.66	0.9999	0.0005		0.999	3.0902
3.21	0.9993	0.0023	3.67	0.9999	0.0005			

REFERENCES

Abramowitz, M. and Stegun, I. A. (Editors), 1964. Handbook of Mathematical Functions. National Bureau of Standards, Washington, D. C., Applied Mathematics Series, 55, Sixth Printing, 1967, 1045 pp.
Aitchison, J. and Brown, J. A. C., 1957. The Lognormal Distribution. Cambridge University Press, 176 pp.
Aitkin, M. A. and Hume, M. W., 1963. Correlation in a singly truncated bivariate normal distribution, II Rank correlation. Biometrika, 52: 639–643.
Alishouse, J. C., Crone, L. J., Flemming, H. E., Vancleef, F. L. and Wark, D. G., 1967. A discussion of empirical orthogonal functions and their application to vertical temperature profiles. Tellus, 19: 477–481.
Anscombe, F. J., 1948. The transformation of Poisson, binomial and negative binomial data. Biometrika, 35: 246–254.
Aroian, L. A., 1941. Continued fractions of the incomplete beta functions. Ann. Math. Stat., 12: 218–223 (correction 30: 1265).
Barnett, V. D., 1966. Evaluation of the maximum-likelihood estimator where the likelihood equation has multiple roots. Biometrika, 53: 151–165.
Bartlett, M. S., 1936. Square-root transformation in the analysis of variance. J. R. Stat. Soc., Suppl., 3: 68–78.
Bartlett, M. S., 1946. On the theoretical specification of sampling properties of autocorrelated time series. J. R. Stat. Soc., Ser. B, 8: 27.
Bartlett, M. S., 1947. The use of transformations. Biometrics, 3: 39–52.
Bartlett, M. S., 1948. Smoothing periodograms from time-series with continuous spectra. Nature, 161: 686–687.
Bartlett, M. S., 1950a. Periodogram analysis and continuous spectra. Biometrika, 37: 1–16.
Bartlett, M. S., 1950b. Tests of significant in factor analysis. Brit. J. Psychol. (Stat. Sect.), 3: 77–85.
Bartlett, M. S., 1951. A further note on tests of significance in factor analysis. Brit. J. Psychol. (Stat. Sect.), 1: 1–2.
Bartlett Jr., R. P. and Provost, L. P., 1973. Tolerances in standards and specifications. Qual. Progr., 6: 14–19.
Bauer, K. G., 1971. Linear Prediction of a Multivariate Time Series Applied to Atmospheric Scalar Fields. Thesis, Univ. of Wisconsin, 180 pp.
Behboodian, J., 1970. On a mixture of normal distributions. Biometrika, 57: 215–217.
Beyer, W. H., 1966. Handbook of Tables for Probability and Statistics. The Chemical Rubber Co., Cleveland, Ohio, 502 pp.
Bhattacharya, C. G., 1967. A simple method of resolution of a distribution into Gaussian components. Biometrics, 23: 115–135.
Birnbaum, Z. W., 1950. Effect of linear truncation on a multi-normal population. Ann. Math. Stat., 21: 272–279.
Birnbaum, Z. W. and Tingey, F. H., 1951. One-sided confidence contours for probability distribution functions. Ann. Math. Stat., 22: 592–596.
Blackman, R. B. and Tukey, J. W., 1958. The Measurement of Power Spectra. Dover Publ., New York, 190 pp.
Blischke, W. R., 1964. Estimating the parameters of mixtures of binomial distributions. J.

Am. Stat. Assoc., 26: 510—528.
Blischke, W. R., 1966. Asymptotic properties of some estimates of quantiles of circular error. J. Am. Stat. Assoc., 61: 618—631.
Bloch, D., 1966. A note on the estimation of the location parameter of the Cauchy distribution. J. Am. Stat. Assoc., 61: 852—855.
Bloomfield, P., 1972. On the error of prediction of a time series. Biometrika, 59: 501—507.
Bloomfield, P., 1973. An exponential model for the spectrum of a scalar time series. Biometrika, 60: 217—226.
Boas, M. L., 1966. Mathematical Methods in the Physical Sciences. Wiley, New York, 778 pp.
Bolgiano, R., Jr., 1959. Turbulent spectra in a stably stratified atmosphere. J. Geophys. Res., 64: 2226—2229.
Bolgiano, R., 1962. Structure of turbulence in stratified media. J. Geophys. Res., 67: 3015—3023.
Bowman, K. O. and Shenton, L. R., 1970. Properties of the maximum-likelihood estimator for the parameter of the logarithmic series distribution. In: G. P. Patil (Editor), Random Counts in Models and Structures. Pennsylvania State Univ. Press, p. 127—150.
Box, G. E. P. and Jenkins, G. M., 1970. Time Series Analysis, Forecasting and Control. Holden Day, San Francisco, 553 pp.
Box, G. E. P. and Pierce, D. A., 1970. Distribution of residual auto-correlations in autoregressive-moving average time series models. J. Am. Stat. Assoc., 65: 1509—1526.
Bradley, J. V., 1973. The central limit effect for a variety of populations and the influence of population moments. J. Qual. Technol., 5: 171—177.
Brandtner, B. and Essenwanger, O. M., 1957. Zur Statistik phaenologischer Daten. Meteorol. Rundsch., 10: 151—156.
British Association for the Advancement of Science, 1952. Tables of the Bessel-Functions, Part II, 10. Functions of positive integer order, 255 pp.
Broadbent, S. R., 1956. Lognormal approximation to products and quotients. Biometrika, 43: 404—417.
Brooks, C. E. P. and Carruthers, N., 1953. Handbook of Statistical Methods in Meteorology. Her Majesty's Stationary Office, London, 412 pp.
Brooks, C. E. P., Durst, C. S. and Carruthers, N., 1946. Upper winds over the World, I. The frequency distribution of winds at a point in the free air. Q. J. R. Meteorol. Soc., London, 72: 55—73.
Brooks, C. E. P., Durst, C. S., Carruthers, N., Dewar, D. and Sawyer, J. S., 1950. Upper winds over the World. Geophys. Mem., 85, London. New Edition, No. 130, 1960, 217 pp.
Brouwer, R. K., 1971. The fitting of time series models. Rev. Int. Stat. Inst., 28: 233—244.
Brunk, H. D., Holstein, J. E. and Williams, F., 1968. A comparison of binomial approximations to the hypergeometric distribution. The Am. Stat., 22: 24—26.
Bryson, R. A. and Kuhn, P. M., 1956. Half hemispheric 500 mbar topography description by means of orthogonal polynomials, 1. Sci. Rep. 4, Dep. Meteorol., Univ. Wisc., 22 pp.
Buell, C. E., 1971. Integral equation representation for factor analysis. J. Atmos. Sci., 28: 1502—1505.
Buell, C. E. and Bundgaard, R. C., 1971. An analysis of winds to 60 km over Battery MacKenzie (Can. Zone). J. Appl. Meteorol., 10: 803—810.
Burns, A., 1963. Power spectra of low-level atmospheric turbulence measured from an aircraft. Royal Aircraft Establishment, Technical Note Structures, 329.
Burns, A. and Rider, C. K., 1965. Project TOP-CAT, power spectral measurements of clear-air turbulence associated with jet streams. Royal Aircraft Establishment, Technical Report 65210.

REFERENCES

Burt, C., 1941. The factors of the Mind; an Introduction to Factor Analysis in Psychology. MacMillan, New York, 509 pp.
Burt, C., 1952. Tests of significance in factor analysis. Brit. J. Psychol. (Stat. Sect.), 5: 109–133.
Cadwell, J. H., 1951. The bivariate normal integral. Biometrika, 38: 475–481.
Carter, M. C., 1973. Multivariate normal integration. NASA-CR-129005, 39 pp.
Cattell, R. B., 1944. 'Parallel proportional profiles' and other principles for determining the choice of factors by rotation. Psychometrika, 9: 267–283.
Cattell, R. B., 1952. Factor-Analysis. Harper, New York, 462 pp.
Cattell, R. B., 1965. The configurative method for surer identification of personality dimensions, notably in child study. Psych. Rep., 16: 269–270.
Cattell, R. B. and Cattell, A. K. S., 1955. Factor rotation for proportional profiles, Analytical solution and an example. Brit, J. Stat. Psychol., 8: 83–92.
Cehak, K., 1962. Orthogonale Polynome in der Meteorologie. Arch. Meteorol. Geophys., Bioklimatol., Ser. A., 12: 40–61.
Chakravarti, I. M., Laha, R. G. and Roy, J., 1967. Handbook of Methods of Applied Statistics, I. Wiley, New York, 460 pp.
Chan, L. K., Chan, N. N. and Mead, E. R., 1971. Best linear unbiased estimates of the parameters of the logistic distribution based on selected order statistics. J. Am. Stat. Assoc., 66: 889–892.
Chapman, D. G., 1956. Estimating the parameters of a truncated gamma distribution. Ann. Math. Stat., 27: 498–506.
Chew, V., 1966. Confidence, prediction, and tolerance regions for the multivariate normal distribution. J. A. Stat. Assoc., 61: 605–617.
Choi, K. and Bulgren, W. G., 1968. An estimation procedure for mixtures of distributions. J. R. Stat. Soc., 30: 444–460.
Chow, V. T., 1954. The log-probability law and its engineering application. Proc. Am. Soc. Civ. Eng. 80: 1–25.
Chow, V. T. (Editor), 1964. Handbook of Applied Hydrology. McGraw-Hill, New York, 1453 pp.
Christensen, W. I., Jr. and Bryson, R. A., 1966. An investigation of the potential of component analysis for weather classification. Mon. Weather Rev., 94: 697–709.
Cleveland, W. S., 1972. The inverse autocorrelation of a time series and their applications. Technometrics, 14: 277–293.
Cohen, A. C., 1950. Estimating parameters of type-III populations from truncated samples. J. Am. Stat. Assoc., 45: 411–423.
Cohen, A. C., 1951. Estimation parameters of logarithmic-normal distributions by maximum likelihood. J. Am. Stat. Assoc., 46: 206–212.
Cohen, A. C., 1954. Estimation of the Poisson parameters from truncated samples and from censored samples. J. Am. Stat. Assoc., 49: 158–168.
Cohen, A. C., 1955. Restriction and selection in samples from bivariate normal distribution. J. Am. Stat. Assoc., 50: 884–893.
Cohen, A. C., 1959. Simplified estimators for the normal distribution when samples are singly censored or truncated. Technometrics, 1: 217–237.
Cohen, A. C., 1961a. Estimating the Poisson parameters from samples that are truncated on the right. Technometrics, 3: 433–438.
Cohen, A. C., 1961b. Tables for maximum-likelihood estimates: singly truncated and singly censored samples. Technometrics, 10: 535–541.
Cohen, A. C., 1964. Estimation in mixtures of two Poisson distributions. NASA-TMX-53189: 104–107.
Cohen, A. C., 1969. A generalization of the Weibull distribution. NASA-CR-61293, 21 pp.
Cohen, A. C. and Fall, L. W., 1967. Estimation in mixed frequency distributions. NASA-TN-D-4033, 99 pp.

Comrie, L. J., 1948. Chambers Six Figure Mathematical Tables. Chambers, Edinburgh I: 576 pp; II: 576 pp.
Cooley, C. G. and Cohen, A. C., 1970. Tables of maximum-likelihood estimating functions for singly truncated and singly censored samples from the normal distribution. NASA-CR-61330, 99 pp.
Court, A., 1949. Separating frequency distributions into two normal components. Science, 110: 500–501.
Court, A., 1957a. Maximum variability level of winds. Sci. Rep. No. 2, Contract AF19(604)-2060, 19 pp.
Court, A., 1957b. Vertical correlation of wind components. AFCRL-TN-57-292, 65 pp.
Court, A., 1962. Applying statistical representation of wind. GRD, AFCRL-62-273, I, No. 140, 49–54.
Craddock, J. M. and Flood, C. R., 1969. Eigenvectors for representing the 500 mbar geopotential surface over the Northern Hemisphere. Q. J. R. Meteorol. Soc., 95: 576–593.
Craig, C. C., 1936. A new exposition and charts for the Pearson system of curves. Ann. Math. Stat., 7: 16–28.
Crutcher, H., 1956. Wind aid from wind roses. Bull. Am. Meteorol. Soc., 37: 391–402.
Crutcher, H., 1957. On the standard vector deviation wind rose. J. Meteorol., 14: 28–33.
Crutcher, H., 1959. Upper wind statistic charts of the Northern Hemisphere. Chief Naval Operations, NAVAER-50-16-535, Vol. I and II, 100 pp.
Crutcher, H. and Moses, H., 1962. A note on ellipsoidal wind distributions. Argonne Nat. Lab. Summary Report, ANL-6769: 181–196.
Daeves, K., 1952. Rationalisierung durch Grosszahl-Forschung. Stahl-eisen, Düsseldorf, 199 pp.
Daeves, K. and Beckel, A., 1948. Grosszahlforschung und Häufigkeitsanalyse. Chemie, Weinheim, 60 pp.
Daniell, P. J., 1946. Discussion on "Symposium on autocorrelation in time-series". J. R. Stat. Soc., Ser. B, 8: 88–90.
Davis, H. T., 1963. Tables of the Mathematical Functions. The principia Press of Trinity University, San Antonio, Texas, I, 401 pp., II, 391 pp., III, 549 pp.
Davis, J. M. and Rappaport, P. N., 1974. The use of time-series analysis techniques in forecasting meteorological drought. Mon. Weather Rev., 102: 176–180.
Day, N. E., 1969. Estimating the components of a mixture of normal distributions. Biometrika, 56: 463–474.
Defrise, P., Sneyers, R., Struylaert, W. and Isacker, J., 1972. Climatologue Aerologique. Inst. R. Meteorol. Belg., Brussels, 41 pp.
Des Raj, 1953. Estimation of the parameters of type-III populations from truncated samples. J. Am. Stat. Assoc., 48: 336–349.
Dick, D. C. and Bowden, D. C., 1973. Maximum-likelihood estimation for mixtures of two normal distributions. Biometrics, 29: 781–790.
Dixon, R., 1969. Orthogonal polynomials as a basis for objective analysis. Meteorol. Office, London, Sci. Pap., 30, 20 pp.
Dixon, R., Spackman, E. A., Jones, I. and Francis, A., 1972. The global analysis of meteorological data using orthogonal polynomial base functions. J. Atmos. Sci., 29: 609–622.
Doberitz, R., 1969. Cross-spectrum and filter analysis of monthly rainfall and wind data in the tropical Atlantic region. Meteorol. Abh., 11, 54 pp.
Doetsch, G., 1928. Die Elimination des Doppler Effekts bei spektroskopischen Feinstrukturen und exakte Bestimmung der Komponenten. Phys., 49: 705–730.
Doetsch, G., 1936. Zerlegung einer Funktion in Gauss'sche Fehlerkurven und zeitliche Zurückverfolgung eines Temperaturzustandes. Math. Z., 41: 283–318.

Durbin, J., 1960. The fitting time-series models. Rev. Int. Inst. Stat., 28: 233–244.
Durbin, J., 1970. Testing for serial correlation in least-square regression when some of the regressions are lagged dependent variables. Econometrica, 38: 410–421.
Dutt, J. E., 1973. A representation of multivariate normal probability integrals by integral transforms. Biometrika, 60: 637–645.
Dwyer, P. S., 1941. The solution of simultaneous equations. Psychometrika, 6: 101–129.
Dwyer, P. S., 1945. The square-root method and its use in correlation and regression. J. Am. Stat. Assoc., 40: 493–503.
Edgeworth, F. Y., 1898. On the representation of statistics by mathematical formulae. J. R. Stat. Soc., 61: 670–700.
Elandt, R. C., 1961. The folded normal distribution: two methods of estimating parameters from movements. Technometrics, 3: 551–562.
Elderton, W. P., 1953. Frequency Curves and Correlation. Harren Press, Washington, D. C., 4th ed., 272 pp.
Epstein, B., 1958. The exponential distribution and its role in life testing. Industr. Qual. Control, 15: 4–9.
Essenwanger, O. M., 1954a. Neue Methode der Zerlegung von Häufigkeitsverteilungen in Gauss'sche Normalkurven und Ihre Anwendung in der Meteorologie. Ber. Dtsch. Wetterd., 10: 11 pp.
Essenwanger, O. M., 1954b. Probleme der Häufigkeitsanalyse. Meteorol. Rundsch. 7: 85–88.
Essenwanger, O. M., 1955a. Zur Häufigkeitsanalyse Meteorologischer Beobachtungen. Z. Meteorol., 9: 258–266.
Essenwanger, O. M., 1955b. Zur Häufigkeitsanalyse von Grosswetterlagen. Meteorol. Rundsch. 8: 55–56.
Essenwanger, O. M., 1955c. Zur Realität der Zerlegung von Häufigkeitsverteilungen in Normalkurven. Arch. Meteorol. Geophys., Bioklimatol., Ser. B, 7: 49–59.
Essenwanger, O. M., 1957. Tafeln zur Häufigkeitszerlegung mit Anwendungsbeispielen. Ber. Dtsch. Wetterd., 39 (Bd. 5), 25 pp.
Essenwanger, O. M., 1959. Probleme der Windstatistik. Meteorol. Rundsch., 12: 37–47.
Essenwanger, O. M., 1961. On defining and computing the mean and the standard deviation for wind directions. AOMC Rep. No. RR-TR-61-1, 75 pp.
Essenwanger, O. M., 1964. The cumulative distribution of wind direction frequencies. Meteorol. Rundsch., 17: 131–135.
Essenwanger, O. M., 1967. The negative binomial distribution applied to atmospheric parameters. Proc. 12th Conf. on Design of Experiments in Army Research, Development and Testing, ARO-D Rep. 67-2, p. 221–242.
Essenwanger, O. M., 1968. On deriving 90 to 99 percent wind and wind-shear thresholds from statistical parameters. Proc. 3rd Natl. Conf. on Aerospace Meteorology. Am. Meteorol. Soc., p. 145–153.
Essenwanger, O. M., 1974a. Elements of Statistical Analysis. Also: World Survey of Climatology, 3, Elsevier, Amsterdam.
Essenwanger, O. M., 1974b. Applied Statistics in Atmospheric Science. B. Smoothing and Filtering, Analysis of Variance, Tests, Meteorological Elements. Elsevier, Amsterdam (in preparation).
Essenwanger, O. M., Horn, L. H. and Bryson, R. A., 1958. Half-hemispheric 500 mbar topography description by means of orthogonal polynomials, 2. Dep. Meteorol. Univ. Wisc., 46 pp.
Essenwanger, O. M., Bradford, R. E. and Vaughan, W. W., 1961. On verification of upper-air winds by vertical shear and extremes, Mon. Weather Rev., 89: 197–204.
Fall, L., 1970. Estimation of parameters in compound Weibull distributions. Technometrics, 12: 399–407.

Feller, W., 1940. On the logistic law of growth and its empirical vertification in biology. Acta Biotheoretica, 5: 51–66.
Feller, W., 1966. An Introduction to Probability and Its Applications, II. Wiley, New York, 669 pp.
Feller, W., 1968. An Introduction to Probability and Its Applications, I. Wiley, New York, 510 pp.
Finney, D. J., 1949. The truncated binomial distribution. Ann. Eugen., 14: 319–328.
Finney, D. J. and Varley, G. C., 1955. An example of the truncated Poisson distribution. Biometrics, 11: 387–394.
Fisher, R. A., 1931. The truncated normal distribution. Brit. Assoc. Adv. Sci., Math. Tables, p. 33–34.
Fisher, R. A., 1950. Statistical Methods for Research Workers. Oliver and Boyd, Edinburgh, 11th ed., 354 pp.
Fisher, R. A. and Yates, F., 1963. Statistical Tables for Biological, Agricultural and Medical Research. Hafner, New York, 6th ed., 146 pp.
Fisher, R. A., Corbet, A. S. and Williams, C. B., 1943. The relation between an animal population. J. Anim. Ecol., 12: 42–58.
Fisz, M., 1963. Probability Theory and Mathematical Statistics. Wiley, New York, 677 pp.
Fox, A. J., 1972. Outliers in time series. J. R. Stat. Soc., B, 35: 350–363.
Francis, P. E., 1972. The possible use of Laguerrs polynomials for representing the vertical structure of numerical models of the atmosphere. Q. J. R. Meteorol. Soc., 98: 662–667.
Freeman, M. F. and Tukey, J. W., 1950. Transformations related to the angular and the square root. Ann. Math. Stat., 21: 607–611.
Frenkiel, F. N., 1951. Frequency distributions of velocities in turbulent flow. J. Meteorol., 8: 316–320.
Friedman, D. G. and Jones, B. E., 1957. Estimation of rainfall probabilities. Rep. Connecticut Agric. Exp. Stat., 3: 21 p.
Fryer, J. G. and Robertson, C. A., 1972. A comparison of some methods for estimating mixed distribution. Biometrika, 59: 639–648.
Garnett, J. C. M., 1919. A certain independent factor in mental measurements. Proc. R. Soc. London, 46: 91–111.
General Electric Company, Defence Systems Department, 1962. Tables of the Individual and Cumulative Terms of Poisson Distribution. Van Nostrand. Princeton, N. J., 202 pp.
Gilliland, D. C., 1962. Integral of the bivariate normal distribution over an offset circle. J. Am. Stat. Assoc., 57: 758–768.
Gilman, D. L., 1957. Empirical orthogonal functions applied to thirty-day forecasting. M.I.T. Dep. Meteorol. Sci. Rp. 1: 129 pp. Clearinghouse for Federal Scientific Technical Information, US Dep. of Commerce, Washington, D. C., AD-117249.
Glahn, R. R., 1962. An experiment in forecasting rainfall probabilities by objective methods. Mon. Weather Rev., 90: 59–67.
Gnanadesikan, R., Pinkham, R. S. and Hughes, L. P., 1967. Maximum-likelihood estimation of the parameters of the beta distribution from smallest order statistics. Technometrics, 9: 607–620.
Godske, C. L., 1962. Contributions to statistical meteorology I, Geophys. Norv, 24: 161–210.
Godske, C. L., 1965. Statistics of meteorological variables. Final Report, Univ. of Bergen, 115 pp.
Godske, C. L., 1966. On the time-dependence of smoothed variables. Tellus, 18: 714–721.
Godske, C. L., 1967. Further studies of statistical meteorology. Final Report. Univ. of Bergen, 71 pp.

Gray, A. and Mathews, G. B., 1966. A Treatise on Bessel Functions and Their Application to Physics. Macmillan, London. 2nd edition by Gray and Macrobert, Dover, New York, 327 pp.

Graybill, F. A., 1961. An Introduction to Linear Statistical Models. McGraw-Hill, New York, 463 pp.

Graybill, F. A., 1969. Introduction to Matrices with Application in Statistics. Wadsworth, Belmont, Cal., 372 pp.

Greenwood, J. A. and Durand, D., 1960. Aids for fitting the gamma distribution by maximum likelihood. Technometrics, 2: 55–65.

Gregor, J., 1969. An algorithm for the decomposition of a distribution into Gaussian component. Biometrics, 25: 79–93.

Grimmer, M., 1963. Space filtering of monthly surface anomaly data in terms of patterns, using empirical orthogonal functions. Q. J. R. Meteorol. Soc., 89: 395–408.

Gross, A. J., 1970. A note on the convolution of Poisson distributions with the zero-class missing. Am. Stat., 24: 42–43.

Gumbel, E. J., 1944. Ranges and midranges. Ann. Math. Stat., 15: 414–422.

Gumbel, E. J., 1961. Bivariate logistic distribution, J. Am. Stat. Assoc., 56: 335–349.

Gupta, A. K., 1952. Estimation of the mean and standard deviation of a normal population from a censored sample. Biometrika, 39: 260–273.

Gupta, S. S. and Gnanadesikan, R., 1966. Estimation of the parameters of the logistic distribution. Biometrika, 53: 565–570.

Gupta, S. S. and Shah, B. K., 1965. Exact moments and percentage points of the order statistics and the distribution of the range from the logistic distribution. Ann. Math. Stat., 36: 907–920.

Gupta, S. S., Qureishi, A. S. and Shah, B. K., 1967. Best linear unbiased estimators of the parameters of the logistic distribution using order statistics. Technometrics, 9: 43–56.

Guss, H., 1951. Zur Struktur von Häufigkeitsverteilung und Bildung von Mittel-und Schwankungswerten der Sichtweite. Ann. Meteorol., 4: 27–39.

Guss, H., 1955. Uber die Bildung typischer Mittel- und Schwankungswerte in der Klimatologie. Ann. Meteorol. 7: 127–133.

Guttman, I., 1956. 'Best possible' systematic estimates of communalities. Psychometrika, 21: 273–285.

Haas, G., Bain, L. and Antle, C. H., 1970. Inferences for the Cauchy distribution based on maximum-likelihood estimators. Biometrika, 57: 403–408.

Hahn, G. J., 1970. Statistical intervals for a normal population. I, Tables, examples and application. J. Qual. Techn., 2: 115–125; II, Formulas, assumptions, some derivations. J. Qual. Techn., 2: 195–206.

Hald, A., 1952. Statistical Tables and Formulas. Wiley, New York, 97 pp.

Hald, A., 1957. Statistical Theory with Engineering Application. Wiley, New York, 783 pp.

Hald, A., 1968. The mixed binomial distribution and the posterior distribution of p for a continuous prior distribution. J. R. Stat. Assoc., 30: 359–367.

Haldane, I. B. S., 1941. The fitting of binomial distributions. Ann. Eugenics, 11: 179–181.

Hamming, R. W., 1971. Introduction to Applied Numerical Analysis. McGraw-Hill, New York, 331 pp.

Hamming, R. W. and Pinkham, R. S., 1966. A class of integration formulas, J. Am. Comput. Mach., 13: 430–438.

Hannan, E. J., 1970. Multiple Time Series. Wiley, New York, 536 pp.

Hannes, G., 1974. Factor analysis of coastal air and water temperatures. J. Appl. Meteorol., 13: 3–7.

Harman, H. H., 1967. Modern Factor Analysis. Univ. of Chicago Press, 2nd ed., 474 pp.

Harter, H. L., 1960. Circular error probabilities. J. Am. Stat. Assoc., 55: 723–731.

Harter, H. L., 1967. Maximum-likelihood estimation for the parameters of a four-param-

eter generalized gamma population from complete and censored samples. Technometrics, 9: 159–165.
Harter, H. L. and Moore, A. H., 1965. Maximum-likelihood estimation of the parameters of gamma and Weibull populations from complete and censored samples. Technometrics, 7: 639–643.
Hartley, H. O., 1949. Task of significance in harmonic analysis. Biometrika, 36: 194–201.
Hartley, H. O. and Fitch, E. R., 1951. A chart for the incomplete beta function and the cumulative binomial distribution. Biometrika, 38: 423–426.
Hassanein, K. M., 1971. Percentile estimators for the parameters of the Weibull distribution. Biometrika, 58: 673–676.
Haugen, D. A. (Editor), 1973. Workshop on Micrometeorology. Am. Meteorol. Soc., Boston, 392 pp.
Hesselberg, T. and Bjorkdal, E., 1929. Über das Verteilungsgesetz der Windunruhe. Beitr. Phys. Freien Atmos., 15: 121–133.
Heyde, C. C., 1963. On a property of the lognormal distribution. J. R. Stat. Soc., 25: 392–393.
Hoel, P. G., 1966. Introduction to Mathematical Statistics. Wiley, New York, 6th printing, 427 pp.
Hogg, R. V. and Craig, A. T., 1965. Introduction to Mathematical Statistics. MacMillan, New York, 6th printing, 1967, 383 pp.
Holmström, I., 1963. On a method for parametric representation of the state of the atmosphere. Tellus, 15: 127–149.
Holmström, I., 1970. Analysis of time series by means of empirical orthogonal functions. Tellus, 22: 638–647.
Ihm, P., 1960. Zur numerischen Integration n-dimensionaler Normalverteilungen. Metrika, 3: 74–78.
Jacobi, K. G. J., 1846. Uber ein leichtes Verfahren, die in der Theorie der Säkularstorungen vorkommenden Gleichungen numerisch aufzulösen. J. Reine Angew. Math., 30: 51–95.
Jaquez, R., 1962. K-Factors for computing tolerance limits for normal distributions. Industr. Qual. Control., 19: 27–28.
John, S., 1968. A central tolerance region for the multivariate normal distribution. J. R. Stat. Assoc., 30: 599–601.
Johnson, N. L., 1949. Systems of frequency curves generated by method of translation. Biometrika, 36: 139–176.
Johnson, N. L., 1962. The folded normal distribution: accuracy of estimation by maximum likelihood. Technometrics, 4: 249–256.
Johnson, N. L. and Kotz, S., 1969. Discrete Distributions. Houghton Mifflin, Boston, 328 pp.
Johnson, N. L. and Kotz, S., 1970a. Continuous Univariate Distribution-1. Houghton Mifflin, Boston, 330 pp.
Johnson, N. L. and Kotz, S., 1970b. Continuous Univariate Distributions-2. Houghton Mifflin, Boston, 306 pp.
Johnson, N. L. and Kotz, S., 1972a. Continuous Multivariate Distributions. Houghton Mifflin, Boston, 333 pp.
Johnson, N. L. and Kotz, S., 1972b. Power transformations of gamma variables. Biometrika, 59: 226–229.
Jones, R. H., 1964. Spectral analysis and linear prediction of meteorological time series. J. Appl. Meteorol., 3: 45–52.
Jones, R. H., 1965. A reappraisal of the periodogram in spectral analysis. Technometrics, 7: 531–542.
Jones, R. H., 1970. Spectrum estimation with unequally spaced observations. Proc., Kyoto Int. Conf. Circuit and System Theory, p. 253–254.

REFERENCES

Jones, R. H., 1971. Spectrum estimation and time-series analysis, a review. Int. Symp. Probability and Statistics in Atmospheric Science, Honolulu, 1971, Am. Meteorol. Soc., p. 35–40.

Jones, R. H., 1972. Aliasing with unequally spaced observations. J. Appl. Meteorol., 11: 245–254.

Joseph, E. S., 1973. Time-series analysis of annual temperatures. Mon. Weather Rev., 101: 501–505.

Julian, P. R., 1971. Some aspects of variance spectra of synoptic scale, tropospheric wind components in mid-latitudes and in the tropics. Mon. Weather Rev., 99: 954–965.

Kamat, A. R., 1962. Some more estimates of circular probable error. J. Am. Stat. Assoc., 57: 191–195.

Kao, S. K. and Woods, H. D., 1964. Energy spectra of meso-scale turbulence along and across the jet stream. J. Atmos. Sci., 21: 513–519.

Kendall, M. G. and Stuart, A., 1961. The Advanced Theory of Statistics, Hafner, New York, I, 433 pp, II, 676 pp.

Kenney, J. F. and Keeping, E. S., 1954. Mathematics of Statistics. Van Nostrand, Princeton, N. J., 3rd ed., I, 348 pp., II, 429 pp.

Khatri, C. G. and Jaiswal, M. C., 1963. Estimation of parameters of a truncated bivariate normal distribution. J. Am. Stat. Assoc., 58: 519–526.

Kirkpatrick, R. L., 1970. Confidence limits on a percent defective characterized by two specification limits. J. Qual. Tech., 2: 150–155.

Knighting, E., 1954. Upper winds over the world, correspondence. Q. J. R. Meteorol. Soc., 80: 239–240.

Koopmans, L. H., Owen, D. B. and Rosenblatt, J. I., 1964. Confidence intervals for the coefficient of variation for the normal and lognormal distributions. Biometrika, 51: 25–32.

Kramer, C. Y. and Jensen, D. R., 1969. Fundamentals of multivariate analysis, I. Inference about means. J. Qual. Tech., 1: 120–133.

Kutzbach, J. E., 1966. Representation and Classification of Fields of Atmospheric Variables. Thesis, Univ. of Wisconsin, 125 pp.

Kutzbach, J. E., 1967. Empirical eigenvectors of sea-level pressure, surface temperature and precipitation complexes over North America. J. Appl. Meteorol, 6: 791–802.

Kutzbach, J. E., 1970. Large-scale features of monthly mean Northern Hemisphere anomaly maps of sea-level pressure. Mon. Weather Rev., 98: 708–716.

Kutzbach, J. E. and Wahl, E. W., 1965. The representation of scalar fields with functions orthogonal in polar coordinates. J. Appl. Meteorol, 4: 542–544.

Lambert, J. A., 1964. Estimation of parameters in the three-parameter lognormal distribution. Aust. J. Stat., 6: 29–32.

Lambert, J. A., 1970. Estimation of parameters in the four-parameter lognormal distribution. Aust. J. Stat., 12: 33–43.

Landsberg, H., Mitchell, J. and Crutcher, H., 1959. Power-spectrum analysis of climatological data for Woodstock College, Maryland. Mon. Weather Rev., 87: 283–298.

Laurent, A. G., 1963. The lognormal distribution and the translation method: description and estimation problems. J. Am. Stat. Assoc., 58: 231–235 (correction 1163).

Lawley, D. N., 1956. Tests of significance for the latent roots of covariance and correlation matrices. Biometrika, 43: 128–136.

Lawley, D. N. and Maxwell, A. E., 1973. Regression and factor analysis. Biometrika, 60: 331–338.

Leone, F. C., Nelson, L. S. and Northingham, R. B., 1961. The folded normal distribution. Technometrics, 3: 543–550.

Levinson, N., 1949. The Weiner RMS Error Criterion in Design, Appendix B in Extrapolation, Interpolation, and Smoothing of Stationary Time Series (N. Wiener). M.I.T. Press, Cambridge, Mass., 163 pp.

Lilliefors, H. W., 1969. On the Kolmogorov-Smirnov test for the exponential distribution with mean unknown. J. Am. Stat. Assoc., 64: 387–389.

Linder, A., 1960. Statistische Methoden für Naturwissenschaftler, Mediziner und Ingenieure. Birkhäuser, Basel, 484 pp.

Lorenz, E. N., 1956. Empirical orthogonal functions and statistical weather prediction. M.I.T. Dep. of Meteorology, Sci. Rep. 1, 49 pp. (Clearinghouse for Federal Sci. and Tech. Information, US Dep. of Commerce, Washington, D. C., AD-110268).

Lyness, F. K. and Badger, E. H. M., 1970. A measure of winter severity. J. R. Stat. Soc., C, 19: 119–134.

Malone, T. F., 1956. Studies in synoptic climatology, Final Rep. N5-ORI-07883, Dep. Meteorol. M.I.T., 47 pp.

Mauchly, J. W., 1940. A significance test for ellipticity in the harmonic dial. Terr. Magn. and Atmos. Electr., 45: 145–148.

McElrath, G. W. and Bearman, J. E., 1959. Some economic considerations of inefficient statistics, Indust. Qual. Control., 16: 10–14.

Mielke, P. W., Jr., 1973. Another family of distributions for describing and analyzing precipitation data. J. Appl. Meteorol., 12: 275–280.

Molina, E. C., 1932. An expansion for Laplacian integrals in terms of incomplete gamma functions, and some applications. Bell System Tech. J. 11: 563–575.

Monin, A. S., 1962. On turbulence spectrum in temperature-inhomogeneous atmosphere. Izv. USSR, Geophys. Ser., 3.

Monin, A. S. and Vulis, I. L., 1971. On the spectra of long-periodic oscillations of geophysical parameters. Tellus, 23: 337–345.

Mood, A. M. F., 1950. Introduction to the Theory of Statistics. McGraw-Hill, New York, 433 pp.

Moranda, P. B., 1960. Effect of bias on estimates of the circular probable error. J. Am. Stat. Assoc., 55: 732–735.

Mostafa, M. D. and Mahmoud, M. W., 1964. On the problem of estimation for the bivariate lognormal distribution. Biometrika, 51: 522–527.

Mosteller, F., 1946. On some useful "inefficient" statistics. Ann. Math. Stat., 17: 377–408.

Muller, F. B. and Clodman, J., 1968. The application of empirical orthogonal functions to the analysis of meso-scale variations of precipitations in storms. Proc. 1st Stat. Meteorol. Conf., Hartford, Conn., Am. Meteorol. Soc., p. 116–123.

Munro, A. G. and Wixley, R. A. J., 1970. Estimators based on order statistics of small samples from a three-parameter lognormal distribution. J. Ann. Stat. Assoc., 65: 212–225.

National Bureau of Standards, 1959. Tables of the Bivariate Normal Distribution Function and Related Functions. National Bureau of Standards, Math. Ser., 50, 258 pp.

Nelder, J. A., 1961. The fitting of a generalization of the logistic curve. Biometrics, 17: 89–109.

Nelson, W., 1967. The truncated normal distribution — with application to component sorting. Ind. Qual. Control, 7: 261–271.

Nelson, W. C. and David, H. A., 1967. The logarithmic distribution: A review. Va. J. Sci., 18: 95–102.

Neumann, C. G., 1867. Theorie der Bessel'schen Funktionen. Ein Analogon zur Theorie der Kugel funktionen. Leipzig, 107 pp.

Neyman, J. and Pearson, E. S., 1933. The problem of the most efficient tests of statistical hypotheses. Philos. Trans. R. Soc. London, Ser. A, 231: 289–337.

Nicholson, C., 1943. The probability integral for two variables. Biometrika, 33: 59–72.

Nordo, J., 1959. Expected skill of long-range forecast when derived from daily forecasts and past weather data. Nor. Meteorol. Inst., Oslo, Sci. Rep., 4, 15 pp.

Nordo, J., 1966. Significance of statistical relations derived from geophysical data. Tellus, 18: 39–53.

Obukhov, A. M., 1960. The statistically orthogonal expansion of empirical functions. Bull. Acad. Sci. USSR Geophys. Ser., 1: 288–291 (Engl. transl.).

Ogawa, J., 1951. Contributions to the theory of systematic statistics I. Osaka J. Math., 3: 175–213.

Owen, D. B., 1956. Tables for computing bivariate normal probabilities. Ann. Math. Stat., 27: 1075–1090.

Owen, D. B., 1962. Handbook of Statistical Tables. Addison-Wesley, Reading, Mass., 580 pp.

Pagano, M., 1972. An algorithm for fitting autoregressive schemes. Appl. Stat., 21: 274–281.

Palmieria, S., 1968. A study of the space relationships of meteorological files over Northern Italy by means of principal components analysis. Rev. Meteorol. Aeron., 28: 5–19.

Panofsky, H. A., Cramer, H. E. and Rao, V. R., 1958. The relation between Eulerian time and space spectra. Q. J. R. Meteorol. Soc., 84: 270–273.

Parzen, E., 1957a. Optimum consistent estimates of the spectrum of a stationary time-series. Ann. Math. Stat., 28: 2.

Parzen, E., 1957b. On choosing an estimate of the spectral density function of a stationary time-series. Ann. Math. Stat., 28: 4.

Parzen, E., 1961. Mathematical considerations in the estimation of spectra. Technometrics 3: 166–190.

Patil, E. P., 1961. Some methods of estimation for the logarithmic series distribution. Biometrics, 18: 68–75.

Pearson, E. S. and Hartley, H. O., 1958. Biometrika Tables for Statisticians, I. Cambridge Univ. Press, 240 pp.

Pearson, K., 1894. Contribution to the mathematical theory of evolution. Philos. Trans. R. Soc. London, Ser. A, 185: 71–110.

Pearson, K., 1895. Contribution to the mathematical theory of evolution. Philos. Trans. R. Soc. London, Ser. A, 186: 343–414.

Pearson, K., 1931. Tables for Statisticians and Biometricians, 2. Cambridge University Press, 214 pp.

Pearson, K., 1956. Tables of the Incomplete Beta Function. Cambridge University Press, London, 2nd ed., 494 pp.

Peizer, D. B. and Pratt, J. W., 1968. A normal approximation for binomial, F. Beta, and other common, related tail probabilities, I. J. Am. Stat. Assoc., 63: 1416–1456.

Pennington, R. H., 1970. Introductory Computer Methods and Numerical Analysis. MacMillan, London, 2nd ed., 496 pp.

Peto, R. and Lee, P., 1973. Weibull distribution for continuous-carcinogenesis experiments. Biometrics, 29: 457–470.

Pierce, D. A., 1971. Distribution of residual autocorrelations in the regression model with autoregressive moving average errors. J. R. Stat. Soc., Ser. B, 33: 140–146.

Pierce, D. A. and Dykstra, R. L., 1969. Independence and the normal distribution. Am. Stat., 23 (4): 39 p.

Pinus, N. Z., Reiter, E. R., Shur, G. N. and Vinnichanko, N. K., 1967. Power spectra of turbulence in the free atmosphere. Tellus, 19: 206–213.

Pratt, J. W., 1968. A normal approximation for binomial, F. Beta, and other common, related tail probabilities, II. J. Am. Stat. Assoc., 63: 1457–1483.

Prescott, P., 1970. Estimation of the standard deviation of a normal population from doubly censored samples using normal scores. Biometrika, 57: 409–419.

Priestly, M. B., 1962. Basic considerations in the estimation of spectra. Technometrics, 4: 551–564.

Ralston, A. and Wilt, H. S. (Editors), 1967. Mathematical Methods for Digital Computers. Wiley, New York, I: 293 pp., II: 287 pp.

Rao, C. R., 1948. Tests of significance in multivariate analysis. Biometrika, 35: 58–79.
Rao, C. R., 1952. Advanced Statistical Methods in Biometric Research. Wiley, New York, 390 pp.
Reeves, J. E., 1972. The distribution of the maximum-likelihood estimator of the parameters in the first-order autoregressive series. Biometrika, 59: 387–394.
Rider, P. R., 1955. Truncated binomial and negative binomial distributions. J. Am. Stat. Assoc., 50: 877–883.
Rider, P. R., 1962. Estimating the parameters of mixed Poisson, binomial, and Weibull distributions by the method of moments. Bull. Int. Stat. Inst., 39: 225–232.
Robertson, C. A. and Fryer, G. G., 1970. The bias and accuracy of moment estimators. Biometrika, 57: 57–65.
Rosenbaum, S., 1961. Moments of a truncated bivariate normal distribution. J. R. Stat. Soc., Ser. B, 23: 405–508.
Rosenthal, G. W. and Rodden, J. J., 1961. Tables of the integral of the elliptical bivariate normal distribution over offset circles. Lockheed Missile and Space Division, Rep. LMSD-800618, 85 pp.
Roughan, J. L. and Evans, H. H., 1970. Editing time series. Austr. J. Stat., 12: 141–149.
Ruben, H., 1960. Probability content of regions under spherical normal distribution. Ann. Math. Stat., 31: 598–618.
Samiuddin, M., 1970. On a test for an assigned value of correlation in a bivariate normal distribution. Biometrika, 57: 461–464.
Sampford, M. R., 1955. The truncated negative binomial distribution. Biometrika, 42: 58–69.
Schafer, R. E. and Sheffield, T. S., 1973. Inferences on parameters of the logistic distribution. Biometrics, 29: 449–455.
Schips, B. and Stier, W., 1972. Autokorrelations Tests und autoregressive Schätzverfahren. Z. Angew. Math. Mech, 52: 273–282.
Schneider-Carius, K. and Essenwanger, O., 1955. Eigentümlichkeiten der Niederschlagsverhältnisse im Norden and Süden der Schweizer Alpen, dargestellt durch die Niederschlagswahrscheinlichkeit von Basel, St. Gotthard und Lugana. Arch. Meteorol., Geophys., Bioklimatol, B, 7: 32–48.
Sellers, W. D., 1957. A statistical-dynamic approach to numerical weather prediction. M.I.T. Dep. of Meteorology, Sci. Rep. 2, 151 pp. (Clearinghouse for Federal Sci. and Tech. Information, US Dep. of Commerce, Washington, D. C., AD-117231).
Selvin, J., 1971. Maximum-likelihood estimation in the truncated, single parameters, discrete exponential family. Am. Stat., 25: 41–42.
Severo, N. C. and Olds, E. G., 1956. A comparison of tests on the mean of a logarithmic normal distribution with known variance. Ann. Math. Stat., 27: 670–686.
Shah, S. M., 1966. On estimating the parameters of a doubly truncated binomial distribution. J. Am. Stat. Assoc., 61: 259–263.
Shapiro, H. S. and Silverman, R. A., 1960. Alias-free sampling of random noise, J. Soc. Indust. Appl. Math., 8: 225–248.
Shenton, L. R., 1965a. Transforming non-normal distributions into nearly normal distributions. 1, Description of the general approach to the problem. Univ. Georgia, College of Agriculture, Tech. Bull., 49, 36 pp.
Shenton, L. R., 1965b. Transforming non-normal distributions into nearly normal distributions. 2, Pearson Type-III distributions. Univ. Georgia, College of Agriculture, Tech. Bull., 50, 48 pp.
Shih Yung-Nien, 1965. An experiment on the quantitative description of climatic element fields by orthogonal functions. Acta Meteorol. Sinica, 35: 343–351.
Sneyers, R., 1962. Sur l'emploi de distributions normales tronquées en climatologie, WMO Tech. Note 71, WMO-No. 178, TP 88, p. 177–183.

REFERENCES

Spearman, C., 1904. General intelligence objectively determined and measured. Am. J. Psychol., 15: 201–293.
Spearman, C., 1927. The Abilities of Man. MacMillan, London, 87 pp.
Stein, C. M., 1962. Confidence sets for the mean of a multivariate normal distribution. J. R. Stat. Soc., Ser. B, 24: 265–296.
Stidd, C. K., 1967. The use of eigenvectors for climatic estimates. J. Appl. Meteorol., 6: 255–264.
Stigant, S. A., 1959. The Elements of Determinants, Matrices and Tensors for Engineers. MacDonald, London, 433 pp.
Stralkowski, C. M., Wu, S. M. and DeVor, R. E., 1970. Charts for the interpretation and estimation of the second-order autoregressive model. Technometrics, 12: 669–685.
Stumpf, F. K., 1937. Grundlagen und Methoden der Periodenforschung. Springer, Berlin, 332 pp.
Stumpf, H., 1972. Zur numerischen Berechnung zweiseitiger Schranken für beliebige vektorielle und tensorielle Feldgrössen elastischer Eigenschwingungszustände, Z. Angew. Math. Mech., 52: 37–44.
Stumpf, H., 1973. Fehlerabschätzung bei der angenäherten Berechnung von Eigenwerten und Eigenzustandsgrössen. Z. Angew. Math. Mech, 53: 182–184.
Susuki, E., 1964. Hyper gamma distribution and its fitting to rainfall data. Pap. Meteorol. Geophys., 15: 31–51.
Tate, R. F. and Goen, R. L., 1958. Minimum variance unbiased estimation for the truncated Poisson distribution. Ann. Math. Stat., 29: 755–765.
Taubenheim, J., 1969. Statistische Auswertung geophysikalischer und meteorologischer Daten. Akademische Verlagsgesellschaft, Leipzig, 386 pp.
Teich, M., 1971. Statitische Analyse der Kältesummen von Berlin. Meteorol. Rundsch., 24: 1–7.
Tennekes, H. and Lumley, J. L., 1972. A First Course in Turbulence. M.I.T. Press, Cambridge, 300 pp.
Thom, H. C. S., 1968. Approximate convolution of the gamma and mixed gamma distributions. Mon. Weather Rev., 96: 883–886.
Thomson, D. H., 1947. Approximate formulae for the percentage points of the incomplete beta function and of the χ^2 distribution. Biometrika, 34: 368–372.
Thöni, H., 1969. A table for estimating the mean of a lognormal distribution. J. Am. Stat. Assoc., 64: 632–636.
Tucker, G. B., 1960. Upperwinds over the world, III. Standard vector deviation of the wind up to the 100-millibar level of the world. Air Ministry, London, Geophys. Mem. 105, 101 pp.
Tukey, J. W., 1949. One degree freedom for non-additivity. Biometrics, 5: 232–242.
Tukey, J. W., 1961. Discussion emphasizing the connection between variance and spectrum analysis. Technometrics, 3: 191–218.
Tukey, J. W., 1962. The future of data analysis. Ann. Math. Stat., 33: 1–67.
Wagle, B., 1968. Multivariate beta distribution and a test for multivariate normality. J. R. Stat. Soc., Ser. B, 30: 511–516.
Wallace, J. M., 1972. Empirical orthogonal representation of time series in the frequency domain, II. Application to the study of tropical wave distrubances. J. Appl. Meteorol., 11: 893–900.
Wallace, J. M. and Dickinson, R. E., 1972. Empirical orthogonal representation of time series in the frequency domain, I. Theoretical considerations. J. Appl. Meteorol., 11: 887–892.
Watson, J. N., 1958. A treatise on the Theory of Bessel Functions. Cambridge Univ. Press, 804 pp.
Weil, H., 1954. The distribution of radial error. Ann. Math. Stat., 25: 168–170.

Weinberger, H. F., 1960. Error bounds in the Rayleigh-Ritz approximation of eigenvectors. J. Res. Natl. Bur. Stand., 64 (B): 217–225.
Weingarten, H. and DiDonato, A. R., 1961. A table of generalized circular error. Math. Comput., 15: 169–173.
White, R. M., Cooley, D. S., Derby, R. C. and Seaver, F. A., 1958. The development of efficient linear statistical operators for the prediction of sea-level pressure. J. Meteorol., 15: 426–434.
Whittaker, E. and Robinson, G., 1944. The Calculus of Observations. Blackie, London, 4th ed., 397 pp.
Wichern, D. W., 1973. The behavior of the sample autocorrelation function for an integrated moving average process. Biometrika, 60: 235–239.
Wilk, M. B., Gnanadesikan, R. and Huyett, M. J., 1962. Probability plots for the gamma distribution. Technometrics, 4: 1–20.
Wilkinson, J. H., 1965. The Algebraic Eigenvalues Problem. Oxford Clarendon Press, 662 pp.
Williams, C. B., 1944. Some applications of the logarithmic series and the index of diversity to ecological problems. Ecology, 32: 1–44.
Williams, C. B., 1952. Sequences of wet and dry days considered in relation to the logarithmic series. Q. J. R. Meteorol. Soc., 78: 91–96.
Wilson, E. B. and Hilferty, M. M., 1931. The distribution of Chi-square. Proc. Natl. Acad. Sci., Wash., 17: 684–690.
Wilt, H. S., 1962. Mathematics for the Physical Sciences. Wiley, New York, 284 pp.
Wise, M. E., 1960. On normalizing the incomplete beta function for fitting to dosage–response curves. Biometrika, 47: 173–175.
Yuan, P. T., 1933. On the logarithmic frequency distribution and the semi-logarithmic correlation surface. Ann. Math. Stat., 4: 30–74.
Yudin, M. I., 1966. Application of natural orthogonal functions to atmospheric dynamics and thermal regime studies. Proc. Symp. Arctic Heat Budget and Atmospheric Circulation. Memo RM-5233-NSF, Rand Corp., Santa Monica, Calif., p. 345–368.
Zelen, M. and Severo, N. C., 1960. Graphs for bivariate normal probabilities. Ann. Math. Stat., 31: 619–624.
Zurmühl, R., 1958. Matrizen. Springer, Berlin, 467 pp.

AUTHOR INDEX

Abramowitz, 102, 117, 118, 196, 205, 287, 323
Aitchison, 27, 31
Aitkin, 91
Alihouse, 275
Anscombe, 323
Antle, 36
Aroian, 69

Badger, 27, 28
Bain, 36
Barnett, 36
Bartlett, M.S., 238, 239, 240, 269, 283, 284, 302, 318, 323
Bartlett, R.P., 9
Bauer, 275
Bearman, 8
Beckel, 150, 160
Behboodian, 159
Beyer, 167, 196, 322
Bhattacharya, 160, 161, 162, 164, 165, 174
Birnbaum, 9, 132
Bjorkdal, 83
Blackman, 236, 240
Blischke, 104, 184, 185
Bloch, 36
Bloomfield, 298, 310, 315
Boas, 381
Bolgiano, 241, 242
Bowden, 160
Bowman, 114
Box, 296, 299, 300, 302, 307
Bradford, 190
Bradley, 14
Brandtner, 152
Broadbent, 27
Brooks, 64, 73, 83, 93, 101, 105, 109, 111
Brouwer, 304
Brown, J.A.C., 27, 31
Brunk, 19
Bryson, 224, 225, 276, 286
Buell, 276, 287
Bulgren, 144, 159
Bundgaard, 276

Burns, 242
Burt, 276, 283, 284

Cadwell, 95, 96
Carruthers, 64, 73, 83, 93, 101, 105, 109, 111
Carter, 86
Cattell, A.K.S., 285
Cattell, R.B., 276, 280, 285, 287
Cehak, 224
Chakravarti, 183
Chan, L.K., 79
Chan, N.N., 79
Chapman, 139, 141, 142
Chew, 86, 105
Choi, 144, 159
Chow, 26, 27, 63
Christensen, 276, 286
Cleveland, 306, 307, 309
Clodman, 255, 270, 271, 272, 273
Cohen, 25, 32, 117, 125, 126, 127, 133, 136, 137, 138, 139, 150, 185, 187
Comrie, 82
Cooley, 125, 126, 127, 255
Corbet, 114
Court, 93, 147
Craddock, 255
Craig, A.T., 83, 89
Craig, C.C., 58
Cramer, 241
Crone, 275
Crutcher, 83, 93, 101, 105, 109, 241, 242

Daeves, 150, 160
Daniell, 240
David, H.A., 115
Davis, H.T., 44, 67
Davis, J.M., 315
Day, 179, 183
Defrise, 14
Derby, 255
Des Raj, 132, 139, 141
DeVor, 300, 307
Dewar, 101
Dick, 160

Dickinson, 274, 275, 315
DiDonato, 98
Dixon, 224
Doberitz, 242
Doetsch, 145, 148, 152, 153, 155, 174, 175
Durand, 119
Durbin, 296, 303, 304
Durst, 83, 101, 105
Dutt, 86
Dwyer, 379
Dykstra, 89

Edgeworth, 317
Elandt, 191, 192, 193
Elderton, 58, 67
Epstein, 114
Essenwanger, 1, 5, 6, 9, 26, 38, 48, 57, 73, 80, 83, 84, 86, 101, 111, 114, 116, 119, 121, 127, 144, 145, 150, 152, 165, 190, 224, 226, 241, 250, 252, 269, 279, 291, 318
Evans, 312, 313, 314, 315

Fall, 185, 187
Feller, 15, 76
Finney, 133, 138
Fisher, 114, 124, 150, 198
Fisz, 312
Fitch, 69
Fleming, 275
Flood, 255
Fox, 310, 311, 312
Francis, A., 224
Francis, P.E., 224
Freeman, 324
Frenkiel, 96
Friedman, 183
Fryer, 145, 150, 160

Garnett, 276
Gilliand, 101, 103, 104
Gilman, 255
Glahn, 255
Gnanadesikan, 45, 47, 79
Godske, 308, 309, 310
Goen, 138
Gray, 251
Graybill, 366, 368, 381
Greenwood, 119
Gregor, 152, 155
Grimmer, 255

Gross, 186
Gumbel, 26, 76
Gupta, A.K., 128, 129
Gupta, S.S., 79, 126
Guss, 22
Guttman, 281

Haas, 36
Hahn, 9
Hald, 96, 97, 124, 150, 167, 184, 322
Haldane, 136
Hamming, 204
Hannan, 294
Hannes, 276
Harman, 276
Harter, 98, 116, 118
Hartley, 69, 196, 237
Hassanein, 118
Haugen, 242
Hesselberg, 83
Heyde, 25
Hilferty, 322
Hoel, 83
Hogg, 83, 89
Holmström, 255, 273, 274
Holstein, 19
Horn, 224, 225
Hughes, 45, 47
Hume, 91
Huyett, 47

Ihm, 86, 105
Isacker, 14

Jacobi, 282
Jaiswal, 129, 130, 131, 132
Jaquez, 9
Jenkins, 299, 300, 302, 307
Jensen, 86
John, 105
Johnson, 17, 49, 69, 71, 83, 183, 191, 193, 318, 320, 321
Jones, B.E., 183
Jones, I., 224
Jones, R.H., 242
Joseph, 315
Julian, 242

Kamat, 98
Kao, 242
Kendall, 15, 58, 79, 83
Keeping, 7, 15, 57, 73

AUTHOR INDEX

Kenney, 7, 15, 57, 73
Khatri, 129, 130, 131, 132
Kirkpatrick, 9
Knighting, 105
Koopmans, 27
Kotz, 17, 49, 69, 71, 83, 183, 321
Kramer, 86
Kuhn, 224, 225
Kutzbach, 225, 244, 249, 251, 255, 276

Laha, 183
Lambert, 31, 32, 33
Landsberg, 241, 242
Laurent, 27
Lawley, 269, 276, 283
Lee, 120
Leone, 190, 192, 193
Levinson, 304
Lilliefors, 30
Linder, 83
Lorenz, 11, 154
Lumley, 242
Lyness, 27, 28

Mahmoud, 105
Malone, 224
Mathews, 251
Mauchly, 88
Maxwell, 276
McElrath, 8
Mead, 79
Mielke, 50, 51
Mitchell, 241, 242
Molina, 47
Monin, 242
Mood, 83
Moore, 118
Moranda, 98
Moses, 105
Mostafa, 105
Mosteller, 8
Muller, 255, 270, 271, 272, 273
Munro, 27

Nelder, 76
Nelson, L.S., 190, 192, 193
Nelson, W., 125
Nelson, W.C., 115
Neumann, 248
Neyman, 27
Nicholson, 95, 96
Nordo, 9, 11, 12, 14, 308

Northingham, 190, 192, 193

Obukhov, 255
Ogawa, 79
Olds, 27
Owen, 27, 95, 101, 104, 167, 322

Pagano, 304, 305
Palmieria, 288
Panofsky, 241, 242
Parzen, 240
Patil, 115
Pearson, E.S., 27, 196
Pearson, K., 4, 41, 42, 46, 47, 49, 57, 64, 69, 95, 117, 144, 145, 146, 318
Peizer, 46
Pennington, 366, 368, 373, 381
Peto, 120
Pierce, 89, 296
Pinkham, 45, 47, 204
Pinus, 242
Pratt, 46
Prescott, 126
Priestly, 238, 239
Provost, 9

Qureishi, 79

Ralston, 381
Rao, C.R., 145, 148, 269
Rao, V.R., 241
Rappaport, 315
Reeves, 292
Reiter, 242
Rider, C.K., 242
Rider, P.R., 133, 134, 184
Robertson, 145, 150, 160
Robinson, 370
Rodden, 104
Rosenbaum, 129, 131
Rosenblatt, 27
Rosenthal, 104
Roughan, 312, 313, 314, 315
Roy, 183
Ruben, 105

Samiuddin, 91
Sampford, 134, 136
Sawyer, 101
Schafer, 78
Schips, 306
Schneider-Carius, 152

Seaver, 255
Sellers, 255
Selvin, 139
Severo, 27, 132
Shah, B.K., 79
Shah, S.M., 134
Shapiro, 197
Sheffield, 78
Shenton, 114, 321, 322
Shih Yung-Nien, 225
Shur, 242
Silverman, 197
Sneyers, 14, 121
Spackman, 224
Spearman, 276
Stegun, 102, 117, 118, 196, 205, 287, 323
Stein, 86, 105
Stidd, 255
Stier, 306
Stigant, 381
Stralkowski, 300, 307
Struylaert, 14
Stuart, 15, 58, 79, 83
Stumpf, H., 273
Stumpff, K., 232, 238
Suzuki, 116, 117, 118

Tate, 138
Taubenheim, 240
Teich, 27, 28, 30
Tennekes, 242
Thom, 183
Thomson, 69
Thöni, 27
Tingey, 9
Tucker, 101

Tukey, 236, 239, 240, 313, 324

Vancleef, 275
Varley, 138
Vaughan, 190
Vinnichanko, 242
Vulis, 242

Wagle, 105
Wahl, 225, 244, 249, 251
Wallace, 274, 275, 315
Wark, 275
Watson, 247, 251
Weil, 99
Weinberger, 273
Weingarten, 98
White, 255
Whittaker, 370
Wichern, 301, 306
Wilk, 47
Wilkinson, 381
Williams, C.B., 114
Williams, F., 19
Wilson, 322
Wilt, 381
Wise, 69
Wixley, 27
Woods, 242
Wu, 300, 307

Yates, 198
Yuan, 24
Yudin, 255

Zelen, 132
Zurmühl, 357

SUBJECT INDEX

Adjoint, *see* matrix
Albrook (Canal Zone), 109, 110
algorithm, 305, 313
aliasing, 197, 224, 237, 241
angular variates, angle, 35, 38, 90
antimode, 64
associative law, *see* matrix
autocorrelation, *see* correlation, time series
—, inverse, *see* time series
autoregression, *see* time series
axis, *see* minor or major

Backward shift operator, *see* times series
bell-shaped, *see* distribution
Berlin, 28
Bernoulli, 244
Bessel function, 99, 244
— —, complex, 245
— —, cylindrical harmonics, 244, 249
— —, first kind, 245, 246
— —, Fourier-Bessel function, 248
— —, potential function, 244
— —, recurrence formula, 246
— —, second kind, 247
— —, spherical harmonic, 244, 249
— —, zeros, 247
beta, *see* distribution
— ratio, 41, 46
Bienaymé, 15
binomial, *see* distribution
biology, 114, 147
bivariate, *see* distribution
—, circular, *see* distribution
—, elliptical, *see* distribution
British Association for the Advancement of Science, 99, 251

Canonical matrix, *see* matrix
Cauchy, *see* distribution
censored sample, *see* distribution
central limit theorem, 14
characteristic equation, *see* matrix
characteristics, 2
—, cumulants, 2, 5
—, mathematical, 2

haracteristics, moments, 2
—, primitive, 2
chi-square, *see* tests
circle with radius, *see* radius
circular, *see* bivariate
— probable error, *see* error
class, boundary, 71
— —, lower, 2
— —, upper, 2
—, central value, 2, 70, 319
— interval, 2
—, marginal, 70
— width, 2
climatic fluctuations, *see* fluctuations
cloud cover, cloudiness, 63, 72, 73, 75
— distribution, *see* U-distribution
coefficients, *see* correlation, polynomials, etc.
cofactor, *see* matrix
cold sums, 27
Colorado, 315
commutative law, *see* matrix
complex conjugate, 297
conditional frequency, *see* frequency
confidence, 6
— interval, 6
— limit, 41
contingency table, 288
conversion, 48
cooling degree days, 28
correlation, autocorrelation, *see* time series
— coefficient, 11, 87, 91, 275, 283
— factor analysis, 279
—, lag, 14
—, linear, *see* coefficient
— matrix, 277
—, multiple, 279
—, partial, 279
correlograms, *see* time series
cosine function, *see* distribution
covariance, *see also* factor analysis
Cramer's rule, *see* matrix
cross-product, *see* matrix, vector
cubic equation, 61, 62
cumulants, 5

cumulants, generating function, 4
— and moments, 2
curve fitting, 195, 210
cycle, 294
cylindrical harmonics, see Bessel

Decomposing, see mixed distribution, individual methods (Gauss)
degrees of freedom, see individual tests, 11
determinant, 331
—, cofactor, see matrix
—, equal, 334
— inversion in permutation, 332
—, minor, 332, 335
—, partition rule, 332, 345
—, permutations, 332
—, product rule, 332
— properties, 333
diagonal matrix, see matrix
difference operator, see time series
di-gamma, see functions
distribution, area integrals (Gaussian), 155
—, bell-shaped, 35, 41
—, beta (= incomplete beta), 17, 40, 48, 49, 56, 57, 64, 323
— —, generalized, 43
— —, negative binomial, 48
— —, ratio, 41, 46
— —, recomputation, 46
—, bimodal, 319
—, binomial, see also truncated, mixed, 17, 18, 19, 20, 49, 324
—, bivariate, 17, 80
— —, circular, 89, 91, 92
— —, cumulative, Gaussian, 93 ff, 108
— —, elliptical, 87, 88, 105, 109, 111
— —, Gaussian, 17, 87, 93
— — —, logistic, see logistic
— —, lognormal, 105
— —, normal, see bivariate Gaussian
—, Cauchy, 17, 35, 37, 38, 39, 40
—, censored, 121, 128, 137
— —, doubly, 137
—, chi-square, see also tests, 183
— —, non-central, 183
—, circular, see bivariate
— cloudiness, see also U-distribution, 63, 71, 72, 74
—, compound, 183
—, cosine function, 56
—, cumulative, see also bivariate, 2, 93, 187
—, elliptical, see bivariate

distribution, exponential, see also mixed, 62, 113
—, extreme value, 17
—, folded Gaussian, 190
—, gamma (= incomplete gamma), see also truncated, 17, 57, 59, 116, 118, 183, 321, 322, 323
— —, three and four parameters, 116, 118
— —, two parameters, 116
—, Gaussian, see also log normal, truncated, mixed, 9, 17, 37, 38, 39, 40, 57, 85, 89, 109
— —, integration of bivariate, 94 ff
— —, joint distribution, 85
— —, multivariate, 85
—, Gumbel, 26
—, hypergamma, 116
—, hypergeometric, 17, 18, 19, 20, 21
—, J-shaped, 41, 66
—, joint distribution, see Gaussian distrib.
—, kappa, 50, 51
—, logarithmic series, 114
—, logistic, 76, 79
— —, bivariate logistic, 76, 80, 109
— —, special form, 79, 82
—, lognormal, logarithmic normal, 22, 27, 29, 30, 32, 318
—, marginal, 89
—, mixed (mixture), 147, 183
— —, binomial, 17, 184
— —, exponential, 186
— —, Gaussian, 147, 164
— — —, Doetsch method = Fourier transform
— — —, Fourier transform method, 152, 153, 155, 156, 157, 159, 174, 175, 179
— — —, graphical method, 160, 165
— — —, maximum-likelihood method, 159
— — —, moments method, 145
— — —, multivariate, 179
— — —, order statistic, 159
— — —, truncation method, 150, 165
— —, Poisson, 17, 185
— —, Weibull, 186 ff
—, multivariate, see also truncated, 83, 85
—, negative binomial, see also truncated, 17, 41, 42, 48
—, normal, see Gaussian
—, Pareto, 17
—, Pearson system, type I, 58
— — —, type II, 58
— — —, type III, 57, 58, 59, 116, 321, 322

SUBJECT INDEX

distribution, Pearson system, type IV, 58, 59
— — —, type V, 58, 60
— — —, type VI, 58, 60
— — —, type VII, 58, 60
— — —, type VIII, 61
— — —, type IX, 61
— — —, type X, 62
— — —, type XI, 62
— — —, type XII, 62
—, Poisson, *see also* truncated, mixed, 17, 19, 20, 103, 183, 323, 324
— —, censored, *see* truncated
—, rectangular, 35, 56
—, truncated, 121
— —, binomial, 133, 135
— —, bivariate, 129
— —, double-sided, 125
— —, gamma, 139, 140
— — Gaussian, 122
— — Gupta, 128
— —, least-squares method, 127
— —, maximum-likelihood method, 126, 135, 181
— —, multivariate, 129
— —, negative binomial, 133, 134, 135
— —, one-sided, 122
— — Poisson, 136, 137, 138
—, unimodal, 319
—, U-type, approximations, 63, 65, 69
— —, shape, 30, 41, 47, 56, 58, 63, 65, 66, 71, 72, 74
— —, symmetric, 65, 68
— —, recomputation, 69
—, V-shape, *see* U-shape, 63
—, Weibull, 17, 57, 59, 116, 118, 119, 120, 189, 317
distributive law, *see* matrix
Doetsch method, *see* distribution
Doppler effect, 152
duration of storms, 114

Eccentricity, 88
ecological problems, 114
editing, *see* time series
efficient, efficiency, *see* estimation
eigenvalues, *see* matrix
eigenvector, *see* matrix
ellipticity, *see* bivariate, distribution
empirical Bayes, 147
— orthogonal functions, *see* polynomials
— polynomials, *see* polynomials
equation, characteristic, *see* matrix

equation, cubic, *see* cubic
—, linear system, *see* matrix
—, second-order, *see* second
error balance, 163
—, circular (probable), 98
—, eigenvalues, *see* matrix
—, Gaussian, 291
—, gross-error, 310, 312, 313
—, random, 10, 253
—, residual, 253, 254
—, standard, 17, 123, 302
—, type I, 6
—, type II, 6
estimation, estimators, 6
—, best linear unbiased, 36
—, consistent, 7
—, efficiency, efficient, 7
— —, asympotic, 36
— —, maximum, 255
—, maximum-likelihood, 7, 19, 36, 44, 193, 292, 319
—, moments, 19, 40
—, residuals, 296
—, spectra, *see* time series
—, sufficient, 7
—, unbiased, 7
Euler, 244, 245
— constant, 45
expectancy, 1, 2
expected number, 2
— value, 15
extrapolation, 196

Factor analysis, 276
—, closed model, 280
—, cofactor rotation, 285
— communalities, 279, 281
—, component analysis, 280
— correlation, 279 ff
—, covariance analysis, 286
—, eigenvector, eigenvalue, 282
—, number of factors, 284
—, open model, 280
—, orthogonal transformation, 286
—, principal factors, 280
— procedure, 283
— rotation, 285
—, simple structure, 285
— structure criterion, 285
filter, filtering, 241, 292
—, linear, 294
—, transfer function, 292
—, stable, 293

Fisher's *F*-test, *see also* tests, 88, 93
fluctations, 294
Forbenius, 245
forward shift operator, *see* time series
Fourier analysis = series
— -Bessel, *see* Bessel
— series, 157, 195, 226, 227, 250, 289
— transform, 86, 299
fractiles, 123
F-ratio, *see* test, Fisher
frequency, *see also* distribution
—, conditional, 89
—, cumulative, 2
— density, 2, 89
— distribution, *see* distribution
—, fundamental, 296
—, Nyquist, *see* timer series
—, relative, 1
function, autocovariance, *see* time series
—, beta, *see* distribution
—, complex, *see* Bessel
—, di-gamma, 44, 47, 117
—, hypergeometric, *see* distribution
—, potential, *see* Bessel
—, Riemann zeta, 45, 77
—, transfer, *see* time series
—, tri-gamma, 119
Gamma (incomplete), *see* distribution
Gaussian distribution, *see* distribution
Gauss-Jordan, *see* matrix
Gauss quadrature, 86, 205, 287
— -Seidel, *see* matrix
General Electric, 103
geometric mean, *see* mean
— series, 295
goodness-of-fit, 73
Gram-Schmidt process, *see* matrix
Gregory's formula, 204, 205, 206, 208, 213, 216
gross-error, *see* error
Grosswetterlage, 162, 164, 166
grouping correction, *see* Sheppard
Gumbel's distribution, *see* distribution

Handbook of Mathematical Functions, 102
harmonic analysis, *see* Fourier series
harmonic wave, *see* time series
Hermitian matrix, *see* matrix
homogeneity, 5
Huntsville, Alabama, 242, 243
hyperbola, rectangular, 317
hyperbolic cosine, 191

hyperbolic secant, 76
— sine, 320
— tangent, 193, 318, 323
— trigonometric functions, 76, 82
hypergeometric, *see* distribution

Inequality, *see* Tchebycheff
information matrix, 31, 33
integration of bivariate, *see* distribution, Gaussian, 94, 155
interpolation, 196
inversion, *see* matrix
iterative method, *see* polynomials

Jacobi method, *see* matrix
Jordan, *see* matrix
J-shape, *see* distribution

Karlsruhe, 162, 164, 166, 176
Kolmogorov-Smirnov, *see* test
Kronecker delta, 278, 288, 339
kurtosis, 4, 43

Lag correlation, *see* correlation
Lagrange multiplyer, 256
Lambert's model, 31, 32, 33
Laplace, 244
— equation, 244
— expansion, 47, 334
— rule, *see* matrix
— transformation, *see* transformation
least-squares solution, 10, 202, 208, 379
left variance, *see* variance
Legendre, *see* polynomials
likelihood ratio, 311
linear equation, *see* matrix
logarithmic scale, progression, 2
— series, *see* distribution
logistic, *see* distribution
lognormal, *see* distribution

Major axis, 91
mapped parameters, 266
marginal class, *see* class
— distribution, *see* distribution
Markov (also Markovian), 6
— chain, 12, 13
—, first-order, 12, 315
—, higher-order, 12
— law, *see* process
— model, *see* time series
—, non-Markovian, 309

SUBJECT INDEX 409

Markov process, *see* time series
— time series, 292, 299, 308, 309
matrix, 325
—, adjoints, 343
—, anti-commute, 329
—, associative law, 326, 327
—, augmented, 374
—, autocorrelation, 315
—, canonical, 337
—, characteristic equation, 257, 258, 266, 348, 350, 351, 359 ff
— — polynomial, 349
— — —, common factor, 367, 368
— — —, complex roots, 363, 365, 369
— — —, computation, 360
— — —, constants of, 361
— — —, multiple roots, 363, 365
— — —, roots (= eigenvalues), 349, 363
—, cofactor, 263, 332
—, commutative, 329
—, commutative law, 326
—, conformable, 326
—, conjugate, 331
—, data, 277
— definitions, 325
— determinant, *see* determinant
—, diagonal, diagonalization, 208, 254, 256, 257, 279, 329, 371
— — and eigenvalues, 352, 371
—, distribution law, 327
—, eigenvalues, 253, 254, 256, 257, 266, 283, 315, 348, 349
— —, calculation, 259, 363, 374
— —, error, 269
— —, explained variance, 261
— —, largest, 358
— —, significance, 269
— —, smallest, 358
—, eigenvector, 253, 254, 257, 266, 278, 315, 341, 348, 351
— —, calculation, 257
— —, determination, 358, 363, 371, 374
— —, iteration, 359
— —, orthogonal, 353, 354
— —, significance, 269, 275
— —, stability, 270
— —, variance, 271
— elements, 325
— —, diagonal, 325
— equivalence, 337
—, factor, 278
—, Gauss-Jordan, 373

matrix, Gauss-Seidel, 373
—, Gram-Schmidt, 225
—, hermetian, 331
—, idempotent, 329
—, identical, *see also* similar matrix, 325
—, identity, 254, 329
—, invariant, 348
—, inverse, 254, 330, 344 ff, 361
— inversion, 332
—, involutory, 330
—, Jacobi method, 259, 262, 282, 283
—, Jordan m, 350, 356, 373
— —, canonical form, 350, 351
—, Laplace expansion, *see* Laplace
— — rule, 335
—, latent roots = eigenvalues
—, linear equations, 348, 379
— — —, Cramer's rule, 380
— — —, elimination method, *see* Gauss-Jordan
— — —, square-root method, *see also* Gauss (matrix), 368
— multiplication, 327
—, nilpotent, 329
—, non-singular, 337, 354
—, normal form, 338, 377
— operations, 326 ff
— order, 325
—, orthogonal, 339, 353
—, partition rule, 332, 345
—, periodic, 329
—, permutations, 332
— rank, 337
—, rectangular, 325
—, scalar factor, 326
— — product, *see* vector
—, similar, 348
—, skewed symmetric, 331
— solution, *see* mixed distribution, Gauss
—, square, 325, 352
— sums, 326
—, symmetric, 254, 330
—, trace, 325
—, transformations, 337, 356
—, transpose, 254, 330
—, triangular, 304, 329
—, vector, 326
— —, latent = eigenvector
— —, normalized, 260
— —, scalar product, 328
— —, standardized, 277, 283
— —, zero, 326
maximum efficiency, *see* estimation

maximum likelihood, *see* truncated, mixed distribution, estimation
Mean, angular variate, 38
—, arithmetic, 3
—, geometric, 23, 117
—, overlapping, 175
median, 319, 320
minor axis, 91
mixed distribution, *see* distribution
— model, *see* time series
mixtures of populations, *see* distribution
moments, 19, 23, 67, 68, 82, 146, 187
—, central, 3
—, factorial, 184
—, general reference, 2
—, generating functions, 4, 49
—, incomplete, 191
—, mixed, 4
—, zero reference, 3, 42
Monte Carlo, 86
Montgomery (Alabama), 108, 111, 189, 223
motor boating, 121
moving average, *see* time series

National Bureau of Standards, 96
negative binomial, *see* distribution
Newton-Raphson, 37, 117, 140, 370
noise, level of, 292
—, white, 292, 294, 296, 304, 315
non-harmonic waves, *see* time series
non-stationary, *see* time series
Nyquist frequency, *see* time series

Observations, spacing, 197
Ohio data, 315
open model, *see* factor analysis
orthogonal, orthogonalization, *see also* polynomial, 13, 195
— matrix, *see* matrix
oscillation, 294
outliers, *see also* quality control, 121, 310, 311, 313
overlapping means, *see* mean

Parabolic relationship, 295
Pareto, *see* distribution
Pearson's system, *see* distribution
percentage reduction, 195, 196, 214, 221, 232, 250, 275
percentile values, 319, 321
periodogram, *see* time series

persistence, 6, 275
— of wind, 105
phase angle, *see* Fourier series, spectral analysis
polar coordinates, 101
polynomials, characteristic, *see* matrix
—, constants, 203
—, continuous type, 197, 224
—, discrete type, 198, 200, 201, 224
—, empirical, 252, 253, 254, 255, 258, 266
— fit, 317
—, general, 195
—, iterative method, 214
—, Laguerre, 224
—, Legendre, 196, 199, 200, 202, 208, 210, 211, 212, 213, 215, 216, 217, 222, 252
—, matrix solution, 208
—, miscellaneous, 224
—, orthogonal, 13, 195, 200, 208, 250
—, Tchebycheff, 196, 197, 202, 210, 211, 212, 213, 215, 216, 222, 252
—, two-dimensional, 224
—, Yates, 198
power spectrum, *see* time series
precipitation, 34
predictand, 263
prediction model, 265, 275
predictors, 252, 253, 255, 265
probability, 1
—, conditional, 89
—, definition, 1
— density function, 2
pseudo-periodic behavior, *see* time series

Quadrature formula, *see* Gauss quadrature
quality control, *see also* time series, 121, 360
quantiles, 2, 31, 32
quartiles, 40
quasi-periodicity, *see* time series

Radius of circle, 100, 101
random sampling, 10
range, 2
rectangular, *see* distribution
recurrence (formula), *see* individual topics
regression curve, 89
— line, 90
— model, 267
residual, 10
— error, *see* error

SUBJECT INDEX 411

residual model, *see* time series
— process, *see* time series
resultant wind vector, 101
Rieman zeta, *see* functions
river floods, 34
Rodrigue's formula, 199, 200
root, characteristic, *see* matrix
—, latent, *see* matrix
Rossby, *see* waves
rotation, *see* factor analysis

Sampling, random, *see* random
satellite data, 73
scalar product, *see* matrix
scale factor, 48
Schuster's criterion, 243
second-order equation, 185
Sheppard's correction, 4
shocks, 292
significance, 6, 9, *see also* eigenvalues
—, level of, 6
—, statistical significance check, 91, 171, 269, 275
Simpson's formula, 204, 205, 207, 208, 213, 216
skewness, 4, 24, 43, 52
smoothing, exponential, 315
spacing, *see* observations
Spearman's rank correlation, 288
spectral density, *see also* time series
spectrum, *see* time series
spherical harmonics, *see* Bessel
standard deviation, 93
— error, *see* error
stationary, *see* time series
statistic, *see also* estimators, 7
—, best, 7
—, consistent, 7, 8
—, efficient, 7
—, inefficient, 8
— order, 8, 40
—, sufficient, 7, 8
—, unbiased, 7
Stirling number, 138
Stoke's method, 247
student, *see* test

Taylor's formula, 197, 370
Tchebycheff's inequality, 15
terminus, 121
testing of a hypothesis, *see also* test, 6
tests, chi-square (χ^2), 12, 30, 73, 74, 96, 296

tests, Fisher's F, 17, 88, 93, 150, 311, 312
—, Kolmogorov-Smirnov, 17, 30, 73, 74
—, power of, 6
—, Student's t, 17, 91, 150
—, two-sided, 9
—, z-transformation, 17, 318
tetrachoric coefficient, 288
Thule, 105, 106
time, interrelationships with space, 10
— series, 10
— —, aliasing, *see* aliasing
— —, autocorrelation, 229, 289, 296, 298, 300, 301, 303, 304, 305
— —, autocovariance, 298
— —, autoregressive model (process), 13, 289, 291, 293, 296, 297, 299, 309
— — — —, first-order, 297
— — — —, second-order, 297, 300, 308
— —, backward difference operator, 294
— —, backward shift operator, 291, 292, 299
— —, correlogram, 228
— —, cross-spectrum, 315
— —, cross-variance, 301
— —, editing, 312
— —, eigenvector, eigenvalue, 315
— — estimation, 237, 238, 239, 302
— — — —, Bartlett, 238
— — — —, Tukey, 239, 240
— —, exponential model, 298
— — — —, filters, *see* filter
— —, forward shift operator, 291, 299
— — filters, *see* filter
— —, Godske model, 308
— —, harmonic wave, 6, 13, 234
— —, identification of model, 302
— —, inverse autocorrelation, 306
— —, mixed model, 293, 296, 298, 300, 301, 315
— —, Markov, *see* Markov
— —, moving average (process), 289, 296, 306
— — — —, model, 292, 293, 300
— — — — —, first-order, 297, 300, 304, 308
— — — — —, second-order, 297, 300
— —, non-harmonic wave, 234, 235, 237
— —, non-stationary, 294, 300, 305
— —, Nyquist frequency, 240
— —, operators, *see* backward, forward
— —, periodogram analysis, 231, 232, 234, 237
— —, power spectrum, *see* spectrum

time series, pseudo-periodic behavior, 300
— —, quality control, 310, 311, 313
— —, quasi-periodicity, 243
— —, recursive formula, 303
— —, residual process, 293, 296
— —, spectral band, 14
— —, spectral density, 289
— —, spectral relationship, 296
— — — —, autoregressive, 297
— — — —, mixed model, 298
— — — —, moving average, 297
— —, spectrum, 226, 230, 231, 234, 235, 237, 289, 290, 295, 297, 315
— —, spectrum analysis, 226, 234
— —, stationary, 269, 290
— —, stationary process, 293, 294, 299
— —, transfer function, 296, 316
— —, trend, 295, 313
— —, truncation, 295, 313
— —, turbulence, 243
— —, vector diagram, 233, 234, 243
trace, see matrix
transformations, 33, 69, 274, 294, 316
—, chi-square, see Shenton, transformation
—, cube-root, 322
—, equation, 201
—, Fisher's, see test, z-transformation
—, Fourier transform, see mixed distribution
—, Gaussian, 317, 321
—, Johnson system, 318, 320
—, Laplace, 321
—, matrix, see matrix
—, orthogonal, see factor analysis
—, Shenton, 321, 322, 323
—, special functions, 316
—, square root, 323
—, summary, 323
—, z, see test
translation method (= transformation), 317
trapezoid integration method, 287

trend, see time series
tri-gamma, see functions
truncation, see distribution, time series

U-shape, type, see distribution

Variance, 3
—, cross-variance, see time series
—, eigenvalues, eigenvectors, see matrix
—, explained, 214, 257, 261, 275
—, left, see also percentage reduction, 195, 196, 214, 218, 219, 220, 265, 269, 275
— matrix, 31, 33
— ratio, see Fisher's F test
—, residual, 11
— stabilization, 324
—, unexplained, 269
vector, see also matrix
— diagram, see time series
— standard deviation, 93
— wind, 83, 101
visibility data, 22
V-shaped, see U-distribution

Waves, gravity, 315
—, harmonic, see Fourier
—, Rossby, 315
—, tropical, 315
weather maps, 266
Weibull, see distribution
white noise, see noise
wind direction, 38, 39
— profile, 223
— speed, 189
—, upper-air data, 121
— vector, 83
winsorization, 313
winter severity, 27

Yates polynomial formula, see polynomial
Yule-Walker equations, 299, 303